CW00723012

HETEROTROPHIC ACTIVITY IN THE SEA

NATO CONFERENCE SERIES

I   Ecology
II   Systems Science
III Human Factors
IV   Marine Sciences
V   Air–Sea Interactions
VI   Materials Science

IV MARINE SCIENCES

Recent volumes in this series

HETEROTROPHIC ACTIVITY IN THE SEA

Edited by

John E. Hobbie
Marine Biological Laboratory
Woods Hole, Massachusetts, USA

and
Peter J. leB. Williams
Department of Marine Microbiology
Gothenburg University
Gothenburg, Sweden

Published in cooperation with NATO Scientific Affairs Division

PLENUM PRESS · NEW YORK AND LONDON

Library of Congress Cataloging in Publication Data

NATO Advanced Research Institute on Microbial Metabolism and the Cycling of Organic Matter in the Sea (1981: Cascais, Portugal)

Heterotrophic activity in the sea.

(NATO conference series. IV, Marine Sciences; v. 15)
"Published in cooperation with NATO Scientific Affairs Division."
Bibliography: p.
Includes index.
1. Bacteria, Heterotrophic—Congresses. 2. Marine bacteria—Congresses. 3. Marine microbiology—Congresses. I. Hobbie, John E. II. Williams, Peter J. leB. III. Title. IV. Series.
QR106.N37 1981   576′.15′09162   84-11601
ISBN 0-306-41724-3

Proceedings of a NATO Advanced Research Institute on Microbial Metabolism and the Cycling of Organic Matter in the Sea, held November 1981 in Cascais, Portugal

©1984 Plenum Press, New York
A Division of Plenum Publishing Corporation
233 Spring Street, New York, N.Y. 10013

Printed in the United States of America

PREFACE

Introduction

This book contains papers given at a NATO Advanced Research Institute (A.R.I.) held at Caiscais, Portugal, in November, 1981. The subject of the A.R.I. was marine heterotrophy; this is defined as the process by which the carbon autotrophically fixed into organic compounds by photosynthesis is transformed and respired. Obviously all animals and many microbes are heterotrophs but here we will deal only with the microbes. Also, we restricted the A.R.I. primarily to microbial heterotrophy in the water column even though we recognize that a great deal occurs in sediments. Most of the recent advances have, in fact, been made in the water column because it is easier to work in a fluid, apparently uniform medium.

The reason for the A.R.I. was the rapid development of this subject over the past few years. Methods and arguments have flourished so it is now time for a review and for a sorting out. We wish to thank the NATO Marine Science Committee for sharing this view, F. Azam, A.-L. Meyer-Reil, L. Pomeroy, C. Lee, and B. Hargrave for organizational help, and H. Lang and S. Semino for valuable editing aid.

Autoecology vs. Synecology

The techniques of microbiology were developed for laboratory use. When they are applied to seawater they also appear to work; at least microbes can be isolated and identified, organic compounds disappear and are oxidized, and oxygen decreases. However, have these techniques identified the important organisms, rates, and controls occurring in nature? This must still be the overriding question for the study of marine heterotrophy. With this emphasis on the process in nature, one concludes that most of the traditional laboratory techniques become unusable: most microbes from seawater do not grow on agar plates, microbes quickly multiply or die in incubation bottles, and incubations in the past have often been carried out for extended periods in order to measure change in substrate concentration. Whereas it may be too extreme to say that most of the measurements in the past have been artifacts, the probability of artifacts caused by experimental conditions is quite high and has to be considered in every measurement.

The great advances in methods in the past decades have been in synecology: the study of organisms as communities or as parts of cycles or processes. The other approach, autecology, or the study of individual species, has been in eclipse because of an inability to relate microbial species, isolated from nature and studied in laboratory cultures, back to their role in natural processes. Synecology has flourished recently in large part due to advances in biochemical and biomedical techniques and instruments and to their adaptation to ecological problems. For example, bacteria may now be counted in natural waters because of greatly improved epifluorescent microscopes and new types of membrane filters, both developed initially for medical and industrial research.

Yet, in spite of new methods and instruments, the problem of how to measure the various aspects of microbial heterotrophy in the sea has not been resolved. This is due in part to the necessity for making measurements in the field and in part to the very low levels of bacterial activity. There is still no single, agreed upon method for measuring bacterial growth or respiration so the literature is filled with arguments about different methods. Many of the arguments would be ended if a single method were available, no matter how laborious or difficult, as a standard for calibration of other easier methods.

The Problem of Production Measurements

Throughout this book and in discussions at conferences and in the literature, the emphasis is on production or growth of microbes. Heterotrophy is also a degradative process and it would appear that the rate and controls of the process could be measured without measuring production. For some questions it is true that a rate measurement would have also been adequate to answer the question. There are also major questions about the pathways of carbon and energy flow in marine food webs that can only be answered with a production measure. For example, if microbes are being fed upon by protozoans and higher filter feeders, then the production rate of microbes is an integral part of the calculation of energy flow.

The techniques for measuring production have been developed for a number of groups of marine organisms. Why then is there a problem for heterotrophic microbes? The general answer is that marine microbes live in a dilute environment where every process is difficult to measure; this is further confounded by the fact that microbes respond rapidly to any change in their surroundings so all incubations may influence results. Some detailed comments about the production problem follow.

1. Microbes degrade and eventually take up, grow on, and respire an enormous variety of organic compounds. At one point in time each species or even each cell may be specializing on one or several compounds or may be simultaneously taking up a large number of compounds. The practical effect is that production cannot be measured by adding a single radioactively labelled compound (e.g., glucose) and measuring the rate of use by microbes. This contrasts wth algal primary production which may be estimated by following the metabolism of a single compound such as oxygen or carbon dioxide.

2. Microbes in the sea are able to increase and decrease their activity over wider ranges than any other group of organisms. If environmental conditions are favorable they grow rapidly. If conditions are unfavorable they may become nearly dormant. The result is that merely demonstrating the presence of 1×10^6 cells per ml in the environment tells us nothing immediately about their production rate. Higher organisms either respire continuously or must enter distinctive resting states. For this reason, we learn a great deal more from their presence when we try to interpret energy flow.

3. The continuous processes that remove microbes - that is, death and grazing - occur at the same time as production. The result is that neither a steady state nor changes in the numbers of bacteria necessarily reflect bacterial production or lack of production. In this context, death may occur as attack by other microbes or as natural lysis. Both are possible but their importance in the sea is unknown.

4. Other organisms compete for substrates with the microbes. While a small flagellate taking up dissolved organic carbon is a part of a heterotrophic process, it can interfere with a measurement of prokaryotic microbial production. It is also known that higher organisms, such as mussels, can remove large amounts of dissolved organic matter from solution and this makes it difficult to infer microbial production from mass balance.

5. Some microbes rapidly respond to changed conditions. When water samples are collected and incubated over time, some microbes flourish and some may die. In a practical sense, only those techniques may be used that have short incubations of minutes or perhaps a few hours.

6. Microbial production is low in most of the sea, at times less than 10 μg C liter^{-1} day^{-1}, so that techniques such as measuring the loss of oxygen or the production of ammonia are analytically very demanding.

The General Questions of Marine Heterotrophy

The discussion above stresses the difficulties of studying marine heterotrophy. In this way we have attempted to explain without being too negative why the workers in this field do not have a better idea of microbial processes in the sea. Yet, in this NATO Advanced Research Institute the participants dealt with the recent progress in measuring and understanding marine heterotrophy; a feeling of accomplishment and optimism prevailed. We now will give an overview of the outcome of the discussions organized around four general questions:

   1. What is the identity and abundance of marine microbial heterotrophs such as bacteria and microflagellates?

   2. What are the rates of heterotrophic processes such as growth, respiration, and mineralization of nitrogen and phosphorus?

   3. What is the role of the marine heterotrophs in the food web of the sea?

   4. What are the controls of the various processes that make up heterotrophy in the sea?

The Identity and Abundance of Heterotrophs

Despite our progress in measuring rates and identifying the role of marine microbes in the heterotrophic process, we are uncertain about the species composition of the community. The classic techniques of plate or dilution cultures yield data on abundance and identity of the organisms we are able to cultivate. Within the last decade it has become clear that the resulting numbers were wrong and that a very small percentage of the marine bacteria grow in enrichment cultures. This leaves us uncertain as to whether the species isolated are representative of the community as a whole. Monoclonal antibody techniques may well resolve this particular question but will, of course, only work with species amenable to culture. If we did know the identity of all the bacteria in a water body, then we could answer the recurring question of whether the sea is populated by microbial specialists or microbial generalists. The incomplete evidence so far would indicate that Pseudomonas-like generalists predominate.

Advances in microscopic techniques in recent years have finally allowed direct counts to be made of microbes in seawater. These somewhat labor intensive methods utilize an epifluorescent microscope to count the bacteria which have been dyed with a fluorescent nuclear stain and filtered onto a membrane. Important findings have included the large number of bacteria (10^5-10^6 ml^{-1}) and the

generally small size of the cells (0.2–0.4 μm in diameter). It
appears that the technique can be automated with computer image
analysis (J. Sieburth, personal communication). With slight modifications, the epifluorescent methods can also be used to count small
flagellates in the 2–20 μm size range and to differentiate between
the autotrophic and heterotrophic forms. An extension of fluorescent
dye techniques, flow cytometer cell sorting, can be used to effect
some physical partitioning of the microbial community. The technique
has been successful for larger bacteria such as nitrifiers.

While it is now relatively easy to obtain numbers of microbes,
it is more difficult to convert these numbers to biomass or carbon.
It is clearly difficult to determine with any precision the exact
dimensions of cells close to the limit of resolution of the light
microscope. Even after a biovolume is obtained, it often needs to
be converted for ecological reasons to units of carbon and these
factors have so far been derived entirely from laboratory cultures
of large bacteria. An alternative approach to the problem has been
to measure some biochemical component of the bacterial cell. Some
of these components, such as ATP, are not specific to bacteria.
Most have the same problem of the conversion factors which might
well vary from population to population and from time to time. For
example, a muramic acid conversion factor will change by almost an
order of magnitude depending on whether the bacteria are gram
negative or gram positive.

In summary, whereas we now know that microbes are abundant and
have a significant biomass relative to other heterotrophs such as the
metazoa, the indigenous species composition is still a closed book.

Rates of Heterotrophic Processes

As a result of growth and activity, heterotrophs are involved
in a number of interrelated processes such as decomposition of
complex molecules, respiration, remineralization, mineralization,
and the uptake of dissolved organic compounds and their conversion
to particulate material. Measurements of any of these has been very
difficult in the past but in the last decade very real advances have
been made.

Bacterial growth, a fundamental property, has been estimated
both directly and indirectly. Direct measures depend upon changes
in biomass (or related parameters) and will only work when grazers
are unimportant or are removed from the system. In practice, it has
proven difficult to remove the grazers, particularly microflagellates, without disrupting the system.

Because of these problems, indirect methods of measuring growth
have been developed. One technique, the frequency of dividing cells,
involves no incubation but instead relies upon determination of the
percentage of the population in the dividing state. The technique

is promising but needs testing in a wide variety of environments.
Another set of techniques involves short incubations in order to
determine the rate of incorporation of labelled precursors into RNA
and DNA. These approaches are in an active state of development at
present and seem to have great potential. However, they still need
testing in a variety of environments and clarification of their
biochemistries. Until these problems are resolved, the usefulness
of the techniques remains in question.

The measurement of a change in concentration of oxygen, carbon
dioxide, or inorganic nutrients is a further way to determine heterotrophic rates. So far, the oxygen change method has been most widely
used for in vitro studies. With recent improvements in instrumental
analysis, it may be expected that both the oxygen technique and the
total carbon dioxide measurements will be more widely used in the
study of overall microbial activity. In situ studies depend both on
analytical sensitivity and on the ability to describe mixing and gas
exchange in the environment. In spite of the long history of this
approach, there are only a few measurements that pertain to the upper
mixed layers of the ocean where the biology is interesting but the
mixing is difficult to measure.

In situations where recycling is important, then isotope
dilution techniques are often used. For example, ^{15}N techniques
allow both uptake and production of ammonia to be calculated.

Another system in which uptake and production are closely
coupled is the production of dissolved organic compounds by algae
and their rapid uptake by bacteria. The algal compounds can be
labelled with ^{14}C and the separation of algae and bacteria attempted
by filtration. However, even after very short incubations, some of
the photoassimilated ^{14}C will have already been respired by the
bacteria so the method underestimates bacterial production from
algal exudates. In addition to the exudates, bacterial growth will
also be sustained by the breakdown of algal particulate matter.

The first useful isotope method for investigating heterotrophy
was the measurement of the uptake and respiration of single organic
compounds, such as glucose, amino acids, or acetate. This approach
gives a relative measure of activity and information on conversion
efficiency of these compounds, but can not be expected to give a
measure of overall microbial metabolism.

A method that stands by itself is the enzymatic measure of the
activity of the electron transport system. This has the great
advantage of not involving an incubation of cells, but in order to
interpret the results in conventional ecological terms either
biochemical assumptions have to be made or the method has to be
cross calibrated against other methods.

In summary, we now have a number of promising techniques for measuring rates of heterotrophic activity and microbial growth in the sea. The early results are encouraging, as there is a measure of agreement between quite independent methods. Further development and intercalibration seem to be necessary.

Role in Food Chains

In contrast to the great amount of work on microbial numbers and species and to the moderate amount on heterotrophic rates, there has been little work on the trophic role of microbes in the food chain. Progress beyond the purely descriptive levels of understanding had to await the development of methods of determining biomass and activity. Among the several microbial processes that are important are the conversion of dissolved to particulate organic carbon, the mineralization of organic nitrogen and phosphorus compounds, and the decomposition of refractory organic compounds. Through the conversion of dissolved organic carbon and the growth of cells, the microbes become the basis of a microheterotrophic food chain and also provide food to supplement metazoan food chains.

At the descriptive level, we now know in a general way the numbers and biomasses of the bacteria and protozoans in the heterotrophic food chains. Much work still needs to be done on the details and on conversion factors but it would appear that the biomass of the microheterotrophic community is on occasion comparable to that of the metazoa. We are in the early stages of understanding the grazing on bacteria of flagellates, ciliates, and metazoa. The indications are that the flagellated protozoans are the principal bacterial grazers. More laboratory work on a variety of protozoans is needed as well as work on grazing within natural communities.

A few studies have also been made of the relative role of microbes in the mineralization of organic nitrogen and phosphorus. Size fractionation studies point to the great importance of the microbial community but have not yet separated mineralization by protozoans from that by bacteria.

Another role suggested for microbes is to enrich particulate material in nitrogen and thus to improve its nutritive value. However, the increase in microbial biomass often does not appear to be large enough to account for the increase in nitrogen. There is evidence for a microbially mediated process of humification that occurs in organic material and this may well explain much of the increase in nitrogen.

Finally, a consideration of whole-system budgets for organic carbon indicates that something in the region of half the photosynthetically produced carbon in the ocean eventually is oxidized by planktonic bacteria. The distribution of biomass within the microbial community is better known than the distribution of activity.

Controls of Heterotrophic Processes

The topic of control of heterotrophy and of heterotrophic organisms is our area of greatest ignorance and uncertainty. This ignorance at the control level is, of course, no different from the situation in other areas of ecology.

In the ocean, there is indirect evidence for tight controls over the standing stocks of organisms and substrates. For example, bacterial numbers are surprizingly predictable and vary little from place to place. The variability is much less than we would expect from such a rapidly reproducing group of organisms. Preliminary evidence also indicates that the heterotrophic flagellates exhibit the same pattern of constant numbers and little variability. A similar situation obtains for the dissolved solutes such as sugars and amino acids which are presumably used by marine bacteria. Concentrations are about the same in very eutrophic and in very oligotrophic waters.

We can speculate that bacterial numbers are controlled by grazing of microflagellates and there is some experimental laboratory evidence that this is so. The demonstration in the field has yet to be made.

The relationship between the rate of solute uptake by bacteria and the concentration of solute is well known in the sea. Uptake is dominated by bacteria and they are capable of turning the substrate over at rates of a few hours to a few days. However, we do not know how the microbes can maintain such a constant substrate concentration. It is possible that there is some threshold concentration at the nanno molar level at which uptake ceases but there is as yet no evidence for this.

Microbial activity may also be controlled in a more sophisticated manner by chemical communicants. For example, cyclic AMP has been found in the ocean and this compound is known to control protein synthesis within the cell.

We expect that improvements in the measurement of heterotrophic biomass, activity, and production in the next few years will allow the study of the control processes. Until these controls are understood, the functioning of the system will remain obscure.

J. E. Hobbie

P. J. leB. Williams

CONTENTS

MICROBIAL PROCESSES IN THE SEA: DIVERSITY IN NATURE AND SCIENCE

Lawrence R. Pomeroy

Institute of Ecology
University of Georgia
Athens, Georgia   30602

INTRODUCTION

Marine microbial ecology is, as its name implies, an interdisciplinary pursuit which draws not from two but from several disciplines, creating a unique and diversified field for research. The
results of this interaction of disciplines have had some impact on
ecological theory. For example, recent advances in marine microbial
ecology have made the trophic level concept almost meaningless,
because microorganisms enter food webs at many trophic levels simultaneously. Also, studies of microorganisms have contributed to the
demise of the belief in a standard "ecological efficiency" or 10
percent per trophic level, and they have done violence to the concept
of a pyramid of numbers, or biomass, in natural communities. At the
same time, recent developments in microbial ecology have been influenced by ecological concepts and methods, and many of the classical
methods of microbiology have been replaced in marine studies by ones
developed by ecologists. The recent results have brought the
realization that microbial activities in the ocean biome are greater
quantitatively and more varied than marine ecologists had realized.

Bacteria are potentially so fast growing and so responsive to
changes in their environment that the microbial populations cultured
from isolates taken from the sea sometimes bear little physiological
resemblance to the populations actually living there. Laboratory
cultures typically are highly enriched, while the ocean, except for
some potentially important microenvironments such as fecal pellets,
is impoverished in many essential nutrients. When organisms which
are adapted to such poverty are presented with the riches of the
laboratory-culture environment, some dramatically change form and
growth state while others simply die from the excess (Morita 1980;

1

Wiebe and Pomeroy 1972). We are still defining the range of natural
environment conditions of marine microorganisms, but it is apparent
that many free-living marine bacteria live in and are physiologically
adapted to conditions of extreme scarcity of resources.

The direct counting of bacteria in sea water has become successful and relatively easy with the development of fluorescent stains
and epifluorescence microscopy. Culture methods, on the other hand,
lead to an underestimation of the numbers and biomass of bacteria in
ocean water, because many marine bacteria do not grow in enriched
media. They also tell little about growth state or metabolic
activity. A number of approaches are being tried to discriminate
among activity states of naturally occurring populations of marine
bacteria. Often what the ecologist wishes to know is the rate of
production or the metabolic rate, and several promising methods have
been introduced recently.

Controversy has arisen over a number of significant roles
bacteria may play in marine ecosystems. Not only do they degrade
many refractory substrates which might otherwise accumulate, but they
also compete very effectively for labile substrates. Therefore, they
may be significant links in marine food chains (Williams 1981).
Questions about these roles are bringing marine microbial ecology
into more direct contact with other areas of marine ecology. Marine
ecologists are recognizing that they must consider the roles and
effects of microorganisms if they are to understand marine food webs
and trophic relationships. At the same time marine microbiologists
are recognizing that microbial activities are interrelated with
those of the eukaryotes at many levels and cannot be considered in
total isolation. That recognition is reflected in the makeup of
this conference.

HISTORICAL PERSPECTIVE

Marine microbiology, like some of the organisms included in its
study, remained small, with a long doubling time, for decades, only
to undergo a pleomorphic change into a large, rapidly growing field.
The preconditions necessary for such a burst of activity developed
long ago, however. Recognition of the global significance of microorganisms goes back more than 50 years (Vernadskii 1926). A number
of early studies implicated microorganisms in the utilization of both
particulate and dissolved organic matter in sea water (Waksman et al.
1933; Keys et al. 1935). In spite of this, much of marine microbial
ecology remained isolated from other aspects of ecology, an isolation
which appears to have been fostered both by microbiologists, whose
approach to ecology has been quite distinctive, and by the specialists in higher organisms, who recognized no central role in marine
ecosystems for microorganisms. Both groups perceived microorganisms

as carrying out the utilization of dilute or refractory substrates but not as serious competitors for the food of higher organisms or the nutrients of higher plants. This view is reflected even in most current textbooks on oceanography. For example, Parsons et al. (1977) devote 4 pages to chemosynthesis and 8 pages to heterotrophic processes, much of the latter an exposition of hyperbolic uptake functions. Since textbooks are usually less than a decade behind the current literature, this is a measure of the rapidity of the change in the perceptions of marine biologists and microbiologists.

The usual experimental method in microbiology involves the isolation of defined organisms and their culture in defined media. A powerful approach, because everything is known and controlled, it is also the procedure against which all work in microbiology is judged. Many marine microbiologists have been understandably reluctant to involve themselves in direct studies of undefined ocean ecosystems, and of natural populations of bacteria, not all of which can be cultured and described fully at this time. For ecologists this is a familiar problem, one with which they have perhaps learned to live all too comfortably. In place of the defined conditions of the microbiologist or the laboratory chemist, the ecologist tends to fall back upon statistical inferences and order-of-magnitude differences. The significance of observations is less assured, but it is virtually all we have when we directly study the dirty, natural world around us. Some chemists and microbiologists are difficult to persuade that the clean, defined systems they study may not behave in the same way in nature that they do in the culture tube. In the case of the ocean, recent events have demonstrated that the systems we create for the study of marine bacteria are not like the ocean in a number of ways. Most of the ocean is impoverished of both organic and some inorganic substrates. Most of it is very cold, quite dark, and does not support photoautotrophs as such. Moreover, it is not a single, uniform environment. Not only does the water column itself vary, but within it are numerous microhabitats for bacteria: living organisms, non-living particulate matter including feces, the surface film, and the bottom.

Science is a world of ideas, and progress in science is limited ultimately by the emergence of new ideas. Sometimes ideas seem to have a life of their own, emerging over and over until they are accepted by the scientific community at large. Often, however, the emergence and acceptance of ideas is limited by our ability to test them. In the case of marine microbial ecology there have been severe limits in methodology. The usual defined-culture methods limited our perception and understanding of the real-ocean ecosystems, but those ecosystems were not readily accessible to us as investigators. In recent years an array of new, powerful methods has changed that, and investigators are trying to catch up with technology which offers new insights into marine microbial ecology.

THE OCEAN'S ROLES IN THE GLOBAL CYCLE OF CARBON

Primary Production

    Very small autotrophs are responsible for a substantial fraction
of total photosynthesis in the ocean, often more than 90% (Malone
1980). Chroococcoid cyanobacteria are now known to be ubiquitous and
abundant in the ocean (Waterbury et al. 1979; Johnson and Sieburth
1979). Substantial populations of small autotrophic flagellates of
various taxa are globally abundant. It is too early to say whether
the flagellates or the cyanobacteria are mainly responsible for
photosynthetic activity in the ocean, or whether dominance shifts
from one to the other under various environmental conditions. On
a global basis organisms < 10 μm are responsible for much of the
fixation of organic carbon in the ocean, with dominance shifting to
larger phytoplankton in some river plumes and upwellings.

    Several gaps in our knowledge of primary production remain to be
filled. A very important one is verification of the validity of the
methods currently in use. There have been many challenges to the ^{14}C
method over the years (Ryther 1956; Odum et al. 1963), and recently
a frequent criticism is that the incubation time is long relative to
the rate of flux of carbon down the food chain. The microcrustacean
food chain is usually eliminated from the bottles in which photosynthetic rate is measured, but the protozoan food chain is not. In the
absence of much data no consensus exists at this time regarding the
rate at which ciliates and heterotrohic flagellates feed on photo-
autotrophic microflagellates and cyanobacteria. The few published
observations suggest considerable variability. The possibility does
exist for substantial cycling of carbon through a protozoan food
chain (Fig. 1), even in a 4-6 hour experiment (Haas and Webb 1979;
King et al. 1980). The DOC released by phytoplankton and organic
carbon respired as CO_2 will not be measured. While we do not have
direct experimental evidence on which to base a good estimate of the
significance of these pathways of carbon flux, we know the biomass of
protozoans in the ocean and their metabolic rate are sufficient to
warrant serious consideration with respect to this problem. Quantification of this food chain may go a long way toward explaining
so-called bottle effects on production of both phytoplankton and
bacteria.

Allochthonous Materials

    Input of organic matter from the land to the sea has been the
subject of speculation and possible over-emphasis, for recent work
on the characterization of humates suggests that most marine organic
matter originates in the sea (Hatcher et al. 1980). Inputs of
particulate organic matter do occur, including logs, leaves, and
garbage introduced by rivers, may be of local importance, and may
even support certain consumer populations in unique ways; river-borne

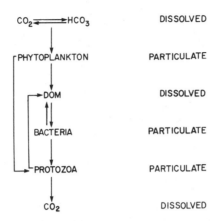

Figure 1. The microbial food chain in a bottle during measurment of photosynthesis by the ^{14}C method. Time to complete the entire pathway is on the order of minutes.

allochthonous organic matter may even be significant in the global cycle of carbon. However, micro-organic matter moving from the land through the atmosphere may be equally significant. Fallout of both inorganic and organic dust of continental origin has been extensively documented (Delaney et al. 1967; Folger 1970). A number of organisms, fungal hyphae, spores, and freshwater diatoms, are common constituents of dust samples taken over the North Atlantic. I once examined some freshly collected samples of atmospheric particles at Bermuda (Bricker and Prospero 1969) and was surprised to see numerous organic particles similar in appearance to the class of organic aggregates called flakes (Riley 1963; Gordon 1970a). Of course, these might have been organic aggregates swept into the atmosphere from the sea surface. However, they were collected in mid-summer, at a time of relatively calm seas, and Folger (1970) reports finding only terrigenous organic particles over the North Atlantic. Folger (1970) examined water samples for particles of terrestrial origin and identified mostly organic ones, largely concentrated in what he called organic aggregates. Probably they were largely fecal aggregates, but in any case he postulated that they would be ingested and would be removed to the deep water as fecal matter. In this connection it is interesting to note that Gordon (1970b) detected a winter maximum in particulate organic matter in the North Atlantic. Gordon showed that at least 20% of this was readily hydrolyzed by proteolytic enzymes, so presumably it had arrived recently and was subject to future microbial transformation. Since the winter maximum corresponds not with the seasonal peak in surface primary production but with winter winds, the winter pulse of POC may be allochthonous terrestrial dust.

Volatile organic materials, such as terpenes and organic

sulfides, move into the atmosphere from both terrestrial and marine sources. The flux of terpenes was crudely estimated by Went (1966) to be 10^9 tons yr^{-1}, which is on the order of 10% of global photosynthesis and 1000 times the input of petroleum to the ocean (Morris et al. 1976). This does not include all volatile organic materials analytically recognized today, and it ignores particulate flux. Current interest in the organic sulfides has focused on their oxidation in the upper atmosphere and their impact on the ozone layer. However, some fraction of these materials must also be washed out in rainfall and dryfall over the ocean, where they may be transformed by bacteria. Rasmussen and Went (1965), and Went et al. (1967) observed that volatile organic matter from terrestrial vegetation forms a significant number of condensation nuclei in the atmosphere. While the nature, source, and residence time of both volatile and particulate organic materials in the atmosphere remains uncertain, a fairly uniform concentration of condensation nuclei has been found over the ocean, on the order of 500 cc^{-1} (Elliott 1976; Ketseridis et al. 1976; Eichmann et al. 1979). Their chemical composition is uniform and is compatible with either an origin from plants, both terrestrial and marine, or from anthropogenic sources. Sooner or later most organic fallout is going to be transformed by microorganisms in soils or in the ocean.

Although allochthonous inputs to the sea are probably small compared to phytoplankton photosynthesis, they may not be trivial. Because of their physiochemical nature, they are probably assimilated in the sea by microorganisms. These inputs and the resulting food web are worth further study. Interactions between continental, atmospheric and oceanic constituents present logistically difficult interdisciplinary problems which often are ignored as a result.

Secondary Production

The significant gaps in our understanding of primary sources of organic carbon in the ocean are small by comparison with the gaps in our knowledge of secondary production. Many marine biologists believe that secondary production is only the production of microcrustacea, but there is also a very substantial production of microorganisms (Fuhrman and Azam 1980; Fuhrman et al. 1980). The credibility of microbial production data has suffered because there is no single, generally accepted method for measuring it in the ocean. Moreover, estimates by the various methods have differed by orders of magnitude. While investigators still do not agree on a single method, the divergence between results by current methods is narrowing. The methods of Karl (1979) and Fuhrman and Azam (1980), which are somewhat similar in approach, appear to be in reasonable agreement with each other but not necessarily with the method of Hagström et al. (1979) which is based on the frequency of dividing cells. However, we can probably have at least as much confidence in the recent data on bacterial production as in the data on primary

production. If so, there can be little doubt that bacterial production and also bacterial consumption of organic carbon in the sea is a
significant part of the total carbon flux. That proposition is not
accepted, however, by many marine biologists. Walsh et al. (1981)
claim to have discovered the "missing carbon" in global models to be
excess production by phytoplankton on continental shelves, based on
simulation modeling which includes no microbial pathways.

In view of all of the uncertainties about the flux of carbon
through marine food webs, one is drawn to the conclusion that
predictive modeling of the global cycle of carbon really is not
possible until there is fundamental agreement on some important
biological fluxes – not only their amount but their very existence.
All models suffer from the condensation necessary to keep them
within the capacity of computing facilities. As a result, each
model tends to emphasize those features of the system which the
modeler believes to be significant. If we can ever agree on what
those features are, modeling of marine food webs may become really
predictive rather than the heuristic device it is today.

Microbial Roles in Carbon Flux

Bacteria have a generally accepted role in the transformation
of particulate organic carbon (POC) to living biomass, but there is
little agreement on how they do it. The evidence, such as it is,
comes from observations of particulate material in the water (Wiebe
and Pomeroy 1972) and from microcosm experiments in which bacterial
tranformations of particulate material of various kinds were observed
(Kranck and Milligan 1980; Tenore 1977; Tenore et al. 1977; Fenchel
1970; Herbland 1975). The reality of these latter observations as
representing what happens in the ocean varies from none to considerable, and most of them suffer from the effects of excessive concentration of organic substrates.

The origin of particulate matter in the open ocean must be
primarily from phytoplankton. Only a small fraction of the particulate matter in ocean water is visually recognizable as algal remains.
However, histochemical tests show that most of the particulate matter
reacts like phytoplankton (Gordon 1970b; Wiebe and Pomeroy 1972), so
the unrecognizable fraction must have passed through some process
which transformed phytoplankton or products of phytoplankton into
what we call particulate organic detritus. The obvious processes
are grazing, predation, and defecation.

Particulate material is also formed as organic aggregates (Riley
1963). Although this term has appeared in many different contexts,
originally and as used here this term is limited to particles which
form authigenically (de novo) in ocean water, from organic matter
derived ultimately from marine photosynthesis. Several mechanisms of
authigenic organic particle formation have been confirmed (Johnson

1976; Johnson and Cooke 1980; Wheeler 1975), but there has not been
any quantification of the rate of formation of aggregates in the
ocean. The relationship of microorganisms to aggregates has been the
subject of debate since the concept of aggregate formation emerged.
Some investigators have suggested that aggregates form only in the
presence of bacteria, either because they are really products of the
bacteria, such as slime or other secretions, or because bacteria in
some way catalyze aggregate formation. Neither of these occurs in
the case of aggregates produced by bubble collapse, but secondary
aggregation of large particles (sea snow) originally produced by
bubbling (e.g., Kranck and Milligan 1980) may involve bacterial
processes in the water.

     Once formed, aggregates presumably are potential substrates
for the growth of bacteria, although here again the observational
evidence is slight. Bacteria are reported to be present on some
aggregates in sea water and not on others (Pomeroy and Johannes
1968; Wiebe and Pomeroy 1972). Observers differ on the frequency of
attached bacteria and their significance. There is no experimental
evidence showing whether organic aggregates are transformed into
bacterial biomass, digested directly by eukaryotes, or both.
Presumably they are utilized in the food web in one way or another,
because they are rare in deep water > 500 m (Wiebe and Pomeroy
1972). Another possibility is that most continue to aggregate with
each other until they become sea snow large enough to sink (Shanks
and Trent 1980).

     One of the difficulties with observations of particulate matter
in samples of ocean water is the lack of agreement among observers
about the nature of the particles they see and describe. Many kinds
of particles, including sea snow, have been described as aggregates.
Only rarely have particles collected in the water been described as
fecal, perhaps because of the widespread belief that all fecal matter
is in the form of pellets which fall rapidly to the bottom. Even in
the case of the microcrustacea this is not true (Hofmann et al.
1981). The early life history stages of microcrustacea produce very
small fecal particles which do not sink rapidly. Other planktonic
organisms produce less compacted feces, often lacking a peritrophic
membrane. The pelagic tunicates, which sweep from the water the
smallest organisms, produce fecal ribbons consisting of a gelatinous
matrix in which boluses of compacted material appear sequentially.
The ribbons are fragile, break apart soon after release, and the
boluses sink rapidly and fall to the ocean's bottom. They have been
collected in sediment traps and unfortunately called fecal pellets
(Honjo 1978; 1980). The boluses are perhaps half of the total fecal
material. The remainder of the fecal ribbon disintegrates into small
(~ 50 µm) fragments which remain in the water. They are rapidly
colonized by bacteria and subsequently by protozoans (Pomeroy and
Deibel 1980). Over several days they disintegrate into very small
fragments which do not appear to be further colonized by bacteria.

In the ocean such particles have a significant chance of being reingested, and they may be actively sought by grazers. Certainly they are a potentially good source of food for either grazers or net feeders as they contain bacteria, protozoa, and little-digested phytoplankton.

The fecal pellets of copepods are also transformed by bacteria and protozoans (Ferrante and Ptak 1978; Turner and Ferrante 1979). Depending on their size and rate of fall through the water, which may be from zero to hundreds of meters per day, fecal pellets in the water column will contain bacteria and protozoa and will be a good source of nutrition for grazers. Considering the scarcity of nutritionally suitable materials in the ocean, it seems probable that there are zooplankton which selectively seek and eat fecal materials. If they do not, they are ignoring a substantial supply of nutrition. Either through direct ingestion or bacterial transfomation, most fecal material does not reach the bottom of the ocean (Bishop et al. 1977, 1978, 1980). This does not in any way conflict with the observations of fecal materials in sediment traps in deep water, but it does contradict the idea that all fecal matter falls to the bottom (e.g., Steele 1974), even in shelf-waters and epicontinental seas.

The efficiency and versatility with which bacteria transform organic substrates makes them different from other components of food webs. Laboratory studies of assimilation efficiency show bacteria to be the most efficient of all organisms (Payne 1970; Ho and Payne 1979). From the viewpoint of microbial ecology it is unfortunate that these studies were done at substrate concentrations higher than those ordinarily found in sea water. While efficiency actually seemed to increase with decreasing substrate concentration, the lowest concentrations were still above natural ones, and one would expect efficiency to begin to drop off at some point. Estimates derived from uptake of defined substrates labeled with ^{14}C suggest that efficiency does not drop and is still 70-80 percent at natural substrate concentrations (Hobbie and Crawford 1969; Williams 1970). If this is so, bacterial transformations of dilute or relatively refractory materials in the ocean may be accomplished with much greater efficiency than ecologists ordinarily assume, and more bacterial biomass than expected may be available to consumers.

Studies of heterotrophic uptake have shown that bacteria do utilize dissolved substrates in microgram and even nanogram per liter concentrations (Wright and Hobbie 1966; Hobbie and Crawford 1969; Azam and Holm-Hansen 1973). There may be lower limits below which bacteria do not remove substrates from the water or shift to another more abundant substrate. Experiments which would verify this do not appear to have been done, although they might be rather straightforward. Bacteria do have different lower concentration limits for the uptake of various compounds, for ATP is taken up at concentrations far below the normal concentrations of glucose or individual amino acids.

Uptake experiments involving defined substrates may not reflect
the rate of uptake of naturally produced organic compounds. In one
test, bacteria utilized undefined DOM from phytoplankton more rapidly
than defined substances (Smith and Wiebe 1976; Wiebe and Smith 1977).
Jacobsen (1981) also found significantly faster uptake of DOM
produced by a diatom culture than of defined substrates at comparable
concentrations. While there may be some odd compound lurking in the
DOM pool which is more significant in the flux of labile DOM than
the compounds people have chosen to study, we have no clue to its
identity. Perhaps a large number of compounds are released by phytoplankton and the collective uptake of all of them, utilizing many
sets of uptake sites, is at least an order of magnitude faster than
uptake of any one. This hypothesis should be testable with the range
of defined substrates now at hand.

Free bacteria in the water have no source of nutrition other
than dissolved substrates, while attached bacteria presumably are
utilizing particles both as habitat and substrate while utilizing
dissolved material as well. Hodson et al. (1981) measured the uptake
of labeled ATP by free and attached bacteria, using a 6 μm Nuclepore
filter to separate free bacteria from larger particles with attached
bacteria. Although there were about 10^3 large, attached bacteria and
10^5 small free bacteria per milliliter, the two populations took up
approximately equal amounts of the labeled substrate. Because of
their size difference, when the uptake rates were expressed per unit
biomass or per unit bacterial surface area, the two populations were
seen to be exposing the same amount of cell surface to the water and
presumably had the same number of uptake sites per unit of surface
area. Thus, despite the difference in numbers the two populations
were equally active and had nearly the same biomass. This may not,
however, be the case throughout the ocean.

Respiration

In evaluating bacteria as movers of energy and materials in the
ocean, we need to measure the respiratory rate of marine bacteria in
their natural state, but it has proven to be difficult. The rate of
respiration per unit volume of natural sea water is too small to be
measured directly. In estuarine and coastal water, direct measurements are now possible with the high-precision Winkler method (Bryan
et al. 1976). Concentration of microorganisms from a large volume of
ocean water in order to make respiratory rate measurements possible,
leads to deactivation or loss of a significant fraction of the activity. In the reverse flow concentration method of Pomeroy and
Johannes (1968) probably the free bacteria were largely lost through
the filters which were used, leaving the attached bacteria and
phytoplankton, which may account for about half of total microbial
respiration. Today Nuclepore filters offer a major improvement, but
there are still serious questions about the effects of concentration

on respiratory rate. Moreover, there are problems in separating the respiration of phytoplankton from that of bacteria. Mechanical separation of the free bacteria probably can be accomplished, but separation of bacteria attached to detritus from the phytoplankton and protozoans is still impossible. A study of the ratio of adenylates to chlorophyll in the upper layers of the ocean suggests that most of the biomass, most of the time, is phytoplankton (Campbell et al. 1979). It may follow that most of the respiration which can be measured in natural sea water is that of phytoplankton. Therefore, we need ultimately not only a sensitive method for measuring respiratory rate in the ocean but also the means to discriminate between the respiration of free bacteria, attached bacteria, phytoplankton, cyanobacteria, and other microorganisms.

Sources and Sinks of Dissolved Organic Carbon

One of the largest standing stocks of carbon on the planet is that of dissolved organic carbon (DOC) in ocean water, which is rather uniformly distributed throughout the ocean (Menzel and Ryther 1968) and has a mean radiocarbon age in deep water on the order of 3000 years (Williams et al. 1969). The DOC is relatively refractory. All or some of this material is believed to be transformed by free-living deep-sea bacteria, which are slow-growing psychrophiles, and this is reasonable, in view of the fact that residence time of DOC in the ocean is orders of magnitude shorter than that of Na^+ or Cl^-. Further clarification of the nature and fate of this material would be worthwhile in view of the size of the DOC pool, although not more than 0.5% of primary production is estimated to enter the DOC pool in deep water (Williams et al. 1969).

In the upper mixed layer of the ocean we find microgram per liter quantities of monosaccharides and amino acids and nanogram per liter quantities of such metabolites as ATP (Azam and Hodson 1977). The residence time of these labile materials varies widely (Azam and Holm-Hansen 1973; Hodson et al. 1981) but is never as long as that of the pool of refractory DOC. We do not know with certainty the source, the rate of production, or the fate of the labile DOC, although it has been generally assumed to be produced by phytoplankton (Fogg 1971; Thomas 1971; Nalewajko 1977). Other secondary sources of DOC may in fact prove to be more significant than primary release from phytoplankton. Some DOC is produced by zooplankton through excretion and defecation (Pomeroy et al. 1963; Lampert 1978). Additional DOC is produced during the transformation of fecal materials by bacteria and protozoans, most of which takes place in the upper 100 meters of the water column. A study of the food web of the California bight suggests that more DOC originates from zooplankton than from phytoplankton (Fuhrman et al. 1980). A shift of attention from phytoplankton to zooplankton as producers of DOC may be appropriate, while bacteria should be viewed both as producers and consumers of DOC.

One of the impediments to understanding oceanic DOC is that all of it has not been described qualitatively, and doing so is a difficult task at present. The refractory DOC which makes up most of the standing stock is presumably humic and fulvic in character, but what is particularly difficult to characterize is the labile material which is probably diverse in origin and chemistry. However, it constitutes 99% of the DOC flux. Those compounds present in the smallest standing stock and therefore most difficult to find are probably the most significant in terms of short turnover time and high flux rate while the less labile materials tend to accumulate.

THE SIGNIFICANCE OF DECOMPOSITION

While no one doubts that marine bacteria are responsible for much decomposition of organic substrates, including DOC, a number of interesting questions remain concerning the long-term global effects of microbial processes, for there are some notable exceptions in decomposer abilities (Alexander 1980). Synthetic organic materials provide a new and sudden challenge to bacterial versatility, and if bacteria were totally efficient in transforming all substrates, there would be no petroleum, coal, or methane in the sedimentary rocks. Would there also be no oxygen in the atmosphere? Although this is widely believed, the evidence is less than compelling.

Fossil Carbon, Recent Oxygen, and Chroococcoid Photosynthesis

The orthodox view of the history of the planet is that the atmosphere lacked oxygen until the rise of oxygen-producing photosynthetic organisms. Once photosynthesis began, according to this view, some reduced carbon was lost to the sediments, excess oxygen was released to the atmosphere, and all of the excess reduced compounds near the surface of the earth, both organic and inorganic, were oxidized. Then, as all exposed reduced materials became oxidized, oxygen began to accumulate in the atmosphere. Such a reasonable and widely accepted paradigm is difficult to challenge, but several investigators have. Van Valen (1971) pointed out that a major source of oxygen is the photolysis of water in the upper atmosphere and the subsequent loss of hydrogen to the solar wind as it streams past the earth.

The large requirement for oxygen over the history of the planet, amounting to 1000 times the present standing stock, could have been supplied by the net difference between planetary photosynthesis and respiration even though the difference was very small. However, regulation of oxygen in the atmosphere appears to be very weak, and the concentration may have varied by a factor of 10 (Cope and Chaloner 1980). Furthermore, there is evidence that photosynthesis of the oxygen-producing kind evolved only after the oxygen content of the atmosphere had gone well above the Pasteur point (Schwartz and Dayhoff 1978). We still do not have a really good estimate of

the rate of production of oxygen in the upper atmosphere and know
even less about the factors which may cause that rate to vary over
time. Therefore, we cannot determine whether the biosphere really
influences the oxygen content of the present atmosphere significantly. With respect to the early atmosphere, oxygen-producing
photosynthesis may have evolved in response to an acute need for
reduced carbon compounds at a time when an oxidizing atmosphere and
rampant oxidative metabolism made photoautotrophy an essential part
of the evolving biosphere. If this were the course of events,
probably the pioneers were the small chroococcoid cyanobacteria.

One of the pervasive mysteries associated with the accumulation
of petroleum is how this accumulation began and how it escaped
bacterial transformation. We look in vain for evidence in the
present ocean of the precursors of hydrocarbon concentrations.
Because we find so many hydrocarbon deposits of many different ages,
is this a rather unusual period in the earth's history? Porter and
Robbins (1981) suggest that conditions for accumulation of organic
matter occur off California, but this is difficult to perceive. The
basins are well oxygenated all the way to bottom. However, near the
shore of Peru under the upwelling region the bottom is anaerobic, or
very nearly so, with green sediments. Fecal pellets may be an
important constituent, and the high concentration of porphyrins
suggested by the color of the sediments is suspicious. No one seems
to have looked at them from that viewpoint. However, it is not clear
how nearshore sediments on an orogenic coast will be conserved. A
piece of the puzzle may still be missing.

The Oligotrophic Ocean

Little organic matter accumulates on the bottom of the ocean,
for what falls has to run the gauntlet of hungry mouths through four
kilometers of water, and it has to fall rapidly enough to escape
bacterial decomposition. What does fall is transformed by the
heterotrophic food web. Considerable layering of activity appears
to be orchestrated by the physical regime of solar illumination
which promotes both photosynthesis and thermal stratification.
Solar penetration is maximal in the blue water of the ocean, but the
ocean would be more productive if there were more phytoplankton to
intercept the light near the surface of the water, as is the case
near shore. Stratification created largely by solar warming appears
to have a great influence on the way the ocean works as a biome, and
most likely the ocean would be a more productive place if it were
more mixed for as we all know, some of the most productive places
are those where water upwells from below the usual depth of the
thermocline. Sverdrup (1955) predicted that this would be so, and
except for that part of the year when the Antarctic is in darkness
and extreme cold, all of the regions of upwelling identified by
Sverdrup have proven to be exceptionally productive. However, this
productivity requires alternate episodes of upwelling and stability.

A continuously mixed ocean is not productive, and there is an optimum mixing intensity (Eppley and Peterson 1979).

Most of the ocean is stratified, however, and the thermocline is a barrier to flux in both directions, although we usually think of it primarily as a barrier to the upwelling of nitrate. Only relatively large (> 100 µm) fecal particles and the occasional dead organism go swiftly through the thermocline toward bottom. So the fallout which is so desperately needed at the bottom is diminished both by animal feeders along the way and by bacteria doing their work both in the upper mixed layer and in the oxygen minimum layer in the thermocline.

Bubnov (1966) and Menzel and Ryther (1968) asserted that the oxygen minimum layers of the major ocean basins are only an artifact of the physical regime; they believe there is no significant biological activity except at the point of origin, said to be an upwelling region, where water of reduced dissolved oxygen content produced beneath the productive upwelling slides across the ocean along a constant-density isopleth. The intensity of the oxygen minimum also decreases with distance from the upwelling, an observation which caused the investigators to suggest that all that is happening, once the water leaves its biologically active origin, is gradual diffusive exchange with the adjacent water masses. There is evidence, however, that some distinctive biological activities do occur in oxygen minima. Karl et al. (1976) reported a large ATP maximum, approaching half the near-surface value, in the oxygen minimum layer of the central North Atlantic, and they assumed that this represented a large population of bacteria. On several cruises in the western North Atlantic and the Caribbean my colleagues and I have found that the number of bacteria in the oxygen minimum layer is greater than that in adjacent water. Moreover, the oxygen minimum layer is the only place in the water column, other than the interior of fecal pellets, where we have seen motile bacteria, a characteristic of an environment of diminished dissolved oxygen (Hobbie et al. 1972). In the northeast Pacific Ocean, Fellows, Karl, and Knauer (1981) found an increase in RNA synthesis in the oxygen-minimum layer. These evidences of bacterial biomass and activity in oxygen minima suggest that the layers receive a sufficient source of energy and are sites of biological activity.

Marine Humus

Litter and humus are important parts of the terrestrial environment where they provide both a habitat and a substrate for microorganisms (Wiegert and Owen 1971). In terresrial environments litter is produced mostly by direct fall of plant materials from the overstory. In the marine biome most litter is microscopic. Some bloom organisms such as Trichodesmium, which are not readily eaten by most zooplankton, do accumulate and become detritus. Movies of zooplankton feeding by Strickler and Paffenhofer show that feeding is

sometimes inefficient, especially when the food is long chains of
diatoms. Predators are also less than perfectly efficient in
consuming planktonic prey (Dagg 1974). So, while Steele's (1974)
assumption in his North Sea model that all phytoplankton are eaten by
zooplankton certainly is not correct, it is possible that the real
value is 80 or 90%, with a large variance. Therefore, the major
source of marine detritus probably is not scraps of phytoplankton
lost by inefficient zooplankton but is fecal material, primarily
from grazers and mucus-net feeders.

The complex detritus food web is difficult to observe in the
real world, because it is microscopic and highly dispersed. When we
concentrate it for observation, we may get a distorted view of what
is really happening. Good observations are still rare and difficult;
laboratory simulations of detritus systems are more common but
probably even more misleading. The ocean is not a hay infusion.

Detritus in the marine biome fills much the same role as litter
does on land; it provides a habitat for microorganisms and small
metazoans and at the same time is a sort of gingerbread house which
its inhabitants eat. One of the significant processes in the
detritus food web is the regeneration of plant nutrients, especially
nitrogen and phosphorus. Both bacteria and protozoans may play a
role in this, and the roles vary with the concentration of dissolved
oxygen in the microenvironment of the detritus particle. In large
fecal pellets there is evidence of anaerobic or reduced oxygen conditions. Under those conditions bacteria do not accumulate polyphos-
phates (Shapiro 1967), but release both ammonia and phosphate to the
surrounding water. However, in smaller (< 100 μm) fecal particles
and other detritus, oxygen diffuses to the center of the particle,
and motile bacteria are not apparent. Under those conditions
bacteria accumulate polyphosphate. They not only use all phosphate
from their particulate substrates but take it up from the surrounding
water. The phosphate will be released only when protozoans consume
the bacteria, releasing excess phosphate. The literature is by no
means unanimous or coherent on this point. The postulate that under
aerobic conditions metazoans and protozoans excrete significant
amounts of phosphate while bacteria do not, was originally proposed
by Pomeroy and Bush (1959), Pomeroy et al. (1963) and Johannes (1964,
1965). Subsequent investigators have both confirmed and denied it.
Beuchler and Dillon (1974) confirmed it, using Paramecium cultures.
Barsdate et al. (1974) and Fenchel (1977) denied that protozoans play
a significant role in nutrient regeneration. This conclusion was
based on the use of experiments with high concentrations of higher
plants in which anaerobic microenvironments were probable and no
retention of phosphate by bacteria would be expected. Moreover, in
reporting the results Barsdate et al. seem to confuse turnover time
with net flux, the former involving transmembrane exchanges which
are irrelevant to the latter. Recently, Kerrick (1981) has shown
that in dilute culture, the marine flagellate, Bodo, eats bacteria

and excretes phosphate while the bacteria retain phosphate. Most of
the work on microbial recycling of nutrients has been done with
phosphate. Bacteria do not seem to have a mechanism for accumulating
nitrogen other than by excess protein synthesis, so there is a more-
or-less steady flux of ammonia being lost from marine bacteria which
grow on a nitrogen-sufficient natural organic substrate.

     In contrast to many terrestrial biomes, organic matter accumu-
lates in the ocean primarily as DOC. Large detrital particles fall
rapidly to the bottom, where they are degraded by the combined
action of invertebrates and bacteria (Sieburth and Deitz 1974).
Although degradation by bacteria alone on the ocean bottom is slow
(Jannasch et al. 1971; Jannasch and Wirsen 1973), it still proceeds
quite rapidly in the presence of invertebrates which break up the
particles. In this repsect the sea bottom is analogous to the
forest floor, where the same synergism (or competition?) between
invertebrates and bacteria occurs (Janzen 1977).

     Dissolved or colloidal humic acids probably have more influence
on the marine environment than is generally appreciated, although
their role in the food web is debatable since they are degraded very
slowly. Humates in fresh water play a major role in metal chemistry,
both as chelators and as zwitterions, oxidizing and reducing sites.
Because of the increase in ion strength in sea water, marine humates
are different in structure from fresh water humates. Nevertheless
they probably play a role in marine chemistry, one which does not
appear to have been explored as thoroughly as it has been in fresh
water. Marine humates are more readily transformed into bacterial
biomass than fresh-water humates because of their lower aromaticity.
Marine humates lack the lignin fraction with its alkylaromatic
esters which are resistant to bacterial transformation (Dereppe et
al. 1980). Coastal and estuarine humates vary in their aromaticity
(Hatcher et al. 1980), reflecting diverse origins.

STRATEGIES FOR SUCCESS IN THE MARINE BIOME

     Any population needs some kind of strategic advantage in order
to compete and survive. While a population may overlap with other
populations, there is a need for some unique niche dimension which
enhances the chance for survival. Bacteria have many potentialities
for unique niche development. Bacteria are metabolically flexible
and versatile to a greater degree than other organisms. They can
transform most organic material, except a few new synthetic compounds
and perhaps some humates, into bacterial biomass. They combine
patience with fast response. They can remain in resting stages or
spores which involve very little expenditure of maintenance energy
for very long periods, variously estimated from tens to millions of
years. Marine bacteria also have shorter lag times and faster
doubling rates than any of the eukaryotes, and they can compete

successfully with other organisms for transitory supplies of substrates and inorganic nutrients, such as phosphate, nitrate, and
ammonia.

Small size offers a number of advantages, and some of the marine
bacteria are extremely small (close to 0.1 μm). Small size carries
with it the potential for low maintenance cost for the genome,
although whether or not maintenance costs are really low will depend
upon the environmental conditions. Small size is also a refuge from
predators. While some organisms, particularly the mucus-net feeders,
are able to concentrate and eat free-living minibacteria (< 0.5 μm
in diameter), most have no means of catching them, and it is hardly
worth their while to do so. An organism can achieve the same
nutrient intake by eating one 2 μm rod attached to a detritus
particle as would be obtained by gathering 20 minibacteria from the
water. Most metazoans will feed on particles rather than small,
free-living bacteria, both because it is easier to do and because
the particles are a richer source of bacterial biomass. Relatively
few metazoans larger than the protozoans have developed a strategy
for eating free-living minibacteria, and probably most protozoans
preferentially swim from particle to particle, munching on the
larger bacteria. Thus the population of minibacteria in the ocean
may not be controlled by grazers but only by the availability of
substrates, and minibacteria can play a waiting game, statistically
safe from enemies until some food comes along. So long as there is
a biosphere, some food will come along, and because the bacteria are
so versatile, almost anything will suffice.

The bacteria on particles obviously have a different strategy.
They change size dramatically in the presence of a suitable substrate, growing from a tiny resting form into a large rod in a very
short time, then doubling until a colony is formed on the substrate
(Wiebe and Pomeroy 1972). These colonies are rapidly found and eaten
by protozoans. The particle-colonizing bacteria therefore switch
from a size-refuge strategy in their resting state to a rapid multiplication strategy in the presence of a substrate. The statistics of
both strategies seem to have been well worked out in an evolutionary
sense.

The strategies of marine microorganisms are those which evolve
in the face of usually impoverished and highly variable conditions.
No group of organisms is more diversified, and so it is appropriate
that scientific approaches to the ecology of marine microorganisms
have been diverse as well.

ACKNOWLEDGMENTS

I thank Janet Pomeroy and W. J. Wiebe for critically reviewing
the manuscript. Work reported from my laboratory was supported by

contract DE-AS09-76EV00639 with the U. S. Department of Energy and grants from the National Science Foundation.

REFERENCES

Alexander, M.   1980.   Helpful, harmful, and fallible microorganisms:
    Importance in transformations of chemical pollutants.
    Microbiology 80: 328-332.
Azam, F., and R. E. Hodson.   1977.   Dissolved ATP in the sea and its
    utilization by marine bacteria.   Nature 267: 696-697.
Azam, F., and O. Holm-Hansen.   1973.   Use of tritiated substrates in
    the study of heterotrophy in seawater.   Mar. Biol. 23: 191-196.
Barsdate, R. J., R. T. Prentki, and T. Fenchel.   1974.   Phosphorus
    cycle of model ecosystems: significance for decomposer food
    chains and effect of bacterial grazers.   Oikos 25: 239-251.
Beuchler, D. G., and R. D. Dillon.   1974.   Phosphorus regeneration
    in freshwater paramecia.   J. Protozool. 21: 339-343.
Bishop, J. K. B., R. W. Collier, D. R. Ketten, and J. M. Edmond.
    1980.   The chemistry, biology, and vertical flux of particulate
    matter from the upper 1500 m of the Panama Basin.   Deep-Sea
    Res. 27: 591-614.
Bishop, J. K. B., J. M. Edmond, D. R. Ketten, M. P. Bacon, and W. B.
    Silker.   1977.   The chemistry, biology, and vertical flux of
    particulate matter from the upper 400 m of the equatorial
    Atlantic Ocean.   Deep-Sea Res. 24: 511-548.
Bishop, J. K. B., D. R. Ketten, and J. M. Edmond.   1978.   The
    chemistry, biology, and vertical flux of particulate matter
    from the upper 400 m of the Cape Basin in the southeast
    Atlantic Ocean.   Deep-Sea Res. 25: 1121-1162.
Bricker, O. P., and J. M. Prospero.   1969.   Airborne dust in the
    Bermuda Islands and Barbados .   (Abstract)   Trans. Am. Geophys.
    Union 50: 176.
Bryan, J. R., J. P. Riley, and P. J. leB. Williams.   1976.   A
    Winkler procedure for making precise measurements of oxygen
    concentration for productivity and related studies.   J. Exp.
    Mar. Biol. Ecol. 21: 191-197.
Bubnov, V. A.   1966.   The distribution pattern of minimum oxygen
    concentrations in the Atlantic.   Okeanologiya 6: 240-250.
Campbell, W. B., T. R. Jacobsen, and L. R. Pomeroy.   1979.
    Heterotrophic-photoautotrophic index: a qualitative parameter
    of microbial interactions applied to a Gulf Stream intrusion.
    Mar. Sci. Commun. 5: 383-398.
Cope, M. J., and W. G. Chaloner.   1980.   Fossil charcoal as evidence
    of past atmospheric composition.   Nature 283: 647-649.
Dagg, M. J.   1974.   Loss of prey body contents during feeding by an
    aquatic predator.   Ecology 55: 903-906.
Delaney, A. C., A. C. Delaney, D. W. Parker, J. J. Griffin, E. D.
    Goldberg and B. E. F. Reimann.   1967.   Airborne dust collected
    at Barbados.   Geochim. Geocosmochim. Acta 31: 885-909.

Dereppe, J. -M., C. Moreaux, and Y. Debyser. 1980. Investigation of marine and terrestrial humic substances by ^{1}H and ^{13}C nuclear magnetic resonance and spectroscopy. Org. Geochem. 2: 117-124.

Eichmann, R., P. Neuling, G. Ketseridis, J. Hahn, R. Jaenicke, and C. Junge. 1979. N-alkane studies in the troposphere. I. Gas and particulate concentrations in North Atlantic air. Atmos. Environ. 13: 587-599.

Elliott, W. P. 1976. Condensation nuclei concentrations over the Mediterranean Sea. Atmos. Environ. 10: 1091-1094.

Eppley, R. W., and B. J. Peterson. 1979. Particulate organic matter flux and planktonic new production in the deep ocean. Nature 282: 677-680.

Fellows, D. A., D. M. Karl, and G. A. Knauer. 1981. Large particle fluxes and the vertical transport of living carbon in the upper 1500 m of the northeast Pacific Ocean. Deep-Sea Res. 28: 921-936.

Fenchel, T. 1970. Studies on the decomposition of organic detritus derived from turtle grass Thalassia testudinum. Limnol. Oceanogr. 15: 14-20.

Fenchel, T. 1977. The significance of bactivorous protozoa in the microbial community of detritus particles, pp. 529-544. In: J. Cairns [ed.], Aquatic Microbial Communities. Garland Publ. Co.

Ferrante, J. G., and D. J. Ptak. 1978. Heterotrophic bacteria associated with the degradation of zooplankton fecal pellets in Lake Michigan. J. Great Lakes Res. 4: 221-225.

Fogg, G. E. 1971. Extracellular products of algae in fresh water. Arch. Hydrobiol. 5: 1-25.

Folger, D. W. 1970. Wind transport of land-derived mineral, biogenic, and industrial matter over the North Atlantic. Deep-Sea Res. 17: 337-352.

Fuhrman, J. A., J. W. Ammerman, and F. Azam. 1980. Bacterioplankton in the coastal euphotic zone: Distribution, activity and possible relationships with phytoplankton. Mar. Biol. 60: 201-207.

Fuhrman, J. A., and F. Azam. 1980. Bacterioplankton secondary production estimates for coastal waters of British Columbia, Antarctica, and California. Appl. Environ. Microbiol. 39: 1085-1095.

Gordon, D. C. 1970a. A microscopic study of particles in the North Atlantic Ocean. Deep-Sea Res. 17: 175-185.

Gordon, D. C. 1970b. Some studies on the distribution and composition of particulate organic carbon in the North Atlantic Ocean. Deep-Sea Res. 17: 233-243.

Haas, L. W., and K. L. Webb. 1979. Nutritional mode of several non-pigmented microflagellates from the York River estuary, Virginia. J. Exp. Mar. Biol. Ecol. 39: 125-134.

Hagström, Å., U. Larsson, P. Horstedt, and S. Normark. 1979. Frequency of dividing cells, a new approach to the determination of bacterial growth rates in aquatic environments. Appl. Environ. Microbiol. 37: 805-812.

Hatcher, P. G., R. Rowan, and M. A. Mattingly. 1980. ^{1}H and ^{13}C NMR of marine humic acids. Org. Geochem. 2: 77–85.

Herbland, H. 1975. Utilization par la flore heterotrophe de la matiere organique naturelle dans l'eau de mer. J. Exp. Mar. Biol. Ecol. 19: 19–31.

Ho, K. P., and W. J. Payne. 1979. Assimilation efficiency and energy contents of prototrophic bacteria. Biotechnol. Bioeng. 21: 787–802.

Hobbie, J. E., and C. C. Crawford. 1969. Respiration corrections for bacterial uptake of dissolved organic compounds in natural waters. Limnol. Oceanogr. 14: 528–532.

Hobbie, J. E., O. Holm-Hansen, T. H. Packard, L. R. Pomeroy, R. W. Sheldon, J. P. Thomas, and W. J. Wiebe. 1972. A study of the distribution and activity of microorganisms in ocean water. Limnol. Oceanogr. 17: 544–555.

Hodson, R. E., A. E. Maccubbin, and L. R. Pomeroy. 1981. Dissolved adenosine triphosphate and its utilization by free-living and attached bacterioplankton. Mar. Biol. 64: 43–51.

Hofmann, E. E., J. M. Klinck, and G.-A. Paffenhofer. 1981. Concentrations and vertical fluxes of zooplankton fecal pellets on a continental shelf. Mar. Biol. 61: 327–335.

Honjo, S. 1978. Sedimentation of materials in the Sargasso Sea at a 5,367 m deep station. J. Mar. Res. 36: 469–492.

Honjo, S. 1980. Material fluxes and modes of sedimentation in the mesopelagic and bathypelagic zones. J. Mar. Res. 38: 53–97.

Jacobsen, T. R. 1981. Autotrophic and heterotrophic activity measurements on the continental shelf of Southeastern U. S. Doctoral dissertation, University of Georgia.

Jannasch, H. W., and C. O. Wirsen. 1973. Deep-sea microorganisms: In situ response to nutrient enrichment. Science 180: 641–643.

Jannasch, H. W., K. Eimhjellen, C. O. Wirsen, and A. Farmanfarmaian. 1971. Microbial degradation of organic matter in the deep sea. Science 171: 672–675.

Janzen, D. H. 1977. Why fruits rot, seeds mold, and meat spoils. Am. Nat. 111: 691–713.

Johannes, R. E. 1964. Phosphorus excretion as related to body size in marine animals: microzooplankton and nutrient regeneration. Science 146: 932–934.

Johannes, R. E. 1965. Influence of marine protozoa on nutrient regeneration. Limnol. Oceanogr. 10: 434–442.

Johnson, B. D. 1976. Nonliving organic particle formation from bubble dissolution. Limnol. Oceanogr. 21: 444–446.

Johnson, B. D., and R. C. Cooke. 1980. Organic particle and aggregate formation resulting fom the dissolution of bubbles in sea water. Limnol. Oceanogr. 25: 653–661.

Johnson, P. W., and J. McN. Sieburth. 1979. Chroococcoid cyanobacteria in the sea, a ubiquitous and diverse phototrophic biomass. Limnol. Oceanogr. 24: 923–935.

Karl, D. M. 1979. Measurement of microbial activity and growth in the ocean by rates of stable ribonucleic acid synthesis. Appl.

Environ. Microbiol. 38: 850-860.

Karl, D. M., P. A. LaRock, J. W. Morse, and W. Sturges. 1976. Adenosine triphosphate in the North Atlantic Ocean and its relationship to the oxygen minimum. Deep-Sea Res. 23: 81-88.

Kerrick, K. H. 1981. Studies on the relationships of marine bacteria with particulate material and with protozoan predators. Doctoral dissertation, University of Georgia.

Ketseridis, G., J. Hahn, R. Jaenicke, and C. Junge. 1976. The organic constituents of atmospheric particulate matter. Atmos. Environ. 10: 603-610.

Keys, A., E. H. Christiensen, and A. Krogh. 1935. The organic metabolism of sea-water with special reference to the ultimate food cycle of the sea. J. Mar. Biol. Assoc. UK 20: 181-196.

King, K. R., J. T. Hollibaugh, and F. Azam. 1980. Predator-prey interactions between the larvacean Oikopleura dioica and bacterioplankton in enclosed water columns. Mar. Biol. 56: 49-57.

Kranck, K., and T. Milligan. 1980. Macroflocs: production of marine snow in the laboratory. Mar. Ecol. Prog. Ser. 3: 19-24.

Lampert, W. 1978. Release of dissolved organic carbon by grazing zooplankton. Limnol. Oceanogr. 23: 831-834.

Malone, T. C. 1980. Size-fractionated primary productivity of marine phytoplankton, pp 301-319. In: P. G. Falkowski [ed.], Primary Productivity in the Sea. Plenum Press, New York.

Menzel, D. W., and J. H. Ryther. 1968. Organic carbon and the oxygen minimum in the South Atlantic Ocean. Deep-Sea Res. 15: 327-337.

Morita, R. Y. 1980. Low temperature, energy, survival, and time in microbial ecology, pp. 62-66. In: R. R. Colwell and J. Foster [eds.], Aquatic Microbial Ecology. Maryland Sea Grant Publ. No. UM-SG-TS-80-03.

Morris, B. F., J. Cadwallader, J. Geiselman, and J. N. Butler. 1976. Transfer of petroleum and biogenic hydrocarbons in the Sargassum community, pp. 235-259. In: H. L. Windom and R. A. Duce [eds.], Marine Pollutant Transfer. Lexington Books.

Nalewajko, C. 1977. Extracellular release in freshwater algae and bacteria: Extracellular products of algae as a source of carbon for heterotrophs, pp. 589-624. In: J. Cairns [ed.], Aquatic Microbial Communities. Garland Publ. Co.

Odum, H. T., R. J. Beyers, and N. E. Armstrong. 1963. Consequences of small storage capacity in nanoplankton pertinent to measurement of primary production in tropical waters. J. Mar. Res. 21: 191-198.

Parsons, T. R., M. Takahashi, and B. Hargrave. 1977. Biological Oceanographic Processes, 2nd Edition. Pergamon Press.

Payne, W. J. 1970. Energy yields and growth of heterotrophs. Ann. Rev. Microbiol. 24: 17-52.

Pomeroy, L. R., and F. M. Bush. 1959. Regeneration of phosphate by marine animals, pp. 893-895. Preprints, Int. Oceanogr. Congr., New York.

Pomeroy, L. R., and D. Deibel. 1980. Aggregation of organic matter
    by pelagic tunicates. Limnol. Oceanogr. 25: 643-652.
Pomeroy, L. R., and R. E. Johannes. 1968. Respiration of ultra-
    plankton in the upper 500 meters of the ocean. Deep-Sea Res.
    15: 381-391.
Pomeroy, L. R., H. M. Matthews, and H. S. Min. 1963. Excretion of
    phosphate and soluble organic phosphorus compounds by zooplank-
    ton. Limnol. Oceanogr. 8: 50-55.
Porter, K. G., and E. I. Robbins. 1981. Zooplankton fecal pellets
    link fossil fuel and phosphate deposits. Science 212: 931-933.
Rasmussen, R. A., and F. W. Went. 1965. Volatile organic material
    of plant origin in the atmosphere. Proc. Nat. Acad. Sci. USA
    53: 215-220.
Riley, G. A. 1963. Organic aggregates in sea water and the dynamics
    of their formation and utilization. Limnol. Oceanogr. 8:
    372-381.
Ryther, J. H. 1956. The measurement of primary production. Limnol.
    Oceanogr. 1: 72-84.
Schwartz, R. M., and M. O. Dayhoff. 1978. Origins of prokaryotes,
    eukaryotes, mitochondria, and chloroplasts. Science 199:
    395-400.
Shanks, A. L., and J. D. Trent. 1980. Marine snow: Sinking rates
    and potential role in vertical flux. Deep-Sea Res. 27A:
    137-143.
Shapiro, J. 1967. Induced rapid release and uptake of phosphate
    by microorganisms. Science 155: 1269-1271.
Sieburth, J. McN., and A. S. Dietz. 1974. Biodeterioration in the
    sea and its inhibition, pp. 318-326. In: R. R. Colwell and R.
    Y. Morita [eds.], Effect of the Ocean Environment on Microbial
    Activities. University Park Press.
Smith, D. F., and W. J. Wiebe. 1976. Constant release of photo-
    synthate from marine phytoplankton. Appl. Environ. Microbiol.
    32: 75-79.
Steele, J. H. 1974. The Structure of Marine Ecosystems. Harvard
    Univ. Press.
Sverdrup, H. U. 1955. The place of physical oceanography in ocean-
    ographic research. J. Mar. Res. 14: 287-294.
Tenore, K. R. 1977. Utilization of aged detritus derived from
    different sources by the polychaete, Capitella capitata. Mar.
    Biol. 44: 51-55.
Tenore, K. R., J. H. Tietjen, and J. J. Lee. 1977. Effect of
    meiofauna on incorporation of aged eelgrass, Zostera marina,
    detritus by the polychaete, Nephthys incisa. J. Fish. Res.
    Board Can. 34: 563-567.
Thomas, J. P. 1971. Release of dissolved organic matter from natur-
    al populations of marine phytoplankton. Mar. Biol. 11: 311-323.
Turner, J. T., and J. G. Ferrante. 1979. Zooplankton fecal pellets
    in aquatic ecosystems. BioScience 29: 670-677.
Van Valen, L. 1971. The history and stability of atmospheric
    oxygen. Science 171: 439-443.

Vernadskii, V. I. 1926. Biosfera. Leningrad.

Waksman, S. A., C. L. Carey, and H. W. Reuszer. 1933. Marine
    bacteria and their role in the cycle of life in the sea. I.
    Decomposition of marine plant and animal residues by bacteria.
    Biol. Bull. 65: 57-59.

Walsh, J. J., G. T. Rowe, R. L. Iverson, and C. P. McRoy. 1981.
    Biological export of shelf carbon is a sink of the global CO_2
    cycle. Nature 291: 196-201.

Waterbury, J. B., S. W. Watson, R. R. L. Guillard, and L. E. Brand.
    1979. Widespread occurrence of a unicellular, marine, plank-
    tonic, cyanobacterium. Nature 277: 293-294.

Went, R. 1966. On the nature of Aitken condensation nuclei. Tellus
    18: 549-556.

Went, F. W., D. B. Slemmons, and H. N. Mozingo. 1967. The nature
    of atmospheric condensation nuclei. Proc. Nat. Acad. Sci. USA
    58: 69-74.

Wheeler, J. R. 1975. Formation and collapse of surface films.
    Limnol. Oceanogr. 20: 338-342.

Wiebe, W. J., and L. R. Pomeroy. 1972. Microorganisms and their
    association with aggregates and detritus in the sea: a micro-
    scopic study. Mem. Ist. Ital. Idrobiol. 29 (Suppl.): 325-352.

Wiebe, W. J., and D. F Smith. 1977. Direct measurement of
    dissolved organic carbon release by phytoplankton and
    incorporation by microheterotrophs. Mar. Biol. 42: 213-223.

Wiegert, R. G., and D. F. Owen. 1971. Trophic structure, available
    resources, and population density in terrestrial vs. aquatic
    ecosystems. J. Theor. Biol. 30: 69-81.

Williams, P. J. leB. 1970. Heterotrophic utilization of dissolved
    organic compounds in the sea. I. Size distribution of popula-
    tions and relationship between respiration and incorporation of
    growth substrates. J. Mar. Biol. Assoc. UK 50: 859-258.

Williams, P. J. leB. 1981 Incorporation of microheterotrophic
    processes into the classical paradigm of the planktonic food
    web. Kiel. Meeresforsch. Sonderh. 5: 1-28.

Williams, P. M., H. Oeschger, and P. Kinney. 1969. Natural radio-
    carbon activity of the dissolved organic carbon in the north-
    east Pacific Ocean. Nature 224: 256-258.

Wright, R. T., and J. E. Hobbie. 1966. Use of glucose and acetate
    by bacteria and algae in aquatic ecosystems. Ecology 47:
    447-464.

STRATEGIES FOR GROWTH AND EVOLUTION OF MICRO-ORGANISMS

IN OLIGOTROPHIC HABITATS

H. van Gemerden[1] and J. G. Kuenen[2]

[1]Department of Microbiology, University of Groningen
Kerklaan 30, 9751 NN HAREN, The Netherlands
[2]Department of Microbiology, Delft University of Technology
Julianalaan 67 A, 2628 BC DELFT, The Netherlands

INTRODUCTION

The strategy of microbes to adapt to a particular environment occurs both at the phenotypical and at the genotypical level. In general phenotypic responses may be needed to cope with temporary changes, whereas genetic adaptations may be needed for long lasting changes in the environment.

The enormous diversity of microorganisms in nature illustrates the equally enormous diversity in ecological niches to harbour these organisms. Within the limits of biological feasibility, organisms have adapted to physical, chemical, and biological changes or stress. Examples are the adaptation to different temperatures, to low nutrient environments, to low light intensities, to the supply of single or mixed substrates, to the continuous or discontinuous availability of nutrients, and to environments with aerobic-anaerobic changes.

The genetic adaptation for these changes may have resulted in the development of organisms which became either highly specialized or remained highly versatile; in other words, evolution led to organisms with either a low or a high potential for phenotypic changes. Irrespective whether we talk about a genetic or a phenotypic change, in each case it is the competition and the subsequent selection in the environment which determines the value of a certain change. A better understanding of competition and selection thus will teach us more of the strategies and mechanisms of adaptations in microorganisms. These selection processes can be assumed to have taken place in both freshwater and marine habitats. In the NATO meeting attention was focussed on the sea, but relevant information can be deduced

25

as well from the many studies with freshwater organisms. So far, no
essential difference with respect to the principles of selection and
competition have been encountered in marine and freshwater bacteria,
and it seems admissible to include studies with both groups of
organisms.

In addition, for a proper judgement of the selection pressure
exerted on organisms living under oligotrophic conditions (i.e., low
population densities, low substrate concentrations, but high turnover
rates) a comparison with more eutrophic habitats may be very reward-
ing. This holds in particular for studies in which substrate limitation occurs. Perturbation of the established balanced growth provides
us with extremely useful information on an organism's capacity to
assimilate excess nutrients. Such culture systems are technically
much easier to handle than strict oligotrophic systems, yet both
systems share a number of characteristics. In the study of selection
and competition between microorganisms the continuous culture
("chemostat") has been proven to be a most useful tool. This culture
method allows us to artificially, and specifically, amplify the
selective pressure exerted on bacteria. In general, organisms show a
saturation-type of growth-rate response to increasing substrate
concentrations. For bacteria, the relation between the specific
growth rate (μ) and the concentration of the substrate (s) under
conditions of balanced growth initially has been described by Monod
(1942) in analogy to the Michaelis-Menten kinetics of enzyme systems.

In Monod's description the specific rate of growth at a given
substrate concentration is determined by the organism's kinetic
parameters μ_{max} and K_s, according to

$$\mu = \mu_{max} \frac{s}{K_s + s}$$

in which μ_{max} is the maximum growth rate attainable in the presence
of excess substrates and K_s is the saturation constant numerically
equal to s at $\mu = 1/2 \, \mu_{max}$.

In later modifications, the maintenance energy concept and the
effects of inhibitory substrates were included in the mathematical
description of balanced growth.

Organisms which are able to grow relatively rapidly at low
nutrient concentrations are usually said to have a high affinity for
the substrate. However, the term affinity is poorly described and
certainly is not only related to the K_s value. Indeed, in many cases
organisms with a high affinity for a substrate were found to have a
low K_s value, but erroneously a low K_s value is often interpreted as
equivalent to a high affinity. An organism with a high μ_{max} value
may have a high K_s value compared to another organism with a lower

μ_{max} value, but still be able to grow faster at low substrate concentrations.

The specific rate of growth at the prevailing substrate concentration is the decisive factor. A mere comparison of K_s values makes sense only for organisms with similar μ_{max} values. In order to compare different organisms with respect to their affinity of the same substrate, the slope of the μ/s curve at a given low concentration of s (not exceeding K_s) can be calculated according to

$$\frac{d\mu}{ds} = \mu_{max} \frac{K_s}{(K_s + s)^2}$$

The organism with the highest $d\mu/ds$ value then is said to have the highest affinity (Zevenboom 1980; see also Healey 1980; Brown and Molot 1980).

When substrate concentrations are low compared to the K_s values, as usually is the case during carbon and energy limitation, the ratio μ_{max}/K_s can be used to compare the affinity for a substrate. However, it should be realized that threshold concentrations, and the minimum growth rate (discussed later) may complicate the interpretation.

In the chemostat, low substrate concentrations can be maintained for long periods of time, and the continuous cultivation thus is an elegant and indispensible tool in the study of competition and selection. When discussing the "strategies" of microorganisms we mean a way of life which by selection has shown survival value during the evolution.

ADAPTATION TO LOW NUTRIENT CONCENTRATIONS

Affinity

Physiologically very similar organisms may occupy different niches on the basis of slight differences in the kinetic parameters K_s and $_{max}$ for a given growth-rate limiting substrate. This can be demonstrated with the aid of continuous cultures inoculated with natural samples; the operation of chemostats at different dilution rates often results in the selective enrichment of different organisms. These phenomena have been extensively studied by Jannasch in Woods Hole, U.S.A. and the group of Veldkamp in Groningen, The Netherlands. Jannasch (1967) demonstrated the reproducible enrichment of several marine bacteria by varying the dilution rate of sea water supplemented with a nutrient (lactate). The dominant species were isolated and grown in chemostats at various dilution rates in order to estimate the kinetic parameters K_s and μ_{max}. Thereafter, competition experiments under similar conditions were performed with

mixtures of the two pure cultures. The outcome of these selection
experiments could be predicted on the basis of the pure culture
responses to various concentrations of the substrate (various
dilution rates) according to Fig. 1.

In another example (Kuenen et al. 1977), two closely related
obligate chemolithotrophic sulfur bacteria (Thiobacillus thioparus
and the spirillum-shaped Thiomicrospira pelophila) were isolated from
the same sample taken from an intertidal mud flat of the Waddensea
(Holland). Their carbon and energy metabolism was remarkably similar
(Kuenen and Veldkamp 1973) but they had different sulfide tolerances
(Kuenen and Veldkamp 1972).

The small spirillum grew better at higher sulfide concentrations
than the rod-shaped Thiobacillus. The magnitude of the sulfide concentrations employed (300 ml seawater on top of 100 ml agar containing 2 to 20 mM sulfide) and the ambiguous results under thiosulfate
limitation suggested that the success of the spirillum at higher
concentrations of sulfide was not explained by a better affinity for
sulfide, but rather by some other factor related to the high sulfide
concentrations. It was postulated that the actual selective factor
was the concentration of iron. At high sulfide concentrations iron
is precipitated as ferrous iron with an extremely low solubility
product. Subsequently, the μ/Fe curves of the two isolates were
determined and found to cross. Competition experiments performed at
dilution rates of 0.05-0.25 h^{-1} resulted in a dominant population of
the spirillum; a further increase in the dilution rate resulted in
the selective enrichment of the Thiobacillus.

Similarly, different chemo-organotrophic bacteria became
dominant in phosphate-limited continuous cultures inoculated with
natural samples and run at either low or high dilution rates (Kuenen
et al. 1977). After five volume changes a spirillum had become
dominant in the culture run at a dilution rate of 0.03 h^{-1}.

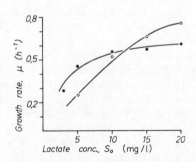

Figure 1. Relation between the specific growth rate of Pseudomonas
(201) (0--0) and Spirillum (101) (●--●) and the substrate lactate.
From Jannasch (1967).

Table 1.   Kinetic parameters of organisms selected in continuous
            culture under phosphate limitation (from Kuenen et al.
            1977).

Organism	μ_{max} (h^{-1})	$K_s(PO_4^{3-})$ (nmol·L^{-1})	Dilution Rate of Enrichment (h^{-1})
Rod-shaped bacterium	0.48	66	0.03
Spirillum sp.	0.24	27	0.3

The parallel experiment at a dilution rate of 0.3 h^{-1} yielded a
dominant culture of a rod-shaped bacterium. When the respective
organisms were isolated by plating and assayed for their kinetic
parameters, they were found to differ in both K_s and μ_{max} value
(Table 1). The extremely low K_s values (10^{-7}, 10^{-8} M) reflect
actual concentrations found in the habitat of the organisms.

Assuming the absence of interactions other than the competition
for phosphate, the spirillum will outcompete the rod-shaped organism
at phosphate concentrations below 0.01 μM (the crossing point). At
that concentration, the specific growth rate of both organisms is
0.07 h^{-1}, well below the maximum specific growth rate of the
spirillum (0.24 h^{-1}). The spirillum thus can be said to be best
adapted to low phosphate concentrations. Interestingly, experiments
with other growth-rate limitations showed that its high affinity is
not restricted to phosphate. It was found that the spirillum
outgrew the rod-shaped bacterium at low dilution rates as well under
limitations of K^+, Mg^{2+}, NH_4^+, aspartate, succinate, and lactate.
For all carbon sources, the specific growth rate of the rod-shaped
bacterium was higher than that of the spirillum. Thus it appears
that spirillum is well adapted to low nutrient concentrations in
general and its good affinity does not seem to be a specific
property related to one or two nutrients.

A very intriguing case of crossing μ/s curves was reported by
Harder (1969; Harder and Veldkamp 1971). An obligately psychrophilic
Pseudomonas and a facultatively psychrophilic Spirillum were both
isolated from samples taken from the North Sea in the winter and the
early spring using lactate as the carbon and energy source (Harder
and Veldkamp 1968). On the basis of μ_{max} determinations of pure cultures and the outcome of competition experiments at various dilution
rates, the μ/s curves at different temperatures as shown in Fig. 2
were derived.

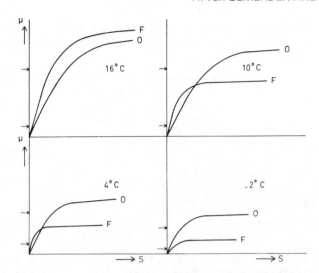

Figure 2. Specific growth rates of an obligately psychrophilic
Pseudomonas sp. (O), and a facultatively psychrophilic Spirillum sp.
(F) as a function of the lactate concentration at different temperatures. The curves are schematic and based on the measurements at the
growth rates indicated. From Harder and Veldkamp (1971).

At 10°C, the obligate psychrophile was unable to outcompete the
Spirillum at low lactate concentrations; lowering the temperature to
4°C restricted the competitive advantage of the Spirillum to even
lower concentrations. A still further reduction in the temperature
to -2° resulted in the competitive exclusion of the Spirillum
irrespective of the dilution rate. The opposite was found true at
a temperature of 16°C. These data suggest that mineralization in
permanently cold ocean waters proceeds through the action of
obligately psychrophilic bacteria.

Surface to Volume Ratio

One question emerging from the results discussed above is why
certain organisms exhibit higher affinities for substrates present
in low concentrations. And also, why can such organisms not express
their supremacy at higher substrate concentrations as well? Or, in
other words, why does an organism not combine a high μ_{max} with a high
affinity?

The first point, the scavenging potential, may be related to
the organism's surface to volume ratio (S/V ratio). In general,
this ratio is higher in organisms which become dominant at lower
substrate concentration. Table 2 shows some relevant data. These
figures suggest that in general a genetic adaptation is required for

Table 2.   Surface to volume ratios of different spiral- and rod-
shaped bacteria grown at limiting substrate concentrations
(adapted from Keunen et al. 1977).

Organism	Limiting Substrate	Specific Growth Rate (h^{-1})	Average Length (μm)	Average Diameter (μm)	Surface/ Volume (μm^{-1})
Pseudomonas sp.[a]	lactate	0.1	2.9	1.1	4.3
Spirillum sp.	lactate	0.1	3.5	0.55	8.0
Pseudomonas sp.[b]	lactate	0.16	2.26	0.80	5.66
Spirillum sp.	lactate	0.16	2.21	0.56	7.82
Unidentified rod[c]	phosphate	0.2	2.8	1.1	4.3
Spirillum sp.	phosphate	0.2	3.8	0.6	7.2
Thiobacillus thioparus[c]	thiosulfate	0.1	2.0	0.4	11
Thiomicrospira pelophila	thiosulfate	0.1	2.5	0.2	21

[a] Harder and Veldkamp unpublished
[b] Matin and Veldkamp 1978
[c] Kuenen et al. 1977

obtaining a high S/V ratio and that such an adaptation is related to
the potential to grow well at low concentrations of the nutrient.

Adaptations to the uptake of nutrients present in low
concentrations by virtue of an increase in the S/V ratio are not
only controlled at the genetic level. A phenotypic response was
reported by Matin and Veldkamp (1978).

A Spirillum and a Pseudomonas, both freshwater isolates, were
shown to have crossing μ/lactate curves, the Pseudomonas outgrew the
Spirillum at dilution rates exceeding 0.29 h^{-1}, but the converse
occurred at lower D values. Both organisms were found to increase
the S/V ratio at low dilution rates, the ratio in Spirillum remained
the highest, however (Table 3). This might explain the success of
this organism at low substrate concentrations.

The formation of prosthecae also increases the S/V ratio. At
very low nutrient concentrations the prosthecae of Ancalomicrobium
are fully developed, whereas the opposite is true under nutrient-
rich conditions. In other species the occurrence of prosthecae is
not so easily explained (Dow and Whittenbury 1980). Oligotrophic

Table 3. Surface to volume ration of two chemoorganotropic
          bacteria at various specific growth rates (from
          Matin and Veldkamp 1978).

| Specific Growth Rate | Surface/Volume (μm^{-1}) | |
	Spirillum	Pseudomonas
0.06	8.05	6.24
0.16	7.82	5.66
0.35(μ_{max})	6.27	
0.64(μ_{max})		4.59

environments oft~n contain microbial forms with very long stalks or
appendages which do not necessarily improve the rate of nutrient
uptake but may instead be a reflection of a reduced growth rate
under near starvation conditions (Dow and Whittenbury 1980).

However, it has been demonstrated in Hyphomicrobium that
organisms with long stalks have a higher ~ffinity for methanol than
organisms with short stalks. Long stalks were observed at low
specific growth rates whereas short stalks were found at high growth
rates (Table 4; Meiberg et al., unpublished). Phenotypic changes
are usually smaller than genetically controlled modifications. The
organisms can apparently cope with small variations in the environment by phenotypic responses, but for larger changes in the S/V
ratio a genetic change is needed.

The second question, why does not an organism combine fast
growth at saturating substrate concentrations (high μ_{max}) with relatively fast growth at low concentrations (high affinity), is hard to
answer. It may be that at the phenotypic level the "elasticity" of
the system is often too small to allow dramatic changes in the S/V
ratio. A genetic adaptation seems to be required to change the S/V
ratio by more than 20-30%. As a consequence, the price to be paid
could be a lower μ_{max}. This may be rationalized by arguing that the
synthesis of cells with relatively large amounts of membrane (high
S/V ratio), cannot proceed at the same high speed as that of cells
which have more protein per unit dry weight (low S/V ratio) available
for biosynthetic processes (growth).

It is cbvious that the phenomenon just described as "limited
elasticity" is one of the constraints put upon the evolution of
microorganisms. It should be realized that much of this "limited

Table 4.   Affinities for methanol of <u>Hyphomicrobium</u> X (DSM 1869)
            at different dilution rates (from Meiberg et al.,
            unpublished).*

Specific Growth Rate (h^{-1})	Stalk Length	Apparent K_s (Methanol) (M)
0.01	long	10^{-7}
0.1	short	10^{-6}

*The organism's μ_{max} is approximately 0.18 h^{-1}.

elasticity" in all probability is to be traced back to the purely
biochemical level, for example to the properties and limitations of
single enzymes, or membranes, or of the total machinery for protein
synthesis.

Growth vs. Storage

The underlying principle of the dilution-rate dependent outcome
of competition experiments is the apparent inconsistency of a high
μ_{max} with a high affinity. As a consequence organisms with different attributes, resulting in crossing μ/s curves, become dominant at
different substrate concentrations. Non-crossing μ/s curves would
have resulted in the competitive exclusion of one of the organisms,
irrespective of the dilution rate imposed. It must be pointed out
that this refers to balanced growth only; that is, the nutrient is
added continuously and at a constant rate. As a rule, such conditions do not occur in nature for long periods of time. In addition,
the decisive influence of the S/V ratio would suggest that uptake
mechanisms always are the rate-limiting factor.

The importance of other parameters can nicely be demonstrated
by the outcome of competition studies between two species of phototrophic bacteria, <u>Chromatium vinosum</u> and <u>Chromatium weissei</u> (van
Gemerden 1974). <u>C. vinosum</u> measures on the average 2 x 4 μm,
resulting in a S/V ratio of 2.4 μm^{-1}; <u>C. weissei</u> measures on the
average 4 x 8 μm, resulting in a S/V ratio of 1.2 μm^{-1}. However,
the μ/sulfide curves of the two organisms do not cross: <u>C. vinosum</u>
exhibits a higher μ_{max} value for sulfide and has a higher affinity
as well. As a consequence, <u>C. weissei</u> is outcompeted at all dilution
rates. However, the addition of substrate to non-growing pure
cultures resulted in a faster uptake by <u>C. weissei</u>, the organism
with the lower S/V ratio.

Under these conditions, the oxidation of sulfide results in part
in the formation of sulfate and in part in the formation of sulfur.
The latter compound is accumulated inside the cells. Also, the
glycogen synthesized from CO_2 is stored intracellularly. Thus, the
substrate sulfide is depleted virtually without the synthesis of
structural cell material.

The specific rate of glycogen synthesis was found to be similar
in the two organisms. However, C. weissei produced less sulfate and
more sulfur than C. vinosum. This hoarding of the substrate is a
competitive advantage under fluctuating conditions. Both sulfur and
glycogen were utilized for growth once sulfide was depleted. In
fact, balanced coexistence of the two organisms was observed under
fluctuating conditions; the relative abundance of the two organisms
was dependent on the amount of sulfide accumulated in the dark period
(Table 5).

A last point of particular importance is that the specific rate
of CO_2 fixation exhibited by C. weissei is higher than its maximum
specific growth rate. C. weissei was found to oxidize sulfide and

Table 5.   Abundancy of Chromatium vinosum and Chromatium weissei
            during balanced coexistence in relation to the light
            regimen (from van Gemerden 1974).*

| Light Regimen | Organism | Relative Abundance (% Biovol.) | |
		Inoculum	Balanced Coexistence
continuous light	C. vinosum	10	100
	C. weissei	90	0
18 h light	C. vinosum	30	100
6 h dark	C. weissei	70	0
6 h light	C. vinosum	20	63
6 h dark	C. weissei	80	37
4 h light	C. vinosum	60	30
8 h dark	C. weissei	40	70

* Sulfide-limited continuous cultures were illuminated as indicated.
  In the dark periods the addition of the substrate was continued
  resulting in the accumulation of sulfide in the culture vessel,
  followed by its rapid oxidation during the next light period.
  Experiments were performed at D = 0.011 h^{-1}.

sulfur at a rate that would permit the organisms to grow at a μ of
0.05 h^{-1}, whereas the organisms μ_{max} is only 0.04 h^{-1}. This implies
that even if the substrate were available at elevated concentrations
for a somewhat longer period of time, it would only be advantageous
for a limited period.

Once the organisms are "stuffed" with sulfur and glycogen, the
rate of sulfide oxidation has to slow down. Thus, a high uptake
capacity is of particular relevance if fluctuating concentrations of
substrates are encountered in the organism's habitat.

Specialism vs. Versatility

It can be deduced from the previous discussion that no organisms
can combine more than a limited number of attributes. This implies
that organisms evolving in one or another direction are under the
continuous challenge to either give up an old property in order to
hold the new one, or to discard the newly acquired trait. This
"give-and-take" principle will be illustrated by comparing a highly
specialized autotroph with a very versatile mixotroph. Both bacteria
are representative of the colorless sulfur bacteria; these organisms
are not encountered in the open ocean, however, the principle may
have a wide application.

The specialist is the obligately chemolithotrophic Thiobacillus
neapolitanus, able to grow rapidly in thiosulfate + CO_2 media. This
organism cannot grow at all heterotrophically and even under mixo-
trophic conditions (thiosulfate + organic compounds) little of the
cellular carbon is derived from the organic carbon supplied in the
medium (Kuenen and Veldkamp 1973).

The versatile organism is Thiobacillus A2, able to grow autotrophically (thiosulfate + CO_2), heterotrophically (20 or more different organic compounds) and mixotrophically (thiosulfate + organic
compounds). Two substrates can be used simultaneously by this
organism, provided both substrates are at limiting concentrations.

In a comparison of these species, we first focus on the μ_{max}
exhibited under various conditions (Table 6). The specialized
Thiobacillus neopolitanus shows a very high μ_{max} in thiosulfate +
CO_2 media and the addition of acetate did not enhance this value.
The versatile Thiobacillus A2 has a much wider choice of carbon and
energy sources. However, it seems that the price to be paid is that
it cannot grow rapidly on any substrate and that its heterotrophic
μ_{max} is not enhanced by the addition of thiosulfate.

Relevant information on the organism's affinity for thiosulfate
could be deduced from the outcome of competition experiments. Under
autotrophic conditions (thiosulfate limitation), Thiobacillus
neapolitanus outcompeted Thiobacillus A2 at dilution rates of

Table 6.   Maximum specific growth rates of two specialists
            (Thiobacillus neapolitanus, Spirillum G7) and a versatile
            bacterium (Thiobacillus A2) on thiosulfate, acetate, or
            both (from Kuenen 1980).

Organism	Thiosulfate $\mu_{max}(h^{-1})$	Thiosulfate and Acetate $\mu_{max}(h^{-1})$	Acetate $\mu_{max}(h^{-1})$
T. neapolitanus	0.35	0.35	no growth
T. A2	0.10	0.22	0.22
Spirillum G7	no growth	0.43	0.43

0.075 h^{-1}, 0.050 h^{-1}, and 0.025 h^{-1}, indicating a superior affinity
for thiosulfate of the specialist. The fact that the versatile
organism was not completely eliminated but maintained at a level of
5-10% of the total cell number is explained by the excretion of
organic compounds (glycollate) by Thiobacillus neapolitanus. The
latter organism is unable to re-assimilate the glycollate, whereas
this compound is efficiently utilized by Thiobacillus A2.

     The low but persistent numbers of the versatile Thiobacillus A2
at dilution rates of 0.025 h^{-1} and higher, point to the importance of
organic compounds for the survival of this facultative chemolithotroph. The inclusion of increasing amounts of glycollate or acetate
in the media resulted in balanced coexistence with increasing numbers
of Thiobacillus A2. The outcome of a series of such experiments performed at a dilution rate of 0.07 h^{-1} is shown in Fig. 3 (Gottschal
et al. 1979). Media containing 7.8 mM acetate or glycollate in
addition to 40 mM thiosulfate resulted in 90% of the versatile
Thiobacillus A2 and 10% of the specialist Thiobacillus neapolitanus
(Gottschal et al. 1979). Similarly, the inclusion of glucose in the
thiosulfate medium resulted in a rapid domination of the versatile
organism and the virtual exclusion of Thiobacillus neapolitanus
which previously made up about 90% of the biomass (Smith and Kelly
1979). These data clearly demonstrate that the capacity for mixo-
trophy is of selective advantage to the versatile organism in its
competition with obligate chemolithotrophic species. The phenomenon
appears to be a general one; similar competition experiments were
conducted between the versatile Thiobacillus A2 and a heterotrophic
specialist, the Spirillum G7 (see Table 6 for μ_{max} values). With
acetate as the sole growth-limiting substrate, Thiobacillus A2 was
excluded at dilution rates of 0.07 h^{-1} and 0.15 h^{-1}. However,
increasing concentrations of thiosulfate (0-10 mM) in the acetate

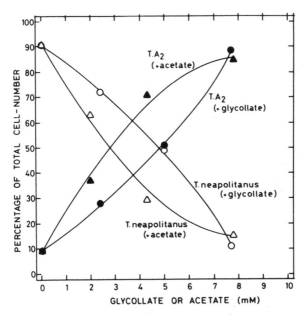

Figure 3. Effect of different concentrations of organic substrate on the outcome of the competition between Thiobacillus A2 and T. neapolitanus for thiosulfate. Chemostats were run at a dilution rate of 0.07 h^{-1}. The inflowing medium contained thiosulfate (40 mM) and either acetate or glycollate at concentrations ranging from 0 to 7 mM. Cell numbers were determined after steady states were established. From Gottschal et al. (1979).

(10 mM) medium eventually led to the elimination of the hetero-trophic specialist (Gottschal et al. 1979) (see Fig. 4).

The general validity of these observations is indicated by the fact that facultatively chemolithotrophic thiobacilli indeed could be enriched from various freshwater environments by feeding continuous cultures with mixtures of acetate and thiosulfate. Surprisingly, similar experiments inoculated with samples from tidal mud flats resulted in a mixed culture of a heterotroph and a specialized chemolithotroph (Gottschal and Kuenen 1980).

Another example of competition between a specialist and a more versatile organism for mixed substrates was reported by Laanbroek et al. (1979). The specialist, Clostridium cochlearium, was found to outcompete the less specialized Clostridium tetanomorphum during growth on glutamate as the only carbon and energy source. Upon the inclusion of glucose in the reservoir solution, the two organisms were able to coexist. Glutamate can be utilized by both organisms but the specialist is unable to grow on glucose. Conceivably, the

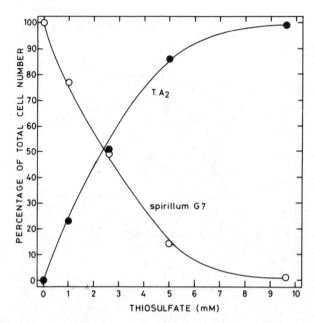

Figure 4. Effect of different concentrations of thiosulfate on the outcome of the competition between Thiobacillus A2 and Spirillum G7 for acetate. Chemostats were run at dilution rates of 0.07 h^{-1}. The inflowing medium contained acetate (10 mM) together with thiosulfate in concentrations ranging from 0 to 9.7 mM. Cell numbers were estimated after steady states were established. From Gottschal et al. (1979).

versatile organism grew on the glutamate in addition to the glucose in the mixed culture.

Summarizing this section, it may be concluded that specialists have an advantage over versatile organisms when the turn-over rate of a single substrate is high. Versatile organisms may be expected to flourish when two (or more) substrates are turned over at comparable speeds. Based on this information, one could expect that versatile organisms are more abundant in natural ecosystems. However, as will be discussed in the next section, an organism's reactivity is of utmost importance when substrates are available at limited and, in particular, at fluctuating concentrations.

It can be argued that in general specialists, not being flexible, have no need for inducible pathways for their carbon sources and energy sources. Such organisms may even have some metabolic "lesions" in pathways which are not functional in the organism's metabolism. For example, in the case of the specialist Thiobacillus neapolitanus, it is clear that it does not need a cyclic oxidation of acetyl-CoA; thus the fact that its TCA cycle is incomplete is not a disadvantage. Being a specialist, depending for growth and survival on one substrate, it can be expected to be equipped with excess capacity to oxidize that substrate.

On the other hand, the versatile organisms cannot permit them-selves to have excess capacity for all the different compounds which might be utilized when the excess capacities are not needed. Instead, the full potential is not switched on in the absence of the corresponding nutrient; in other words these organisms will show a strong adaptation to different growth conditions. Thus, Thiobacillus A2 can fix CO_2 when necessary; its ability to oxidize thiosulfate is substrate-inducible and is repressed by organic compounds (Gottschal et al. 1981b). Similarly, in Rhodopseudomonas capsulata, a versatile phototroph, the utilization of acetate results in a low potential rate of sulfide oxidation; this is in contrast to the more special-ized Chromatium vinosum (Wijbenga and van Gemerden 1981; Beeftink and van Gemerden 1979).

These phenomena have two consequences, both of which lead to a lower ability of the versatile organisms to react when compared to the specialists. First of all, in an environment where the concen-tration of substrates fluctuate - which rather is a rule than an exception - the versatile organisms may be unable to respond quickly to a sudden deviation from zero concentration because its potential is repressed. In the resulting lag period, its specialist competitor with a constitutive system is strongly favored. Secondly, when there is a sudden increase in the supply of a nutrient that has been avail-able in low concentrations, the nutrient will be oxidized more slowly by the versatile organism because even under fully induced conditions it has a lower capacity than the specialist. Under fluctuating conditions of organic and inorganic compounds, the versatile Thiobacillus A2 is at a competitive disadvantage compared to the specialist Thiobacillus neapolitanus and the specialist Spirillum G7. Its continuous adaptation causes the versatile organism to be slow in its response to a single substrate (Gottschal et al. 1981a). Likewise, under fluctuating conditions the specialist C. vinosum has a competitive advantage over the versatile R. capsulata (Wijbenga and van Gemerden, unpublished). Some characteristic differences between metabolically versatile organisms and highly specialized organisms are listed in Table 7.

In conclusion, the price to be paid for being versatile is to be less reactive; in other words, a strategy directed at versatility is incompatible with a strategy aimed at reactivity. To some extent we can understand in a functional sense why such a choice can or even must be made. Again, one should realize that this strategy applies not only to intact organisms, but is also a part of the basic properties of subcellular structures.

Table 7.   Differences between metabolically versatile and special-
            ized organisms extrapolated from studies with
            Thiobacillus species.

Specialist	Versatile
1. Few substrates utilized	1. Many substrates utilized
2. High specific growth rate on a single substrate	2a. Low specific growth rate on a single substrate
	2b. Relatively high specific growth rate on mixed substrates
3. High excess capacity to respire substrates	3. Low excess capacity
4. Constitutive enzymes for energy and carbon metabolism	4. Inducible enzymes
5. High reactivity for few substrates	5. Low reactivity for many substrates
6. Metabolic lesions	6. Many pathways, often over-lapping
7. Low endogenous respiration	7. High endogenous respiration
8. Very resistant to starvation	8. Probably less resistant

Adaptations of Enzyme Systems

     One way for an organism to be able to continue to grow at
decreasing concentrations of a nutrient, is to increase the content
of the critical enzyme. Another way is to modulate the enzyme's
activity, for example, by allosteric control. A third way is the
de novo synthesis of special low-affinity enzymes.

     The effect of the dilution rate on the synthesis of various
enzymes in carbon-limited continuous cultures has recently been
summarized by Matin (1979). Out of 51 responses reported, the
majority showed an increased activity with decreasing dilution rate.
Such a response was observed for almost all the catabolic enzymes
examined. An increased activity with increasing dilution rate was
found only in enzymes which are involved either in biosynthetic
reactions or with the respiratory chain. The simplest explanation

for the latter phenomenon is that induction increases with the increasing substrate concentration encountered at higher dilution rates. Likewise, increasing enzyme synthesis with decreasing dilution rates can be explained by a release of catabolite repression as the result of decreasing intracellular pools of metabolites in organisms grown at lower dilution rate (Matin 1979).

The increased synthesis of catabolic enzymes in slower growing organims is understandable if one realizes that the steady-state substrate concentrations are well below the K_m of the enzymes involved. If however such enzymes are inducible rather than derepressible, a low substrate concentration would not, or not sufficiently, initiate the synthesis of that enzyme. This shortage of enzyme would automatically result in a rise in the substrate concentration unless another organism were present that contained the enzyme constitutively. Thus, at low substrate concentrations, organisms with inducible enzyme systems can be expected to be out-competed by those that contain such an enzyme constitutively. Alternatively, the enzyme can be derepressable rather than inducible so that at low metabolite concentrations it can be present at sufficient levels. The competitive exclusion of E. coli wild type by mutants which are constitutive for the hydrolyzing enzyme β-galactosidase under lactose-limitation may illustrate this phenomenon (Matin 1979).

Another way to cope with low nutrient concentrations is the preferential synthesis of enzymes with improved affinity. The best example is the combined action of glutamine synthetase (GS) and glutamine-oxoglutarate-amidotransferase (GOGAT) at low concentrations of NH_4^+ which replaces the glutamate dehydrogenase (GDH) operative at high concentrations of $NH4^+$ (Tempest et al. 1970). In Aerobacter aerogenes the $K_mNH_4^+$ of GDH is about 10 mM, while the corresponding value of GS is about 0.5 mM NH_4^+; however, in the latter system 1 ATP is consumed in the synthesis of 1 glutamic acid. In many organisms these two systems are inducible and can be repressed. However, in a species of Caulobacter studied by Ely et al. (1978) GDH is lacking and GS is present at a constitutive level. The obvious advantages for the organism are that it does not have to carry the genetic information for other assimilation pathways, and that NH_4^+ can be taken up without any time lag. The disadvantage of such a "strategy" is that at higher levels of NH_4^+ the organism still must use the system which requires ATP, and, even worse, is inhibited at high concentrations of NH_4^+. The latter phenomenon may ultimately even result in the complete inhibition of growth. This may be considered an extreme case of adaptation (specialization) to low nutrient concentrations.

Generally, at higher concentrations, caulobacters are inhibited by a wide variety of organic and inorganic nutrients. This nutrient-concentration sensitivity is probably related to the balance of

nutrients in general and not to the concentration of a single
compound (Poindexter 1981).

The existence of low and high affinity enzyme systems is not
restricted to the assimilation of NH_4^+ and such systems have been
reported as well for organic nutrients as glycerol and glucose
(Neijssel et al. 1975; Whitting et al. 1976).

In addition to the adaptations mentioned, organisms have
developed mechanisms to avoid low nutrient environments, or to
"escape" from predation. Such adaptations are beyond the scope of
the present paper. For a detailed discussion the reader is referred
to Dow and Whittenbury (1980).

Maintenance, Survival, and the Minimum Growth Rate

Organisms, even when not growing, require a certain amount of
energy per unit time and biomass. This is required first of all for
the maintenance of a proton motive force (p.m.f.) over the membrane
(Konings and Veldkamp 1980) which is needed for the performance of
such life-sustaining processes as osmotic regulation, control of
internal pH, synthesis of ATP, and turnover of essential macromolecules (Pirt 1975; Tempest 1970; Dawes 1976). The energy
required for all processes not directly related to growth is
collectively called the energy of maintenance. As a consequence,
the curve describing the relation between the specific rate of
growth and the concentration of the limiting source of energy does
not pass through the origin. In such a plot, the ordinate intercept
of the extrapolated curve has been described as the specific
maintenance rate, μ_e (Powell 1967). The same principle applies to
the growth rate dependency of phototrophic bacteria on the intensity
of light. A high μ_e value of organism A compared to organism B then
indicates that organism A requires a higher irradiance to maintain
its cellular integrity.

In the continuous cultivation of microorganisms, the specific
rate of growth, μ, is usually considered identical to the dilution
rate, D, once a steady state is established. Also, the magnitude of
the energy of maintenance is considered small compared to the energy
required for growth processes. Actually, however, under steady-state
conditions μ is always somewhat higher than D because the viability
is somewhat lower than 100% (Sinclair and Topiwala 1970; Veldkamp
1976). Also, the steady-state cell densities are somewhat lower due
to the fact that some of the (limiting) energy source is used for
maintenance purposes. In most studies the dilution rates employed
are such that neglecting the maintenance rate, μ_e, and the death
rate, γ, can be justified. However, with decreasing dilution rate
these parameters become increasingly important. It can be argued,
therefore, that it is not justified to apply the original Monod
equation to growth under oligotrophic conditions.

The increasing importance of the energy for maintenance at decreasing dilution rates becomes noticeable in cell densities which are lower than expected from the yields at higher growth rates. This phenomenon has been observed in a number of bacteria and yeasts, and can be assumed to hold for all organisms whose growth rate is limited by the concentration of the energy source. In other words, under conditions which only allow growth at a very low specific rate, the size of the standing crop is low compared to the supply of nutrients. Therefore, under oligotrophic conditions, a substantial amount of the limiting energy source must be expected to be utilized for maintenance purposes.

Conceivably, at very low dilution rates the energy source could eventually be utilized exclusively for maintaining the cell integrity, thus resulting in a specific rate of growth of zero. However, there is evidence that μ cannot be any given value between μ_{max} and zero, but has a certain minimal value, μ_{min}. Progressively decreasing viabilities were observed by Tempest et al. (1967) upon a decrease in the dilution rate of carbon- and energy- (glycerol) limited cultures of Aerobacter aerogenes. At a D of 0.04 h^{-1} (μ_{max} = 1.0 h^{-1}), the viability was still 90%, but decreased to less than 40% at a dilution rate of 0.004 h^{-1}. As a consequence of the fact that steady states could be established, the specific rate of growth of the viable population must have been higher than the dilution rate (compare Fig. 5). It was estimated that the minimum value of μ was 0.009 h^{-1}, which is approximately 1% of μ_{max}. Considering the nature of the growth limitation, this might be attributed to a shortage of energy. However, the comparison with ammonia-limited cultures receiving the energy source in excess throughout, revealed that energy is not the (only) decisive factor. The minimum rate of growth of the viable population under ammonia-limitation was found to be about 0.007 h^{-1}.

It is to be expected that organisms with lower μ_{max} values will have correspondingly lower μ_{min} values. As mentioned earlier, competition studies between the specialist Thiobacillus neapolitanus and the versatile Thiobacillus A2, performed under thiosulfate limitation at dilution rates of 0.025 h^{-1}, 0.050 h^{-1}, and 0.075 h^{-1}, resulted in a dominant population of T. neapolitanus. The sustained presence of T. A2 (5-10%) was attributed to the utilization of glycollate excreted by the specialist. However, similar experiments at a D of 0.004 h^{-1} resulted in increased T. A2 numbers (up to 70% after five volume changes), concomitant with more drastically reduced viabilities of T. neapolitanus (40% viable) compared to T. A2 (85% viable). A dilution rate of 0.004 h^{-1} is only 1.1% of the μ_{max} of T. neapolitanus, but is still 4% of the μ_{max} of T. A2.

From the considerations mentioned above, it may be expected that organisms which are adapted in such a way that growth under oligotrophic conditions is possible are slow growing organisms even at higher substrate concentrations (low μ_{max}). The fact that cells

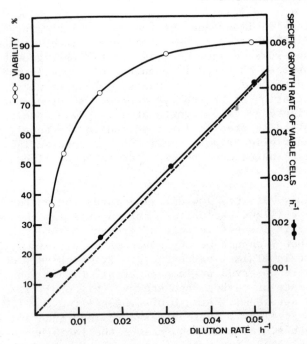

Figure 5.   Steady-state viability (● -- ●) and specific growth
rate of the viable cells (■ -- ■) of glycerol-limited <u>Aerobacter</u>
<u>aerogenes</u>. Dashed line represents the specific growth rate assum-
ing a 100% viability. Redrawn from Tempest et al. (1967).

geared to an extremely accurate control over biosynthesis at very
low rates have low maximum specific growth rates, may well be
related to the properties of enzymes (see S/V ratio). Again the
principle of the "limited elasticity" is applicable.

It is obvious that organisms thriving in oligotrophic habitats,
conceivably often facing starvation conditions, will not have high
endogenous rates of metabolism. Such a "strategy" would result in
even less energy available for growth. Genuine oligotrophs such as
<u>Arthrobacter</u>, <u>Caulobacter</u>, and <u>Hyphomicrobium</u> may be expected to
show low endogenous growth rates (Poindexter 1981; Boylen and Ensign
1970a, b). An interesting observation on this respect was made by
Meiberg and Harder (unpublished) in starving <u>Hyphomicrobium</u> cells.
After a week the O_2-uptake rate became undetectably low but all
cells remained viable; even after 3 months of starvation methanol
was respired immediately.

In most heterotrophic organisms the energy source also serves as
the source of carbon; thus under oligotrophic conditions many organisms can be expected to be both carbon- and energy-limited. However,
in phototrophic bacteria the effect on survival of a low rate of

energy supply (low irradiance) can be separated from the effect of low fluxes of carbon. Such studies have been performed with a Chromatium strain in the presence of both an organic (acetate) and an inorganic (CO_2) carbon source (van Gemerden 1980). In the studies the relation between the specific growth rate and the light intensity was estimated, and by extrapolation of the data to zero growth rates, the light intensity required for maintenance purposes was deduced. Indeed, at this light intensity Chromatium maintained a 100% viability for 10 days without visible growth. In the corresponding dark cultures which also contained the carbon sources, the viability decreased rapidly. This was confirmed with ATP data. It is to be expected that similar observations can be made for green algae, cyanobacteria (Gons and Mur 1975), and green bacteria. In this respect it is of interest to note that green sulfur bacteria appear to require less light for the maintenance purposes than purple bacteria (Biebl and Pfennig 1978; van Gemerden 1980). This can be ecologically significant. In layered ecosystems green sulfur bacteria usually are found below purple sulfur bacteria; thus they will receive less light. The ability of the purple sulfur bacteria to outcompete the green sulfur bacteria at levels closer to the surface may be related to the fact that growth of the green sulfur bacteria is completely inhibited by traces of oxygen. This is not the case for the purple sulfur bacteria who can grow chemolithotrophically under microaerophilic conditions (Kämpf and Pfennig 1980). Although habitats which support the exuberant growth of phototrophic bacteria are usually characterized as being (highly) eutrophic, this is to be considered a misconception with respect to light.

The differences just described with respect to the intensity of the light required for the maintenance between the purple and the green sulfur bacteria are similar to those found between the green algae and cyanobacteria (Gons and Mur 1975; van Liere 1979). It was suggested by Raven and Baerdall (1981) that the lower μ_e value observed in the green sulfur bacteria and the cyanobacteria is related not only to the quality and quantity of the pigment of these organisms, but also to the spatial distribution of the (bacterio) chlorophylls within the cells. For instance, the Chlorobium vesicles would prevent an excessive leakage of protons to the external medium. Without the vesicles, an additional amount of energy would be required to compensate for this leakage; this would raise the maintenance requirements. In addition, the high pigment content of these organisms enables them to scavenge the available photons more effectively. Such organisms thus can best be described as well adapted to oligophototrophic conditions. In the green sulfur bacteria such a strategy may well have been developed during evolution as a result of the obligate anaerobic growth requirements which forced them to inhabit the deeper layers to avoid the toxic oxygen produced at the surface layers. At the same time, their most common electron donor, sulfide, usually is produced in the underlying sediments.

THE "MODEL" OLIGOTROPH

Thus far we have described differences between organisms. From these facts we have speculated about how the properties we observe allowed the organisms to survive.

Another way to proceed is to first try to design an ideally equipped organism for a certain job and then by looking at examples found in nature see how well evolution has been able to design such an organism. In the present context we would like to discuss the profile of the "model" oligotroph on the basis of the stimulating thoughts and ideas put forward by Poindexter in Hirsch et al. (1979).

Oligotrophs are defined as those organisms known to be able not only to survive but particularly to multiply under conditions of extremely low and often discontinuous supply of nutrients. In other words, organisms adapted to low and irregular fluxes of substrates. One might wonder what properties such organisms should have to survive optimally and what the consequences would be. Table 8 (Hirsch et al. 1979) summarizes the characteristics of such an organism with respect to the uptake of nutrients, and Table 9 (Hirsch et al. 1979) does the same with respect to the utilization of substrates.

a. A high surface/volume ratio would enable such organisms to scavenge substrates, due to the relatively high proportion of uptake sites. In addition, the internal diffusion pathway would be reduced. The "model" oligotroph thus can be expected to be small, elongated, and may have cellular appendages.

Studies with bacteria isolated from the same habitat generally show that bacteria with comparatively low surface-volume ratios are outcompeted at low substrate concentrations. Examples obeying this statement are summarized in Table 2.

The occurrence under nutrient limitation of fully expressed prosthecae in Ancalomicrobium and the long stalks in Hyphomicrobium also are in acccordance with these expectations.

b. The preferential utilization of energy for the uptake of nutrients can be expected to be of advantage under conditions of low nutrient supply. The continued uptake of substrates at concentrations which are either too low to permit growth at $\mu > \mu_{min}$ or are unbalanced, would lead to the synthesis of reserve materials. These substances could later, when conditions have changed, be used to replenish the nutrients which are available at that time and in this way allow growth. Storage material could even be synthesized under conditions enabling growth. At first glance, it seems unlikely that reserve materials are synthesized under energy limitation, nevertheless it has been reported to occur. Matin and Veldkamp (1978) observed increased amounts of PHB with decreasing dilution rate in

Table 8.   Characteristics of nutrient uptake and their consequences
for model oligotrophs (from Hirsch et al. 1979).

Characteristic	Consequence
High surface/volume ratio	a) cells small and/or elongated b) morphology possibly unusual c) relatively high proportion of uptake sites among surface components d) decreased diffusion path from surface to internal sites of utilization
Metabolic energy used preferentially for nutrient uptake, especially when not actively growing and reproducing. Energy not dissipated in irrelevant activities.	a) relatively high proportion of endogenous metabolic energy would be consumed for uptake b) possession of specific electron donors that could be quickly tapped for the generation of the proton gradient allowing transport. The proton gradient could be coupled to inhibition mechanisms for other gradient-consuming processes.
Constantly capable of uptake	Could not afford to produce resting stage cells impermeable to nutrients or unable to perform transport (except possibly for dispersal).
Uptake systems of high affinity; possibility for simultaneous uptake of mixed substrates	a) highly sensitive to toxic solutes of biotic and abiotic origin b) during nutrient enrichment, a relative excess of one nutrient might interfere with uptake of others c) A nutrient present in relative excess will be stored; the ability to store a variety of reserve materials would be needed
Uptake would lead to reserve accumulation, particularly when the rate of uptake was submaximal and metabolic pools were not saturated due to extremely low substrate concentrations, but possibly also when substrates were adequate for balanced growth.	Relatively low maximum growth rates. In laboratory cultures, this could result in burdensome accumulation of reserve materials.

Table 9. Characteristics of nutrient utilization and their
          consequences for model oligotrophs (modified from
          Hirsch et al. 1979).

Characteristic	Consequence
Low endogenous metabolic rate with energy used principally to support uptake constituents	a) slow degeneration of reserve materials and cellular b) prolonged viability in absence of exogenous nutrient supply
Increase in metabolic rate to a level supportive of net bio-synthesis would occur only when both the uptake systems and the metabolic pools were saturated	Slow response to nutrient enrichment (as measured by net biosynthesis or overall multiplication); the oligo-troph would, therefore, not be among the first flush of microbial multipli-cation following nutrient enrichment
Biosynthetic rate would drop rapidly in response to decline decline of uptake rate due to exhaustion of substrates below a concentration that allowed maximal rate of transport	Growth rate would not be maintained during depletion of exogenous nutri-ents, although uptake rate would be maintained and nutrient accumulation would continue
Catabolically efficient and versatile	a) would posess at least average genetic complexity b) would by oxybiontic
Large proportion of catabolic enzymes would be inducible, while carriers would be constitutive	a) the usual economy of specific protein synthesis b) reduction in variety of enzymes available for endogenous metabolisr c) slow response to nutrient enrichment d) even with constitutive carriers, low specificity of uptake might cause interference by a relatively excessive nutrient with uptake of others
Minimal level of catabolite repression	Ability to utilize several carbon- and energy-sources simultaneously
Ability to store a variety of reserve materials	a) relatively large proportion of cell volume could become committed to storage, reducing capacity for other activities b) growth (as net, balanced biosyn-thesis) could result from unbal-anced, single-nutrient enrichment in the environment of the well-stocked cell

lactate (carbon and energy!) limited continuous cultures of a
Spirillum sp. A similar phenomenon has been observed in
Hyphomicrobium (Meiberg and Harder, unpublished).

c. Low endogenous rates of metabolism would result in a slow
degradation of the reserve materials or essential macromolecules. It
would enable an organism to grow faster under supporting conditions,
or to remain viable for longer periods of time under starvation
conditions.

Low endogenous rates have indeed been reported to occur in such
genuine oligotrophs as Arthrobacter, Caulobacter and Hyphomicrobium.

d. The possession of inducible carriers under conditions of low
and discontinuous supply of nutrients can be expected to be fatal.
Similarly, the production of resting cells impermeable to nutrients
can be concluded to be highly disadvantageous. Thus, the "model"
oligotroph does not form spores and has constitutive uptake carriers.
The possession of non-specific carriers would be most favorable.

In addition, the uptake systems must have a high affinity
towards the substrate(s). In Caulobacter crescentus a high affinity
uptake system for NH_4^+ has indeed been reported to be present on a
constitutive level (Ely et al. 1978). According to Hirsch et al.
(1979; see Table 8), the "model" oligotroph should have the
possibility to utilize a number of substrates simultaneously.
This is correct, but seems not be a special feature of oligotrophs;
all organisms studied so far in this respect can simultaneously
utilize different substrates, provided the concentration of the
nutrients is low.

e. Under conditions where not all requirements for growth are
met, uptake should continue and result in the accumulation of
reutilizable storage materials. Balanced net biosynthesis (growth)
should not occur unless both the uptake systems and the internal
metabolic pools were saturated. As mentioned earlier, the synthesis
of storage materials has been observed even when conditions are such
that balanced growth is possible.

As a consequence, the "model" oligotroph can be expected to
have a relatively low μ_{max}, and in addition shows a slow growth
response upon a sudden increase in the concentration of nutrients.

f. A continuation of growth under conditions where uptake rates
decline (as the result of decreasing substrate concentrations), would
result in the exhaustion of intracellular reserves. The "model"
oligotroph may therefore be expected to show a rapid decline in
growth rate in response to a decreased uptake rate. Consequently,
an internal nutrient enrichment is to be expected upon the continued
uptake even if the rate of uptake would be somewhat reduced as a

result of the decreased concentrations.

g. Being a specialist, and handling available energy sources
inefficiently, cannot be regarded a "wise" strategy under conditions
of low and irregular nutrient availability. Thus, the utilization of
nutrients by "model" oligotrophs can be expected to be catabolically
versatile and with a high efficiency of energy conversion.

In general this is indeed found in most of the cases.
Caulobacter sp. and Arthrobacter sp. are best described as versatile
organisms. On the other hand, Hyphomicrobium species are mainly
restricted to growth on C_1-compounds. One might wonder why the
latter organism, even under oligotrophic conditions, has been able
to specialize on so few substrates. It may mean that in many
environments a continuous supply of C_1-compounds is available. For
example, air invariably contains volatile C_1-compounds other than
CO_2, and the same is true for environments where CH_4 is formed.

h. Under oligotrophic conditions, a delay in the uptake can be
a disadvantage of capital importance in the presence of other organisms. Therefore, the "model" oligotroph does contain its uptake
carriers on a constitutive level. On the other hand, having a wide
variety of catabolic enzymes constitutively at high levels is meaningless under conditions of low and irregular substrate availability.
Thus, most catabolic enzymes can be expected to be inducible, or
possibly constitutive but then at low levels. Therefore, upon
nutrient enrichment, the "model" oligotroph would show a fast
response with respect to uptake, but a slow response with respect
to growth.

i. From the foregoing it is obvious that under conditions where
the rates of uptake and growth are not comparable, the capability to
synthesize reserve materials becomes essential. The ability to store
a variety of reserve materials enhances the chances for a full and
complete utilization of the uptake potential under the conditions
encountered.

In conclusion, it can be stated that known oligotrophs to some
extent fulfill the expectations. In some other cases at least the
biochemical mechanisms for the "model" oligotroph are available.

However, some of the properties mentioned are observed in
organisms which cannot be regarded as oligotrophs at all. The
secret of the sustained life under oligotrophic conditions would
then lie in the combination of as many of the properties listed as
possible. Thus, not necessarily all properties would be encountered
in all oligotrophic organisms. It may very well be that there does
not exist a single organism possessing all these properties, but
rather that a whole spectrum of oligotrophs exist which possess one
or more of the properties to a greater or lesser extent. The

educated guess of Poindexter (Hirsch et al. 1979) may help us to look specifically for those characteristics which can be expected to be of particular relevance for the strategy of the oligotrophic mode of life.

ACKNOWLEDGMENT

We thank Professor W. Harder for many stimulating discussions.

REFERENCES

Beeftink, H. H., and H. van Gemerden. 1979. Actual and potential rates of substrate oxidation and product formation in continuous cultures of Chromatium vinosum. Arch. Microbiol. 121: 161-167.

Biebl, H., and N. Pfennig. 1978. Growth yields of green sulfur bacteria in mixed cultures with sulfur and sulfate reducing bacteria. Arch. Microbiol. 117: 9-16.

Boylen, C. W., and J. C. Ensign. 1970a. Long-term starvation survival of rod and spherical cells of Arthrobacter crystallopoites. J. Bacteriol. 103: 569-577.

Boylen, C. W., and J. C. Ensign. 1970b. Intracellular substrates for endogenous metabolism during long-term starvation of rod and spherical cells of Arthrobacter crystallopoites. J. Bacteriol. 103: 578-587.

Brown, E. J., and L. A. Molot. 1980. Competition for phosphorus among phytoplankton. Abstract Int. Symp. Microb. Ecol. 2: 163.

Dawes, E. A. 1976. Endogenous metabolism and the survival of starved prokaryotes. Symp. Soc. Gen. Microbiol. 26: 19-53.

Dow, C. S., and R. Whittenbury. 1980. Prokaryotic form and function. Proc. Int. Symp. Microb. Ecol. 2: 391-417.

Ely, B., A. B. C. Amarasinghe, and R. A. Bender. 1978. Ammonia assimilation and glutamate formation in Caulobacter crescentus. J. Bacteriol. 133: 225-230.

Gons, H. J., and L. C. Mur. 1975. An energy balance for algal populations in light-limiting conditions. Verh. Int. Verein Limnol. 19: 2729-2733.

Gottschal, J. C., and J. G. Kuenen. 1980 Selective enrichment of facultatively chemolithotrophic Thiobacilli and related organisms in continuous culture. FEMS Microbiol. Lett. 7: 241-247.

Gottschal, J. C., H. Nanninga, and J. G. Kuenen. 1981. Growth of Thiobacillus A2 under alternating growth conditions in the chemostat. J. Gen. Microbiol. 126: 85-96.

Gottschal, J. C., A. Pol, and J. G. Kuenen. 1981. Metabolic flexability of Thiobacillus A2 during substrate transitions in the chemostat. Arch. Microbiol. 129: 23-28.

Gottschal, J. C., S. de Vries, and J. G. Kuenen. 1979. Competition between the facultatively chemolithotrophic Thiobacillus A2, an

obligately chemolithotrophic Thiobacillus and a heterotrophic
Spirillum for inorganic and organic substrates. Arch.
Microbiol. 121: 241-249.

Harder, W. 1969. Obligaat psychrofiele mariene bacteriën. Ph.D.
thesis, Univ. Groningen, The Netherlands.

Harder, W., and H. Veldkamp. 1968. Physiology of an obligately
psychrophilic marine Pseudomonas species. J. Appl. Bacteriol.
31: 12-23.

Harder, W., and H. Veldkamp. 1971. Competition of marine psychrophilic bacteria at low temperatures. Antonie van Leeuwenhoek
J. Microbiol. Serol. 37: 51-63.

Healey, F. P. 1980. Slope of the Monod equation as an indicator of
advantage in nutrient competition. Microb. Ecol. 5: 281-286.

Hirsch, P. 1979. Life under conditions of low nutrient concentrations, pp. 357-373. In: M. Shilo [ed.], Strategies of Microbial
Life in Extreme Environments. Berlin Dahlem Konferenz.

Jannasch, H. W. 1967. Enrichments of aquatic bacteria in continuous culture. Arch. Mikrobiol. 59: 165-173.

Kämpf, C., and N. Pfennig. 1980. Capacity of Chromatiaceae for
chemotrophic growth. Specific respiration rates of Thiocystis
violaceae and Chromatium vinosum. Arch. Microbiol. 127:
125-135.

Konings, W. N., and H. Veldkamp. 1980. Phenotypic responses to
environmental change. Proc. Int. Symp. Microb. Ecol. 2:
161-191.

Kuenen, J. G. 1980. "Recycling" door micro-organismen, pp. 51-85.
In: H. Veldkamp [ed.], Oecologie van Micro-organismen. Pudoc,
Wageningen.

Kuenen, J. G., J. Boonstra, H. G. J. Schröder, and H. Veldkamp.
1977. Competition for inorganic substrates among chemoorganotrophic and chemolithotrophic bacteria. Microb. Ecol. 3:
119-130.

Kuenen, J. G., and H. Veldkamp. 1972. Thiomicrospira pelophila,
nov. gen. nov. sp. a new obligately chemolithotrophic colour-
less sulfur bacterium. Antonie Van Leeuwenhoek J. Microbiol.
Serol. 38: 241-256.

Kuenen, J. G., and H. Veldkamp. 1973. Effects of organic compounds
on growth of chemostat cultures of Thiomicrospira pelophila,
Thiobacillus thioparus and Thiobacillus neapolitanus. Arch.
Mikrobiol. 94: 173-190.

Laanbroek, H. J., A. J. Smit, G. Klein Nulend, and H. Veldkamp.
1979. Competition for L-glutamate between specialised and
versatile Clostridium species. Arch. Microbiol. 120: 61-66.

Matin, A. 1979. Microbial regulatory mechanisms at low nutrient
concentrations as studied in chemostat, pp. 323-339. In: M.
Shilo [ed.], Strategies of Microbial Life in Extreme
Environments. Berlin Dahlem Konferenz.

Matin, A., and H. Veldkamp. 1978. Physiological basis of the
selective advantage of a Spirillum sp. in carbon-limited
environment. J. Gen. Microbiol. 105: 187-197.

Monod, J. 1942. Recherches sur la croissance des cultures bactériennes. Ph.D. thesis, Paris.
Neijssel, O. M, S. Hueting, K. J. Crabbendam, and D. W. Tempest. 1975. Dual pathway of glycerol assimilation in Klebsiella aerogenes. Arch. Microbiol. 104: 83-87.
Pirt, S. J. 1975. Principles of Microbe and Cell Cultivation. Blackwell Sci. Publ., Oxford, England.
Poindexter, J. V. 1981. The Caulobacters: ubiquitous unusual bacteria. Microbiol. Rev. 45: 123-179.
Powell, E. O. 1967. The growth rate of microorganisms as a function of substrate concentration, pp. 34-55. In: E. O. Powell, C. G. T. Evans, R. E. Strange, and D. W. Tempest [eds.], Microbial Physiology and Continuous Culture. Proc. Third Int. Symp., HMSO Porton Down.
Raven, J. A., and J. Beardall. 1981. The intrinsic permeability of biological membranes to H^+: the efficiency of low rates of energy transformation. FEMS Microbiol. Lett. 10: 1-5.
Sinclair, C. G., and H. H. Topiwala. 1970. Model for continuous culture which considers the viability concept. Biotechnol. Bioeng. 12: 1069-1079.
Smith, A. L., and D. P. Kelly. 1979. Competition in the chemostat between an obligately and a facultatively chemolithotrophic Thiobacillus. J. Gen. Microbiol. 115: 377-384.
Tempest, D. W. 1970. The continuous cultivation of micro-organisms. I. Theory of the chemostat, pp. 259-277. In: J. R. Norris and D. W. Ribbons [eds.], Methods in Microbiology. Academic Press, New York.
Tempest D. W., D. Herbert and P. J. Phipps. 1967. Studies on the growth of Aerobacter aerogenes at low dilution rates in a chemostat, pp. 240-253. In: E. O. Powell, C. G. T. Evans, R. E. Strange, and D. W. Tempest [eds.], Microbial Physiology and Continuous Culture. Proc. Third Int. Symp., HMSO Porton Down.
Tempest, D. W., J. L. Meers, and C. M. Brown. 1970. Synthesis of glutamate in Aerobacter aerogenes by a hitherto unknown route. Biochem. J. 117: 405-407.
van Gemerden, H. 1974. Coexistence of organisms competing for the same substrate: an example among the purple sulfur bacteria. Microb. Ecol. 1: 104-119.
van Gemerden, H. 1980. Survival of Chromatium vinosum at low light intensities. Arch. Microbiol. 125: 115-121.
Van Liere, E. 1979. On Oscillatoria agardhii Gomont, experimental ecology and physiology of a nuisance bloom-forming cyanobacterium. Ph.D. thesis, Univ. Amsterdam.
Veldkamp, H. 1976. Continuous Culture in Microbial Physiology and Ecology. Meadowfield Press, Shildon Co., Durham.
Whitting, P. H., M. Midgley, and E. A. Dawes. 1976. The role of glucose limitation in the regulation of the transport of glucose, gluconate and 2-oxogluconate and of glucose metabolism in Pseudomonas aeruginosa. J. Gen. Microbiol. 92: 304-310.
Wijbenga, D. J., and H. van Gemerden. 1981. The influence of

acetate on the oxidation of sulfide by <u>Rhodopseudomonas</u>
<u>capsulata</u>. Arch. Microbiol. 129: 115–118.
Zevenboom, W. 1980. Growth and nutrient uptake kinetics of
<u>Oscillatoria</u> <u>agardhii</u>. Ph.D. thesis, Univ. Amsterdam.

PHYSIOLOGICAL AND BIOCHEMICAL ASPECTS OF MARINE BACTERIA

W. J. Wiebe

Department of Microbiology and Institute of Ecology
University of Georgia
Athens, Georgia 30602

INTRODUCTION

This paper is not intended as a review. There is an enormous
literature dealing directly with bacteria isolated from the sea and
an even larger literature indirectly important to questions of marine
bacterial physiology and biochemstry. In this paper I want to call
attention to the contributions physiological and biochemical studies
of pure cultures have made to marine ecology.

In recent years there has been a marked shift away from consideration of pure culture studies in ecology, and it might be well to
briefly review the history of "marine microbiology" in order to find
out how this change in approach occurred. Early bacteriologists were
forced to study the activities of mixed bacterial populations because
of the difficulties of obtaining pure culture. Koch simplified the
pure culture technique and, with this single stroke, transformed
emphasis in microbial investigations to the study of pure cultures.
This approach became firmly entrenched as the modus operandi for all
microbiologists. The emphasis continued in the marine field until
the mid-1960's. [See Benecke (1933) and Zobell (1946) for reviews
of early work and Wood (1965) for an update to the mid-1960's.]

There was a revolution developing however. In 1962, Strickland
and Parsons proposed a technique to measure directly the in situ
metabolic activity of marine microheterotrophs. This entailed adding
a radioactive substrate directly to a water sample and measuring the
amount of label incorporated into the particulate material after
incubation. In principle the technique was analagous to the method
of Steemann Nielsen (1952), who developed the use of $^{14}C-NaHCO_3$ to
measure primary production in oceanic water. Wright and Hobbie

(1965, 1966) investigated many aspects of this "heterotrophic potential technique" and expanded the theoretical basis for its application. Their efforts stimulated the wide-scale application of the
technique. It has proven to be a powerful tool for examination of
the in situ activity of microheterotrophs.

There was another, more subtle, reason for the revolution:
ecological research was entering a new age and investigators were
starting to consider the function of entire ecosystems [e.g., coral
reefs (Odum and Odum 1955), salt marshes (Teal 1962)]. At this level
the importance of microbes and their activities became obvious and
this stimulated interest in microbial ecology. However, because
most microbiologists were committed to pure culture approaches, the
major advances in ecology were achieved by zoologists and botanists.

One effect of this revolution, with some notable exceptions, has
been to take emphasis away from the study of the physiology and biochemistry of axenic and mixed gnotobiotic populations for ecological
research. In this paper then, I would like to consider the value of
the more "classical" (i.e., organism oriented) approaches when they
are applied to current marine ecological problems. Specifically, I
wish to consider five topics: growth yields, microbial size, substrate utilization, versatility of bacteria for substrates, and long
term cell survival.

GROWTH YIELDS

Definition and Techniques

The subject of growth yields of bacteria is one more often
associated with industrial application, such as the production of
single cell protein, than with ecology. Yet as van Niel (1949)
stated, "Growth is the expression par excellence of the dynamic
nature of living organisms." Growth represents the integrated
response of an organism to all of the biotic and abiotic factors in
the environment. In many natural systems and in batch pure culture
these factors change with time, providing not a "steady state"
environment but a changing or fluctuating one. The yield of cells
which one can expect under specified conditions is an important
concern for ecologists interested in system dynamics and energetics.
For example, Pomeroy (1974) stated that it was impossible to
distinguish whether bacteria in the open ocean serve as important
links in planktonic food webs or represent energy sinks, with most
of the reduced substrates being converted to carbon dioxide via
respiration. While progress has been made in resolving this question
(e.g., Pomeroy 1979), we are a long way from understanding the
dynamics involved.

Growth yields of heterotrophic microorganisms and the techniques

used in their assessment were reviewed in 1978 by Payne and Wiebe. A
distinction should be made between three types of energy metabolism:
aerobic respiration, where the electron acceptor is oxygen; anaerobic
respiration, where the electron acceptor is an oxidized inorganic
compound; and fermentation, where an organic compound serves as the
electron acceptor. Much effort has gone into developing predictive
equations for the microbial conversion of substrates to biomass in
pure cultures. Bauchop and Elsden (1960) demonstrated that about
10.5 g (dry weight) of cells resulted from optimal fermentative
growth at the expense of each mole of adenosine-triphosphate (ATP)
generated (Y_{ATP} = 10.5). While this equation holds for specific
conditions, it cannot be considered a general property of cells.
Mayberry et al. (1968) related yield of aerobes to the number of
available electrons (ave^-) in a compound, i.e., the electrons
transferred to oxygen during combustion of a mole of substrate.
Payne (1970, 1972) subsequently found that prototrophic heterotrophs
cultured under optimal conditions in carbon-limited, minimal media
had a mean Y_{ave}^- of 3.07 g (dry weight) of cells for each ave^-
removed from the medium. Some values since then have been reported
that are significantly lower (e.g., Abbott et al. 1973) and higher
(e.g., Linton et al. 1975) than the mean value, but these data
appear to result from unique problems associated with the type of
substrate or the growth conditions. For example growth on hydrocarbons yields a mean Y_{ave}^- of only 2.08 (Payne 1972). This low
value is thought to be due to internal pooling of substrate in
vacuoles, yielding an over-estimate of substrate utilization.

Of perhaps more interest to ecologists is the prediction of the
grams (dry weight) of cells obtained per kilocalorie (kcal) of
energy removed in any way by growing microorganisms. Mayberry et
al. (1968) proposed the following equation:

$$Y_{kcal} = \frac{Y_{ave}^-}{kcal/ave^-}$$

They found experimentally that there were 26.5 kcal/ave^- for organic
compounds. Thus, with Y_{ave}^- = 3.14, Y_{kcal} would be 0.11 kcal/g dry
weight. Payne (1972) showed that the average value, taken from 41
experimental results using nine different microorganisms and many
substrates, was 0.11 kcal/g. This information was used to calculate
the energy assimilation efficiency (Payne and Wiebe 1978):

$$\% \text{ Energy efficiency} = 100 - \frac{\text{Initial Energy} - \text{Assimilated Energy}}{\text{Initial Energy}} \times 100$$

The solution for microorganisms, grown in minimal media and without
excessive maintenance expenditure, is 62%. An assimilation:respira-
tion ratio of about 60:40 was experimentally noted for bacteria
(Payne 1970; Servizi and Bogan 1964) and even bleached Euglena

(Danforth 1961) prior to the theoretical calculation. There have been reports calculated in various ways that assimilation can be much greater or less than the theoretical 62% and as a general statement:

> "...these results suggest(ed) that the 'classical thermo-dynamics' approach (accounting for first and last states but not intervening events) should give way more and more to 'statistical thermodynamics' continuous accounting for every relevant event during growth." (Payne and Wiebe 1978)

One factor that can affect growth efficiency is the substrate concentration. Bacteriologists almost always incubate pure cultures with high concentrations of both inorganic and organic nutrients. Ho and Payne (1979) examined the assimilation efficiencies of several bacterial species. They found one group whose assimilation effici-ency increased as substrate concentration decreased. The threshold substrate concentration was between 1 and 2 mg/ml for most species. Many increased assimilation efficiency to great than 80%. Unfortun-ately, the lowest concentration of substrate tested was 0.2 mg/ml (or 100 mg/l), so whether high assimilation efficiencies are maintained in natural marine environments is unknown. Ho and Payne (1979) found a second group of bacteria that exhibited no change in assimilation efficiency over the entire range of substrate concentrations tested. The examination of growth under low substrate conditions is one that deserves further investigation; we will return to this topic later.

Under conditions of anaerobic respiration, the yields of cells per gram of substrate are reduced from those incubated under aerobic conditions: Pseudomonas denitrificans shows only 60% the assimilation of substrate under anaerobic conditions as under aerobic conditions (Koike and Hattori 1974). These results and others (e.g., Stouthamer 1973) have yet to be explained satisfactorily.

Growth yields of natural marine populations have not been meas-ured, except by indirect methods, because bacterial populations and activity in ocean water are too low to use conventional techniques. The theoretical basis for this work is uncertain, for we do not often know the growth history and the precise mixture of species in samples. Furthermore it is often difficult to separate the bacteria from the other microflora and fauna. Williams (1970) found growth efficiencies from heterotrophic potential studies using glucose and amino acids were often greater than 60%. Crawford et al. (1974) examined a large number of substrates in a North Carolina estuary for over a year. There was some systematic variation between substrates, with five showing incorporation rates always greater than 70% and five always less than 70%. One problem with this method is that the ratio of $^{12}CO_2$:$^{14}CO_2$ respired for short time periods is uncertain and thus the values may not reflect true growth efficiency.

In summary there is evidence that growth efficiency in pure cultures can be substrate concentration dependent. In addition there may be some effect on substrate composition. There are, however, no clear rules as to how to relate these data to in situ growth.

Rate of Growth

The rate of growth is related to the intrinsic growth properties of an organism and the physical and chemical conditions of incubation. Escherichia coli under ideal conditions can divide every 20 minutes; Vibrio natriegen has been documented to divide in less than 10 min (Eagon 1962). At the other extreme Mycobacterium tuberculosis grows at 24 h/generation under optimal culture conditions.

There is limited evidence in nature concerning growth rates. Wiebe and Pomeroy (1972) reported that individual cells observed under slide culture conditions from coral reefs and salt marsh waters had generation times of 22 minutes, but only after an initial induction period. Sheldon and Sutcliffe (1978) reported rapid growth of small particles and an increase in ATP in Sargasso Sea samples that were incubated in small bottles. They suggested that the reason for the lack of population explosions in nature was the presence of active grazing organisms. Hagström et al. (1979) (see also discussion in present volume) proposed using frequency of dividing cells in natural waters to estimate in situ bacterial growth rates. Using this technique, Newell and Christian (1981) found that bacterial populations in estuarine samples had 12-68 h generation times. Sieburth (1976) used diffusion chambers to estimate bacterial growth rates and found a doubling time of 4 h during daylight but only 3 doublings/day in Narragansett Bay waters. Christian et al. (1975) calculated an average generation time greater than 50 days for the benthic bacteria in a salt marsh, based on the number of cells present (as estimated by ATP) and the amount of organic matter available on a sustained basis.

While the rate of growth can vary greatly for both single and mixed populations there are two lines of evidence concerning in situ growth rates which seem directly opposed. Some workers claim that open ocean marine bacteria grow rapidly (e.g., Azam this volume; Sieburth 1976; Sheldon and Sutcliffe 1978). Others have evidence that population densities are constant (Watson and Hobbie 1979; Ferguson and Rublee 1976) and that most cells are small and in a starved state or dormant (Novitsky and Morita 1978a, b; Stevenson 1978). I will return to this topic in the section on microbial size.

Control of Growth and Yield

There are many factors that can influence growth rates and yield, including temperature, substrate concentration, and substrate composition. Aspects of these and other factors will be mentioned throughout this paper; here I want to provide some examples of these effects.

Christian and Wiebe (1974) examined the growth rate and respiration of an obligate psychrophile isolated from the Antarctic Ocean. The generation time decreased as temperature rose from 1 to 4°C, plateaued from 4 to 7°C, and then increased from 7 to 11°C, its maximum growth temperature. Respiration, however, increased over the entire temperature range. Thus, the temperature of growth clearly affected the assimilation efficiency. The uncoupling of growth rate from respiration has been noted for many cultures of cells [see also Brown (1957) and Harder and Veldkamp (1967)], but its significance, ecologically, has not been fully evaluated. For example, is slow growth more tightly coupled and efficient than rapid growth?

Substrate concentration also controls growth rate and yield. Sheehata and Marr (1971) found that the growth rates of E. coli in glucose minimal medium was substrate dependent only in the 0.01-5 μM range. They also found that cell volume changes occurred at different substrate concentrations. The reason for altered growth rate may be partly explained on the basis of changes in biochemical transport mechanisms (e.g., Taylor and Costilow 1977), and this might also explain the cell volume changes, as will be discussed below.

Substrate composition can affect growth and yield. Williams (1973) speculated that complex media could exert a sparing effect on substrate assimilition by bacteria. Payne and Williams (1976) documented this effect; they showed that "investment expenditure" responses were measurable only if most of the required supplements were present. Wiebe and Chapman (1968a, b) showed that changing substrate concentration (but not composition) or temperature of incubation greatly affected the fine structure of marine Pseudomonas and Acinetobacter species. While these results point out some possible responses, we lack enough information to form a coherent theory. Too few organisms and conditions have been examined to permit us to predict the effects of altering conditions in pure cultures, let alone natural populations.

MICROBIAL SIZE

While our understanding of the conditions that affect growth yields is very incomplete, as we have just discussed, there is a great deal known about the conditions that determine the size and shape of microorganisms in culture and nature, perhaps because this information can be obtained directly. In this section I discuss the evidence for the distribution of different size cells in culture and nature and consequences implied by such distributions.

ASPECTS OF MARINE BACTERIA

   Bacteriologists most often concentrate on the examination of
cells that are in the logarithmic state of growth and ignore other
states. There is, however, ample evidence that the size and function
of cells change throughout the entire growth cycle. As Konings and
Veldkamp (1980) pointed out in an excellent review of "Phenotypic
Responses to Environmental Changes", the surface to volume ratio is
a function of the specific growth rate. At decreasing rates the
surface to volume ratio increases and generally small-size bacteria
grow under low nutrient conditions. [But see Hodson et al. (1981)
on heterotrophpic activity of small vs. large cells.] In broth
cultures lag phase cells are generally largest in volume; in log
phase they decrease to a small and generally uniform shape. Upon
reaching stationary phase, the size and shape of cells becomes
highly variable. On solid media, which may be a model for growth on
surfaces, the evidence is less clear. For example, Wiebe and Pomeroy
(1972) showed that a marine <u>Vibrio</u> sp. had a highly pleomorphic
morphology at the growing edge of micro-colony. There are also
alterations of the fine structure in different phases of growth and
under different incubation conditions (Wiebe and Chapman 1968b).
Unfortunately, there has been only limited work on this topic, and
full extent of morphological plasticity is unknown.

Is There an Optimal Size for Planktonic Bacteria?

   Until the early 1970's the notion was firmly established in the
literature that most bacteria in the sea were attached to particles,
where substrates could be concentrated. This was an extension of
the "bottle effect" hypothesis (Zobell 1946). At present, evidence
is overwhelming that small, round cells, less than 0.4 μm diameter,
predominate in seawater (e.g., Ferguson and Rublee 1976; Watson and
Hobbie 1979; Azam and Hodson 1977; Wiebe and Pomeroy 1972) and that
larger cells are found mostly in enriched waters, at physical
interfaces or attached to particles.

   Small cells can transform into larger ones under appropriate
conditions. Wiebe and Pomeroy (1972) reported that on initial
microscopic examination, bacteria from the waters of a variety of
marine habitats most often were small and unattached. In slide
culture with non-nutrient agar these cells enlarged over a period of
hours and then begin dividing as "typical" cultured cells (see also
Novitsky and Morita 1976).

   Williams (1970) and Derenbach and Williams (1974) reported that
most of the heterotrophic activity resided in the fraction of
filtered water that passed through a 3.0 μm filter. Azam and Hodson
(1977) confirmed these results using 0.6 and 1.0 μm filters. There
have been numerous papers since then that support these findings.
Hodson et al. (1981) found that while most of the heterotrophic
activity occurred in the "particle free" fraction, the activity/cell
was much greater for cells attached to particles. However, when

activity was normalized to cell surface area, not cell number, most of the difference between the two fraction disappeared. This leads to the conclusion that large cells are <u>not</u> intrinsically more active than small cells. We will return to this topic subsequently.

Let us consider now some possible reasons for the observations. What advantage does a small round cell have, floating free in water? One proposed advantage is refuge. Grazing by filter feeders is active in the upper, and probably lower, water column (Sheldon and Sutcliffe 1978). Large particles may be preferably scavenged and the cells on them digested. Individual small cells would be more likely to escape this pressure. The observation that small cells in the water column under other conditions enlarge and divide suggests that in nature they are either not dividing very often or that their division strategy is different than that observed after capture.

Novitsky and Morita (1976, 1978a,b) have produced small cells identical in size to those in nature by starving pure cultures. They have shown that these cells can survive long periods in this state and that when presented substrate, they again enlarge and divide. Thus, one plausible explanation for the occurrence of small, round bacteria in oceanic water columns is that the cells are living under very low nutrient conditions; they are free of particles either by chance or because of predation pressure. Interestingly, Marshall (1979) found that the first organisms to attach to fresh surfaces were "mini-cells."

Another possible explanation for the occurrence of small, round cells in the water column is that such cells have a higher surface to volume ratio than large round cells and this could facilitate substrate utilization. We do not know why round cells become rod-shaped upon culturing. Do round forms have a rigid cell wall or is rigidity somehow impaired? There is recent evidence (e.g., Maca and Fitzgerald 1979) that some enzymes are cell wall associated. In a low nutrient environment bacteria, with, as we shall see subsequently, highly versatile metabolic capabilities, could place many enzymes or binding proteins into their cell walls, where they would be closest to external substrates. Is it possible that this loading of non-wall proteins could reduce rigidity?

Small, round cells have some inherent disadvantages in terms of growth potential. Surface area, which could be used to encounter nutrients, is reduced to the theoretical minimum value of surface to volume. Laws (1975) suggested that diatoms, which are large phytoplankters, have better survival characteristics than small dinoflagellates because diatoms, by virtue of size, can store more material and reduce their direct need on the external environment. Here, however, we have the opposite case: the small bacteria dominate in low nutrient systems.

The actual role of the small cells in nature is unclear.
Ferguson and Rublee (1976) and Watson and Hobbie (1979) present
convincing evidence that the numbers of cells at a particular depth
in several oceanic zones are constant, while Sheldon and Sutcliffe
(1978) and Sieburth (1976) show that there is a great deal of microbial activity in similar locations. If small cells are dormant, then
the former data could be explained by physical factors spreading a
very slow-growing population. If they are growing, as implied by
Sheldon and Sutcliffe (1978) and Sieburth (1976), they can only be
maintained in reasonably constant densitites by intensive grazing,
or they do not constitute the population they measure as growing.
We do not have enough information to resolve this dilemma, but the
solution is certainly necessary for our understanding of marine
biological dynamics.

ATTACHED VERSUS "FREE" CELLS

The roles of free and attached bacteria are brought up throughout this paper. In this section I want to examine some general
aspects of attachment.

Marshall (1980a) defines four general states regarding
bacterial attraction to surfaces. The weakest attractive forces
are reversible, with cells sorbed and desorbed rapidly. In theory
this should be of limited occurrence in sea water where both
bacteria and particles are negatively charged. There are three
types of adhesion. The first is temporary. Gliding bacteria, some
diatoms, and blue-green bacteria form a polymeric bond with surfaces
but can move over the surface. Second is specific irreversible
attachment. This is restricted to organism-organism interactions.
The third is a non-specific attachment in which a polymer is the
cement. There is little known about the chemical structure of these
polymers. Marshall (1980b) has found that most often very small
cells, or "dwarf" bacteria, are the primary colonizers. They form
non-specific attachment polymers that spread over the surface. Thus
there are potentially several modes and strengths of attachment.

Detrital Attachment

In discussing the question of attached vs. free cells in the
ocean it is necessary to specify location. In most zones of the
coastal and open ocean Wiebe and Pomeroy (1972) found few attached
bacteria on particles and Azam and colleagues (see refs.) have
demonstrated that most of the heterotrophic potential activity
resides in a water fraction that can pass through a 1 μm screen.
In estuarine waters the situation is less clear. Hanson and Wiebe
(1977) found that most of the heterotrophic potential activity in
estuarine water at Sapelo Island, Georgia was associated with large
particles. Goulder (1977) reported similar findings in a British

estuary which had abundant suspended solids. However, Harvey and
Young (1980) found from 24-44% of the bacteria attached to particles
in the surface film vs. 1.3 to 4.6% attached in subsurface waters of
the Sippewissett, Massachusetts and Palo Alto, California marshes
respectively.

In most recent papers few bacteria are reported to be attached
to flocculant or flake-like particles. For example, Jannasch (1973)
found sparse numbers of bacteria on masses of particles encountered
in tropical surface waters; he concluded that this indicated they
were of recent origin, i.e., they had not had time to be effectively
colonized. Pomeroy and Deibel (1980) examined tunicate faeces and
found that both flocculant and plate-like aggregates were produced.
They examined the sequence of microbial succession and found that
for up to 18 hours few bacteria were visible; from 18-36 hours
bacterial numbers increased greatly; by 48-96 hours protozoan
populations arose. Subsequently, the particles fragmented and few
attached microorganisms were seen. From these results particles
with few bacteria could be fresh or old.

Recently, the role of the "rain" of large particles into the
deep ocean, e.g., algae, faeces, whales, has been investigated.
Sieburth (1976) and Sheldon et al. (1972) pointed out that while a
few large particles could supply deep ocean energetic requirements,
it would be difficult to capture them with conventional gear. Wiebe
et al. (1976) and Fellows et al. (1981) among others have suggested
that particles greater than 35 μm diameter may constitute the major
flux of material to the deep sea, but that with a concentration of
from 0.1-4 per liter they would not be routinely sampled. Fellows
et al. (1981) trapped large, sinking particles and found that they
constituted an important carbon source. Further, their steady-state
concentration represented less than 1% of the flow-through. Large
particles were particularly rich in ATP compared to adjacent water
samples. Eppley and Peterson (1979) estimated that 20-30% of the
organic carbon produced by phytoplankton is found in faeces which
can sink rapidly and that "new" production equals the loss via
faecal pellets to the deep ocean.

In summary, most small particles harbor few bacteria. The
particles probably are derived from faeces of grazing zooplankton
and for short periods may contain an abundant microflora. Large,
rapidly sinking particles may routinely contain a large microflora.

Attachment to Living Organisms

As many investigators have discovered, and has been elegantly
documented by Sieburth (1979), large numbers of bacteria can be
associated with the surfaces of marine animals and plants. For
example, Castille and Lawrence (1979) examined glucose and mannitol
uptake by postlarval peneid shrimp and found that bacteria on the

carapace accounted for most of the uptake. Provasoli and Pinter
(1980) have shown that the attached bacteria on the green alga,
Ulva, induce polymorphism. Filtrates of cells were ineffective, and
the bacteria in culture lost the ability. It was regained only
after several generations of growing with Ulva. This illustrates
that close and important relationships can exist between microbes
and large organisms, but we have little information about these
phenomena.

SUBSTRATES THAT CAN BE USED BY MICROBES

     Virtually all substrates can be metabolized by bacteria, either
singly or in consortia. In many cases combinations of organisms
increase the rate of substrate utilization (e.g., cellulose digestion
to methane). Studies of the physiology and biochemistry of marine
bacterial isolates have provided us with information on the range of
substrates used but only limited information regarding activity
under in situ conditions. In this section I want to examine some of
the factors that control substrate utilization in sea water and look
at the evidence for the utilization of some exotic and man-made
compounds. In the following section I will look more closely at the
metabolic versatility of marine bacteria.

Physical Factors that Affect Substrate Utilization

     Bacteria can metabolize substrates from about -10°C to > 90°C
and from fresh water to saturated salinities. These factors influence the rates of activity but do not prevent substrate utilization.
The most important factors in the open ocean that control substrate
utilization appear to be the mechanisms of dispersal and concentration. Jannasch (1970) for example showed that pure cultures grown
in chemostats under one set of conditions had thresholds below which
a substrate could not be incorporated rapidly enough to sustain
growth; changing some physical condition (in this case O_2 tension)
altered the threshold.

     It is uncertain what concentration of substrates exist around
individual or clumps of cells. Several mechanisms have been proposed
that would result in concentration gradients of substrates in sea
water. Zobell (1946) proposed a model whereby dissolved substrates
were attracted to surfaces, and these provided a concentrated mileau
for bacteria; Goldman and McCarthy (1978) speculated that phytoplankton may live in much richer nutrient microzones than one would
suspect from the bulk concentration measurements. However calculations by Jackson (1980) and Williams and Muir (1981) have cast doubt
on the importance of microzones in this instance. (See also Azam
and Fuhrman, this volume.) Bacteria in pure culture have been shown
to concentrate substrate in the periplasmic space, between the cell
wall and cell membrane. It has also been suggested that much of the

microbial growth and activity in water could take place in or on the immediate vicinity of phytoplankters or zooplankters, where substrate concentrations would be higher than in the average ambient mileau.

There is also evidence that anaerobic metabolism occurs in oxic ocean water. Oremland (1979) has reported methane generation from incubated zooplankton, and methane has been found in open ocean water columns. In oxygen minimum zones, N_2O, the penultimate product of denitrification, is found (Yoshinari 1976). Thus even in the open ocean there can be considerable diversity of habitats.

Whatever the case, the accumulated evidence (e.g., Pomeroy 1979, 1980) is that microbial activity in ocean water is responsible for a large amount of the substrate processing.

Substrate Concentration and Utilization

As discussed in the section on growth yields there is evidence that substrate concentrations and mixtures can control not only the assimilation of compounds but also the efficiency at which they are used (Ho and Payne 1979; Williams 1973; Payne and Williams 1976). Law and Button (1977) provided important confirmation of this idea by examining the growth rate kinetics of a marine coryneform bacterium grown on single and multiple carbon substrates. They found using chemostat cultures that this bacterium had a growth threshold of 0.21 mg glucose/l when only glucose was supplied. Subsequently, they found that total substrate concentration of 0.11 mg C/l would support normal growth if the "right" mixture of substrates was provided and that 0.3 µg glucose/l could function as a nutrient source at reduced growth rates. This latter concentration is in the range found in sea water. Fuhrman and Azam (1980) concluded that not only were bacterioplankton (< 3.0 µm) quantitatively important in coastal marine food webs, but their data..."also supported the novel conclusion that net growth of free-living bacteria does occur at a significant rate." Thus one might argue that ocean water may be a good growth medium in spite of the low total concentration of available substrates. It is clear that we have not heard the end of this argument.

"Exotic" and Man-made Compounds

There is no evidence that organic matter is accumulating in the sea, and thus most substrates produced must be utilized fairly rapidly. But what of the situation with compounds, either natural or synthesized, that are added by man? A variety of workers have demonstrated the complete or partial utilization of many hydrocarbons and organocides (e.g., Payne et al. 1970) such as DDT may be only partially decomposed initially and may require special conditions for further oxidation. Spain et al. (1980) examined the biodegradation rates of methyl parathion and p-nitrophenol in estuarine and river

water and sediments. Exposure of as little as 60 µg p-nitrophenol/l
in river samples greatly increased subsequent rates of degradation.
Estuarine microflora did not show a similar adaptive response at any
concentration. They concluded that prior adaptation of the ecosystem
microflora must be considered when looking at xenobiotic decay.
Horvath (1972) and others have examined the decomposition of
substrates in which no utilization of energy is derived from the
oxidation to support microbial growth. This phenomenon, termed
co-metabolism, may account for the decomposition of many exotic
compounds.

Perhaps our best understanding of this topic comes from oil
spills. Decomposition rates have been measured for various
components and their ultimate fate documented. There has been some
effort to find active oil decomposers and to encourage their growth
by addition of inorganic nutrients to oil spills in order to increase
oxidation rates. This enthusiasm appears premature, since some
major components of oil can be incorporated directly into cells,
structurally unaltered (Scott and Finnerty 1976). One could equally
argue that antibiotics should be spread on oil spills to prevent
bacterial metabolism and permit maximum dilution of spills. Whatever
the final outcome, it is clear that even these massive spills of
exotic compounds can be metabolized by bacteria.

METABOLIC VERSATILITY VERSUS SPECIALIZATION OF MARINE BACTERIA

In the previous section I examined some of the factors that
control substrate utilization. Here I examine the evidence for
microbial metabolic versatility.

Growth Studies Using Pure Cultures

As discussed in the Introduction, marine bacteriologists have
been wedded to pure culture techniques until very recently. While
it is fashionable now to ignore these studies, they are relevant to
"modern" microbial ecology.

Konings and Veldkamp (1980) recently reviewed microbial
strategies for the utilization of energy sources. They pointed out
that among non-exacting chemoorganotrophic bacteria isolated from
several terrestrial and aquatic environments, the most striking
features were 1) the wide range of substrates that could be used as
sole carbon and energy sources and 2) that few fresh isolates have
precisely the same suite of substrate capabilities. In one cited
study (Sepers 1979, not seen) 83% of the aquatic ammonifying bacteria
isolated could use 21-35 of the 42 substrates tested, although of the
100 isolates examined only two had identical capabilities. Wiebe
and Liston (1972) found a similar pattern: isolates from near-shore
sediments grouped taxonomically into a few well-defined species

clusters, while those from continental slope sediments did not cluster at the species level because of diverse substrate utilization by
individual strains. Almost all isolates used substrates that are
commonly found (free or combined) in organisms, e.g., glucose,
ribose, proline, alanine, glycine, galactose. They were separated
taxonomically by their use of "rare" compounds, e.g., melizotose,
methionine, etc. "Rare" compounds collectively, however, constitute
a significant percentage of the total available substrate. Wiebe and
Liston (1972) postulated that for organisms to remain numerically
significant in an ecosystem they must use some of these rare substrates. If they used none or few, they would be non-competitive;
there would be no pressure to use them all.

Almost all investigators who have examined the taxonomy of chemo-
organotrophic bacteria in the marine environment have commented on
the lack of clear speciation among fresh isolates except near-shore.
This latter effect may result from pulsed inputs of substrates
causing rapid growth of a few types. Konings and Veldkamp (1980)
answered the question, "How can so many overlapping types coexist?",
by concluding that heterogeneity of the environment yields rapid
growth of different organisms for short periods. Taylor and Williams
(1975) examined, in a theoretical treatment, the coexistence of
competing species under continuous-flow conditions and concluded that
"it is necessary that there are at least as many growth-limiting substrates as there are different species." Heterogeneity on a broader
scale also exists. For example Austin et al. (1979) compared the
microflora in Chesapeake Bay and Tokyo Bay. They found intrinsic
differences in the types of microflora in the two bays and concluded
that estuaries, for a variety of physical and biological reasons,
cannot be assumed to contain microbial communities of similar taxa.

Notwithstanding the difficulties in interpreting such studies,
they do provide insight to microbial metabolic function and community
structure.

Biochemical Study of Pure Cultures

Many questions posed by heterotrophic potential analyses (see
below) are beginning to be answered by using pure cultures, e.g.,
the kinetics of uptake, the metabolic activity of small cells, the
utilization of nanomolar substrate concentrations.

Solute is transported across the cytoplasmic membrane by two
major mechanisms. In secondary transport systems the transport is
driven by electrochemical gradients, and the solute is translocated
in an unmodified form. This may either be a passive process or
facilitated by specific protein carriers. Passive transport would
appear to be unimportant in low substrate environments, while
facilitated transport may involve only a specific protein carrier
of, in the case of active transport, energy (ATP). Many carriers

are constitutive in the membrane but space is limited.

The second transport mechanism is group translocation. The
solute is the substrate for a specific enzyme system in the membrane;
chemical modification of the substrate occurs before release into
the cytoplasm. The phosphoenolpyruvate-sugar phosphotransferase
system (PTS) is the only well-established mechanism in this group.
It is not ATP dependent, and because the substrate is transported
and phosphorylated it is more efficient than facilitated transport.

Geesey and Morita (1979) measured L-arginine uptake in a marine
psychrophile. Kinetics differed for nanomolar and micromolar concentrations. Uptake occurred without added substrate, even after pre-
incubation in starvation media. At 70 nM L-arginine there was a 300
times concentration by cells after 60 seconds, and the uptake was
relatively specific; L-lysine or L-ornithine in 20 times excess
reduced L-arginine uptake by only 35%. They proposed two transport
systems: 1) high-affinity, active transport at low concentrations and
2) reduced affinity and chemotaxis at high substrate concentrations.
The uptake was energy dependent; dinitrophenol completely inhibited
uptake, even though exogenous energy was not required.

Hayasaka and Morita (1978) examined galactose permease in a
marine psychrophilic vibrio and found that both constitutive and
inducible enzymes were present. Hodson and Azam (1980) found that
glucose uptake in Serratia marinorubra was mediated by a single
constitutive enzyme via the PTS. Uptake was not ATP dependent.
Interestingly, glycerol uptake (a non-PTS facilitated substrate) was
not regulated by the presence of glucose as it is in Escherichia
coli and Salmonella typhimurium. The interpreted their results as
an adaptation to low, mixed substrate growth. They also speculated
that the high specificity of the system would be of advantage where
the relative concentrations of substrate vary greatly; this would
prevent substrate competition. PTS may be widespread among marine
bacteria, having been found in Mn oxidizers, Flavobacterium,
Pseudomonas, Bacillus, Photobacterium, and Benekea. PTS has been
found in a wide variety of obligate and facultative anaerobes but
not in obligate aerobes, although low nutrient environments have not
been examined extensively. (Will we find other group translocation
transport systems in these bacteria?)

These few examples demonstrate the value of pure culture studies
for gaining insight into in situ processes. One final example
concerns another way in which pure culture enzymatic studies can help
augment in situ observations. Whitt et al. (1979) examined tryptophanase from a marine vibrio. Tryptophanase is formed only by
bacteria, and all known bacteria that have the enzyme can use animal
intestines as a habitat. Thus, enzymatic analysis could give clues
not only to strategies of substrate uptake, but also possible habitat
preferences.

Strategies for Increasing the Versatility of Substrate Utilization

There are several major problems that bacteria face when attempting to use some compound. They must have the appropriate uptake mechanism to capture and transfer the substrate across the cytoplasmic membrane, the substrate must be in the appropriate concentration for these mechanisms to function, and they must have the enzymes to utilize the substrate. If these conditions are not fulfilled, there are several responses that can be made. An individual bacterium can induce an enzyme or change genetically, consortia may form to facilitate substrate decomposition, or the population structure can change.

Perhaps the simplest mechanism for increasing the capacity of a bacterial population to use a substrate is by species succession, as suggested by Konings and Veldkamp (1980). Numerous examples of enzyme induction have been published and need not be discussed here. One potentially important mechanism, but one largely unexplored in marine environments, is genetic change within a species. Episomal transfer is well documented for E. coli in polluted ocean waters. Antibiotic resistance factors (R-factors) can be transferred if population densitites and substrate concentrations are high. Slater and Bull (1979) showed that changes of metabolic capabilities can take place especially rapidly in mutator mutants. These mutants increase the rate of mutation over wild type E. coli, and in mixed strain experiments these strains can dominate. In the guts of animals, bacterial transformation has been demonstrated to yield new strains.

A major problem marine bacteria may face is adapting to transiently variable substrate concentrations. Wright and Hobbie (1966) observed two types of kinetics, one for low and one for high substrate concentrations. They suggested that kinetics for low concentrations were caused by bacteria and those for high ones by algae. Recently, Azam and Hodson (1981) found non-linear kinetics in size fractioned samples, where the algae had been physically removed. They concluded that high Kms are "normal" if substrate concentrations are high. They also found that the turnover time (T_t) increased as substrate concentration increased. For glucose the T_t for nanomolar concentrations was about 9.5 h, while at 10^{-6} to 10^{-3} molar the T_t = 320–600 h. Thus, episodic bursts of substrate will only be utilized after diffusion to low bulk concentrations. However, sustained zones of high concentration will result in high Kms and high V_{max}, and in this case substrate will be used rapidly. They also suggested, as have Gocke et al. (1981), that if there are significant microzones in sea water, T_t may be underestimated when the concentration, measured chemically, is used to calculate the specific activity in radioactive tracer studies. (But see previous discussion in Physical Factors that Affect Substrate Utilization.) As mentioned previously, Geesey and Morita (1979) found that a

marine psychrophile had two specific Kms for substrate uptake,
depending on the ambient concentration. Pure culture studies of
substrate capture mechanisms need to be continued and expanded.

Evidence of in situ Metabolic Versatility by Bacteria

Many investigators have examined the uptake of substrates in
natural waters. Virtually all tested substrates appear to be used,
albeit sometimes slowly. Of particular interest is the relationship
of bacterial activity to algal production of dissolved organic carbon
(PDOC). Duursma (1963) noted that dissolved organic carbon (DOC)
during an algal bloom did not increase. Since then a number of
investigators have found that PDOC produced by planktonic algae is
utilized rapidly (Schelyer 1980; Bell and Sakshaug 1980; Wiebe and
Smith 1977; Smith and Wiebe 1976). Chemically, PDOC is a heterogen-
ous mixture of substrates (Wiebe and Smith 1978), and thus one can
conclude that bacteria are generally present that can use any of
these substrates. Ittekkot et al. (1981), however, distinguished
free dissolved organic carbohydrates from combined dissolved organic
carbohydrates and found that the latter increased greatly at the end
of a bloom. This may represent an example of the episodic introduction of substrate discussed by Azam and Hodson (1981) where the
substrate swamps the system and T_t increases.

LONG TERM SURVIVAL AND STARVATION

In previous sections I have discussed the evidence for the
bacterial growth and starvation in nature. Here I examine the
mechanisms that permit bacteria either to survive for long periods
without growth or cause them to die.

Pure Culture Studies

Bacteria have a variety of mechanisms to survive for long
periods under poor or non-growth conditions. Some form spores which
resist even boiling. Spores of some species of the genus Bacillus
have been shown to remain viable for over 100 years. Some cells form
cysts of enlarged capsules surrounding their walls, while others form
modified resting cells. But some cells survive starvation without
special structures. In culture some populations survive via cryptic
growth, where the lysed remains of one cell serves as substrate for
another. And we have seen that small cells of some microorganisms
have remarkable survival characteristics (e.g., Novitsky and Morita
1977). Other bacteria, however, show little capacity for survival
under non-growth conditions. E. coli goes quickly into decline, and
Pseudomonas cultures often lose total viability after a few weeks of
storage.

The ability to survive in culture is also related to the

particular growth state of the organism. Log phase and lag phase
cells are more sensitive to changes in conditions than are stationary
phase cells. For example Yetka and Wiebe (1974) showed that several
antibiotics affected bacterial respiration more in lag and early log
phases than subsequently. The effect of limiting substrates has
also been shown to be greater on log than stationary phase cells;
thymineless death (Stanier et al. 1970) is but one example.

Recently, Novitsky and Morita (1976, 1977, 1978a, 1978b)
produced small cells of a psychrophilic marine vibrio, ANT 300 by
nutrient starvation. They found (1978a) that even after 70 weeks of
starvation not only were viable cells present but that they could
culture 15 times more cells than were present at the start of the
experiments (see below). Cells starved for one week (Novitsky and
Morita 1978a) were much more barotolerant (100% survival) than
actively growing cells (75% killed) and they concluded that starvation is a survival mechanism for cells subjected to high pressure.
Novitsky and Morita (1976) also found that the fine structure of
their starved small cells was identical to normal cells, except
that the small cells had enlarged periplasmic spaces. It is still
uncertain, however, if small cells found in situ function in a
manner similar to ANT-300 or if they are derived by starvation.

In situ Studies

The evidence concerning the growth state of bacteria in the sea
is conflicting. The presence, indeed numerical dominance, of small,
spherical cells has led some investigators to consider that they are
in a poor nutritional state, resting, or dead. However, as we have
seen, there is some evidence (e.g., Fuhrman and Azam 1980; Hodson et
al. 1981; Azam and Fuhrman, this volume) that these cells are active
components of the community.

There have been some attempts to measure directly the activity
of in situ populations. We previously discussed the attempts to
distinguish growth rates and states by measuring the number of
dividing cells in a population, the production of DNA and RNA, and
the heterotrophic potential of selectively screened portions of the
community. Another technique that has been used to estimate growth
state is the measurement of the community adenylate energy charge
(E.C.). This technique had been applied to a wide variety of pure
cultures of organisms in known growth states (e.g., Chapman et al.
1971). Wiebe and Bancroft (1975) proposed the adaptation of the
technique to natural microbial populations. They showed that microorganisms were at least in the equivalent of a stationary phase in
water columns in a transect from the Bahamas to Puerto Rico. Energy
charge values were considered minimum since most of the recognized
errors in the technique lower the observed charge. Subsequently,
Karl and Holm-Hansen (1977) reported similar results for water
columns in the Black Sea, with interface zones showing higher E.C.

ratios. Karl (1980) has thoroughly reviewed the field of nucleotide measurements as applied to microbial ecology. In most areas of the ocean below the euphotic zone, the microbial populations have an E.C. of around 0.6-0.7, equivalent to a lag or stationary phase in a pure culture.

One problem in the interpretation of E.C. data is that they sum the entire community; if a few large zooplankters are included, values in theory could be biased heavily. Fellows et al. (1981) examined ATP, total adenylates (A_t) and E.C. ratios of large particles collected in traps from the surface to greater than 1500 m. They found that ATP and A_t had to be carefully corrected for "swimmers" but that the E.C. ratio was not affected. A more serious problem with the technique has been reported by Karl and Craven (1980). Alkaline phosphatase (APase) was detected in a variety of oceanic zones and its occurrence in nucleotide extracts affected the concentrations of all nucleotides. They concluded ..."the occurrence of APase in nature is dependent upon micro-scale inorganic phosphate limitations of the autochthonous microbial communities." A phosphoric acid extraction procedure circumvents this problem, but APase effects could affect all of the "old" data and unfortunately not by some fixed value. The corrections should not lower the E.C.; higher values would mean that microbial populations are even more active than previously reported.

Hobbie et al. (1977) attempted to examine in situ activity of bacteria by direct observation. They proposed using the fluorescent color given off by bacteria, when stained with acridine orange and viewed with a specific combination of filters, as an index of whether a cell was active or not. They found that most cells were not active. The validity of this method, however, is still unknown.

Attempts have been made to measure radioisotopic incorporation of substrate by individual cells (Hoppe 1976). However, one cannot be certain either whether label is incorporated without cell growth or whether an unlabelled cell is dead, resting, or simply does not use the added substrate.

Yetka and Wiebe (1974) added antibiotics to estuarine water samples and examined changes in the respiration rates. They concluded that the natural microflora responded more like lag or early log phase cells.

Novitsky and Morita (1978a) in commenting on the increased number of marine vibrios after long term starvation stated:

"It therefore appears that the increase in numbers is due to the efficient utilization of reserve material... Efficient replication may permit the cells to increase in numbers without a significant increase in biomass."

We are left with three very different conclusions about the activity of bacteria in the sea: 1) most bacteria are actively growing, 2) most bacteria are in a dormant state, and 3) bacteria can replicate but do not increase in total biomass. It is quite clear that we have only begun to understand the relevant mechanisms involved in permitting bacteria to survive and grow in the sea.

CONCLUSIONS

1. Growth yields. Techniques are available to measure growth yields in pure cultures and to a limited extent in natural systems. Growth efficiency appears to be dependent upon both substrate type and concentration. So far, however, there are no clear rules as to how to relate these observations to in situ growth.

2. Rate of growth. In culture under optimal conditions generation times may vary from a few minutes to > 24 h. Many morphologically recognizable types have never been successfully cultured, so there is still much to learn about conditions necessary for microbial growth. In nature the few rate measurements that have been made range from several to hundreds of hours.

3. Control of growth and yield. Substrate type and composition appear to be the major controllers of cell growth and yield. It might be argued that even though substrate concentrations most often are low, seawater is a good growth medium because of the large number of substrates present.

4. Microbial size. As a rule, at decreasing growth rates the surface to volume ratio increases, and it is most often high in bacteria that grow in low nutrient regimes.

5. Is there an optimal size for planktonic bacteria? Size appears to be related to nutrient conditions but we still do not know why exactly; we also do not know why they are often round. If their metabolism/unit surface area is the same as large cells, then what is the advantage of being small? More work is needed on the enzymes of these cells, particularly on their physical location in the inner and outer membranes.

6. Attached versus free cells. In most of the open ocean, most of the bacteria are free of detritus, but there are specific zones in which large numbers are attached. Whether sparsely colonized particles are fresh or old needs investigation. Clearly both conditions are possible as Pomeroy and Deibel (1980) have established. The relationship of bacteria to living organisms shows some intriguing possiblities as Provasoli and Pinter (1980) demonstrated with Ulva and bacteria.

7. Substrates that can be used by bacteria. The evidence is that
    bacteria are responsible for much of the substrate processing
    in the sea. Work on cultures has helped elucidate questions of
    substrate thresholds and uptake kinetics, but here are large
    questions concerning the direct applicability of this work to
    in situ conditions. Specifically, we know little about
    decomposition rates once a substrate is greatly diluted.

8. Metabolic versatility or specialization. For pure cultures the
    evidence is that individuals isolated from water on complex
    media are highly versatile; they do not often have identical
    capabilities. There are some theories as to why this is, but
    little direct proof. It is here that cultures will be of great
    value in examining strategies of substrate utilization.

9. Long term survival and starvation. Pure cultures have a
    variety of mechanisms to permit them to survive unfavorable
    conditions. It is more difficult to examine this question in
    situ, but Morita and his colleagues and others have begun
    research in this area. It poses one of the major puzzles for
    present day marine microbiologists.

REFERENCES

Abbott, B. J., A. I. Laskin, and C. J. McCoy. 1973. Growth of
    Acinetobacter calcoaceticus on ethanol. Appl. Microbiol. 25:
    787-792.

Austin, B., S. Garges, B. Conrad, E. E. Harding, R. R. Colwell, U.
    Simidu, and N. Taga. 1979. Comparative study of the aerobic,
    heterotrophic bacterial flora of Chesapeake Bay and Tokyo Bay.
    Appl. Environ. Microbiol. 37: 704-714.

Azam, F., and J. A. Fuhrman. Measurement of bacterioplankton growth
    in the sea and its regulation by environmental conditions. In:
    J. E. Hobbie and P. J. leB. Williams [eds.], Heterotrophic
    Activity in the Sea. Plenum Press, New York.

Azam, F., and R. E. Hodson. 1977. Size distribution and activity
    of marine microheterotrophs. Limnol. Oceanogr. 22: 492-501.

Azam, F., and R. E. Hodson. 1981. Multiphasic kinetics for
    D-glucose uptake by assemblages of natural marine bacteria.
    Mar. Ecol. Prog. Ser. 6: 213-222.

Bauchop, T., and S. R. Elsden. 1960. The growth of microorganisms
    in relation to their energy supply. J. Gen. Microbiol. 23:
    457-469.

Bell, W. H., and E. Sakshaug. 1980. Bacterial utilization of algal
    extracellular products. 2. A kinetic study of natural popula-
    tions. Limnol. Oceanogr. 25: 1021-1033.

Benecke, W. 1933. Bacteriologie des Meeres. Abderhalden's Hand.
    der Biol. Arbeitsmethoden IX: 717-854.

Brown, A. D. 1957. Some general properties of a psychrophilic

pseudomonad: the effects of temperature on some of these
properties and the utilization of glucose by this organism and
Pseudomonas aeruginose. J. Gen. Microbiol. 17: 640-648.

Castille, F. L., Jr., and A. L. Lawrence. 1979. The role of
bacteria in the uptake of hexoses from seawater by post larval
peneid shrimp. Comp. Biochem. Physiol. 64A: 41-48.

Chapman, A. G., L. Fall, and D. E. Atkinson. 1971. Adenylate
energy charge in Escherichia coli during growth and starvation.
J. Bacteriol. 108: 1072-1086.

Christian, R. R., and W. J. Wiebe. 1974. The effects of temperature upon the reproduction and respiration of a marine obligate
psychrophile. Can. J. Microbiol. 20: 1341-1345.

Christian, R. R., K. Bancroft, and W. J. Wiebe. 1975. Distribution
of microbial adenosine triphosphate in salt marsh sediments at
Sapelo Island, Georgia. Soil Sci. 119: 89-97.

Crawford, C. C., J. E. Hobbie, and K. L. Webb. 1974. The utilization of dissolved free amino acids by estuarine microorganisms.
Ecology 55: 551-563.

Danforth, W. R. 1961. Oxidative assimilation of acetate by Euglena.
Carbon balance and effects of ethanol. J. Protozool. 8:
152-158.

Derenbach, J. B., and P. J. leB. Williams. 1974. Autotrophic and
bacterial production: fractionation of plankton populations by
differential filtration of samples from the English Channel.
Mar. Biol. 25: 263-269.

Duursma, E. K. 1963. The production of dissolved organic matter in
the sea, as related to the primary gross production of organic
matter. Neth. J. Sea Res. 2: 85-94.

Eagon, R. G. 1962. Pseudomonas natriegens, a marine bacterium with
a generation time of less than 10 minutes. J. Bacteriol. 83:
736-737.

Eppley, R. W., and B. J. Peterson. 1979. Particulate organic matter
flux and planktonic new production in the deep ocean. Nature
282: 677-680.

Fellows, D. A., D. M. Karl, and G. A. Knauer. 1981. Large particle
fluxes and vertical transport of living carbon in the upper 1500
M of the northeast Pacific Ocean. Deep-Sea Res. 28A: 921-936.

Ferguson, R. L., and P. Rublee. 1976. Contribution of bacteria to
standing crop of coastal plankton. Limnol. Oceanogr. 15: 14-20.

Furhman, J. A., and F. Azam. 1980. Bacterioplankton secondary
production estimates for coastal waters of British Columbia,
Antarctica and California. Appl. Environ. Microbiol. 39:
1085-1095.

Geesey, G. G., and R. Y. Morita. 1979. Capture of arginine at low
concentrations by a marine psychrophilic bacterium. Appl.
Environ. Microbiol. 38: 1092-1097.

Gocke, K., R. Dawson, and G. Liebezeit. 1981. Availability of
dissolved free glucose to heterotrophic microorganisms. Mar.
Biol. 62: 209-216.

Goldman, J. C., and J. J. McCarthy. 1978. Steady state growth and ammonium uptake of a fast-growing marine diatom. Limnol. Oceanogr. 23: 695-703.

Goulder, R. 1977. Attached and free bacteria in an estuary with abundant suspended solids. J. Appl. Bacteriol. 43: 399-405.

Hagström, Å., U. Larrson, P. Horstedt, and S. Normark. 1979. Frequency of dividing cells, a new approach to the determination of bacterial growth rates in aquatic environments. Appl. Environ. Microbiol. 37: 805-812.

Hanson, R. B., and W. J. Wiebe. 1977. Heterotrophic activity associated with particulate size fractions in a Spartina alterniflora salt-marsh estuary, Sapelo Island, Georgia, U.S.A. and the Continental Shelf waters. Mar. Biol. 42: 321-330.

Harder, W., and H. Veldkamp. 1967. A continuous culture study of an obligately psychrophilic marine Pseudomonas species. J. Appl. Bacteriol. 31: 12-23.

Harvey, R. W., and L. Y. Young. 1980. Enrichment and association of bacteria and particulates in salt marsh surface water. Appl. Environ. Microbiol. 39: 894-899.

Haysaka, S. S., and R. Y. Morita. 1978. Salinity and nutrient effects on the induction of the galactose permease system in a psychrophilic marine vibrio. Mar. Biol. 49: 1-6.

Ho, K. P., and W. J. Payne. 1979. Assimilation efficiency and energy contents of prototrophic bacteria. Biotechnol. Bioeng. 21: 787-802.

Hobbie, J. E., R. J. Daley, and S. Jasper. 1977. Use of nuclepore filters for counting bacteria by fluorescence microscopy. Appl. Environ. Microbiol. 33: 1225-1228.

Hodson, R. E., and F. Azam. 1980. Occurrence and characterization of a phosphoenolpyruvate glucose phosphotransferase system in a marine bacterium, Serratia marinorubra. Appl. Environ. Microbiol. 38: 1086-1091.

Hodson, R. E., A. E. Maccubbin, and L. R. Pomeroy. 1981. Dissolved adenosine triphosphate utilization by free-living and attached bacterioplankton. Mar. Biol. 64: 43-51.

Hoppe, H.-G. 1976. Determination and properties of actively metabolizing heterotrophic bacteria in the sea, investigated by means of autoradiography. Mar. Biol. 36: 291-302.

Horvath, R. S. 1972. Microbial co-metabolism and the degradation of organic compounds in nature. Bacteriol. Rev. 36: 146-155.

Ittekkot, V., U. Brockmann, W. Michaelis, and E. T. Degens. 1981. Dissolved free and combined carbohydrates during a phytoplankton bloom in the northern North Sea. Mar. Ecol. Prog. Ser. 4: 299-305.

Jackson, G. A. 1980. Phytoplankton growth and zooplankton grazing in oligotrophic oceans. Nature 284: 439-441.

Jannasch, H. W. 1970. Threshold concentrations of carbon sources limiting bacterial growth in seawater, pp. 321-328. In: D. W. Hood [ed.], Organic Matter in Natural Waters. Inst. of Mar. Sci. (Alaska) Occas. Publ. 1.

Jannasch, H. W. 1973. Bacterial content of particulate matter in off-shore surface waters. Limnol. Oceanogr. 18: 340-342.

Karl, D. M. 1980. Cellular nucleotide measurements and applications
in microbial ecology. Microbiol. Rev. 44: 739-796.

Karl, D. M., and D. B. Craven. 1980. Effects of alkaline phosphatase activity on nucleotide measurementsin aquatic microbial
communities. Appl. Environ. Microbiol. 40: 549-561.

Karl, D. M., and O. Holm-Hansen. 1977. Adenylate energy charge
measurements in natural seawater and sediment samples, pp. 141-
169. In: G. A. Brown [ed.], Second Bi-Annual ATP Methodology
Symp. SAI Technol. Co., San Diego, California.

Koike, I., and A. Hattori. 1974. Growth yield of a denitrifying
bacterium, Pseudomonas denitrificans, under aerobic and
denitrifying conditions. J. Gen. Microbiol. 88: 1-10.

Konings, W. N., and H. Veldkamp. 1980. Phenotypic responses to
environmental changes, pp. 161-191. In: D. C. Wilwood, J. N.
Hedger, M. J. Latham, J. M. Lynch, and J. H. Slater [eds.],
Contemporary Microbial Ecology. Academic Press, New York.

Law, A. T., and D. K. Button. 1977. Multiple-carbon-source-limited
growth rate kinetics of a marine coryneform bacterium. J.
Bacteriol. 129: 115-123.

Laws, E. A. 1975. The importance of respiration losses in controlling the size distribution of marine phytoplankton. Ecology
56: 419-426.

Linton, J. D., D. E. F. Harrison, and A. T. Bull. 1975. Molar
growth yields, respiration and cytochrome patterns of Beneckea
natriegens when grown at different medium dissolved oxygen
tensions. J. Gen. Microbiol. 90: 237-246.

Maca, H. W., and J. W. Fitzgerald. 1979. Evidence for a periplasmic
location in Comamonas terrigena of the inducible tyrosine
sulfate sulfohydrolase. Can. J. Microbiol. 25: 275-278.

Marshall, K. C. 1979. Growth at interfaces, pp. 281-290. In: M.
Shilo [ed.], Strategies of Microbial Life in Extreme Environments. Berlin: Dahlem Conferenzen.

Marshall, K. C. 1980a. Bacterial adhesion in natural environments.
Chapter 9. In: R. C. W. Berkeley et al. [eds.], Microbial
Adhesion to Surfaces. Ellis Horwood Ltd., Chichester.

Marshall, K. C. 1980b. Microorganisms and interfaces. BioScience
30: 246-249.

Mayberry, W. R., G. J. Prochozka, and W. J. Payne. 1968. Factors
derived from studies of aerobic growth in minimal media. J.
Bacteriol. 96: 1424-1426.

Newell, S. Y., and R. R. Christian. 1981. Frequency of dividing
cells as an estimator of bacterial productivity. Appl. Environ.
Microbiol. 42: 23-31.

Novitsky, J. A., and R. Y. Morita. 1976. Morphological characterization of small cells resulting from nutrient starvation of a
psychrophilic marine Vibrio. Appl. Environ. Microbiol. 32:
617-622.

Novitsky, J. A., and R. Y. Morita. 1977. Survival of a psychrophilic marine vibrio under long-term nutrient starvation.

Appl. Environ. Microbiol. 33: 635-641.

Novitsky, J. A., and R. Y. Morita. 1978a. Possible strategy for survival of marine bacteria under starvation conditins. Mar. Biol. 48: 289-295.

Novitsky, J. A., and R. Y. Morita. 1978b. Starvation-induced baro-tolerance as a survival mechanism of psychrophilic marine vibrio in the waters of the Antarctic Convergence. Mar. Biol. 49: 7-10.

Odum, H. T., and E. P. Odum. 1955. Trophic structure and productiv-ity of a windward coral reef community on Eniwetok Atoll. Ecol. Monogr. 25: 291-320.

Oremland, R. S. 1979. Methanogenic activity in plankton samples and fish intestines: a mechanism for in situ methanogenesis in oceanic surface waters. Limnol. Oceanogr. 24: 1136-1141.

Payne, W. J. 1970. Energy yields and growth of heterotrophs. Ann. Rev. Microbiol. 24: 17-52.

Payne, W. J. 1972. Bacterial growth yields, pp. 57-73. In: L. J. Guarraia and R. K. Ballentine [eds.], The Aquatic Environment: Microbial Transformations and Water Management Implications. USEPA, Washington, D.C.

Payne, W. J., R. R. Christian, and W. J. Wiebe. 1970. Assays for biodegradability essential to unrestricted usage of organic compounds. BioScience 20: 862-865.

Payne, W. J., and W. J. Wiebe. 1978. Growth yield and efficiency in chemosynthetic microorganisms. Ann. Rev. Microbiol. 32: 155-183.

Payne, W. J., and M. L. Williams. 1976. Carbon assimilation from simple and complex media by prototrophic heterotrophic bacteria. Biotechnol. Bioeng. 18: 1653-1655.

Pomeroy, L. R. 1974. The oceans food web, a changing paradigm. BioScience 24: 499-504.

Pomeroy, L. R. 1979. Secondary production mechanisms of continental shelf communities, pp. 163-188. In: R. J. Livingston [ed.], Ecological Processes in Coastal and Marine Systems. Plenum Press, New York.

Pomeroy, L. R. 1980. Microbial roles in aquatic food webs, pp. 85-109. In: R. R. Colwell and J. Foster [eds.], Aquatic Microbial Ecology. Maryland Sea Grant Publ., College Park.

Pomeroy, L. R., and D. Deibel. 1980. Aggregation of organic matter by pelagic tunicates. Limnol. Oceanogr. 25: 643-652.

Provasoli, L., and I. J. Pinter. 1980. Bacteria induced polymorph-ism in an axenic laboratory strain of Ulva lactuca (Chloro-phyceae). J. Phycol. 16: 196-201.

Schleyer, M. H. 1980. A preliminary evaluation of heterotrophic utilization of a labelled algal extract in a subtidal reef environment. Mar. Ecol. Prog. Ser. 3: 223-229.

Scott, C. C. L., and W. R. Finnerty. 1976. A comparative analysis of the ultrastructure of hydrocarbon-oxidizing microorganisms. J. Gen Microbiol. 94: 342-350.

Servizi, J. A., and R. H. Bogan. 1964. Thermodynamic aspects of

biological oxidation and synthesis. J. Water Pollut. Control
    Fed. 36: 607-618.

Sheehata, T. E., and A. G. Marr. 1971. Effect of nutrient concen-
    tration on the growth of Escherichia coli. J. Bacteriol. 107:
    210-216.

Sheldon, R. W., and W. H. Sutcliffe, Jr. 1978. Generation times of
    3 h for Sargasso Sea microplankton determined by ATP analysis.
    Limnol. Oceanogr. 23: 1051-1055.

Sheldon, R. W., A. Prakash, and W. H. Sutcliffe, Jr. 1972. The size
    distribution of particles in the ocean. Limnol. Oceanogr. 17:
    327-340.

Sieburth, J. McN. 1976. Bacterial substrates and productivity in
    marine ecosystems. Ann. Rev. Ecol. Syst. 7: 259-285.

Sieburth, J. McN. 1979. Sea Microbes. Oxford Univ. Press, New York.

Slater, J. H., and A. T. Bull. 1979. Biochemical basis of microbial
    interactions. Ann. Appl. Biol. 82: 149-151.

Smith, D. F., and W. J. Wiebe. 1976. Constant release of photosyn-
    thate from marine phytoplankton. Appl. Environ. Microbiol. 32:
    75-79.

Spain, J. C., P. H. Pritchard, and A. W. Bourquin. 1980. Effects
    of adaptation on biodegradation rates in sediment/water cores
    from estuarine and freshwater environments. Appl. Environ.
    Microbiol. 40: 726-734.

Stanier, R. Y., M. Douderoff, and E. A. Adelberg. 1970. The
    Microbial World. Prentice-Hall, Inc., Englewood Cliffs, New
    Jersey.

Steemann Nielsen, E. 1952. The use of radioactive carbon (^{14}C) for
    measuring organic production in the sea. J. Cons. Int. Explor.
    Mer. 18: 117-140.

Stevenson, L. H. 1978. A case for bacterial dormancy in aquatic
    systems. Microb. Ecol. 4: 127-133.

Stouthamer, A. H. 1973. A theoretical study on the amount of ATP
    required for synthesis of microbial cell material. Antonie van
    Leeuwenhoek J. Microbiol. Serol. 39: 545-565.

Strickland, J. D. H., and T. R. Parsons. 1962. On the production
    of particulate organic carbon by heterotrophic processes in sea
    water. Deep-Sea Res. 8: 211-222.

Taylor, D. C., and R. N. Costilow. 1977. Uptake of glucose and
    maltose by Bacillus popilliae. Appl. Environ. Microbiol. 34:
    102-104.

Taylor, P. A., and P. J. leB. Williams. 1975. Theoretical studies
    on the coexistence of competing species under continuous-flow
    conditions. Can. J. Microbiol. 21: 90-98.

Teal, J. M. 1962. Energy flow in the salt marsh ecosystem of
    Georgia. Ecology 43: 614-624.

van Niel, C. B. 1949. Bacterial growth, pp. 92-105. In: A. K.
    Parpart [ed.], The Chemistry and Physiology of Growth.
    Princeton Univ., Princeton, New Jersey.

Watson, S. W., and J. E. Hobbie. 1979. Measurement of bacterial
    biomass as lipopolysaccharide, pp. 82-88. In: J. W. Costerton

and R. R. Colwell [eds.], Native Aquatic Bacteria: Enumeration, Activity and Ecology. ASTM STP 695.

Whitt, D. D., M. J. Klug, and R. D. DeMoss. 1979. Tryptophanase from a marine bacterium, Vibrio K-7. Synthesis, purification and some chemical properties. Arch. Microbiol. 122: 169-175.

Wiebe, P. H., S. H. Boyd, and C. Winget. 1976. Particulate matter sinking to the deep-sea floor at 2000 M in the Tongue of the Ocean, with a description of a new sediment trap. J. Mar. Res. 34: 341-354.

Wiebe, W. J., and K. Bancroft. 1975. Use of the adenylate energy charge ratio to measure growth state of natural microbial communities. Proc. Nat. Acad. Sci. USA 72: 2112-2115.

Wiebe, W. J., and G. B. Chapman. 1968a. Fine structure of selected marine pseudomonads and achromobacteria. J. Bacteriol. 95: 1862-1873.

Wiebe, W. J., and G. B. Chapman. 1968b. Variation in the fine structure of marine Achromobacter and Pseudomonas grown under selected nutritional and temperature regimes. J. Bacteriol. 95: 1874-1886.

Wiebe, W. J., and J. Liston. 1972. Studies on the aerobic, heterotrophic, non-exacting marine bacteria of the benthos, pp. 281-312. In: D. L. Alverson [ed.], Bioenvironmental Studies of the Columbia River Estuaries and Adjacent Ocean Region. Univ. of Washington Press, Seattle.

Wiebe, W. J., and L. R. Pomeroy. 1972. Microorganisms and their association with aggregates and detritus in the sea: A microscopic study. Mem. Ist. Ital. Idrobiol. Spec. Vol. 29 (Suppl.): 325-352.

Wiebe, W. J., and D. F. Smith. 1977. Direct measurement of dissolved organic carbon release by phytoplankton and incorporation by microheterotrophs. Mar. Biol. 42: 213-223.

Wiebe, W. J., and D. F. Smith. 1978. ^{14}C-labelling of the compounds excreted by phytoplankton for employment as a realistic tracer in secondary productivity measurements. Microb. Ecol. 4: 1-8.

Williams, P. J. leB. 1970. Heterotrophic utilization of dissolved organic compounds in the sea. I. Size distribution of population and relationship between respiration and incorporation of growth substrates. J. Mar. Biol. Assoc. UK 50: 859-870.

Williams, P. J. leB. 1973. On the question of growth yields of natural heterotrophic populations. Bull. Ecol. Res. Comm. Stockholm 17: 400-401.

Williams, P. J. leB., and L. R. Muir. 1981. Diffusion as a constraint on the biological importance of microzones in the sea, pp. 209-218. In: J. C. J. Nihoul [ed.], Ecohydrodynamics. Elsevier Scientific Publ. Co., Amsterdam.

Wood, E. F. J. 1965. Marine Microbial Ecology. Chapman and Hall, London.

Wright, R., and J. E. Hobbie. 1965. Uptake of organic solutes by bacteria in lake water. Limnol. Oceanogr. 10: 22-28.

Wright, R. T., and J. E. Hobbie. 1966. Use of glucose and acetate

by bacteria and algae in aquatic ecosystems. Ecology 47:
447-468.

Yetka, J., and W. J. Wiebe. 1974. Antibiotic effects on respiration
of bacterial cultures. Appl. Microbiol. 28: 1033-1039.

Yoshinari, T. 1976. Nitrous oxide in the sea. Mar. Chem. 4:
189-202.

Zobell, C. E. 1946. Marine Microbiology. Chronica Botanica,
Waltham, MA.

SUBSTRATE CAPTURE BY MARINE HETEROTROPHIC BACTERIA

IN LOW NUTRIENT WATERS

Richard Y. Morita

Department of Microbiology, School of Oceanography
Oregon State University
Corvallis, Oregon

INTRODUCTION

Except in coastal areas and some isolated marine environments seawater can be classified as oligotrophic or even ultraoligotrophic. If the nearshore values for the dissolved organic carbon (DOC) are ignored, the DOC ranges from 0.3 to 1.2 mg $C \cdot liter^{-1}$ and in the deep sea, ca. 0.5 mg $C \cdot liter^{-1}$ (Riley and Chester 1971). Other values are cited by various authors in review articles dealing with organic matter in seawater (Menzel 1974, Williams 1975; Duursma 1965; Wangersky 1978) and in each instance, the values for DOC are low. The amount of particulate organic carbon (POC) is exceeded by the DOC by a factor of 10 to 20 (Riley and Chester 1971) and its value will depend on various factors. Approximately 1/5 of the particulate organic fraction is composed of living organisms. Although the amount of DOC and POC is not very great per unit volume, the total amount in the entire oceans is extremely huge. The reservoir of DOC and POC in the ocean is placed at 655×10^9 tons and 14×10^9 tons respectively, assuming the values of 0.5 mg $C \cdot liter^{-1}$ and 10 µg $C \cdot liter^{-1}$ respectively (Menzel 1974). The turnover time is 3,300 years (Menzel 1974).

Dating of the ^{14}C in the DOC by Williams et al. (1969) indicates that it is approximately 3,400 years old and the stability of the DOC has been confirmed by Williams and Gordon (1970).

According to Craig (1971), the rate of oxygen utilization in the Pacific Ocean is 0.004 $ml \cdot liter^{-1} \cdot yr^{-1}$ and this value would require the oxidation of 0.0012 mg $C \cdot liter^{-1} \cdot yr^{-1}$. This would indicate a residence time of 2,000 years.

The resistant nature of the organic matter to decomposition in
the sea was demonstrated by Menzel (1970), Menzel and Goering (1966),
and Menzel and Ryther (1970). When organic matter in seawater was
concentrated and allowed to incubate with the indigenous microflora,
Barber (1968) found virtually no decomposition of the material.
However, if POC is hydrolyzed, it will serve as a good substrate for
microbial decomposition (Seki et al. 1968). The hydrolyzable
material, according to Gordon (1970), represents refractory proteins
that are slowly being converted into bacterial cellular material.

Only one organism has been measured in terms of its growth rate
in the oligotrophic water of the open ocean benthos. This organism
was Tindaria callistiformis which took approximately 50 to 60 years
for gonad development and 100 years for it to reach 8 mm size
(Turekian et al. 1975).

Williams (1970) found that 49% of radioactive glucose was taken
up by organisms passing through a 1.2 μm pore size filter, 68% of the
substrate taken up by organisms passing through a 3 μm filter, and
80% of the substrate taken up by organisms passing through a 8 μm
filter. He concluded that bacteria were mainly responsible for most
of the uptake of free glucose as well as amino acids. However, slow
rates of oxidation of amino acids and glucose were detected at 400
and 600 m in the Mediterranean Sea and no detectable oxidation by
organisms in water collected at 2000 m. According to Azam and Hodson
(1977), 90% of the heterotrophic activity results from bacteria
passing through a 1 μm membrane filter. Because of the low number
of bacteria that could be cultured from the waters of the open ocean,
it has been difficult to accept the traditional view that bacteria
were the most important users of detrital and dissolved organic
materials in the sea (Williams 1975). However, two elements must be
added to help resolve this dilemma. Time (as evidenced by residence
time of water) and the fact that many bacteria in the marine environment do respire but do not form colonies on media (Hoppe 1976, 1978)
should be taken into consideration.

The foregoing data (low concentrations of organic matter, age of
the organic matter, slow rate of oxygen utilization coupled with rate
of organic matter utilization, turnover time, residence time of the
organic matter, and slow growth of Tindaria callistiformia) are the
main reasons for studying substrate capture by microorganisms in low
nutrient environments. The importance of substrate capture in low
nutrient environment lies in the reservoir of the organic matter in
the oligotrophic waters.

In discussing substrate capture in this paper, I will
concentrate on amino acids. Unfortunately, there are no data on
this subject for employing oligotrophic bacteria and therefore these
organisms will not be discussed in this presentation. Nevertheless,
many heterotrophic bacteria from oligotrophic waters can be isolated

on seawater media and the presentation will center around these
heterotrophs.

NITROGENOUS MATERIAL IN SEAWATER

The total amount of soluble organic nitrogen ranges from 5 to
300 µg N·liter^{-1} (Strickland and Parsons 1968). The amino acid
concentration varies according to geographic area, and Williams
(1975) summarizes the values obtained by various investigators. The
range of total amino acid is from 6 to 125 µg·liter^{-1}, whereas
individual amino acids range from 0 to 38 µg N·liter-1 (Strickland
and Parsons 1968). Most of the values cited are in nearshore
environments. However, Lee and Bada (1975) indicate that dissolved
combined amino acids in Equatorial surface waters away from the nearshore environment range from 40 to 50 nmoles·liter^{-1} but the amount
of dissolved free amino acids is considerably lower. Deeper waters
have still lower concentrations (Fig. 1). Bada and Lee (1977) state
that it is possible that the dissolved combined amino acids are not
actually dissolved compounds but simply very small biological
detritus which passes through the glass-fiber filter used in processing the seawater samples. However, it is known that bacteria can
pass through a Whatman GF/C filter. Lott and Morita (unpublished
data) demonstrated that approximately 20% of the bacteria in Yaquina
Bay water pass through a Whatman GF/C glass fiber filter. If oceanic
water is used, then it is possible that a greater percentage of the
bacteria could pass through the filter since it is known that starved
heterotrophic bacteria can form ultramicrocells (Novitsky and Morita
1976). In fact, miniaturization of bacterial cells taken from the
open ocean takes place in the absence of nutrients (Amy and Morita,
unpublished data). This point becomes important since it is now
recognized that many of the bacterial cells in seawater are ultramicrocells (Zimmermann and Meyer-Reil 1974; Daley and Hobbie 1975;
Morita 1977; Zimmermann 1977; Watson et al. 1977; Kogure et al.
1979; Torrella and Morita 1981; Tabor et al. 1981; MacDonell and
Hood, personal communication).

Although free amino acids occur in the DOC, they occur mainly
combined as peptides (Riley and Chester 1971). Amino acids are
necessary for all organisms, especially if they are essential amino
acids for the organism. The importance of methionine, as an example
of an essential amino acid, in oceanic and deep waters is discussed
by Morita (1980). This amino acid is essential in the translation
process for protein synthesis and is one of the amino acids not
reported by Lee and Bada (1975).

The foregoing emphasizes the fact that the amino acid concentration in the oligotrophic waters of the oceans is very low and
that the amount reported as dissolved may actually be several fold
too high.

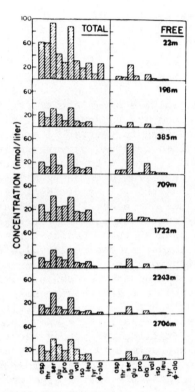

Figure 1. Concentration of individual dissolved amino acids at
station 1 (2°15.5'N, 10°35.3'W). (The extremely high serine content
for the 385 m dissolved free amino acids indicates that the sample
was contaminated.) After Lee and Bada (1975).

MICROBIAL LIFE UNDER LOW NUTRIENT CONDITIONS

    The growth of Caulobacter in distilled water lines remains the
classical example of life under low nutrient conditions. Bacteria
can be isolated in the oligotrophic waters of the oceans as well as
in pelagic red clay of the abyssal plains of the ocean (Morita and
ZoBell 1955). Sieburth (1971) presents a good picture of the
distribution of bacteria in the oligotrophic waters of the sea.
There are numerous reports concerning the heterotrophic activity of
bacteria in oligotrophic waters of the ocean. The growth of bacteria
in low nutrient concentrations has been addressed in the Dahlem
Konferenzen (Shilo 1979) by numerous investigators. However, the
chemostat experiments mentioned cannot operate at below the threshold
levels of substrate for growth and the oligotrophic waters of the
marine environment are below the threshold levels used by many
investigators. What then is the mechanism(s) by which organisms
obtain their substrate? Time must enter into consideration in the

utilization of substrates as evidenced by residence time of water
masses, utilization of oxygen in oligotrophic waters coupled with
the oxidation of organic matter, and the low concentration of
organic matter.

UPTAKE AFFINITIES BY MARINE BACTERIA

The ability of bacteria to function at low substrate levels has
been credited to the affinity of various uptake systems. The affinity constants for various substrates of different microorganisms,
including some high and low affinity uptake systems, by different
investigators has been compiled by Tempest and Neijssel (1978). All
the Km values are in the μM range.

For marine bacteria, it has been shown by Geesey and Morita
(1979) that Ant-300 (a marine vibrio) possess a high and low affinity for arginine uptake (Fig. 2). Even when the cells are depleted
of their endogenous energy reserves, they were found to be capable
of taking up arginine (Fig. 3).

There are many more data in the microbiological literature
concerning uptake affinities but none of the data were collected in
experiments performed at the low concentration of amino acids in the
oligotrophic waters of the ocean. This becomes especially important
when microbial growth rates in natural low nutrient waters appear to
be too slow to be reproducible in steady state cultures (Jannasch
1979).

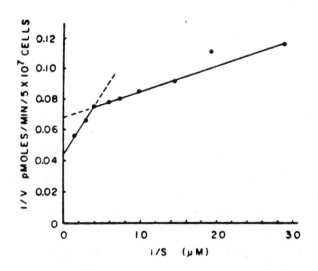

Figure 2. Double-reciprocal plot of initial rates of L-arginine
uptake by cells of Ant-300. After Geesey and Morita (1979).

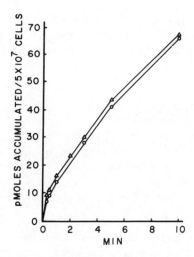

Figure 3. Uptake of L-arginine by cells in the presence (Δ) and
absence (0) of glucose. Cells were exposed to L-[U-^{14}C] arginine
(0.070 μM) for increased periods of time. Glucose was added to a
final concentration of 20 mM. After Geesey and Morita (1979).

Starved cells more closely represent bacteria in oligotropic
waters than freshly cultivated cells, mainly due to the lack of
available substrates in the natural enviornment. The Km values
obtained for glutamic acid uptake by Ant-300 cells starved for 2, 48
and 217 hr and 20 days were 1.8×10^{-7} M, 2.3×10^{-7} M, 1.5×10^{-7} M
and 0.9×10^{-7} M, respectively. The same range of values were found
with arginine, leucine and proline as the substrate (Glick 1980).

Although these affinity uptake values are of interest, the basic
mechanism for affinity towards substrate by the bacteria must still
be answered. In addition, it is doubtful whether these values would
apply when the substrate levels are as low as those found in the
oligotrophic oceanic and deep sea water.

BINDING OF SUBSTRATES TO MICROBIAL CELLS

All organisms have binding proteins. Since bacteria must
obtain their nutrients from their environment, the binding proteins
were probably formed through evolution so that they could scavenge
their nutrients in a nutrient poor system which probably existed
during the early part of the Precambrian.

New binding proteins are continually being discovered; Table 1
represents a partial list. Many of the binding proteins are involved
in substrate transport of the cell. However, it should be mentioned
that most of the binding proteins have been elucidated from

Table 1.   Partial list of some of the binding proteins elucidated.

Binding protein	Reference
Sulfate-binding protein*	Pardee and Prestidge 1966
Phosphate-binding protein*	Medveczky and Rosenberg 1969
Galactose-binding protein*	Anraku 1967
L-arabinose-binding protein*	Hogg and Englesberg 1969; Schleif 1969
Ribose and maltose-binding proteins*	Hazelbauer and Alder 1971
Thiamine-binding protein*	Iwashima et al. 1971
Cyanocobalamine-binding protein	Taylor et al. 1972
Leucine-binding protein*	Piperno and Oxender 1966
Glutamine-binding protein*	Weiner and Heppel 1971
Glutamic acid-binding protein*	Barash and Halpern 1971
Phenylalanine-binding protein*	Kuzuwa et al. 1971
Histidine-binding protein*	Rosen and Vasington 1970; Ames and Lever 1970

*Involved in transport mechanisms of the cells.

Escherichia coli. According to Rosen and Heppel (1973), "it appears entirely reasonable to us that the binding proteins are the carrier proteins for shock-releasable transport systems and they contain the initial recognition site for the substrate."

In 1974, Griffiths et al. suggested that amino acids that bind to the cells before transport into the cell may explain the ability of bacteria to survive and grow in low nutrient aquatic environments. The concept of amino acid pools in bacteria has been accepted for many years but no attempt has been made to separate the amino acids bound by the binding proteins from those in pools inside the cell. The routine method for defining the amino acid pool as outlined by

Britten and McClure (1962), makes use of a TCA treatment. Griffiths
et al. (1974) demonstrated that the ability to retain ^{14}C-amino
acids was primarily a function of osmotic pressure rather than an
ionic function. Figure 4 illustrates the effect of three different
salts on the retention of radioactive glutamic acid in terms of
molarities, osmotic pressure, and ionic strength. By use of an
osmotic shock (at 0.15 M NaCl or its equivalent) they were able to
demonstrate that the amino acid pool consists of two components in

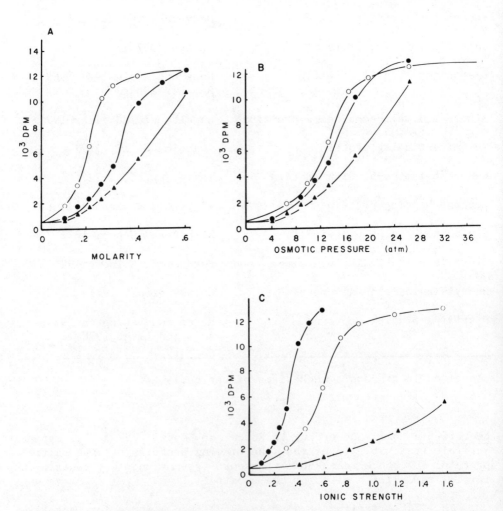

Figure 4. A. The effects of three salts at various molarities on the
retention of labelled material by MP-1. The three salts compared in
this study were NaCL(●), MgCl$_2$ (O), and MgSO$_4$ (△). B. The same
data as presented in 4A expressed in terms of atmospheres. C. The
same data as presented in 4A expressed in terms of ionic strength.
After Griffiths et al. (1974).

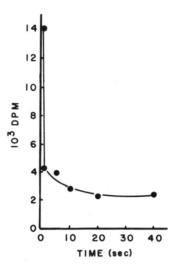

Figure 5.   Time course of glutamate removal from MP-1. The labelled cells were osmotically shocked with 0.15 M NaCl wash for the periods of time indicated.   After Griffiths et al. (1974).

the marine organisms, Vibrio marinus MP-1.   The first pool of amino acids was referred to as AAP-1 since these amino acids were quickly lost upon osmotic shock as illustrated in Fig. 5, whereas a more tightly bound amino acid pool not liberated by the osmotic shock was termed AAP-2.   The use of an osmotic shock provides a technique to demonstrate that amino acids can be loosely bound to the cell by the binding proteins.   Baross et al. (1975) were able to identify the cellular components which incorporated the radioactive substrates (especially when the heterotrophic activity method was employed) as well as to identify the loosely bound amino acid.

As stated previously, starved cells more closely represent bacteria in oligotrophic water than freshly grown cells.   Glick (1980) found with starved cells that there was an osmotically shocked releasable pool (with 0.15 M NaCl) which generally increased with starvation time.   For glutamic acid, over 80% of the amino acids were osmotically shock releasable even when the cells had been starved for 720 hours.

The ability of bacteria to "see" a low substrate concentration is illustrated in Table 2.   Uptake was obtained at concentrations of picograms per liter, which are lower than concentrations found in oligotrophic waters.   It should also be mentioned that the lower limit that bacterial cells can "see" the substrate is not known.   The experiments in Table 2 were not carried out at lower concentrations due to the specific activity of the radioactive glutamate.

Table 2. Uptake of ^{14}C-glutamic acid at low concentration by 50×10^8 72 day starved cells at 5°C. Total volume of incubation mixture was 50 ml. Amy and Morita, unpublished data.

Period of incubation	10^{-9} M (110,000 dpm added)		10^{-10} M (11,000 dpm added)		10^{-11} M (1,100 dpm added)		10^{-12} M (110 dpm added)	
	CO_2 (dpm)	Cells (dpm)	CO_2 (dpm)	Cells (dpm)	CO_2 (dpm)	Cells (dpm)	CO_2 (dpm)	Cells (dpm)
4 h	15,811	4,308	1,699	395	--	--	35	9
7 h	38,141	6,198	3,621	586	508	83	--	--
20 h	--	--	--	--	--	--	81	14
23 h	41,881	7,761	6,080	984	638	99	--	--
32 h	--	--	--	--	--	--	80	0

All values corrected for controls. All samples run in triplicate.

CHEMOTAXIS AS EVIDENCE FOR BINDING OF SUBSTRATE

In the diagram below (redrawn after Anraku 1968) the relationship between the chemoreceptor and its response to the substrate is
indicated. In many cases the chemoreceptor is a binding protein.

                                                        Response

        Transport   Chemoreception

             Other components   Other components

                        Binding protein

Kalckar (1971) states that chemotaxis is primarily a food-finding
device but it is studied mainly to determine the basic nature of a
chemoreception and response. From an evolutionary view point, binding proteins probably evolved to a point where they became the chemo-
receptors which led to a response mechanism (chemotaxis). The value
of chemotaxis in ecology has not been investigated to any great
degree. Chemotaxis helps bring the organism to the most favorable
position for growth, genetic recombination, or dispersal (Carlile
1980). Chemostaxis has been demonstrated in marine bacteria (Bell
and Mitchell 1972). All the amino acids elicit a chemotactic
response in Ant-300 (Torrella and Morita, unpublished data). The
chemotactic response of Ant-300 is illustrated in Fig. 6. Although
the threshold response of Ant-300 appears to be around 10^{-7} M, it
should be recognized that even if no chemotaxis occurs with the cells
because of low substrate concentration, the cells still possess the
chemoreceptor sites for the substrate. Hence, below the threshold
level for chemotaxis, the ability to capture remains functional and
probably helps scavenge substrates from a low nutrient environment.
However, it appears that good chemotaxis does not occur with Ant-300
cells until the cells have been starved at least 24 hours. Better
response occurs if the cells are starved for 48 hours indicating
that, in this organism, active chemotaxis does not occur unless
starvation takes place.

BINDING OF ARGININE BY CELLULAR COMPONENTS

The binding of arginine to cellular components of Ant-300 was
reported by Geesey and Morita (1981). By osmotically shocking the
cells, it was possible to release cellular components from the cell
membrane. When Ant-300 cells were subjected to a SE shock solution
(20% sucrose-1 mM ethylenediaminetetraacetic acid -0.33 M Tris-HCl,
pH 7.5) three percent of the cellular protein was released into the
supernatant fluid. The supernatant fluid can be concentrated by
ultrafiltration yielding a non-dialyzable fraction which bound
arginine when assayed by equilibrium dialysis (Table 3). SE treated

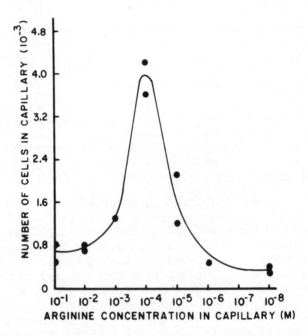

Figure 6.   Concentration-response curve for L-arginine by Ant-300
cells.   Cells were exposed to attractant at 5°C for 1 hr.   After
Geesey and Morita (1979).

cells suspended in reduced osmolarity and increased pH (T-Mg treatment, 1 mM $MgCL_2$ -0.33 M Tris-HCl, pH 8.0) resulted in the release of
an additional 27% protein. Photomicrographs indicated an extensive
damage to the cell envelope (Geesey and Morita 1981). The binding
ability of the T-Mg released components are also given in Table 3.

IMMUNOLOGICAL EVIDENCE FOR BINDING OF SUBSTRATE

    Column fractions (3 and 4) (see Table 3) of Ant-300 shock
released components that exhibited arginine binding activity were
combined with Freund incomplete adjuvant and employed as the
antigen.   This was injected into a rabbit to produce a specific
IgG component(s). Employing the indirect fluorescent staining
procedure of Ward and Perry (1980), Geesey and Morita (1981) were
able to demonstrate rabbits immunized with either fractions 3 or 4
produced an IgG component(s). When this component was incubated in
the presence of intact cells of Ant-300 and fluorescein isothio-
cyanate-labelled goat anti-rabbit IgG, it produced an intense
fluorescence around the bacteria (Fig. 7). For further details see
Geesey and Morita (1981).

Table 3.   Arginine binding activity of components released from
                Ant-300 cells. After Geesey and Morita (1981).

Preparation	Arginine-binding activity	
	pmol bound per ml	Specific activity[a]
SE treatment[b]	144	45
T-Mg supernatant[c]	1,411	185
Column fraction 3 of SE supernatant	192	198
Column fraction 4 of SE supernatant	1,180	303

[a] Specific activity is expressed as units of binding activity per
milligram of protein. One unit of binding activity is 1 pmol of
arginine bound per ml.

[b] A 14-g (wet weight) amount of cells was harvested from a culture
(19 liters) and subjected to SE treatment. Lyophilized SE solution was reconstituted to obtain a final protein concentration of
3.2 mg/ml.

[c] Cells from 11 liters of culture were subjected to T-Mg treatment,
and 500 ml of cell-free T-Mg was concentrated to 51 ml. The final
protein concentration was 7.6 mg/ml.

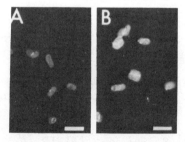

Figure 7. Epifluorescent micrograph of intact cells of Ant-300
suspended in artificial sea water in the presence of fluorescein
isothiocyanate coupled with rabbit IgG raised against (A) column 3
and (B) column fraction 4. Bar, 10 μm. After Geesey and Morita
(1981).

CONCLUSION

It is proposed that bacteria are able to scavenge nutrients in the oligotrophic waters of the oceans by use of their binding proteins. The amount of substrate captured by microorganisms may be sufficient for their energy of maintenance. If time is taken into consideration, then it is possible for erratic growth of the microorganisms to occur. For instance, the doubling time of bacteria from deep water was given by Carlucci and Williams (1978) as 210 hr and they gave a rate of organic utilization of 0.1 mg $C \cdot m^{-3} \cdot yr^{-1}$. No other concept for the capturing of substrate has been proposed. The capture of substrates by the cell's binding proteins has a good scientific basis. Substrate capture should function in all aquatic environments.

ACKNOWLEDGMENTS

This paper was supported in part by the National Science Foundation Grant OCE 8108366. Published as Technical Paper No. 6204, Oregon Agricultural Experiment Station.

REFERENCES

Ames, G. F. and J. Lever. 1970. Components of histidine transport: Histidine binding proteins and his P protein. Proc. Nat. Acad. Sci. 66: 1096-1103.

Anraku, Y. 1967. The reduction and restoration of galactose transport in osmotically shocked cells of Escherichia coli. J. Biol. Chem. 242: 793-800.

Anraku, Y. 1968. Transport of sugars and amino acids in bacteria. II. Properties of galactose- and leucine-proteins. J. Biol. Chem. 243: 2123-3127.

Azam, F., and R. E. Hodson. 1977. Size distribution and activity of marine microheterotrophs. Limnol. Oceanogr. 22: 492-501.

Bada, J. F., and C. Lee. 1977. Decomposition and an alteration of organic compounds in seawater. Mar. Chem. 5: 523-534.

Barasch, H., and Y. S. Halpern. 1971. Glutamate-binding protein and its relation to glutamate transport in Escherichia coli K-12. Biochem. Biophys. Res. Commun. 45: 681-699.

Barber, R. T. 1968. Dissolved organic carbon from deep waters resists microbial oxidation. Nature 220: 274-275.

Baross, J. A., F. J. Hanus, R. P. Griffiths and R. Y. Morita. 1975. Nature of incorporated ^{14}C material retained by sulfuric acid fixed bacteria in pure culture and in natural aquatic populations with trichloroacetic acid precipitates from the sample. J. Fish. Res. Board Can. 32: 1876-1879.

Bell, W., and R. Mitchell. 1972. Chemotactic and growth responses of marine bacteria to algal extracellular products. Biol. Bull. 143: 265-277.

Britten, R. J., and F. T. McClure. 1962. The amino acid pool in
    Escherichia coli. Bacteriol. Rev. 26: 292–335.
Carlile, M. H. 1980. Positioning mechanisms -- The role of motil-
    ity, taxis and trophism in the the life of microorganisms, pp.
    54–74. In: D. C. Ellwood, N. J. Hedger, M. J. Latham, M. J.
    Lynch, and J. H. Slater [eds.], Contemporary Microbial
    Ecology. Academic Press.
Carlucci, A. F., and P. M. Williams. 1978. Simulated in situ
    growth rates of pelagic marine bacteria. Naturwissenschaften
    65: 541–542.
Craig, H. 1971. The deep metabolism: oxygen consumption in abyssal
    ocean water. J. Geophys. Res. 76: 5078–5086.
Daley, R. J., and J. E. Hobbie. 1975. Direct count of aquatic
    bacteria by a modified epifluorescent technique. Limnol.
    Oceanogr. 20: 875–881.
Duursma, E. K. 1965. The dissolved organic constituents of sea-
    water, pp. 433–475. In: J. P. Riley and G. Skirrow [eds.],
    Chemical Oceanography, Vol. I. Academic Press.
Geesey, G. G., and R. Y. Morita. 1979. Capture of arginine at low
    concentrations by a marine psychrophilic bacterium. Appl.
    Environ. Microbiol. 38: 1092–1097.
Geesey, G. G., and R. Y. Morita. 1981. Relationship of cell
    envelope stability to substrate capture in a marine psychro-
    philic bacteria. Appl. Environ. Microbiol. 42: 533–540.
Glick, M. A. 1980. Substrate capture, uptake, and utilization of
    some amino acids by starved cells of a psychrophilic marine
    Vibrio. M.S. thesis, Oregon State Univ., Corvallis, Oregon.
Gordon, D. C. 1970. Some studies on the distribution and composi-
    tion of particulate carbon in the North Atlantic Ocean.
    Deep-Sea Res. 17: 233–243.
Griffiths, R. P., J. A. Baross, F. J. Hanus, and R. Y. Morita. 1974.
    Some physical and chemical parameters affecting the formation
    and retention of glutamate pools in a marine psychrophilic
    bacterium. Z. Allg. Mikrobiol. 14: 359–369.
Hazelbauer, G. L., and J. Adler. 1971. Role of galactose binding
    protein in chemotaxis of Escherichia coli toward galactose.
    Nat. New Biol. 230: 101–104.
Hogg, R. W., and E. Englesberg. 1969. L-arabinose binding protein
    from Escherichia coli B/r. J. Bacteriol. 100: 423–432.
Hoppe, H. G. 1976. Determination and properties of actively
    metabolizing heterotrophic bacteria in the sea, investigated by
    means of microautoradiography. Mar. Biol. 36: 291–302.
Hoppe, H. G. 1978. Relations between active bacteria and hetero-
    trophic potential in the sea. Neth. J. Sea Res. 12: 78–98.
Iwashima, A., A. Matsumura, and Y. Nose. 1971. Thiamine-binding of
    Escherichia coli. J. Bacteriol. 108: 1419–1421.
Jannasch, H. W. 1979. Microbial ecology of aquatic low nutrient
    habitats, pp. 243–260. In: M. Shilo [ed.], Strategies of
    Microbial Life in Extreme Environments. Dahlem Konferenzen,

Verlag Chemie, Weinheim and New York.

Kalckar, H. M. 1971. The periplasmic galactose binding protein of *Escherichia coli*. Science 174: 557-565.

Kogure, K., U. Simidu, and N. Taga. 1979. A tentative direct microscopic method for counting living marine bacteria. Can. J. Microbiol. 25: 415-420.

Kuzuwa, H., K. Bronwell, and G. Gurloff. 1971. The phenylalanine-binding protein of *Comamonas* sp. (ATCC 11299a). J. Biol. Chem. 246: 6371-6380.

Lee, C., and J. L. Bada. 1975. Amino acids in equatorial Pacific Ocean. Earth Planet. Sci. Lett. 26: 61-68.

Medeczky, N., and H. Rosenberg. 1969. The binding and release of phosphate by a protein isolated from *Escherichia coli*. Biochim. Biophys. Acta 192: 369-371.

Menzel, D. W. 1970. The role of *in situ* decomposition of organic matter on the concentration of non-conservative properties in the sea. Deep-Sea Res. 17: 751-764.

Menzel, D. W. 1974. Primary productivity, dissolved and particulate organic matter, and the sites of oxidation of organic matter, pp. 659-678. *In*: E. D. Goldberg [ed.], The Sea, Vol. 5. Wiley-Interscience Publ., John Wiley and Sons.

Menzel, D. W., and J. J. Goering. 1966. The distribution of organic detritus in the ocean. Limnol. Oceanogr. 11: 333-337.

Menzel, D. W., and J. H. Ryther. 1970. Distribution and cycling of organic matter in the oceans, pp. 31-54. *In*: D. W. Hood [ed.], Organic Matter in Natural Waters. Inst. Mar. Sci. Publ., College, Alaska.

Morita, R. Y. 1977. The role of microbes in the marine environment, pp. 445-456. *In*: N. R. Anderson and B. J. Zahurance [eds.], Ocean Sound Scattering Prediction. Plenum Press, New York.

Morita, R. Y. 1980. Microbial life in the deep sea. Can. J. Microbiol. 21: 1375-1385.

Morita, R. Y., and C. E. ZoBell. 1955. Occurrence of bacteria in pelagic sediments collected during the Mid-Pacific Expedition. Deep-Sea Res. 3: 66-73.

Novitsky, J. A., and R. Y. Morita. 1976. Morphological characterization of small cells resulting from nutrient starvation of a psychrophilic marine vibrio. Appl. Environ. Microbiol. 32: 617-622.

Pardee, A. B., and L. S. Prestige. 1966. Cell-free activity of a sulfate binding site involved in active transport. Proc. Nat. Acad. Sci. USA 55: 189-191.

Piperno, J. R., and D. L. Oxender. 1966. Amino acid-binding protein released from *Escherichia coli* by osmotic shock. J. Biol. Chem. 241: 5732-5734.

Riley, J. P. and R. Chester. 1971. Introduction of Marine Chemistry. Academic Press.

Rosen, B. P., and L. A. Heppel. 1973. Present status of binding proteins that are released from Gram-negative bacteria by osmotic shock, pp. 209-239. *In*: L. Lieve [ed.], Bacterial

Membranes and Walls. Marcel Dekker, Inc., New York.

Rosen, B. P., and F. D. Vasington. 1970. Relationship of the histidine binding protein and the histidine permease system in S. typhimurium. Red. Proc. Fed. Am. Soc. Exp. Biol. 29: 342.

Schleif, R. 1969. An L-arabinose binding protein and arabinose permeation in Escherichia coli. J. Mol. Biol. 46: 185-196.

Seki, H., J. Skelding, and T. R. Parsons. 1968. Observations on the decomposition of marine sediment. Limnol. Oceanogr. 13: 440-447.

Shilo, M. 1979. Strategies of Microbial Life in Extreme Environments. Dahlem Konferenzen, Verlag Chemie, Weinheim and New York.

Sieburth, J. McN. 1971. Distribution and activity of oceanic bacteria. Deep-Sea Res. 18: 1111-1121.

Strickland, J. D. H., and T. R. S. Parsons. 1968. A practical handbook of sea water analysis. Fish. Res. Board Can. Bull. 167.

Tabor, P. S., K. Ohwada and R. R. Colwell. 1981. Filterable marine bacteria in deep sea: Distribution, taxonomy and response to starvation. Microb. Ecol. 7: 67-83.

Taylor, R. T., S. A. Norrell and M. L. Hanna. 1972. Uptake of cyanocobalamin by Escherichia coli B: Some characteristics and evidence for a binding protein. Arch. Biochem. Biophys. 148: 366-381.

Tempest, D. W., and O. M. Neijssel. 1978. Eco-physiological aspects of microbial growth in aerobic nutrient-limited environments. Adv. Microb. Ecol. 2: 105-153.

Torrella, F., and R. Y. Morita. 1981. Microcultural study of bacterial size changes and microcolony and ultramicrocolony formation by heterotrophic bacteria in seawater. Appl. Environ. Microbiol. 41: 518-527.

Turekian, K. K., J. K. Cochran, D. P. Kharkar, R. M. Cerrato, J. R. Vaisnys, H. L. Sanders, J. F. Grassle, and J. A. Allen. 1975. Slow growth rate of a deep-sea clam determined by ^{228}Ra chronology. Proc. Nat. Acad. Sci. 72: 2829-2832.

Wangersky, P. J. 1978. Production of dissolved organic matter, pp. 115-220. In: O. Kinne [ed.], Marine Ecology, Vol. IV. John Wiley and Sons.

Ward, B. B., and M. J. Perry. 1980. Immunofluorescent assay for marine ammonium-oxidizing bacterium Nitrosococcus oceanus. Appl. Environ. Microbiol. 39: 913-918.

Watson, S. W., T. J. Novitsky, H. L. Quinby and F. W. Valois. 1977. Determination of bacterial number and biomass in the marine environment. Appl. Environ. Microbiol. 33: 940-946.

Weiner, J. H., and L. A. Heppel. 1971. A binding protein for glutamine and its relation to active transport in Escherichia coli. J. Biol. Chem. 246: 6933-6941.

Williams, P. J. leB. 1970. Heterotrophic utilization of dissolved organic compounds in the sea. I. Size distribution of population and relationship between respiration and incorporation of growth substrates. J. Mar. Biol. Assoc. UK 50: 839-870.

Williams, P. J. leB. 1975. Biological and chemical aspects of dissolved organic material in sea water, pp. 301-364. In: J. P. Riley and G. Skirrow [eds.], Chemical Oceanography, Vol. 2. Academic Press.

Williams, P. M., and L. I. Gordon. 1970. Carbon-13:carbon-12 ratios in dissolved and particulate organic matter in the sea. Deep-Sea Res. 17: 19-28.

Williams, P. M., H. Oeschger and P. Kinny. 1969. Natural radio-carbon activity of dissolved organic carbon in the Northeast Pacific Ocean. Nature 224: 256-258.

Zimmermann, R. 1977. Estimation of bacterial numbers and biomass by epifluorescence microscopy and scanning electron microscopy, pp. 103-120. In: G. Rheinheimer [ed.], Microbial Ecology of a Brackish Water Environment. Springer-Verlag.

Zimmermann, R., and L.-A. Meyer-Reil. 1974. A new method for fluorescence staining of bacterial populations on membrane filter. Kiel. Meeresforsch. 30: 24-27.

INPUTS INTO MICROBIAL FOOD CHAINS

Jonathan H. Sharp

College of Marine Studies
University of Delaware
Lewes, Delaware   19958

INTRODUCTION

There is little question or disagreement with the contention that phytoplankton contribute the vast majority of the input of organic matter to the sea. There are, however, major controversies today on the actual rate of primary productivity in the sea and on the routes by which organic matter arrives in the dissolved pool.[1]

There is the traditional viewpoint that phytoplankton growth in the ocean is slow with rates on the order of 0.1 to 0.2 doublings per day. In contrast to this are recent claims of phytoplankton exhibiting maximum growth rates (presumably more like 1 to 2 doublings per day). A second controversy has as the more prevalent older viewpoint that phytoplankton excrete a significant portion of the photoassimilated carbon as organic losses. This is contrasted by recent claims that phytoplankton under steady state conditions do not suffer large carbon losses.

Recent evidence (e.g., Mopper and Lindroth 1982; Sellner 1982) of large variations in amino acid and carbohydrate concentrations in ocean water and suggestions of large nutritional requirements of

---

[1] Dissolved organic matter (the soluble pool) is functionally defined by microporous filters; it is somewhat misleading since it contains colloidal particles (Sharp 1973) and even some microorganisms. This discrepancy is more than semantic since it bears on the dynamics of production and utilization of extra-organismic organic matter. However, in this paper the functional definition will be used and dissolved organics will describe anything passing a filter of about 0.5 - 1 micron.

marine microheterotrophs (e.g., Pomeroy 1974; Williams 1981b) seem to support the need for high phytoplankton growth rates and extensive excretion. However, an alternative explanation to excretion by actively growing phytoplankton is organic inputs into microbial food chains from zooplankton grazing spillage, zooplankton excretion, and death of phytoplankton.

In this paper, an attempt is made to reassess the mechanism of organic input into the sea by phytoplankton. After a quantification of allochthonous and autochthonous inputs, primary productivity will be discussed with conderation of: growth rates, carbon budgets, and organic excretion. Then discussion will focus on debris from herbivore feeding and on death of phytoplankton as possible sources to the organic pool.

ORGANIC INPUTS

A general picture of organic inputs to the sea is shown in Fig. 1. Two autochthonous inputs are primary productivity by phytoplankton and by macrophytes; the latter including both benthic algae and spermatophytes. All other inputs are considered allochthonous. This latter category is subdivided into five sources: rivers, ground waters, volatiles, aerosols, and rain. Table 1 gives values for the inputs from these sources; volatiles and aerosols are combined for a total dry deposition.

Of the _in situ_ inputs, macrophyte production has often been overlooked or underestimated in the past. Recent reconsiderations (e.g., Woodwell et al. 1978; Nienhuis 1981; Smith 1981) show that this source is much greater than was otherwise thought. A considerable portion of the macrophyte production might be released in surrounding waters as particulate and dissolved organics (Robinson

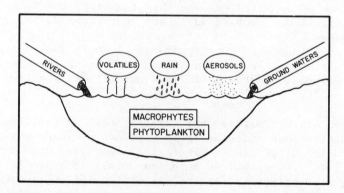

Fig. 1 Inputs of organics to the sea from _in situ_ processes (in boxes), liquid import (in cylinders), and atmospheric import (ovals).

et al. 1982). Some estimates have shown these losses to be minimal
in young tissue, but probably they are significant since most macroscopic algae and spermatophytes have a considerable portion of their
biomass in older tissue (see Mann et al. 1980). Thus macrophyte
dissolved organic inputs may be quite important locally in coastal
waters. Also, as will be considered in the next section, phytoplankton production may have large errors associated with its
estimation and thus could be up to an order of magnitude higher than
the value given in Table 1.

Considerable variation exists with estimates of most of the
allochthonous inputs as well. The river input is from a recent
revised estimate (Richey et al. 1980) and is two to three times the
size of most previous ones; it in itself has been challenged
(Meybeck 1981). However, this estimate is based upon the somewhat
novel considerations of seasonal variations in water flow and
organic content and of oxidation of organic matter (Richey et al.
1980). When compared to other calculations (Meybeck 1981), Richey's
value appears to be reasonable to accept. Ground water input is
based upon somewhat limited estimates on both water volume and
organic concentration, but this is apparently a relatively small

Table 1. Inputs of organics to the sea.

Input	10^{15} g C/yr	% of Total
Primary Production		
Phytoplankton (Woodwell et al. 1978)	23.1	84
Macrophytes (Woodwell et al. 1978)	1.7	6.2
Liquid Import		
Rivers (Richey et al. 1980)	1.0	3.6
Ground Waters (Garrels and Mackenzie 1971)	0.08	0.3
Atmospheric Import		
Rain (Duce and Duursma 1977)	1.0	3.6
Dry Deposition (Duce and Duursma 1977)	0.5	1.8

source. Both rainwater and atmospheric volatile organics entering
the ocean are relatively large and poorly quantified inputs.
Aerosol dry deposition of particulate material is also poorly
defined, but possibly is a smaller source than the other two
atmospheric inputs. The atmospheric inputs are also difficult to
separate (rain scavenges volatiles and aerosols) and to attribute to
sources (much of the input is actually recycling from the sea
surface rather than new input from land). The large international
research project SEAREX is nearing a point at which much better
estimates of these inputs should be available shortly (R. A. Duce,
personal communication); for the present, the estimates given in
Table 1 will have to suffice.

The prevailing thinking, until very recently, has been that the
only in situ production of organic matter in the sea could come from
photoautotrophy. The recent discoveries of chemosynthetic bacteria
in deep hydrothermal vents completely changes the total picture. It
is possible that a relatively large amount of newly fixed carbon
might enter the deep sea organic pool in this fashion. Further
analysis is needed of rates of carbon fixation and extent of deep
chemosynthetic environments before any quantitative assessment can
be made.

Overall, the phytoplankton obviously are the major source of
organic matter for the sea, but contributions from other sources may
be considerably larger than was thought a few years ago (e.g., see
Handa 1977; Williams 1975) and they should not be ignored.

HOW FAST DO PHYTOPLANKTON GROW?

The prevailing view on phytoplankton growth rates in the open
ocean, until a few years ago, was of doubling times on the order of
0.1 to 0.2 per day (Eppley and Strickland 1968; Eppley et al. 1973).
Recently, several rather independent experimental approaches have
given rise to the suspicion of much higher growth rates. Measurements of bacterial demand (Sorokin 1978; Sieburth 1977) suggest
greater primary production rates than previously estimated.
Estimates of increases of total particle volume and ATP in incubated
bottles suggests much higher microbial doubling times (Sheldon et
al. 1973; Sheldon and Sutcliffe 1978). Ammonium nitrogen uptake by
phytoplankton is inferred to be very rapid in the open ocean
(McCarthy and Goldman 1979) and if this rate were sustained for
phytoplankton to maintain the Redfield Ratio (Redfield et al. 1963)
of C, N, and P, then phytoplankton must grow rapidly and near their
maximal growth rate (Goldman et al. 1979). Thus, it has been argued
that phytoplankton are growing in the ocean at rates closer to 1 to
2 doublings per day.

This controversy has been re-evaluated for C, N, and P uptake in the central North Pacific Gyre with experimental checks (Sharp et al. 1980a; Perry and Eppley 1981). Central within the controversy is the evidence for significant underestimates in production using the ^{14}C method (Gieskes et al. 1979) which has led to suggestions for much more tedious and careful procedures (Carpenter and Lively 1980; Goldman 1980). Peterson (1980) has written an excellent review on problems associated with the ^{14}C method. There has been a full array of claims from the ^{14}C method being fairly accurate (Williams et al. 1979; Williams 1981a) to it underestimating productivity somewhat (Peterson 1980), to it grossly underestimating (Gieskes et al. 1979) to it overestimating productivity (Peterson 1978). Probably better calibration of the method is needed since it is not used in any standard format.

It is important to realize that if phytoplankton do indeed produce at rates up to ten times that of our conventional thinking, then other aspects of marine ecology must be consistent with this new thinking. Eppley (1980) has pointed out the discrepancy that claims of higher productivity could make for light harvesting efficiency of phytoplankton. He, on the other hand (Eppley 1980), has suggested that sediment trap experiments may support the idea of higher production. This latter point requires resolution of differences between sediment trap experiments and deep carbon cycling in the ocean (Fiadeiro 1980). One must also take into account the implications to zooplankton and bacterial utilization of the primary production (Jackson 1980; Williams 1981b).

This controversy has led to a PRPOOS study, headed by Eppley, to examine phytoplankton rate processes in the oligotrophic ocean. This study should hopefully provide much information for finally resolving the controversy. However, one cannot help but feel that possibly some of the controversy is blown out of proportion and Yentsch (1980) has probably correctly reminded us that both rapid and slow production can be found in the open ocean depending upon when and where one looks. Thus, one must consider that phytoplankton probably do not exist under steady state conditions in the open ocean for long but rather display large, seemingly random, ups and downs in their metabolic behavior. This subject will be further considered below.

CARBON BUDGETS

Once phytoplankton have fixed carbon in photosynthesis, some may be lost as carbon dioxide in respiration and some as organic matter in excretion. Excretion is considered separately in the next section, the main concern here is the overall cellular carbon budget and respiratory losses.

The older oxygen light-dark productivity method provided in

principle values for gross and net productivity;[2] the ^{14}C method is less clear-cut and is usually considered to give "something between net primary and gross productivity." It is possible that with 24-hour incubations the ^{14}C method gives a fair estimate of net primary productivity (Eppley and Sharp 1975). However, if this is the case and if respiratory losses are large and variable, short-term incubations may not give a good assessment of real productivity. Indeed, there is recent evidence of large and variable respiratory losses by phytoplankton (Eppley and Sharp 1975; Smith 1977; Peterson 1978) or by the full microbial community (Devol and Packard 1978). Hence one should be very cautious in interpreting the results of short-term incubations in the field. On the other hand, long incubations are subject to large potential errors from bacterial effects, zooplankton excretion, and physiological stress.

In the laboratory, where better control can be achieved, it is possible to assess carbon losses and compare productivity from ^{14}C and losses to growth. This has been done with continuous cultures of a marine diatom with the result that growth estimated from ^{14}C uptake matched the growth rate estimated from the continuous culture dilution rate (Sharp et al. 1980b). In this same experimental setup the exponential phase growth rate in a batch culture estimated by ^{14}C uptake matched that from changes in cell counts. Also from this work, a good correlation was found between carbon budget components of an exponential phase batch culture and those from a continuous culture (Sharp et al. 1980b). Li and Goldman (1981) found in 3 out of 5 cases with marine phytoplankton from nitrogen-limited continuous cultures that short term ^{14}C estimates of growth rate did not match those from the culture parameters; the discrepancy occurred as both under and overestimates.

The respiratory losses in the experiment discussed above were very low (< 3% of assimilated carbon) during exponential phase and in continuous culture, but soared in non-exponential phases (as high as 85%). As discussed above, recent field studies have shown much larger and variable respiratory losses. In all cases above only mitochondrial respiration has been discussed, photorespiration is a quite different phenomenon and of poorly understood significance for marine phytoplankton (Burris 1980). Perhaps photorespiration does not represent an error in ^{14}C experiments since carbon is rapidly recycled and never really a part of gross production as organic matter.

Another carbon budget term that must be considered is that of

---

[2] Net productivity implies the net amount from the microbial community and includes the effect of bacterial respiration; net primary productivity includes only algal respiration (and excretion), not bacterial losses.

organic loss not directly related to carbon assimilation. In continuous cultures of marine diatoms at high density this has been seen as a large increase in total dissolved organic carbon but not in short-term organic ^{14}C (Sharp 1977). The same loss was seen by Sharp et al. (1980b) in non-exponential phase of a batch culture of a marine diatom. This loss is probably attributable to algal death and will be considered further below.

Very careful carbon budget experiments have been done in continuous cultures of marine flagellates and diatoms on light-dark cycles with uniform labelling (Laws and Wong 1978; Underhill 1981). In the first case, three species of marine phytoplankters were considered and studied at one temperature over varying growth rates (Laws and Wong 1978); in the second case, a single species of marine diatom was studied over varying growth rates at three temperatures (Underhill 1981). In both cases, dark losses were almost entirely from mitochondrial respiration; excretion was minimal. Carbon loss were considerable greater (26-30%) at low growth rates than at near maximum growth rates (about 10% loss). This growth rate effect and the minimal excretion levels can be seen in Table 2. In the Laws and Wong work, different species appeared to have different percentage losses with flagellates showing greater losses at low growth rates than the diatom. In Underhill's work, higher percentage respiratory losses were found at higher temperatures.

Another point from Underhill's (1981) work is that with light-dark cycles and nitrogen limitation of his culture, he found C/N ratios approximating the Redfield Ratio at growth rates considerably below (< 50%) the maximum. This is in direct opposition with the results of Goldman et al. (1980) who used continuous illumination for their cultures.

PHYTOPLANKTON EXCRETION

The subject of excretion of organic matter by phytoplankton is controversial and has elicited considerable discussion. In two recent reviews of this phenomenon in both culture and field studies, it was concluded that healthy cells do not appear to have excessive organic carbon losses (Sharp 1977; Harris 1978). I will not attempt here to rehash the literature covered previously, but instead will consider more recent publications and attempt to evaluate the subject in a slightly different light.

An area of importance is that of functional excretion. Some compounds are produced by phytoplankton that have ecological significance even if they do not constitute a major carbon loss from the cell (Aaronson 1978; Sharp 1978). Effects can be made on other phytoplankton (e.g., allelopathy; Sharp et al. 1979) on a broad array of other biological targets (see Gauthier and Aubert 1981) or

Table 2.   Respiration and excretion losses from marine phytoplankton
           grown in uniformly labelled media in continuous cultures
           under light/dark cycles at varying growth rates. Data for
           calculations from Laws and Wong (1978) from their Table 1.
           Data from Underhill (1981) from his Table 2.

Dilution rate (d^{-1})	Dark losses as percent of daytime net production	
	Respiration Loss	Excretion Loss
Thalassiosira allenii, 20°C [a]		
0.11	19	1.6
0.36	13	1.3
0.69	10	0.7
1.04	10	1.0
1.42	13	0.9
1.55	11	0.6
Monochrysis lutheri, 20°C [a]		
0.14	35	1.3
0.30	27	0.8
0.47	24	1.0
0.59	21	0.6
0.82	13	0.9
0.95	10	1.0
Dunaliella tertiolecta, 20°C [a]		
0.13	31	1.9
0.22	26	3.1
0.38	29	1.0
0.52	27	1.1
0.66	26	1.4
0.85	21	0.4
1.06	12	0.9

	Dark losses as percent of daytime net production	
Dilution rate (d^{-1})	Respiration Loss	Excretion Loss
Skeletonema costatum, 6°C [b]		
0.22	20	2.3
0.35	18	1.8
0.50	18	2.0
0.57	18	1.2
0.68	18	0.8
0.79	13	1.7
Skeletonema costatum, 16°C [b]		
0.26	24	2.1
0.53	19	1.4
0.75	17	1.9
1.07	17	1.4
1.38	14	1.2
1.72	12	0.6
Skeletonema costatum, 24°C [b]		
0.37	34	2.1
0.51	32	2.0
0.72	25	1.6
1.05	20	0.8
1.39	17	0.9
2.08	10	1.2

[a] Laws and Wong 1978
[b] Underhill 1981

for alteration of the environment (e.g., Fisher and Fabris 1982). However, functional excretion is probably of no consequence in actual carbon budgets since the amount of organic matter produced is usually not a significant portion of the carbon fixed.

In this section, laboratory studies will be discussed first, then field studies on phytoplankton, followed by field studies of specific organic compounds.

Laboratory Studies

The notorious organic producer Dunaliella sp. has been studied because of its use of glycerol for osmoregulation. In somewhat classical work, Ben Amotz (1975) considered glycerol release from D. parva to not increase in higher molarity salt solutions. Jones and Galloway (1979) found the opposite of this with D. tertiolecta, namely more excretion in higher molar NaCl media; thus, suggesting that glycerol production is not purely an osmoregulatory function. As in other studies, they found considerably more glycerol production in stationary phase than in exponential (Jones and Galloway 1979) indicating cellular leakage. Enhuber and Gimmler (1980) confirmed that glycerol production in D. parva was partially due to salt stress.

Other indications of shock effects have been confirmed with light conditions (Mague et al. 1980) and changing of medium by dilution (Naylor 1979). An excellent way to avoid potential shock problems and thus arrive at results from steady state conditions is to use uniform labelling in continuous cultures. The works of Laws and Wong (1978) and Underhill (1981) using this approach have already been mentioned. Neither of these authors were willing to give much meaning to their excretion results. However, in both cases excretion losses were always less than 5% of day-light carbon production and as can be seen in Table 2, these losses consistently appear to be greater at low growth rates than at higher ones.

The present conclusion from laboratory studies is that under steady state conditions, especially at relatively high growth rates, phytoplankton do not lose much of their photoassimiliated carbon. Under non-steady state conditions and when given environmental shock, this is not the case.

Field Studies of Phytoplankton

An early reason for considering field results to be influenced by carbon excretion was the work of Antia et al. (1963) in which there was a serious discrepancy between the ^{14}C and dissolved oxygen measures of production. This has been confirmed by other workers since that time but the recent review of Williams et al. (1979) casts a new light on the subject. They found good agreement between the

two methods in a very careful comparison and suggested that previous
poor agreements were due to inaccurate assumptions in the use of the
photosynthetic quotient. Their higher PQ values arise from taking
the effect of the nitrogen source into account.

Earlier field work in which organic excretion was estimated
with the ^{14}C method has been criticized on the basis of inadequate
assessment of background blanks (Sharp 1977). Recently, considerably
lower values have been reported with this method (e.g., Williams and
Yentsch 1976; Mague et al. 1980; Shifrin 1980; Sellner 1982). In
fact, the conclusion of Mague et al. (1980) was that extracellular
release is probably not a significant loss of oceanic primary
production but may be quite important when viewed in context of
biological cycling of organic matter. Storch and Saunders (1978)
concluded that extracellular release, even if calculated at maximum
rates, did not significally influence seasonal organic cycles in a
lake. However, this may be begging the question since their lake
had appreciable allochthonous input. Furthermore, the labile
fraction of the organic pool even though it is small may turn over
very rapidly in the ocean. There have been a number of recent
attempts to account for direct rapid heterotrophic utilization of
phytoplankton products (e.g., Smith et al. 1977; Wiebe and Smith
1977). This subject will be addressed partially in the next section
and was also discussed by others in this conference.

Several recent works in estuarine and coastal waters have also
addressed excretion by natural populations using the ^{14}C method.
Lancelot (1979) found fairly high extracellular release (20-60%).
However, other recent estuarine studies (Simek 1981; Joint and
Pomeroy 1981) showed very low release (< 5%). This discrepancy is
quite pronounced and it is possible that the environmental situation
studied by Lancelot (late in a spring bloom) is not typical of the
other studies.

Overall, from field studies there is some evidence that important, but not quantitatively significant, excretion probably takes
place. By this, I mean that there is undoubtedly an important rapid
cycling of labile organics starting with phytoplankton excretion.
However, there is still no compelling evidence that significant
portions of the photoassimilated carbon are lost by phytoplankton
normally. The subject of "normally" will be discussed further
below. It should be acknowledged that any extracellular production
measured in the ^{14}C method is a net community production and the
actual amount produced by phytoplankton, before bacterial use, may
be greater.

Field Studies on Organic Molecules

The logical place to look for labile organic molecules that are
rapidly cycled is in common biochemical pathways. The most

successful recent attempts at analytical organic chemistry in marine systems has been aimed at expected biochemical intermediates (Dawson and Liebezeit 1981). The compounds most studied are those with recent breakthroughs in analytical sensitivity, amino acids (Lindroth and Mopper 1979) and carbohydrates (Mopper et al. 1980).

Brockman et al. (1979) made a study of carbohydrates and amino acids in a tank of enclosed seawater for 28 days. They found maximum carbohydrate in the water coincident with stationary phase of the dominant diatom Lauderia borealis. They suggested that there was also some loss of carbohydrates from zooplankton grazing. Amino acid concentrations appeared to perhaps be more directly related to phytoplankton excretion.

Ittekkot et al. (1981) followed carbohydrate concentrations in waters of the North Sea. They found large amounts of carbohydrates released into seawater as a consequence of the phytoplankton spring bloom; a major portion of this release took place toward the end of the bloom. They concluded that the release was mainly from cells under stress induced through changes in available nutrient concentrations and from lysing cells and planktonic debris suspended at the thermocline. Most of the carbohydrate found during the bloom was in combined form, monomers did not appear to be major excretion products. Ittekkot (1982) also followed amino acids in this same study. Amino acids appeared to be more associated with earlier stages of the bloom and to decrease later; this was in contrast to the sequence for carbohydrates.

Mopper and Lindroth (1982) were able to demonstrate pronounced diel variations in dissolved free amino acids in natural waters. In their work, they found variations from 30-40 nM to 200-400 nM amino acid concentrations over the period of several hours. Burney et al. (1980) had previously shown strong diel variations in carbohydrates. In Mopper and Lindroth's (1982) work, maximum concentrations were found in the basic amino acids and ammonium in the evenings and was therefore attributed to grazer activities. Burney et al. (1981) were able to show pronounced diel variations of carbohydrates with increases in the evening; this is also indicative of non-photosynthetic release of phytoplankton organics.

Sellner (1982) studied day/night differences in carbohydrates and amino acids in several oceanic environments. He found apparent production of both groups of organics during the daytime and apparent consumption of this organic matter at night. His most pronounced results were from an area of active upwelling.

From all of these works it is obvious that measureable amounts of labile organics are produced by phytoplankton and rapidly appear in seawater. It is not clear, however, how much of the release is from active photosynthesis and how much results from senescent cells or as a result of herbivore grazing. It is possible that much of this release is from the latter source and that active excretion by phytoplankton is slight. Certainly, larger compounds (e.g., poly-saccharides) primarily would appear to originate largely as debris (Brockman et al. 1979). Studies with short time interval sampling of carbohydrates and amino acids are an exciting new area of much promise. These studies do suggest very rapid recycling of organics since amino acids and carbohydrates seem to disappear rapidly.

ZOOPLANKTON, DEATH, AND OTHER HYDROGRAPHIC CONSIDERATIONS

Zooplankton can serve as indirect avenues for dissolved organic inputs from phytoplankton in two fashions, spillage from phytoplank-ton cells during grazing and zooplankton excretion. When zooplankton graze on phytoplankton, some of the contents of the phytoplankton are spilled into the water. This phenomenon was first demonstrated cleverly by Hellebust (1967) and has since been confirmed (Copping and Lorenzen 1980; Conover and Huntley 1980). Much of the apparent input of organic compounds seen with time course sampling (previous section) could be due to this grazing effect rather than to direct excretion by the phytoplankton.

Zooplankton excretion is also a source of dissolved organics indirectly coming from phytoplankton. This phenomenon has been much studied (e.g. Webb and Johannes 1967; Corner and Newell 1967; Conover and Francis 1973). Recent work (Szyper et al. 1976; Lampert 1978) confirms that it is indeed an important source of organics. This again could be confused with phytoplankton excretion in studies of organic compounds.

To sort out direct phytoplankton production of dissolved organic matter from zooplankton transferred production, carefully controlled experiments are needed. Perhaps diel amino acid and carbohydrate monitoring of continuous cultures with and without a zooplankton grazer is needed. This approach should prove quite illuminating.

Laboratory studies of algal cultures in steady state conditions and field studies with the ^{14}C method do not show that phytoplankton "normally" lose much of their photoassimilated carbon by excretion. However, the question should be raised of how often are "normal" phytoplankton and steady state conditions encountered in nature. Jackson (1980) suggests, based on considerations of zooplankton populations and grazing rates, that phytoplankton in the open ocean do have low growth rates as was traditionally thought. To accomodate claims that phytoplankton must grow at near maximum rates, μ_{max}, to satisfy physiological data (Goldman et al. 1979), Jackson (1980) suggests that these phytoplankton have probably evolved low rates. This might be so, but a possible alternative is that oceanic phyto-

plankton do not always live at steady state, but instead undergo rapid
fluctuations in growth rates. If this were the case, stationary and
declining phases may be common for phytoplankton and appreciable
portions of phytoplankton populations could often become inputs to
the detrital particulate and dissolved organic pools.

Considerable evidence is becoming available for extensive physical variability in the ocean ranging from large features like Gulf
Stream core rings to features of considerably smaller geographic
dimensions. Occasional extensive blooms of pelagic organisms (e.g.,
Tricodesmium, Phaeocystis) are noticed in the open ocean and are
impressive for their seemingly incongruous existence. Extensive
oceanic blooms of phytoplankton have been observed in oligotrophic
waters (e.g., Venrick 1974; Bowman and Lancaster 1965). In the bloom
observed by Venrick (1974), a large population of a diatom in seemingly moribund condition was found on top of the summer thermocline.
The large amount of phytoplankton material found had a small vertical
scale (a few meters) but very large lateral area (hundreds of square
kilometers). Similar scales in oxygen anomalies have been found
(Kester et al. 1973). Observations like these have lead to current
interest in studies such as the Gulf Stream rings. However, a point
that is not often discussed in relation to such observations is that
they suggest extensive death of phytoplankton as an organic input.

I have previously shown that very large organic inputs can be
found in cultures with apparent decomposition of some of the phytoplankton (Sharp 1977; Sharp et al. 1980b). I would like to extend
that argument now by suggesting that death of phytoplankton (senescent and declining growth phases) is a common phenomenon in the ocean.
Menzel (1974) has differentiated between the terrestrial environment
where less than 10% of plant matter is eaten and the remainder
directly enters the decomposition cycle and the marine environment
where greater than 90% of the plant matter is eaten. Perhaps he
is wrong. As Williams (1981b) has pointed out, about 60% of the
primary production is required to go to microheterotrophs for their
nutritional requirements. I suggest that much of this comes from
phytoplankton death but not from active excretion by phytoplankton.

REFERENCES

Antia, N. J., C. D. McAllister, T. R. Parsons, K. Stephens, and J. D.
     H. Strickland. 1963. Further measurements of primary produc-
     tion using a large-volume plastic sphere. Limnol. Oceanogr. 8:
     166-183.
Ben-Amotz, A. 1975. Adaptation of the unicellular alga Dunaliella
     parva to a saline environment. J. Phycol. 11: 50-54.
Bowman, T. E., and L. J. Lancaster. 1965. A bloom of the planktonic
     blue-green alga, Trichodesmium erythraeum, in the Tonga Islands.
     Limnol. Oceanogr. 10: 291-293.

Brockmann, U. H., K. Eberlein, H. D. Junge, E. Maier-Reimer, and D. Siebers. 1979. The development of a natural plankton population in an outdoor tank with nutrient-poor sea water. II. Changes in dissolved carbohydrates and amino acids. Mar. Ecol. Prog. Ser. 1: 283-291.

Burney, C. M., K. M. Johnson, D. M. Lavoie, and J. McN. Sieburth. 1979. Dissolved carbohydrate and microbial ATP in the North Atlantic: concentrations and interactions. Deep-Sea Res. 26: 1267-1290.

Burney, C. M., P. G. Davis, K. M. Johnson, and J. McN. Sieburth. 1982. Dependence of dissolved carbohydrate concentrations upon small scale nanoplankton and bacterioplankton distributions in the western Sargasso Sea. Mar. Biol. 65: 289-296.

Burris, J. E. 1980. Respiration and photorespiration in marine algae, pp. 411-432. In: P. G. Falkowski [ed.], Primary Productivity in the Sea. Plenum Press, New York.

Carpenter, E. J., and J. S. Lively. 1980. Review of estimates of algal growth using ^{14}C tracer techniques, pp. 161-178. In: P. G. Falkowski [ed.], Primary Productivity in the Sea. Plenum Press, New York.

Conover, R. J., and V. Francis. 1973. The use of radioactive isotopes to measure the transfer of materials in aquatic food chains. Mar. Biol. 18: 272-283.

Conover, R. J., and M. E. Huntley. 1980. General rules of grazing in pelagic ecosystems, pp. 461-485. In: P. G. Falkowski [ed.], Primary Productivity in the Sea. Plenum Press, New York.

Copping, A. E., and C. J. Lorenzen. 1980. Carbon budget of a marine phytoplankton-herbivore system with carbon-14 as a tracer. Limnol. Oceanogr. 25: 873-882.

Corner, E. D. S., and B. S. Newell. 1967. On the nutrition metabolism of zooplankton. IV. The forms of nitrogen excreted by Calanus. J. Mar. Biol. Assoc. UK 47: 113-120.

Dawson, R., and G. Liebezeit,. 1981. The analytical methods for the characterization of organics in seawater, pp. 445-496. In: E. K. Duursma and R. Dawson [eds.], Marine Organic Chemistry. Elsevier, New York.

Devol, A. H., and T. T. Packard, 1978. Seasonal changes in respiratory enzyme activity and productivity in Lake Washington microplankton. Limnol. Oceanogr. 23: 104-111.

Duce, R. A., and E. K. Duursma. 1977. Inputs of organic matter into the ocean. Mar. Chem 5: 319-340.

Enhuber, G., and H. Gimmler. 1980. The glycerol permeability of the plasmalemma of the halotolerant green alga Dunaliella parva (volvocales). J. Phycol. 16: 524-532.

Eppley, R. W. 1980. Estimating phytoplankton growth rates in the central oligotrophic ocean, pp. 231-242. In: P. G. Falkowski [ed.], Primary Productivity in the Sea. Plenum Press, New York.

Eppley, R. W., E. H. Renger, E. L. Venrick, and M. M. Mullin. 1973. A study of plankton dynamics and nutrient cycling in the central gyre of the north Pacific Ocean. Limnol. Oceanogr. 18: 534-551.

Eppley, R. W., and J. H. Sharp. 1975. Photosynthetic measurements
    in the central North Pacific: The dark loss of carbon in 24-h
    incubations. Limnol. Oceanogr. 20: 981-987.
Eppley, R. W., and J. D. H. Strickland. 1968. Kinetics of marine
    phytoplankton growth, p. 23-62. In: M. R. Droop and E. J. F.
    Wood [eds.], Advances in Microbiology of the Sea. Academic
    Press, New York.
Fiadeiro, M. 1980. Carbon cycling in the ocean, p. 487-496. In:
    P. G. Falkowski [ed.], Primary Productivity in the Sea. Plenum
    Press, New York.
Fisher, N. S. and J. G. Fabris. 1982. Complexation of Cu, Zn, and
    Cd by metabolites excreted from marine diatoms. Mar. Chem. 11:
    245-255.
Garrels, R. M., and F. T. Mackenzie. 1971. Evolution of Sedimentary
    Rocks. Norton.
Gauthier, M. J., and M. Aubert. 1981. Chemical telemediators in
    the marine environment, pp. 225-257. In: E. K. Duursma and R.
    Dawson [eds.], Marine Organic Chemistry. Elsevier, New York.
Gieskes, W. W. C., G. W. Kraay, and M. A. Baars. 1979. Current [14]C
    methods for measuring primary production: gross underestimates
    in oceanic waters. Neth. J. Sea Res. 13: 50-78.
Goldman, J. C. 1980. Physiological processes, nutrient availabil-
    ity, and the concept of relative growth rate in marine phyto-
    plankton ecology, pp. 179-194. In: P. G. Falkowski [ed.],
    Primary Productivity in the Sea. Plenum Press, New York.
Goldman, J. C., J. J. McCarthy, and D. G. Peavey. 1979. Growth
    rate influence on the chemical composition of phytoplankton in
    oceanic waters. Nature 279: 210-215.
Handa, N. 1977. Land sources of marine organic matter. Mar. Chem.
    5: 341-359.
Harris, G. P. 1978. Photosynthesis, productivity and growth: The
    physiological ecology of phytoplankton. Arch. Hydrobiol. Beih.
    Ergebn. Limnol. 10: 1-171.
Hellebust, J. A. 1967. Excretion of organic compounds by cultured
    and natural populations of marine phytoplankton, pp. 361-366.
    In: G. H. Lauff [ed.], Estuaries. AAAS Sci. Publ. No. 83,
    Washington.
Ittekkot, V. 1982. Variations of dissolved organic matter during a
    plankton bloom: qualitative aspects, based on sugar and amino
    acid analysis. Mar. Chem. 11: 143-158.
Ittekkot, V., U. Brockmann, W. Michaelis, and E. T. Degens. 1981.
    Dissolved free and combined carbohydrates during a phytoplankton
    bloom in the northern North Sea. Mar. Ecol. Prog. Ser. 4:
    299-305.
Jackson, G. A. 1980. Phytoplankton growth and zooplankton grazing
    in oligotrophic oceans. Nature 284: 439-441.
Joint, I. R., and A. J. Pomroy. 1981. Primary production in a
    turbid estuary. Est. Coastal Mar. Sci. 13: 303-316.
Jones, T. W., and R. A. Galloway. 1979. Effect of light quality
    and intensity on glycerol content in Dunaliella tertiolecta

(chlorophyceae) and the relationship to cell growth/osmoregulation. J. Phycol. 15: 101-106.

Kester, D. R., K. T. Crocker, and G. R. Miller. 1973. Small scale oxygen variations in the thermocline. Deep-Sea Res. 20: 409-412.

Lampert, W. 1978. Release of dissolved organic carbon by grazing zooplankton. Limnol. Oceanogr. 23: 831-834.

Lancelot, C. 1979. Gross excretion rates of natural marine phytoplankton and heterotrophic uptake of excreted products in the southern North Sea, as determined by short-term kinetics. Mar. Ecol. Prog. Ser. 1: 179-186.

Laws, E. A., and D. C. L. Wong. 1978. Studies of carbon and nitrogen metabolism by three marine phytoplankton species in nitrate-limited continuous culture. J. Phycol. 14: 406-416.

Li, W. K. and J. C. Goldman. 1981. Problems in estimating growth rates of marine phytoplankton from short-term ^{14}C assays. Microb. Ecol. 7: 113-121.

Lindroth, P., and K. Mopper. 1979. High performance liquid chromatographic determination of subpicomole amounts of amino acids by precolumn fluorescence derivatization with phthaldialdehyde. Anal. Chem. 51: 1667-1674.

McCarthy, J. J., and J. C. Goldman. 1979. Nitrogenous nutrition of marine phytoplankton in nutrient-depleted waters. Science 203: 670-672.

Mague, T. F., E. Friberg, D. J. Hughes, and I Morris. 1980. Extracellular release of carbon by marine phytoplankton; a physiological approach. Limnol. Oceanogr. 25: 262-279.

Mann, K. H., A. R. O. Chapman, and J. A. Gagne. 1980. Productivity of seaweeds: the potential and the reality, pp. 363-380. In: P. G. Falkowski [ed.], Primary Productivity in the Sea. Plenum Press, New York.

Menzel, D. W. 1974. Primary productivity, dissolved and particulate organic matter, and the sites of oxidation of organic matter, pp. 659-678. In: E. D. Goldberg [ed.], The Sea, Vol. 5.

Meybeck, M. 1981. River transport of organic carbon to the ocean, pp. 219-269. In: G. E. Likens, F. T. Mackenzie, J. E. Richey, J. T. Sedell, and K. K. Turekian [eds.], Flux of Organic Carbon by Rivers to the Oceans. U. S. Dept. Energy Publication CONF-8009140, Washington, D.C.

Mopper, K., R. Dawson, G. Liebezeit, and V. Ittekot. 1980. The monosaccharide spectra of natural waters. Mar. Chem. 10: 55-66.

Mopper, K., and P. Lindroth. 1982. Diel and depth variations in dissolved free amino acids and ammonium in the Baltic Sea determined by shipboard HPLC analysis. Limnol. Oceanogr. 27: 336-347.

Naylor, S. 1979. Phytoplankton exudation and culture shock. M.Sc. thesis, Southampton University. Cited in Williams 1981b.

Nienhuis, P. E. 1981. Distribution of organic matter in living marine organisms, pp. 31-69. In: E. K. Duursma and R. Dawson [eds.], Marine Organic Chemistry. Elsevier Scientific Publishing Company.

Perry, M. J., and R. W. Eppley. 1981. Phosphate uptake by phyto-
    plankton in the central north Pacific Ocean. Deep-Sea Res. 28:
    39-49.
Peterson, B. J. 1978. Radiocarbon uptake: Its relation to net
    particulate carbon production. Limnol. Oceanogr. 23: 179-184.
Peterson, B. J. 1980. Aquatic primary productivity and the $^{14}C-CO_2$
    method: A history of the productivity problem. Ann. Rev. Ecol.
    Syst. 11: 359-385.
Pomeroy, L. R. 1974. The ocean's food web, a changing paradigm.
    Bioscience 24: 499-504.
Redfield, A. C., B. H. Ketchum and F. A. Richards. 1963. The influ-
    ence of organisms on the composition of sea water, pp. 26-77.
    In: M. N. Hill [ed.], The Sea, Vol. II. Interscience, New York.
Richey, J. E., J. T. Brock, R. J. Naiman, R. C. Wissmar, and R. E.
    Stallard. 1980. Organic carbon: oxidation and transport in the
    Amazon River. Science 207: 1348-1351.
Robinson, J. D., K. H. Mann, and J. A. Novitsky. 1982. Conversion
    of the particulate fraction of seaweed detritus to bacterial
    biomass. Limnol. Oceanogr. 27: 1072-1079.
Sellner, K. G. 1981. Primary productivity and the flux of DOM in
    several marine environments. Mar. Biol. 65: 101-112.
Sharp, J. H. 1973. Size classes of organic carbon in sea water.
    Limnol. Oceanogr. 18: 441-447.
Sharp, J. H. 1977. Excretion of organic matter by marine phyto-
    plankton: Do healthy cells do it? Limnol. Oceanogr. 22:
    381-339.
Sharp, J. H. 1978. Reply to comment by Aaronson. Limnol Oceanogr.
    23: 839-840.
Sharp, J. H., M. J Perry, E. N. Renger, and R. W. Eppley. 1980a.
    Phytoplankton rate processes in the oligotrophic waters of the
    central north Pacific Ocean. J. Plankton Res. 2: 335-353.
Sharp, J. H., P. A. Underhill, and A. C. Frake. 1980b. Carbon
    budgets in batch and continuous cultures: How can we understand
    natural physiology of marine phytoplankton? J. Plankton Res.
    2: 213-222.
Sharp, J. H., P. A. Underhill, and D. J. Hughes. 1979. Interaction
    (allelopathy) between marine diatoms: Thalassiosira pseudonana
    and Phaeodactylum tricoruntum. J. Phycol 15: 353-362.
Sheldon, R. W., A. Prakash, and W. H. Sutcliffe. 1973. The produc-
    tion of particles in the surface waters of the ocean with
    particular reference to the Sargasso Sea. Limnol. Oceanogr.
    18: 719-733.
Sheldon, R. W., and W. H. Sutcliffe. 1978. Generation times of 3 h
    for Sargasso Sea microplankton determined by ATP analysis.
    Limnol. Oceanogr. 23: 1051-1055.
Shifrin, N. S. 1980. The measurement of dissolved organic carbon
    released by phytoplankton. Estuaries 3: 230-233.
Sieburth, J. McN. 1977. International Helgoland Symposium:
    Convenor's report on the informal session of biomass and produc-

tivity of microorganisms in planktonic ecosystems. Helogol.
Wiss. Meeresunters 30: 697-704.

Simek, E. 1981. Phytoplankton production in a marsh-dominated
estuary. Ph.D. dissertation, University of Delaware.

Smith, S. V. 1981. Marine macrophytes as a global carbon sink.
Science 211: 838-840.

Smith, W. O. 1977. The respiration of photosynthetic carbon in
eutrophic areas of the ocean. J. Mar. Res. 35: 557-565.

Smith, W. O., R. T. Barber, and S. A. Huntsman. 1977. Primary
production off the coast of northwest Africa: excretion of
dissolved organic matter and its heterotrophic uptake.
Deep-Sea Res. 24: 35-47.

Storch, T. A., and G. W. Saunders. 1978. Phytoplankton extracellular release and its relation to the seasonal cycle of
dissolved organic carbon in a eutrophic lake. Limnol.
Oceanogr. 23: 112-119.

Szyper, J. P., J. Hirota, J. Caperon, and D. A. Ziemann. 1976.
Nutrient regeneration by the larger net zooplankton in the
southern basin of Kaneohe Bay. Pac. Sci. 30: 363-372.

Underhill, P. A. 1981. Steady state growth rate effects on the
photosynthetic carbon budget and chemical composition of a
marine diatom. Ph.D. dissertation, Univ. of Delaware.

Venrick, E. L. 1974. The distribution and significance of Richelia
intracellularis Schmidt in the North Pacific Central Gyre.
Limnol. Oceanogr. 19: 437-445.

Webb, K. L., and R. E. Johannes. 1967. Studies of the release of
dissolved free amino acids by marine zooplankton. Limnol.
Oceanogr. 12: 376-382.

Wiebe, W. J., and D. F. Smith. 1977. Direct measurements of
dissolved organic carbon release by phytoplankton and
incorporation by microheterotrophs. Mar. Biol. 42: 213-223.

Williams, P. J. leB. 1975. Biological and chemical aspects of
dissolved organic material in sea water, pp. 301-362. In: J.
P. Riley and G. Skirrow [eds.], Chemical Oceanography, 2nd
Edition, Vol. 2. Academic Press.

Williams, P. J. leB. 1981a. Microbial contribution to overall
marine plankton metabolism: direct measurements of respiration.
Oceanol. Acta. In press.

Williams, P. J. leB. 1981b. Incorporation of microheterotrophic
processes into the classical paradigm of the planktonic food
web. Kiel. Meeresforsch. 5: 1-28.

Williams, , P. J. leB., R. C. T. Raine, and J. R. Bryan. 1979.
Agreement between the ^{14}C and oxygen methods of measuring
phytoplankton production; reassessment of the photosynthetic
quotient. Oceanol. Acta 2: 411-416.

Williams, P. J. leB., and C. S. Yentsch. 1976. An examination of
photosynthetic production, excretion of photosynthetic proudcts,
and heterotrophic utilization of dissolved organic compounds
with reference to results from a coastal, subtropical sea.
Mar. Biol. 35: 31-40.

Woodwell, G. M., R. H. Whittaker, W. A. Reiners, G. E. Likens, C. C.
    Delwich, and D. B. Botkin. 1978. The biota and the world
    carbon budget. Science 199: 141-146.
Yentsch, C. S. 1980. Phytoplankton growth in the sea. A coales-
    cence of disciplines, pp. 17-32. In: P. G. Falkowski [ed.],
    Primary Productivity in the Sea. Plenum Press, New York.

DYNAMICS OF POOLS OF DISSOLVED ORGANIC CARBON

Richard T. Wright

Department of Biology
Gordon College
Wenham, Massachusetts

INTRODUCTION

A Trophic Approach

Bacteria living and growing in the plankton of marine waters are clearly doing so at the expense of organic matter in the dissolved state. In recent years our understanding of this process has improved greatly. This paper will explore the interaction of bacteria and dissolved organic matter from an ecological viewpoint. In energy flow terms, the DOM (dissolved organic matter) of natural waters can be construed as a trophic level, the base of a food chain involving only heterotrophs. The next (higher) trophic level is occupied by the heterotrophic microorganisms dependent on DOM, and there is good evidence now that in planktonic ecosystems the bacteria uniquely occupy this trophic level. They in turn contribute energy to higher trophic levels when they are fed on by planktonic and benthic animals. From this point the energy flow joins that of the food chain based on particulate matter such as algae and detritus.

This sort of ecological approach is possible only because of the recent availability of methods for measuring the total bacteria population and for estimating their growth and metabolic activity. In short, it is now possible to deal with the planktonic bacteria as a functional assemblage. This is a radical departure from most of classical microbiology, based as it is on isolating, culturing and measuring biochemical attributes of bacteria in pure culture. These two approaches may converge some time in the future, but for the time being there is much that can be learned of the dynamics of both the bacterial trophic level and the DOM on which they are dependent.

121

Interaction Between the Bacteria and Dissolved Organic Carbon Pool

The total DOC (dissolved organic carbon) pool is large in
comparison with the standing stock of bacteria; open ocean euphotic
zone values for DOC range from 0.6 to 1.0 mg·liter^{-1} (Williams 1975).
Bacteria numbers in these waters range from 0.2 to 1.5 x 10^6 ml^{-1},
giving an equivalent carbon content of ca. 2 to 20 µg C·liter^{-1} if
we assume an average volume of 0.1 µm^3 for a planktonic bacterium
(Ferguson and Rublee 1976) and a conversion factor of 121 x 10^{-9} µg
C·µm^{-3} (Furhman and Azam 1980). It is customary to consider the DOC
pool as two major fractions: labile, meaning capable of being
metabolized in a fairly short time frame, and refractory, meaning
quite resistant to breakdown and usually of a complex nature and
large molecular weight. By definition the labile pool is the most
dynamic one; an understanding of its trophic importance to the
bacteria must be based not only on concentration data but also on
knowledge of rates of production and use.

The definition of the labile pool indicates that the bacteria
will have a decisive impact on that pool. Evidence will be presented
that reveals a close coupling of the planktonic bacteria and low M.
W. organic solutes like the amino acids and monosaccharides. Transport systems found in marine bacteria function efficiently in the
same concentration range of these organic solutes (Wright 1979).
Little is known, however, of the relationship between the bacteria
and larger M. W. organic solutes.

At least two kinds of trophic interactions are conceivable.
The first is readily understood: dissolved organic compounds are
produced or released and bacteria take them up and respond by an
increase in biomass or numbers or both (bacterial production). A
second interaction is less well understood but possibly quite
important: dissolved organic compounds are produced slowly in small
amounts and are used by the bacteria (which, it now appears, are
always present at significantly high numbers) which use the DOC for
maintenance. In this case, little or no growth is detectable by
methods presently available. A third interaction between the
bacteria and DOC, although not strictly speaking trophic, may be
important. Some compounds may not be useful for maintenance or
growth, either because they are in very low concentrations or can
not be coupled to useful biochemical pathways, yet they will still
be metabolized. This process is called co-metabolism and may be
significant in the biodegradation of pollutants (Alexander 1981).
In summary, the DOC pool in marine waters, by its composition and
concentration, will reflect the impact of bacterial activity.

Origins of DOC Available to Marine Bacteria

It is also to be expected that the DOC pool in marine waters

will reflect its origins. Williams (1975) has reviewed this and
other aspects of the DOM of marine waters, and has classified the
origins of DOM according to external or internal sources. The major
external sources are the atmosphere and rivers, and Williams estimates that these two sources contribute approximately equal amounts
of DOM to the oceans, each equivalent however to only about 1% of
marine net primary production. The contribution from rivers no doubt
produces some important effects in coastal areas. It is important
to note that both atmosphere and rivers may also contribute synthetic
organic chemicals, such as pesticides and PCB's, which can impact the
marine food chain despite their low concentrations (Williams 1975).

The internal sources are better understood and no doubt much
more important from the viewpoint of trophic dynamics. These
sources include: (1) excretion or exudation as a result of algal
photosynthesis; Williams (1975) suggests that 10% of net production
of offshore phytoplankton enters the DOM pool by this route. (2)
exudation and abrasion of benthic seaweeds (Sieburth and Jensen
1970). As much as 25% of annual production may be released as
mucilage, which has a large impact on DOC in surrounding waters
(Linley et al. 1981). (3) exudation by salt marsh macrophytes,
especially Spartina (Turner 1978). This may be the major source
of DOM for estuarine and coastal waters bacteria in some areas.
(4) release of DOM as a result of zooplankton feeding and from
zooplankton fecal material (Williams 1975). (5) death, lysis of
cells, and decomposition. These processes produce both labile and
resistant DOM and are quantitatively the major sources of DOM.

If it can be established that the rate of supply of DOM as
substrate is a major factor controlling bacterial numbers and
activity, then it becomes important in turn to understand those
factors that control the supply of DOM. For example, Fuhrman and
Azam (1980) found a strong correlation of chlorophyll a with
bacterioplankton numbers in coastal waters; this implies direct
dependence of bacteria on DOM coming from phytoplankton. In this
environment, the algal community is controlled by upwelling and
mixing so it can be said that these latter factors ultimately
control the heterotrophic bacteria. As another example, temperature
strongly affects measured bacteria heterotrophic activity (Wright
and Hobbie 1966; Gillespie et al. 1976), and in a study on the
annual cycle of numbers and activity of bacteria in an estuary in N.
Massachusetts there was a strong and highly significant relationship
between temperature and both numbers and activity (Wright and Coffin
1983). However, the evidence indicated that the strong temperature
dependence of the estuarine bacteria was a consequence of substrate
supply from Spartina photosynthate, which is strongly temperature
dependent. Temperature dependence of coastal water bacteria in the
vicinity was much weaker.

Various Approaches to Studying DOC Dynamics

The dynamics of DOC pools and the interaction of DOC with
bacteria represent an extremely complex system to study, one which
must be broken down into manageable units and approached with a
variety of methods. Some of the different approaches are: the
identification and concentration of specific dissolved organic
compounds; the measurement of total DOC and subfractions of this
such as dissolved carbohydrates or free amino acids; the determination of rates of turnover of specific DOC compounds due to bacterial
use; and the use of experimental studies to probe the interaction of
the bacteria with DOC.

DETERMINATION OF CHEMICAL CONCENTRATIONS OF DOC COMPOUNDS

Methods and Their Limitations

The individual dissolved organic compounds are present at great
dilution in seawater compared with the inorganic salts. Most
chemical analyses for specific compounds require concentration or
extraction and desalting, procedures which have the potential to
change the relative composition of organic compounds. There is also
the great potential for contamination from a variety of sources. As
a result of these problems, many if not most of the reported
concentrations of individual organic solutes should be viewed with
some skepticism. Low values are usually more accurate than high
ones in any comparison of results. Theoretically, the best
techniques are those which can measure individual organic compounds
or groups of compounds without de-salting or pre-concentration, such
as the shipboard, HPLC method of Lindroth and Mopper (Lindroth and
Mopper 1979; Mopper and Lindroth 1982). Employing fluorometry as a
detection method, this procedure requires no sample preparation and
performs an analysis in 30 minutes with detection limits well below
nanomolar concentrations.

There are actually three basic approaches that have been
employed in the study of the natural DOC compounds of the sea. The
first, mentioned above, is the identification and concentration of
specific organic solutes. The second approach involves analysis of
a class of compounds, such as the dissolved free amino acids (DFAA),
or the total dissolved carbohydrates (TDCHO). The third approach
measures the gross organic fraction associated with the organic
elements C, P and N (i.e., DOC, DON, DOP). Since most reported
methods have employed significant sample treatment, the accuracy of
some of these approaches has been questioned (Williams 1975).

However, the chemical analyses are probably now within the
proper order of magnitude of real values, and are useful in showing
vertical, horizontal, geographical and seasonal differences in

concentration that may then lend themselves to meaningful interpretation.

Pool Sizes and Fractions: General Indications

The DOC pool in sea water is a large reservoir of potential energy. Most measurements away from coastal influence lie in the range of 0.6 to 1.0 mg C·liter^{-1} for the upper 100 m, and about two-thirds of that level for deep waters (Williams 1975). DOC levels are higher in coastal waters and estuaries (Wheeler 1976), as much as an order of magnitude above those found in the open ocean. Gardner and Stephens (1978) found nearshore levels of DOC to be three times those found 100 km offshore from Georgia, and were able to plot a linear decrease of DOC with increasing salinity. They attributed the relationship to the terrestrial derivation of most of the inshore DOC.

The composition of the DOC of seawater is certainly heterogeneous. Williams (1975) shows a detailed breakdown of organic compounds by general class and, for some, as individual substances. Much of the remainder of this paper deals with specific compounds or classes of compounds that together make up the DOC. An alternative to chemical analysis that also indicates heterogeneity in the DOC is separation by molecular weight. Ogura (1977) demonstrated a range in M. W. from < 500 to > 100,000 daltons in Japanese waters, with a range of 8 to 45% of the total DOC represented by the > 100,000 daltons fraction. Billen et al. (1980) summarized the available data on the low molecular weight pool and concluded that substances in the < 500 dalton range (the size ordinarily available to bacterial transport) represent 10% or less of the total DOC. According to Degens (1970), the bulk of the DOC falls in the range 3000 to 5000 daltons. Below 300 m in stratified ocean waters, the DOC is considered highly refractory, the unused fraction of DOC input from years past (Menzel and Ryther 1970). Radiocarbon dating of DOC from deep Pacific Ocean water gave a mean age of 3400 years (Williams 1975).

Some studies have shown seasonal variations in DOC in coastal waters. Banoub and Williams (1972) found a summer increase of 300 μg C·liter^{-1}, and attributed such increases to the decay of phytoplankton blooms. Diel variations in DOC have been reported by Johnson et al. (1981) for estuarine and open ocean waters, with maximum levels in evening hours indicating a gradual accumulation during the photoperiod, and minimum levels found in the early morning. The magnitude of these diel changes ranged from 1.5 mg·liter^{-1} in estuarine waters to ca. 50 μg·liter^{-1} in Caribbean Sea waters. The observation of seasonal and diel changes in DOC levels indicates the dynamic nature of the DOC pool in euphotic marine waters. The remarkable uniformity in the amounts of DOC in the sea (as contrasted with inorganic nutrients) suggests that production and use are in balance, on a larger time and spatial scale. The heterogeneity of solutes and variations in coupling of input and use suggest that attempting to understand the importance of DOC by simple concentration measurements alone will lead to oversimplification and potential error.

Two classes of compounds that make up the DOM of natural waters have received recent attention by research workers: the amino acids and the carbohydrates. Summarizing the situation for amino acids, Williams (1975) concluded that the concentrations of individual amino acids were usually found at concentrations of 1-10 $\mu g \cdot liter^{-1}$, and the total dissolved free amino acid pool ranged from 5 to 80 $\mu g \cdot liter^{-1}$. It was also noted that concentrations were not much higher in estuaries and coastal waters, nor did they show any seasonal variation. Dawson and Gocke (1978) presented evidence that the DFAA levels found with chemical measurements were substantially higher than the substrate levels apparently available to the bacteria, and if used in flux calculations the data resulted in impossibly high rates of microbial growth and activity in open Baltic Sea waters. In summary, the DFAA fraction is a very small part of the total DOC, and all the evidence to date (much to be cited later) indicates that this fraction is important for bacterial energy and growth and hence there is tight coupling between production and use of this class of compounds.

The dissolved carbohydrates (DCHO) represent a much more heterogeneous class of compounds. Typical values for total CHO show a range of 200-600 $\mu g \cdot liter^{-1}$; individual monosaccharides rarely show values above 20 $\mu g \cdot liter^{-1}$, so most of the carbohydrates must be polymers (Williams 1975). In the North Sea, Ittekot et al. (1981) measured carbohydrate levels during the course of a phytoplankton bloom. They found substantial vertical and temporal changes especially in the dissolved combined carbohydrate fraction, with vertical maxima close to the seasonal thermocline. The authors suggested that large amounts of carbohydrates are rapidly released from stressed cells, from lysing cells, and from planktonic debris at the end of a phytoplankton bloom. Because they found little change in the dissolved free carbohydrates during the bloom, Ittekot et al. (1981) concluded that these substances are not released in quantity by excretion or exudation, although it is possible that heterotrophs are using the dissolved CHO as rapidly as it is produced. These and other data thus indicate that there is tight coupling between the bacterial use and the production of low M. W. CHO but that it might be possible to observe temporal and spatial variations in more complex carbohydrates as a result of sudden pulses of release or of their greater resistance to bacterial uptake and metabolism.

Limitations of This Approach

Accuracy and precision of chemical methods are clearly a major limitation of the chemical approach. However, the most serious limitation is also the most obvious: chemical measurements give an indication only of the instantaneous concentration of a solute or class of solutes. If there are differences in concentration between locations or times, then at best a correlation may be seen with some producer or decomposer organisms. Usually, this is not possible and the author is forced to resort to speculation in interpreting the data. If no differences in concentrations are seen, the reason could be found in completely opposite explanations: the substance is used as rapidly as it is produced, or the substance does not participate significantly in production or uptake processes. For example, Fuhrman et al. (1980) examined a large number of factors suspected of affecting bacterial abundance and activity in the euphotic zone of the South California Bight. Some correlations emerged (e.g., bacterial abundance and glucose turnover rates, or bacterial growth and phytoplankton standing stock), but total DOC and primary amines showed no correlations with any other measurements (bacteria numbers, bacterial growth, glucose turnover, Chl a, primary production, phytoplankton excretion, etc.). The authors concluded that the key to the question of importance of DOC compounds lies in knowing the flux, or turnover, of the compounds in question.

TURNOVER OF DOC

Methods Available

As with any scientific study, the conclusions reached concerning turnover of dissolved organic compounds are strongly influenced by the methods employed. Two different approaches to turnover are used. One approach could be called "intrusive sampling"; samples are removed from the environment and measurements are made over time of rates of use of solutes or growth of microbes in response to substrate present. Usually, radioisotopes are used as tracers, measurements begin immediately after sampling, and incubations are short. The other approach may be called "strategic sampling". Samples are taken for chemical or biomass measurements over close time intervals; any differences are converted to rates on the basis of known biological processes. This approach requires a definite temporal separation of processes producing and using organic solutes. Usually this separation is the diel cycle but it can also be large scale processes such as the growth and decay of a major phytoplankton bloom (Ittekkot et al. 1981).

Intrusive Sampling

The use of radiolabelled organic solutes to measure the activity
of heterotrophic bacteria is described in Wright and Burnison (1979).
The objectives of research employing labelled organic solutes to
measure bacterial activity are usually one or more of the following:
(1) some relative measure of bacterial activity that can be quantified and related to other variables; (2) the actual rate at which
the bacteria are using a given compound or group of compounds; and
(3) the turnover of a given compound in the system in question as a
result of production and consumption rates. The latter objective can
be assessed in two forms. Turnover time is the theoretical time for
complete removal of the substrate, and is the ratio of the concentration to the ambient rate of use. The inverse of turnover time, sometimes called "relative uptake rate" (Azam and Holm-Hansen 1973) or,
more commonly, turnover rate (Odum 1971, p. 17; Wright 1978), is an
expression of the fraction of substrate used per unit time. If the
natural substrate concentration is measured independently, the actual
rate of use of the substrate can be determined from a measurement of
turnover.

The two most commonly employed methods for accomplishing the
above objectives are the tracer approach (Williams and Askew 1968;
Azam and Holm-Hansen 1973) and the kinetic approach (Parsons and
Strickland 1962; Wright and Hobbie 1966). The tracer approach
involves adding labelled substrate at concentrations well below
those known to be found in the natural environment and using the
measured fraction of available labelled substrate taken up to
calculate turnover rate or turnover time (Wright and Burnison 1979).
The kinetic approach assumes that the response of a natural microbial
population to a range of concentrations of added substrate can be
used to describe natural heterotrophic activity. With an equation
derived from Michaelis-Menten enzyme kinetics, it is possible to
generate the parameter V_{max}, a theoretical maximum uptake velocity
for the substrate and microbial population in question, the turnover
time for the substrate, and a combined parameter including natural
substrate and a constant analogous to the Michaelis constant (Wright
and Hobbie 1966). Gocke (1977) compared the turnover of a variety
of solutes with the two methods and obtained similar turnover times
from estuarine and coastal waters. However, oligotrophic waters
consistantly yielded shorter turnover times with the tracer
technique. Gocke attributed the discrepancy to the limitations of
the kinetic approach where microbial activity is low and the kinetics
often diverge from the expectations of the Michaelis-Menten scheme,
a problem explored in theory by Williams (1973). Both of the above
approaches have their strengths and weaknesses, as discussed in
Wright (1973) and Wright and Burnison (1979). However, the two
methods have been successfully employed to generate valuable new
information on the natural activity of the heterotrophic bacteria,
as shown later in this paper.

In recent years some interesting variations on the kinetic themes have appeared. Dietz et al. (1977) presented an "alternative kinetic approach" which was reportedly capable of yielding substrate concentrations as well as natural turnover times and uptake. However, Wright and Burnison (1979) showed that this approach has a serious logical inconsistency and therefore is invalid. Bell (1980) presented a kinetic model modified so as to deal with the uptake of a pool of labelled solutes (algal exudate), and presented two cases where a linear analysis (critical to the validity of the kinetic approach) is theoretically possible. Wiebe and Smith (1977) utilized a "compartmental kinetic tracer analysis" to analyze data on heterotrophic use of a highly labelled algal exudate. None of these variations have been widely adopted, however, and their usefulness has basically been limited to specific research needs or to a confirmation of the validity of the kinetic approach.

It is possible, of course, to bypass kinetic considerations by accounting for the mass transfer of labelled compounds from one planktonic compartment to another. Derenbach and Williams (1974) and Hagström et al. (1979) used this approach to measure the use of naturally occurring labelled algal exudate by heterotrophs.

Strategic Sampling

The basic paradigm for applying the "strategic sampling" approach to turnover of dissolved organic compounds is to calculate rates of solute release or uptake based on the changes in concentration found in the water column between sampling intervals (Burney et al. 1981; Mopper and Lindroth 1982; Ittekkot et al. 1981). In considering the validity of this approach, some basic qualifying assumptions would seem to be: (1) the chemical technique yields a true and precise measurement of the organic solute(s) in question; (2) chemical changes over time are being measured in the same water masses; and (3) the chemical changes observed are the result of the biological processes to which they are ascribed. Because both production and removal of a solute are probably occurring simultaneously (or at least to some extent during the sampling period), the results are expressed as net rates and are considered to err on the low side.

A variety of factors may be measured in conjunction with studies of organic solute turnover (see below). However, no set of factors is more important than measurements of bacterial biomass and growth. Our knowledge of these factors is firm enough to be useful in setting limits on the presumed activities of bacteria in using dissolved organic solutes.

Studies of Natural Turnover of Dissolved Organic Compounds

Williams (1975) found several patterns in his survey of studies

of natural turnover of dissolved organic solutes as determined with
isotopically labelled solutes. First, turnover times increase in a
seaward direction, with estuaries showing turnover times of 1 day,
coastal waters 1 to 10 days, and oceanic waters 10 to 100 days.
Second, the depth distribution shows a similar pattern: if the
surface has a turnover rate of 10–50% day^{-1}, depths of 400 m show
< 1% day^{-1} and at 2000 m turnover of organic solutes is undetectable.
Third, seasonal changes always occur where there are seasonal changes
in phytoplankton activity. Turnover rates for individual compounds
ranged from undetectable levels to 5 μg·liter^{-1}·day^{-1}, usually in
the range 0.1 to 1 μg·liter^{-1}·day^{-1}. Williams pointed out that the
general level of dissolved organic carbon input from phytoplankton
was likely to be 15 μg C·liter^{-1}·day^{-1}, therefore the two methods are
in general agreement. It also follows that studies of the concentration of dissolved organic carbon will not reveal the dynamics of
production and use because of the large amount of refractory DOC and
the (relatively) small changes occurring on a daily basis.

With a few notable exceptions, the general patterns cited in
Williams (1975) have been supported by recent research. One of the
most interesting new perspectives has been provided by some work in
Antarctic waters by Gillespie et al. (1976) and Hodson et al.
(1981). These studies have revealed an active bacterial population
with numbers as high as 1 x 10^6 ml^{-1} in very cold waters (0°C).
There is a psychrophilic bacterial population in Antarctic regions
where primary production is occurring (e.g., east side of McMurdo
Sound) with leucine and glucose turnover times in the range of 1 to
6 days (Hodson et al. 1981). These kinds of turnover are not found
in temperate oceans under winter conditions, or at depths below 500
m in the oceans, and the obvious conclusion is that temperature
alone does not limit heterotrophic activity and growth in bacterial
populations. Where there is production of dissolved organic
substrate, there will be active bacteria.

Another intriguing perspective has been given by Larsson and
Hagström (1979), in attempting to measure the direct transfer of
organic matter from algal photosynthetic exudate to the bacteria.
After a $^{14}CO_2$ incubation in the light, the plankton were separated
by Nuclepore filtration. Radioisotope activity in the 0.2 to 3 μm
fraction was assumed to represent net bacterial use of exudate. They
found, during a spring bloom, rates of net incorporation as high as
2.3 μg C·liter^{-1}·hr^{-1}. In one experiment, measured carbon flow to
the bacterial fraction was ca. 1 μg C·liter^{-1}·hr^{-1}. During a 6 hr
experiment, net growth of the bacteria population of 2 x 10^9 liter^{-1}
occurred. Based on a general figure of 12 μg C per 10^9 bacteria,
this suggests that the bacteria obtained about one fourth of their
nutrition directly from algal exudate. Applying this approach to a
year's cycle of production measurements, Larsson and Hagström found
that an average of 27% of the labelled organic carbon appeared in
the bacterial fraction, and that an estimated bacterial production

of 29 g $C \cdot m^{-2} \cdot yr^{-1}$ was due to the use of algal exudate.

The continued refinement of techniques for measuring the
dissolved free amino acids (DFAA) in seawater has stimulated
considerable research into amino acid flux. Dawson and Gocke
(1978) measured heterotrophic uptake of an amino acid mix (total
0.2 μg $C \cdot liter^{-1}$) with a tracer approach, and combined the turnover
data with measured amino acid concentrations to calculate daily amino
acid flux in Baltic Sea waters. They observed that concentrations
showed little sample to sample variations, hence variations in flux
were due to variations in heterotrophic turnover times. Their values
were in agreement with earlier work; gross uptake rates ranged from
4 to 9 μg $C \cdot liter^{-1} \cdot day^{-1}$ averaged over the water column. They
concluded that rate of supply or production rather than the actual
substrate concentrations of dissolved organic solutes controlled rate
of heterotrophic removal. Billen et al. (1980) came to the same
conclusion from work on amino acids and other substrates over the
course of a year and over a range from estuarine to open ocean
systems. They found no significant seasonal or geographic variations
in concentrations of amino acids, glucose, lactate, glycolate and
acetate, but did find major differences in turnover rates. The rates
for estuaries were an order of magnitude higher than rates offshore
while coastal waters were in between.

Using a new shipboard technique for amino acid analysis that
avoids desalting and concentrating seawater samples, Mopper and
Lindroth (1982) examined the amino acid spectrum and concentrations
at short time intervals on station in the Baltic Sea. They observed
distinct diel changes; evening concentrations ranged from 200 to 400
n$\underline{M}$ (20 to 40 μg DFAA), while morning concentrations ranged from 30
to 50 n$\underline{M}$. Based on these concentration changes, Mopper and Lindroth
calculated apparent net heterotrophic uptake as high as 3 to 6 μg
DFAA·$liter^{-1} \cdot hr^{-1}$ (1 to 2 μg $C \cdot liter^{-1} \cdot hr^{-1}$), an order of magnitude
higher than rates calculated from studies using labelled amino acids.

Sieburth and co-workers have developed a spectrophotometric
method for carbohydrate analysis and have also found a discrepancy
with isotope studies. At some mid-oceanic stations (Burney et al.
1981) they found a diel pattern of carbohydrate accumulation during
the daytime and disappearance at night. Calculations of net apparent
heterotrophic usage gave rates up to 5 μg $C \cdot liter^{-1} \cdot hr^{-1}$, two orders
of magnitude higher than previous values. Estuarine diel differences
were even greater and yielded rates of apparent net uptake in the
range of 30 μg $C \cdot liter^{-1} \cdot hr^{-1}$. These findings are similar to the
diel cycle of dissolved carbohydrates found by Walsh (1965) in a
coastal salt pond. Rates of use calculated from concentration
differences range around 50 μg $C \cdot liter^{-1} \cdot hr^{-1}$. The weight of
evidence indicates that both amino acids and dissolved carbohydrates
reveal diel differences when they are looked for. If the rates of
apparent net heterotrophic use are taken at face value, the two

classes of organic solutes indicate ambient rates of heterotrophic
uptake in the range of 1 to 10 µg $C \cdot liter^{-1} \cdot hr^{-1}$ for offshore oceanic
surface waters and much higher rates in enriched systems. Isotope-
based studies indicate that rates are this rapid only for estuarine
or highly enriched marine systems (Crawford et al. 1974).

A very carefully executed set of isotope-based measurements by
Gocke et al. (1981) indicated one possible souce of confusion in
these considerations. The authors pointed out that past workers have
often found chemically measured natural substrate concentration (S_n)
to be higher than the kinetic approach parameter ($K_t + S_n$) allows
(e.g., Burnison and Morita 1974; Dawson and Gocke 1978). When they
measured glucose concentrations chemically and compared them with
kinetic parameters, they found the same phenomenon. However, when
they added unlabelled glucose to the microbial system they got excellent "recovery" as ($K_t + S_n$) with no change in uptake rate, and
therefore felt that the problem was not simply a matter of procedure.
They concluded that "free" sugars (and this would apply to amino
acids) may not all be free -- some may be adsorbed to colloidal
systems or complexed with metallic ions. Alternatively, chemical
modifications during the analysis may result in the hydrolysis of
some polysaccharides or shift equilibria allowing more to be released
(Gocke et al. 1981). They concluded that either only a fraction of
the "total glucose" is available to microorganisms for the above
reasons, or that for some reason the ^{14}C-glucose taken up is under-
represented in the glucose pools that are really available to the
microbes.

Another source of confusion is illustrated by some interesting
work of DeLattre et al. (1979). These workers set up 50 liter tanks
filled with fresh seawater and measured bacteria increases in diffusion chambers suspended in the tanks. Bacterial production was compared with glucose uptake via the kinetic approach. Rates obtained
with the diffusion chambers were convertible to those with glucose
net uptake by multiplying by a factor of 16. Commenting on this, the
authors noted that bacterioplankton production therefore may be much
greater than previously indicated by ^{14}C assays of heterotrophic
production. However, in view of the fact that only one substrate was
measured, it could just as well be concluded that the results are
entirely consistent and simply indicate that in this system glucose
contributed about 6% of the substrate nourishing the bacteria.

At this point it might be useful to summarize the basic
parameters of this controversy. The chemical, or "strategic
sampling" approach to the measurement of turnover rates has
indicated apparent rates of heterotrophic use of organic solutes
in surface ocean waters in the range 1 to 10 µg $C \cdot liter^{-1} \cdot hr^{-1}$
(24 to 240 µg $C \cdot liter^{-1} \cdot day^{-1}$). Radioisotope based studies,
employing single substrates or amino acid mixtures, indicate much
lower rates. Taking into consideration a large variety of solutes,

the overall rates may be as high as 10 µg C·liter^{-1}·day^{-1} (Williams 1975). A set of explanations offered for this discrepancy by Burney et al. (1981) includes: (1) serious doubts about the applicability of the kinetic approach; (2) a suspicion that measurements made on microorganisms enclosed in containers introduces serious artifacts that yield abnormally low responses. Burney et al. (1981) expressed confidence that direct chemical analysis over diel cycles provided the best ("more ecologically meaningful") approach to substrate flux.

An alternative set of explanations for the discrepancy might include the following: (1) sampling is a serious potential source of discrepancy in the chemical approach. In the studies cited there seems to be little control over possible changes in concentration between sampling times due to movements of different water masses; (2) the method of measurement of carbohydrate could allow colloidal carbohydrate to be included in the dissolved fraction. The method certainly lacks the chemical precision which is a consequence of the use of purified isotopically labelled organic compounds; (3) organic flux data obtained with radioisotopes measure bacterial use of one or several substrates. It has never been claimed by proponents of this approach that the approach yields anything close to total use of dis-solved organic carbon. (4) The analysis of the data by proponents of the chemical approach is often uncritical and sometimes appears to start from the assumption of very high rates of production. Thus, if no change in concentration is found, this is interpreted as a balance of production and consumption. Every change in concentration is used to calculate a rate of production or consumption, whereas some changes could be due to errors in measurement or to movement of water masses. In some papers, for example Sieburth et al. (1977), a few data points are extrapolated to question all previous information on microbial activity. Recognizing that there are these discrepancies, a cautious interpretation should be used until the microbial activity and the chemistry are better understood.

As indicated earlier, the most important parameters to measure in connection with flux studies are those dealing with bacterial population density and growth. The following assumptions set some limits on bacterial use of organic substrates: (1) the mean volume of natural marine bacterial is 0.1 µm^3; (2) the factor for conversion to carbon is 121 fg C·(µm^3)$^{-1}$; (3) there are 0.2 to 1.5 x 10^9 bacteria liter^{-1} in the open ocean surface water; (4) the carbon content, therefore, is 2 to 20 µg C·liter^{-1}; (5) the growth yield is 35% of the substrate taken up (Billen et al. 1980). Given these assump-tions, the chemical approach indicates bacteria turnover times in the range of 6 hours (assuming that bacteria density is correlated with turnover rates), and the radioisotope approach indicates bacteria turnover times of several days. Bacterial production has proven difficult to measure; Fuhrman and Azam (1980) compared bacterial production rates calculated from ^{3}H-thymidine incorporation and from growth in samples passed through a 3 µm pore-size filter

and incubated. The two methods yielded comparable results in CEPEX
enclosures; values for net production based on observed growth ranged
from 10 to 32 μg C$\cdot$liter$^{-1}\cdot$day^{-1} or a generation time of 16 hrs to
2.5 days while the ^{3}H–thymidine times ranged from 16 to 26 hrs. In
view of the relative enrichment of the CEPEX enclosures compared with
the open ocean, these values support the radioisotope derived data
for bacterial growth and production rather than data derived from
chemical measurements.

If we assume a steady state concentration of bacteria, there
should be removal of bacteria by grazing or death to offset the
apparent growth of the bacterial population. These are the subject
of other papers in this publication and will not be commented on
directly here. Instead, attention will now be directed towards some
experimental studies involving the interaction of bacteria populations and a variety of dissolved organic compounds. Hopefully, these
studies will shed some light on the disagreement over flux of organic
solutes and bacterial activity.

EXPERIMENTAL STUDIES

Enrichment Experiments

How does a marine bacterial population respond to enrichment
with organic solutes? An experimental approach to answering this
question can potentially shed some light on several problems. One
relates to the capacity of a freshly collected natural population to
take up substrate. How much can they handle? When and at what
level does substrate concentration clearly have an effect on the
population? Another problem relates to the physiological state of
all or most of the bacterial population. What is the degree of
activity or inactivity, often referred to as dormancy (Novitsky and
Morita 1978; Stevenson 1978)? Another problem relates to the
important question of naturally occurring substrate that can be
readily used by bacteria. How much is present in a given water
body? How closely are bacteria and substrate production coupled?

A good beginning in this direction was made by Vaccaro (1969),
who explored the adaptations of natural bacterial populations to
organic substrates with the kinetic approach. He demonstrated that a
natural population adjusted to enrichment of 200 μg glycine-C$\cdot$liter^{-1}
by increasing the rate of incorporation from very low levels up to
8 μg C$\cdot$liter$^{-1}\cdot$hr^{-1} within 24 hours (see Fig. 1). His work pointed
for the first time to the phenomenon of inactivity of microbes in
open ocean samples with respect to labile substrates like glycine.
Enrichment produced a response within 24–36 hrs; uptake followed
Michaelis–Menten kinetics and the substrate (200 μg C$\cdot$liter^{-1}) was
consumed within 48 hrs. Williams and Gray (1970) extended this
approach by incubating at a number of concentrations. They found

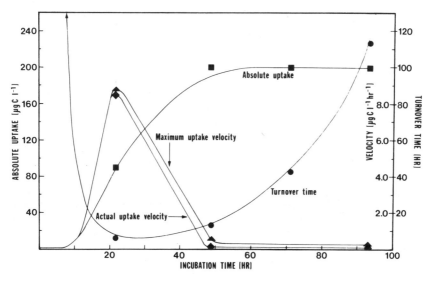

Figure 1.   Changes in kinetic uptake parameters with time for the Woods Hole Harbor water enriched with 200 µg glycine-C·liter^{-1}. From Vaccaro (1969).

that the rate of respiration of amino acids increased as a function of concentration (not directly proportional, however). Also, samples enriched at 500 µg·liter^{-1} amino acids and above showed a marked increase in rate usually occurred within 24 hours, was proportional to the increase in substrate concentration, and resulted in rates as high as 1-2 mg·liter^{-1}·day^{-1} by the second day. The authors attributed this phenomenon to the induction of enzyme systems in the existing population and subsequent microbial growth, and presented theoretical calculations based on Michaelis-Menten kinetics and cell growth which agreed well with the empirical data.

More recently, Hollibaugh (1976, 1978, 1979) measured nitrogen regeneration from amino acid degradation in seawater. He enriched at 10 µM concentrations (between 1 and 2 mg·liter^{-1} of amino acids), and observed a response similar to that reported in Williams and Gray (1970). Although he found that antibiotics inhibited the typical increase in uptake and that transitory increases in ATP coincided with the rate increases, Hollibaugh suggested that the observed phenomena were due to activation of a dormant population or to enzyme induction rather than to a biomass increase in the population. Perhaps it should also be noted here that Williams and Gray (1970) and Hollibaugh (1979) remarked on the lack of any short-term bottle effect. Enclosure in a bottle did not stimulate marked increases in the bacterial population in the first day or two of the experiments.

These papers have indicated that there is uncertainty as to the

nature of the response of the microbial community to the enrichment
regimes employed. It could be due to: (1) activation of a dormant
segment of the natural population; (2) induction of enzymes and
transport systems (overlap with 1); (3) increase in biomass with
growth of individual cells and multiplication of the population.
The following is some work from our laboratory in Massachusetts
which may clear up this ambiguity as well as some of the earlier
questions raised.

Two samples were taken, one from the Essex Estuary and the
other from 6 km offshore in the Gulf of Maine, and filtered at low
vaccuum through a 3 μm Nuclepore filter. Six replicates of each
sample were added to seawater rinsed plastic containers (Whirl-Pak,
1 liter capacity) and incubated at 20°C. To five of the samples
^{14}C-labelled and unlabelled glutamic acid were added to give five
orders of magnitude of glutamate concentrations, from 1 μg·liter^{-1}
to 10 mg·liter^{-1}. One sample was left unenriched (control). The
following parameters were sampled: acridine orange direct counts

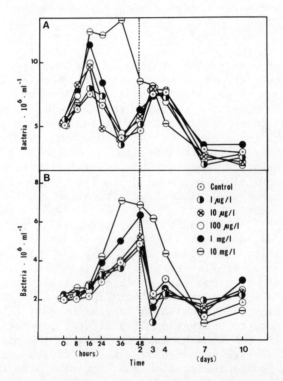

Figure 2. Time course of bacterial AODC in samples prefiltered
through 3 μm Nuclepore filters and enriched with glutamic acid at 1,
10, and 100 μg·liter^{-1} and 1 and 10 mg·liter^{-1}. A: estuarine water;
B: coastal water.

(AODC) on 0.2 and 0.6 μm Nuclepore filters (Hobbie et al. 1977),
total uptake of glutamate (assimilation and respiration), colony
forming units (CFU), and active bacteria via INT counts (Iturriaga
and Rheinheimer 1975). Figure 2 shows the time course of AODC (0.2
μm counts) in the estuarine (2A) and coastal water (2B) samples.
Figure 3 plots the fraction of glutamate used over the course of the
experiment in the 5 enriched subsamples, for estuarine water (3A) and
coastal water (3B). The data from the estuarine samples indicate the
following course of events: (1) within 8 hours, all the glutamate
was used when the initial concentraton was 100 μg·liter^{-1} and lower;
within 24 hours, the glutamate was exhausted in all samples; (2) all
samples showed an increase in AODC that reached a peak in 16 hours in
all but the 10 mg·liter^{-1} sample. The AODC declined after this peak
until the 36-hour sampling but then there was a slow increase and
final decline; (3) AODC counts on the different sized filters indicated that the initial rapid increase and decline in samples with the
lower concentrations of glutamate was due to a proliferation and then
disappearance of smaller bacteria (< 0.6 μm). On the other hand, the
increase in the highest concentration sample (10 mg·liter^{-1}) was due
to large cells.

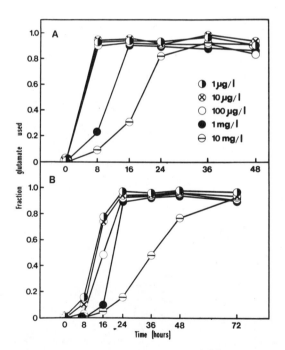

Figure 3. Total uptake and metabolism of ^{14}C-labelled glutamic acid
in enrichment experiment, presented as fraction of available glutamic
used over time. Samples prefiltered through 3 μm Nuclepore filters
and enriched with glutamic acid at 1, 10 and 100 μg·liter^{-1} and 1 and
10 mg·liter^{-1}. A: estuarine water; B: coastal water.

   This experiment points up the significance of two unknown variables pertaining to any such work with natural population. First,
the bacteria in a given sample range from dormant and inactive to
highly active and rapidly growing. Second, they have been removed
from most sources of substrate; therefore, a variable amount of
readily useable substrate is present and additional amounts of
substrate are made slowly available through cell death and from
natural enzymatic breakdown of polymers. The measured responses to
enrichment have the potential to yield some information on these
unknowns. Thus, in the above experiment, it is apparent that the
estuarine bacteria were quite active as measured by their capacity
to use substrate and the immediate cell increases that occurred.
Figure 4 is a plot of the rate of substrate use vs. concentration for
the two water types for all samples where substrate exhaustion had
not yet occurred. The estuarine population was clearly capable of

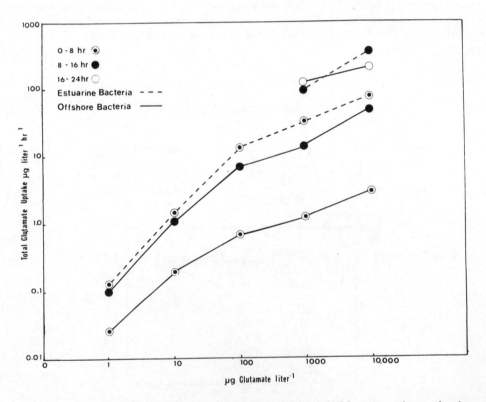

Figure 4. Rates of glutamic acid uptake at different times during
the enrichment experiment, plotted as rate of uptake versus starting
concentration. Rates only plotted if substrate not yet exhausted
(see Fig. 3).

using substrate at rapid rates even during the first eight hours. However, it is apparent that only concentrations above 100 $\mu g \cdot liter^{-1}$ had a substantial impact on the actual course of growth and division of the population. Therefore, the observed changes in numbers and cell size in all lower concentrations can be interpreted on the basis of their response to naturally occurring substrate or the lack thereof.

The results suggest that the estuarine population was very active but closely tied to substrate supply. When the supply was cut off (removal from the natural environment), they divided once (within 16 hours) and then a large proportion of those new small cells apparently died and lysed. Their death and other unknown processes supplied more substrate and allowed the second increase (Fig. 2A) to occur during the second and third days of the experiment.

As Figs. 2B and 3B show, the bacteria in waters 6 km offshore showed different responses to enrichment and enclosure. There was a delayed response in rate of use of all concentrations of substrate; exhaustion of even the lowest concentration did not occur until 24 hours, and it took over two days to exhaust the 10 $mg \cdot liter^{-1}$ sample. Cell numbers showed little increase until 16 hours, but cell growth began occurring between 8 and 16 hours as indicated by an increase in larger cells and a decrease in smaller cells (not shown) in all samples. The data suggest that bacteria in these waters were relatively dormant (inactive), a conclusion that is supported by data reported in Wright (1978) indicating that such bacteria are much less active in heterotrophic uptake on a per cell basis than estuarine bacteria.

These experiments demonstrate the same phenomenon as the enrichment experiments referred to earlier: the capacity to consume a low molecular weight organic solute in a relatively (and variably) short time for added concentrations up to 10 $mg \cdot liter^{-1}$. Response to the substrate, however, is a function both of the concentrations added relative to available natural substrate, and of the recent history of the bacterial population with respect to activity or dormancy. The concentration response is, as noted by Williams and Gray (1970) and shown in Fig. 4, strongly evident. The kind of saturation predicted by the kinetic approach is not evident here. Use is more or less directly dependent on concentration, and, interestingly, not dependent on induced growth or multiplication of the bacteria. Returning to the three possibilities referred to earlier, it is evident that all three occur if the population is initially dormant, while the response of an active population would involve only the last two. Thus, dormant bacteria are "activated" (whatever that means); enzymes and transport are induced (this may be activation); and growth and division of the bacteria occur. It should be evident that the latter is a sensitive indicator of substrate available,

whether natural or added. Our work has indicated that only concentrations of added substrate in the vicinity of 50 to 100 μg·liter^{-1}
(= 20 to 40 μg C·liter^{-1}) will produce a measurable growth and
multiplication response in estuarine and coastal populations.
Rates of substrate use in lower concentrations are, however, quite
respectable: 12 μg C·liter^{-1}·hr^{-1} in the estuarine sample with 100
μg liter^{-1} added glutamate. It is not unreasonable to assume that
this is within the normal range of total substrate use in these
waters. In fact, one possibility is that the population in the
estuary is so closely coupled to the continued supply of substrate
at this rate that they undergo a starvation-induced death when
removed from the environment. This death involves a proliferation
of small cells followed by rapid disappearance (lysis?).

Other Experimental Approaches

     One very obvious drawback to the enrichment studies described
above is the fact that substrates and concentrations used do not
necessarily correspond with those encountered by the bacteria under
natural conditions. Bacteria in coastal and pelagic waters no doubt
are confronted by an array of different substrates supplied
(produced) at various rates which may be changing in tidal or diel
cycles. One substrate source for marine bacteria is phytoplankton
exudates. Iturriaga and Hoppe (1977) prepared natural ^{14}C-labelled
algal exudates and incubated the water with natural samples
containing bacteria. They found uptake ranging from 8 to 12% of the
exudate per hour in the Kiel Fjord and reasoned that at this rate
the bacteria could completely metabolize algal exudate over a 24 hr
day. Bell (1980) and Bell and Sakshaug (1980) reported experiments
using labelled exudate from an algal culture and both cultured and
natural bacteria populations. The uptake compared well with rates
derived from the kinetic approach in spite of the use of an
obviously heterogeneous substrate. They found rates of use as high
as 0.8 μg C·liter^{-1}·hr^{-1} with a natural population using prepared
Skeletonema exudate. Natural phytoplankton exudate production was
lower - ca. 2 μg C·liter^{-1}·day^{-1}. Apparently, the natural bacteria
population was able to rapidly consume algal exudate and was in fact
probably dependent on other substrates as well.

     Some other sources of substrate have been explored. Turner
(1978) used ^{14}C-labelled Spartina exudate to show that planktonic
microorganisms were quite capable of assimilating and respiring
dissolved organic matter released by this salt marsh grass.
Velimirov (1980) demonstrated substantial growth of natural bacterioplankton in water containing "foam" prodced by an interaction of
kelp and phytoplankton in coastal waters. Two interesting papers
(Lucas et al. 1981; Linley et al. 1981) demonstrated the ability of
natural bacteria to grow on dissolved mucilage extracted from kelp.
Eppley et al. (1981) reasoned that since the flux of organic carbon
to bacteria appears to be in the order of 25% of primary production

(Fuhrman and Azam 1980), processes in addition to exudation must be involved. They added extra macrozooplankton to natural plankton incubated with $^{14}CO_2$ and looked for extra $DO^{14}C$ as evidence for "sloppy feeding" by zooplankton. They did not find extra $DO^{14}C$ but did find increased thymidine incorporation by bacteria; thus grazing apparently resulted in an enhanced carbon flow from algae to bacteria. However, no significant increases in bacteria (direct count) were observed, and it must be said that this process is still only hypothetically important.

It is possible to obtain some useful information and to design some interesting experiments using only the substrate naturally occurring in a sample. Figure 5 shows data from an enclosure of 3 µm Nuclepore-filtered populations from an estuary (E-10), from 1 km

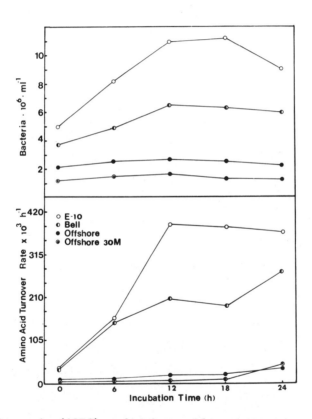

Figure 5. Bacteria (AODC) and amino acid turnover rate vs. incubation time in samples prefiltered through 3 µm Nuclepore filters and incubated in 1 liter Whirl-Paks at 18°C. Turnover rate measured on subsamples amended with ^{14}C-labelled amino acid mix to give a concentration of 6.8 nM, and incubated 1 hr. Samples: E-10 = Essex estuary, Bell = 1 km̄ offshore, Offshore = 10 km offshore, both 1 m and 30 m depths.

offshore (Bell), and from 1 and 30 m depth 10 km offshore. No
enrichment was involved; the response was to naturally occuring substrate.   Growth and activity obtained from the estuary and immediate
coastal water sample for the first 12 hours yielded net production
rates of 12 and 5 µg C·liter^{-1}·hr^{-1}.   The offshore rates were in the
range of 0.8 to 1 µg C·liter^{-1}·hr^{-1}.   Fuhrman and Azam (1980) used a
similar approach to estimate bacterial use of natural substrate, and
found production rates of 10 to 32 µg C·liter^{-1}·day^{-1} for coastal
bacteria populations.   They also found reduced growth after 24 hrs
which they attributed to exhaustion of readily available substrate;
Figure 5 indicates the same phenomenon.   These results are consistent
with the general picture that has already been outlined; there is a
broad range of activity that corresponds with potential access to
substrate (Wright, 1978).   The ability to undergo rapid increase in
the first 12 hrs of incubation is, therefore, indicative of an active
population capable of and indeed accustomed to rapid use of organic
solutes.   Although cell density is usually high for such a population
when sampled, removal from the environment and enclosure without
enrichment often results in rapid decline following the proliferation
stage.   Figure 6 shows such a population from 20°C water incubated at
three temperatures.   The results suggest that the role of temperature
here was to slow down the processes of proliferation and death, but
not to prevent them.

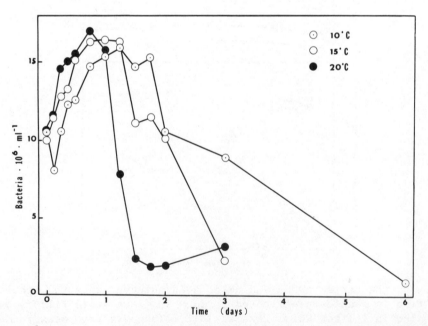

Figure 6.   Active estuarine bacteria population (initial AODC = 10 x
10^6 ml^{-1}) prefiltered through 3 µm Nuclepore filter and incubated at
3 temperatures (10, 15 and 20°C) in 2 liter subsamples with no added
substrates.

This pattern of increase and decline does not always occur with estuarine and coastal bacteria. Figure 7 shows data from two estuarine populations which were prefiltered through 3 μm Nuclepore filters and split into 3 subsamples before incubation. Two were resuspended in 500 ml Whirl-Paks in some sample water that was filtered successively through glass fiber and then GS Millipore (0.2 μm) filters. Bacteria populations were reduced by 50% and 90%. As the AODC data show, no rapid increases or declines occurred in the two control samples during the first 24 hrs. It is noteworthy that the diluted bacteria populations increased within 24 to 36 hrs to the level of the undiluted controls. This suggests that sufficient substrate was

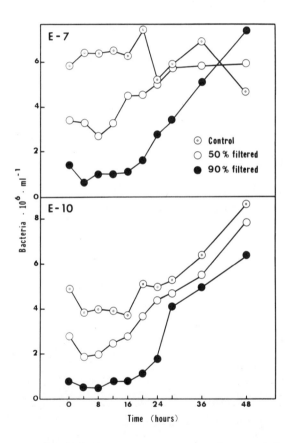

Figure 7. Bacteria (AODC) from two estuarine stations prefiltered through 3 μm Nuclepore filters and incubated at 20°C in 500 ml Whirl-Paks. Bacteria in two of three samples from each station reduced in concentration 50% and 90% respectively by mixing with sample water filtered through glass fiber and GS Millipore filters (0.2 μm).

present in the sample water to either maintain a large population at
steady state for 24 hrs or to allow for growth of a smaller population.

Another experiment was conducted to explore the differences
between estuarine and offshore (10 km) bacterial populations.
Samples were collected on June 20, 1979 from 1 m depth: estuarine
salinity was 31.8 ppt, the offshore sample 31.9 ppt. All water from
both stations was 3 µm Nuclepore filtered with gentle vacuum,
removing eucaryotes but leaving behind ca. 90% of the bacteria.
Subsamples were filtered successively and gently through a glass
fiber and a GS Millipore filter to remove all organisms. The six
experimental 1 liter Whirl-Paks were set up as follows: (1) 100% 3 µm

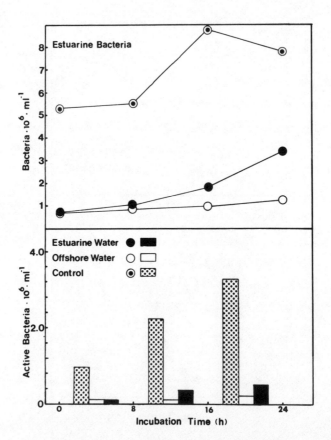

Figure 8. Estuarine bacteria AODC (upper) and active bacteria
(lower) prefiltered through 3 µm Nuclepore filters and treated as
follows: Control: no change; Offshore water: 10% estuarine sample
added to 90% offshore water filtered free of bacteria; Estuarine
water: 10% estuarine sample added to 90% estuarine water filtered
free of bacteria. Samples incubated at 18°C.

filtered estuary water; (2) 10% 3 μm filtered estuary water, 90% GS
filtered estuary water; (3) 10% 3 μm filtered estuary water, 90% GS
filtered offshore water. Samples 4, 5 and 6 were set up similarly,
with the offshore 3 μm filtrate as the test bacterial population.
Samples were incubated in the dark at 18°C, and subsamples taken
every 8 hrs for AODC, ^{14}C-glutamate assimilation, INT activity, and
autoradiography using ^{3}H-glutamate. Figures 8 and 9 show the first
24 hours of data for AODC and autoradiographic determination of
active bacteria. The data show clearly the relative effects of
natural substrate present in the two samples vs. the relative physiological state of the populations. For the first 24 hrs, estuarine
bacteria: (1) in the control increased rapidly and exhausted substrate and began to decline by 24 hrs; (2) diluted and resuspended in
estuarine water showed a substantial net growth, and; (3) suspended
in 90% offshore water showed little growth. The offshore bacteria
population showed no significant growth in any of the subsamples for
the first 24 hours. Twenty-four hour net bacterial production in
the estuary sample was estimated at 84 and 80 μg C·liter^{-1} for the

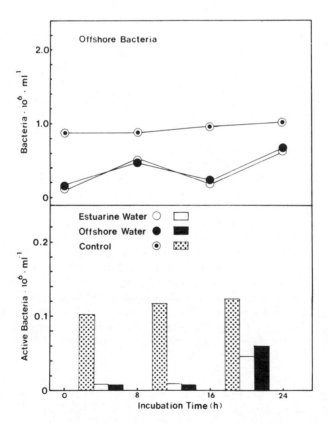

Figure 9. Bacteria from 1 m depth 10 km offshore: AODC (upper),
active bacteria (lower), treated as in Fig. 8 sample.

control and diluted sample, and 12 µg C·liter^{-1} in offshore water.
Corresponding 24 hr values for the offshore samples are 5 and
12 µg C·liter^{-1} for control and diluted offshore samples and
13 µg C·liter^{-1} in estuarine water. The most logical interpretation
of this experiment is that the offshore bacteria were relatively
dormant and were existing in water with essentially no excess
substrate. The estuarine bacteria, on the other hand, were actively
growing as a response to the presence of a substantial amount of
substrate. The autoradiography, INT and glutamate turnover data all
support these conclusions.

PERSPECTIVES ON THE INTERACTION OF THE BACTERIA AND DISSOLVED
ORGANIC SOLUTES

The Capacity of Natural Bacteria Populations to Use Available Pools

     The evidence indicates that the natural bacteria are easily
capable of using algal exudate as rapidly as it is released (Larsson
and Hagström 1979; Bell and Sakshaug 1980). However, it also appears
that algal exudate is not the only or perhaps even the most important
source of useful organic matter. Many workers have concluded that
algal exudate alone could not support either measured flux rates for
low molecular weight solutes or calculated bacterial growth rates.
Although ultimately all bacterial growth in the oceans depends on
algal production, they suggest that bacterial growth in the oceans
depends more on the death or consumption of phytoplankton than on
excretion.

     In the discussion of their work on amino acid flux, Dawson and
Gocke (1978) were concerned that the flux values for amino acids in
Baltic Sea surface waters calculated from radioisotope turnover and
chemical concentrations were too high - that the existing bacteria
population could not account for the flux they measured. Although
they encountered rates as high as 76 µg·liter^{-1}·day^{-1}, most values
ranged from 2 to 10 µg·liter^{-1}·day^{-1}. Bacteria density ranged from
1 to 2 x 10^9 liter^{-1}, but they assumed only 4 µg C per 10^9 bacteria
and so their estimates of the potential for this population density
to consume substrate are low, probably by 60-70%. The 76 µg·liter^{-1}
of amino acids is equivalent to 30 µg C, and their measured bacteria
population was 1.76 x 10^9 liter^{-1} or 21 µg C·liter^{-1} (based on 12 µg
C per 10^9 bacteria). Assuming that 50% of the amino acids are
respired, this population need turn over only once in 34 hrs to
account for the highest observed flux.

     The enrichment experiments cited above indicate the capacity of
active bacterial population to use substrate supplied to them artificially. In fact, their use of added low molecular weight organic
solutes generates a growth response only above some threshold level,
in the range of 40 to 400 µg C·liter^{-1}. This threshold level is

probably a reflection of available natural substrate, which suggests
that during summer months the estuarine and near coastal bacteria
populations are accustomed to daily substrate production rates at
those levels. Thus, the apparent uptake rates for total carbohydrate
reported in Burney et al. (1981) for a salt marsh estuary – daily
rates of 85 to 520 µg $C \cdot liter^{-1} \cdot day^{-1}$ seem to be in the correct order
of magnitude. Net bacterial production rates cited earlier confirm
that bacteria populations in estuarine and immediate coastal waters
in the summer are sufficiently active and concentrated to be assimilating substrate at summer rates of at least 1 to 10 µg $C \cdot liter^{-1} \cdot$
hr^{-1}. It is also clear from the experimental evidence that bacterioplankton are capable of adjusting rapidly to pulses of readily usable
organic solutes.

Mathematical Model of Billen

     Billen et al. (1980) have constructed a simple but elegant
mathematical model which incorporates as variables substrate use,
substrate production, and bacteria population density (Fig. 10).
Constants in the model are $V_{max} \cdot cell^{-1}$, Km, growth yield, and death
rate. The model is particularly useful in helping to define the
behavior of the variables under steady state conditions or with minor
perturbations. At steady state, the model predicts that bacterial
density is a direct function of the rate of substrate production.
The model also predicts that ambient substrate concentrations are
basically independent of production rates and are a function of
physiological properties of the bacteria (especially Km, a measure
of substrate affinity). In a computer test run of the model, the
authors show that the effect of an increased rate of production is,
after one turnover of the bacteria population, to return substrate
concentration to its original value and to adjust the bacteria population to a new, higher steady state density. If production rate is
allowed to vary sinusoidally (as in diel effects), daily variations
will occur in bacteria density and substrate concentration, but they

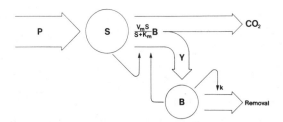

Figure 10. Schematic representation of mathematical model
incorporating substrate production (P), substrate concentration
(S), substrate use (kinetic equation), bacteria population size
(B), growth yield (Y), and a constant (K) to represent removal by
death and predation. From Billen et al. (1980).

will be limited. Billen et al. (1980) have produced an extremely useful model which may serve as the basis for future considerations. It requires further refinement, especially in the evaluation of predation on bacteria and natural death rate. However, it is a good start in the direction of realistic, quantitative modeling of the role of bacteria in the flux of DOC in the oceans.

Conceptual Model

Based on a consideration of the effects of substrate limitation, grazing, and dormancy on bacteria populations, I postulate three ideal bacterioplankton communities. Each community is assumed to be a mixture of many species of bacteria with a finite breadth of biochemical capabilities and varying proportions of dead, dormant, and active cells. Each community is also assumed to be in a steady state level of numbers and heterotrophic activity.

Active, grazer-controlled community. The existence of active, growing heterotrophic bacteria and enough substrate to support substantial continued growth suggests that this community is not resource-limited. The most logical alternative is control by grazing. Some regions of estuaries are well populated by benthic bivalves; also, microzooplankton are often abundant in the water column (e.g., protozoa) and could increase rapidly. The combined grazing pressure from the animals in the water column and the benthos (Wright et al. 1982) could hold bacteria densities well below levels set by available dissolved organic matter released continuously from plants such as Spartina and phytoplankton. Continuous grazing would also tend to prevent the accumulation of dormant or otherwise inactive bacteria. This community is found in estuarine and other systems with a substantial input of labile dissolved organic matter.

The characteristics of this community can be summarized as: (1) numbers moderately high (1 to 6 x 10^6 ml^{-1}); (2) high specific activity (Wright 1982) and percent active cells; (3) plate counts high (10^4 to 10^5 ml^{-1}); (4) ratio of small to large cells is 2 to 3; (5) substantial increase in numbers (20% to > 100%) if 1 μm filtrate incubated for 24 hrs; (6) evidence of grazer populations able to feed on bacteria.

Active, substrate-limited community. In the absence of grazer control, the bacteria will tend to increase rapidly to a higher steady state level where they use substrate as fast as it is supplied. The effects of nutrient starvation may appear because of variations in nutrient supply and the high growth potential of bacteria. Since the large bacteria community represents an unexploited food resource, this may be a transient state - either spatial or temporal - unless other environmental constraints prevent the buildup of a grazer community.

The characteristics of this community are: (1) numbers are high
(3 to 12 x 10^6 ml^{-1}; (2) specific activity, percent active cells
relatively high, but can vary; (3) plate count high; (4) ratio of
small to large cells is variable, may range from 2 to 8; (5) lower
percent increase in numbers if 1 μm filtrate incubated for 24 hours
(10% to 50%); may show rapid decline after short increase.

Dormant, substrate-limited community. This community is characteristic of oligotrophic waters or of strata in the water column
with little dissolved organic matter input. The numbers are not
high enough to support grazers specific for bacteria and instead the
population is controlled by general grazers and slow, starvation-
induced death and the slow growth and increase of other bacteria.
The majority of the cells are inactive; those that are active tend
to be the larger cells (> 0.6 μm).

The characteristics of this community are: (1) numbers are low
(< 1 x 10^6 ml^{-1}); (2) specific activity and percent active cells low;
(3) plate counts very low (10^2 to 10^3 ml^{-1}); (4) ratio of small to
large cells high (4 to 8); (5) no significant increases in first 24
hrs if enriched or incubated or both.

In essence, dormancy prevents numbers (hence, bacterial biomass)
from precipitous declines as substrate becomes and remains unavail-
able and guarantees a relatively rapid response to substrate when it
appears again or if it becomes available in small patches of the
plankton. Dormancy thus sets a limit on the low end of the observed
seasonal and spatial variations in numbers, while grazing may
function to set an upper limit on numbers in most systems. This may
explain the remarkably small seasonal range in numbers reported now
by many workers. On the other hand, these two phenomena foster a
much greater range in heterotrophic activity (uptake and mineralization of substrates), as dormancy creates a large proportion of
inactive cells while continued grazing guarantees a large proportion
of heterotrophically active cells. This scheme is consistent with
most of my data and with the recent literature. Further work is
required in developing mathematical expressions for the model communities and in pursuing some additional problems. In particular, what
does the seasonal temperature effect do to the model? How can an
enrichment response best be measured? Where do the detritus-attached
bacteria, very important in some estuaries, fit into the scheme?
The three community types can probably best be considered as finite
units on what is best viewed as a continuum. Any given sample may
be in transition from one stage to another, or may be the result of a
recent mixture of water masses. How can this kind of variability be
represented in the model? At the present time, this model represents
only a first attempt to use the data I have collected and to deal
with the three major phenomena of dormancy, substrate limitation, and
grazing as they influence bacteria numbers and activity. The model
suggests that most of the variations in bacterioplankton numbers and

activity can be understood as a consequence of these three phenomena.

ACKNOWLEDGMENTS

This work was supported by the National Science Foundation under Grant No. OCE-7925368.

REFERENCES

Alexander, M. 1981. Biodegradation of chemicals of environmental concern. Science 211: 132-138.

Azam, F., and O. Holm-Hanson. 1973. Use of tritiated substrates in the study of heterotrophy in seawater. Mar. Biol. 23: 191-196.

Banoub, M., and P. J. leB. Williams. 1972. Measurements of microbial activity and organic material in the Western Mediterranean Sea. Deep-Sea Res. 19: 433-443.

Bell, W. H. 1980. Bacterial utilization of algal extracellular products. I. The kinetic approach. Limnol. Oceanogr. 25: 1007-1020.

Bell, W. H., and E. Sakshaug. 1980. Bacterial utilization of algal extracellular products. 2. A kinetic study of natural populations. Limnol. Oceanogr. 25: 1021-1033.

Billen, G., C. Joiris, J. Wijnant, and G. Gillain. 1980. Concentration and microbiological utilization of small organic molecules in the Scheldt Estuary, the Belgian coastal zone of the North Sea and the English Channel. Est. Coastal Mar. Sci. 11: 279-294.

Burney, C. M., K. M. Johnson, and J. McN. Sieburth. 1981. Diel flux of dissolved carbohydrate in a salt marsh and a simulated ecosystem. Mar. Biol. 63: 175-187.

Burnison, B. K., and R. Y. Morita. 1974. Heterotrophic potential for amino acid uptake in a naturally eutrophic lake. Appl. Microbiol. 27: 488-495.

Crawford, C. C., J. E. Hobbie, and K. L. Webb. 1974. The utilization of dissolved free amino acids by estuarine microorganisms. Ecology 55: 551-563.

Dawson, R., and K. Gocke. 1978. Heterotrophic activity in comparison to the free amino acid concentrations in Baltic Sea water samples. Oceanol. Acta. 1: 45-54.

Degens, E. T. 1970. Molecular nature of nitrogenous compounds in sea water and recent marine sediments, pp. 77-106. In: D. Hood [ed.], Organic Matter in Natural Waters. Occas. Publ. #1, Institute of Mar. Science, Alaska.

DeLattre, J. M., R. Delesmont, M. Clabaux, C. Oger, and H. Leclerc. 1979. Bacterial biomass, production and heterotrophic activity of the coastal seawater at Gravelines (France). Oceanol. Acta 2: 317-324.

Derenbach, J., and P. J. leB. Williams. 1974. Autotrophic and bacteria production; fractionation of plankton populations by differential filtration of samples from the English Channel. Mar. Biol. 25: 263-269.

Dietz, A. S., L. J. Albright, and T. Tuominen. 1977. Alternative model and approach for determining microbial heterotrophic activities in aquatic systems. Appl. Environ. Microbiol. 33: 817-823.

Eppley, R. W., S. G. Horrigan, J. A. Fuhrman, E. R. Brooks, C. C. Price, and K. Sellner. 1981. Origins of dissolved organic matter in Southern California coastal waters: Experiments on the role of zooplankton. Mar. Ecol. Prog. Ser. 6: 149-159.

Ferguson, R. L., and P. Rublee. 1976. Contribution of bacteria to standing crop of coastal plankton. Limnol. Oceanogr. 21: 141-145.

Fuhrman, J. A., J. W. Ammerman, and F. Azam. 1980. Bacterioplankton in the coastal euphotic zone: Distribution, activity and possible relationships with phytoplankton. Mar. Biol. 60: 201-207.

Fuhrman, J. A., and F. Azam. 1980. Bacterioplankton secondary production estimates for coastal waters of British Columbia, Antarctica, and California. Appl. Environ. Microbiol. 39: 1085-1095.

Gardner, W. S., and J. A. Stephens. 1978. Stability and composition of terrestrially derived dissolved organic nitrogen in continental shelf surface waters. Mar. Chem. 6: 335-342.

Garrasi, C., E. T. Degens, and K. Mopper. 1979. The free amino acid composition of seawater obtained without desalting and preconcentration. Mar. Chem. 8: 71-85.

Gillespie, P. A., R. A. Morita, and L. P. Jones. 1976. The heterotrophic activity for amino acids, glucose and acetate in Antarctic waters. J. Oceanogr. Soc. Japan 32: 74-82.

Gocke, K. 1977. Comparison of methods for determining the turnover times of dissolved organic compounds. Mar. Biol. 42: 131-141.

Gocke, K., R. Dawson, and G. Liebezeit. 1981. Availability of dissolved free glucose to heterotrophic microorganisms. Mar. Biol. 62: 209-216.

Hagström, Å., U. Larsson, P. Horstedt, and S. Normark. 1979. Frequency of dividing cells, a new approach to the determination of bacterial growth rates in aquatic environments. Appl. Environ. Microbiol. 27: 805-812.

Hobbie, J. E., R. Daley, and S. Jasper. 1977. Use of Nuclepore filters for counting bacteria by fluorescence microscopy. Appl. Environ. Microbiol. 33: 1225-1228.

Hodson, R. E., F. Azam, A. F. Carlucci, J. A. Fuhrman, D. M. Karl, and O. Holm-Hansen. 1981. Microbial uptake of dissolved organic matter in McMurdo Sound, Antarctica. Mar. Biol. 61: 89-94.

Hollibaugh, J. T. 1976. The biological degradation of arginine and glutamic acid in seawater in relation to the growth of phytoplankton. Mar. Biol. 36: 303-312.

Hollibaugh, J. T.   1978.   Nitrogen regeneration during the degrada-
    tion of several amino acids by plankton communities collected
    near Halifax, Nova Scotia, Canada.   Mar. Biol. 45: 191-201.
Hollibaugh, J. T.   1979.   Metabolic adaptation in natural bacterial
    populations supplemented with selected amino acids.   Est.
    Coastal Mar. Sci. 9: 215-230.
Ittekkot, V., U. Borckmann, W. Michaelis, and E. T. Degens.   1981.
    Dissolved free and combined carbohydrates during a phytoplankton
    bloom in the northern North Sea.   Mar. Ecol. Prog. Ser. 4:
    299-305.
Iturriaga, R., and G. Rheinheimer.   1975.   Eine einfache methode zur
    auszahlung von bakterien mit aktivem eletronentranspotsystem in
    wasser und sedimentproben.   Kiel. Meeresforsch. 31: 83-86.
Iturriaga, R., and H. G. Hoppe.   1977.   Observations of heterotrophic
    activity on photoassimilated organic matter.   Mar. Biol. 40:
    101-108.
Johnson, K. M., C. M. Burney, and J. McN. Sieburth.   1981.   Enigmatic
    marine ecosystem metabolism measured by direct diel CO_2 and O_2
    flux in conjunction with DOC release and uptake.   Mar. Biol.
    65: 49-60.
Larsson, U., and Å. Hagström.   1979.   Phytoplankton exudate release
    as an energy source for the growth of pelagic bacteria.   Mar.
    Biol. 52: 199-206.
Lindroth, P. and K. Mopper.   1979.   High performance liquid chromato-
    graphic determination of subpicomole amounts of amino acids by
    precolumn fluorescence derivatization with o-phthaldialdehyde.
    Anal. Chem. 51: 167-174.
Linley, E. A. S., R. C. Newell, and S. A. Bosma.   1981.   Heterotroph-
    ic utilisation of mucilage released during fragmentation of
    kelp (Ecklonia maxima and Laminaria pallida).   I. Development
    of microbial communities associated with the degradation of
    kelp mucilage.   Mar. Ecol. Prog. Ser. 4: 31-41.
Lucas, M. I., R. C. Newell, and B. Velimirov.   1981.   Heterotrophic
    utilisation of mucilage released during fragmentation of kelp
    (Ecklonia maxima and Laminaria pallida).   II. Differential
    utilisation of dissolved organic components from kelp mucilage.
    Mar. Ecol. Prog. Ser. 4: 43-55.
Menzel, D. W., and J. H. Ryther.   1970.   Distribution and cycling of
    organic matter in the oceans, pp. 31-54.   In: D. Hood [ed.],
    Organic Matter in Natural Water.   Occas. Publ. #1, Inst. of
    Mar. Science, Alaska.
Mopper, K., and P. Lindroth.   1982.   Diel and depth variations in
    dissolved free amino acids and ammonium in the Baltic Sea
    determined by shipboard HPLC analysis.   Limnol. Oceanogr. 27:
    336-347.
Novitsky, J. A., and R. Y. Morita.   1978.   Possible strategy for the
    survival of marine bacteria under starvation conditions.   Mar.
    Biol. 48: 289-295.
Odum, E. P.   1971.   Fundamentals of Ecology.   W. B. Saunders,
    Philadelphia.

Ogura, N. 1970. High molecular weight organic matter in seawater.
    Mar. Chem. 5: 535-549.
Parsons, T. R., and J. D. M. Strickland. 1962. On the production
    of particulate organic carbon by heterotrophic processes in sea
    water. Deep-Sea Res. 8: 211-222.
Sieburth, J. McN., and A. Jensen. 1970. Production and transforma-
    tion of extracellular organic matter from littoral marine algae:
    A resume, pp. 203-224. In: D. Hood [ed.], Organic Matter in
    Natural Water. Occas. Publ. #1, Inst. of Mar. Science, Alaska.
Sieburth, J. McN., K. M. Johnson, C. M. Burney, and D. M. Lavoie.
    1977. Estimation of in situ rates of heterotrophy using diurnal
    changes in dissolved organic matter and growth rates of pico-
    plankton in diffusion culture. Helgol. Wiss. Meeresunters. 30:
    565-574.
Stevenson, H. L. 1978. A case for bacterial dormancy in aquatic
    systems. Microb. Ecol. 4: 127-133.
Turner, R. E. 1978. Community plankton respiration in a salt marsh
    estuary and the importance of macrophytic leachates. Limnol.
    Oceanogr. 23: 442-451.
Vaccaro, R. F. 1969. The response of natural microbial populations
    in seawater to organic enrichment. Limnol. Oceanogr. 14:
    726-735.
Velimirov, B. 1980. Formation and potential trophic significance
    of marine foam near kelp beds in the Benguela Upwelling
    system. Mar. Biol. 58: 311-318.
Walsh, G. E. 1965. Studies on dissolved carbohydrate in Cape Cod
    waters. II. Diurnal fluctuation in Oyster Pond. Limnol.
    Oceanogr. 10: 577-582.
Wheeler, J. 1976. Fractionation by molecular weight of organic
    substances in Georgia coastal water. Limnol. Oceanogr. 21:
    846-852.
Wiebe, W. J., and D. F. Smith. 1977. Direct measurement of
    dissolved organic carbon release by phytoplankton and
    incorporation by microheterotrophs. Mar. Biol 42: 213-223.
Williams, P. J. leB. 1973. The validity of the application of
    simple kinetic analysis to heterogenous microbial populations.
    Limnol. Oceanogr. 18: 159-164.
Williams, P. J. leB. 1975. Biological and chemical dissolved
    organic material in sea water, pp. 301-364. In: Riley and
    Skirow [eds.], Chemical Oceanography, Vol. 2, Second edition.
    Academic Press, New York.
Williams, P. J. leB., and C. Askew. 1968. A method of measuring
    the mineralization by micro-organisms of organic compounds in
    sea-water. Deep-Sea Res. 15: 365-375.
Williams, P. J. leB., and R. W. Gray. 1970. Heterotrophic utiliza-
    tion of dissolved organic compounds in the sea. II. Observation
    on the responses of heterotrophic marine populations to abrupt
    increases in amino acid concentration. J. Mar. Biol. Assoc. UK
    50: 871-881.
Williams, P. M., H. Oescher, and P. Kinney. 1969. Natural radio-

carbon activity of the dissolved organic carbon in the
Northwest Pacific Ocean. Nature 224: 256-258.

Wright, R. T. 1973. Some difficulties in using ^{14}C-organic solutes
to measure heterotrophic bacterial activity, pp. 199-217. In:
Stevenson and Colwell [eds.], Estuarine Microbial Ecology.
Univ. of S. Carolina Press, Columbia.

Wright, R. T. 1978. Measurement and significance of specific activity in the heterotrophic bacteria of natural waters. Appl.
Environ. Microbiol. 36: 297-305.

Wright R. T. 1979. Natural heterotrophic activity in estuarine and
coastal waters, pp. 119-134. In: Borquin and Pritchard [eds.],
Proc. Workshop: Microbial Degradation of Pollutants in Marine
Environments. EPA-600 9-79-012.

Wright, R. T., and B. K. Burnison. 1979. Heterotrophic activity
measured with radiolabelled organic substrates, p. 140-155. In
Costerton and Colwell [eds.], Native Aquatic Bacteria: Enumeration, Activity and Ecology. American Society for Testing and
Materials, STP 695, Philadelphia.

Wright, R. T., and J. E. Hobbie. 1966. The use of glucose and
acetate by bacteria and algae in aquatic ecosystems. Ecology
47: 447-464.

Wright, R. T., R. B. Coffin, C. P. Ersing, and D. Pearson. 1982.
Field and laboratory measurements of bivalve filtration of
natural marine bacterioplankton. Limnol. Oceanogr. 27: 91-98.

Wright, R. T., and R. B. Coffin. 1983. Planktonic bacteria in
estuaries and coastal waters of northern Massachusetts: spatial
and temporal distribution. Mar. Ecol. Prog. Ser. In press.

SINKING OF PARTICULATE MATTER FROM THE SURFACE WATER OF THE OCEAN

Barry T. Hargrave

Marine Ecology Laboratory
Bedford Institute of Oceanography
Dartmouth, Nova Scotia, Canada B2Y 4A2

INTRODUCTION

Concentrations of suspended particulate matter in the ocean do not change over short periods of time (days) if processes of supply are balanced by processes of removal. Thus, the amount of organic matter available for utilization within the water column can be calculated as the difference between phytoplankton net production and the loss of particulate organic matter by gravitational settling. Advective loss is assumed to be balanced by advective supply. Then, the concentration of organic matter available for heterotrophic consumption in surface waters is that produced and not removed by sedimentation.

Past studies of particulate organic matter loss through utilization within the water column and gravitational deposition have been based on four approaches which can be identified on the basis of methods employed. Mass balance studies utilize empirically determined distributions of dissolved oxygen, inorganic nutrients, or radioactive elements adsorbed to particulate matter to calculate residence time, loss, or consumption rates within the water column. The other three approaches employ observations of particulate matter concentrations, either suspended, sedimented in traps, or accumulated in bottom deposits to quantify the downward transfer of material through the water column.

All of these approaches have been used to provide estimates of the loss of suspended matter from the water column by consumption or sedimentation or both. Each method differs in assumptions and limitations inherent in sampling and analytical techniques. Estimates of vertical flux in various types of study are also not directly comparable because different time and space scales are involved in the different measurements. This review recognizes these limitations while attempting to summarize rates which have been calculated. Rates of vertical flux of particulate organic carbon and photosynthetic production are compared to illustrate the rapid recycling of material which occurs in the surface of the ocean.

Table 1. Summary of rates of oxygen consumption, phosphorus regeneration, average concentration of suspended particulate organic carbon and sedimentation of particulate organic carbon at various depths in the North Atlantic Ocean.

Depth[1] (m)	Oxygen Uptake $\mu l\ O_2 \cdot liter^{-1} \cdot y^{-1}$	Phosphorus Regeneration mg-at $P \cdot liter^{-1} \cdot y^{-1}$	Particulate Suspended[2] $\mu g\ C \cdot liter^{-1}$	Organic Carbon Sedimented[2] $g\ C \cdot m^{-2} \cdot y^{-1}$
200	210	93	10.4	18.5
280	80	40	9.4	12.8
370	50	24	8.7	9.5
510	55	5	7.9	6.7
700	35	3	7.2	4.7
1000	13	4	6.5	3.2
1250	5	4	6.1	2.5
1500	1.6	2.9	5.8	2.1
2500-4000	0.13	0.1	4.3	0.7

[1] Depths are averages for density surfaces used to calculate oxygen consumption and phosphorus regeneration (Riley 1951).

[2] Data taken from Gordon et al. (1979) and Suess (1980). Suspended (POC) and sedimented (SOC) particulate organic carbon is expressed as a function of depth (z in meters). Equations derived by least square regression analysis (n = number of observations);

$$POC\ (\mu g \cdot liter^{-1}) = 49.3\ z^{-0.29} \qquad (n = 13)\ (r^2 = 0.68) \qquad (1)$$

$$SOC\ (g\ C \cdot m^{-2} \cdot y^{-1}) = 5889\ z^{-1.09} \qquad (n = 40)\ (r^2 = 0.74) \qquad (2)$$

DISCUSSION

Material Balance Studies

     Hutchinson (1938) calculated that approximately 3% of the seston
suspended in the surface water of three lakes would have to settle
through the thermocline each day to account for observed rates of
oxygen depletion. Riley (1951) used a similar mass balance approach
to calculate deep-water oxygen consumption and phosphorus regeneration in the Atlantic Ocean. Density gradients were used to estimate
the horizontal velocity and this was combined with profiles of oxygen
and phosphorus to quantify rates of oxygen depletion and phosphorus
regeneration on isoentropic surfaces by a finite difference method.
Values of oxygen consumption in the northern North Atlantic below
200 m varied from 17 to 41 liter·m^{-2}·y^{-1} (summarized in Table 1 from
Riley 1951). A supply of 7 to 15 g C·m^{-2} of oxidizable organic
matter would have to settle from the surface to allow this consumption. Although controversy still surrounds estimates of oceanic
phytoplankton production, rates are generally believed to be at
least ten times this amount. Riley concluded that 90% of the
organic carbon produced by phytoplankton was utilized above 200 m.
Rates of oxygen uptake, phosphorus regeneration and organic carbon
sedimentation below 200 m clearly show that maximum values occur
near the surface (Table 1).

     Studies of the oxygen minimum layer have shown that combined
effects of physical and biological variables determine vertical
profiles of dissolved oxygen in the ocean. The position of the
thermocline which slows the downward settling of organic particles,
the concentration of migrating organisms under the thermocline and
variable rates of vertical mixing all contribute to the shape of the
profile. Smoothed curves which depict a general decrease in in situ
consumption over depth (Wyrtki 1962; Craig 1971) do not reflect these
different processes. However, Wyrtki (1962) fitted an exponential
curve of the form

$$R = R_o e^{-\alpha z} \qquad\qquad\qquad (3)$$

to Riley's (1951) data where R (recalculated as μl O$_2$·liter^{-1}·y^{-1})
is a negative exponential function of depth z with R_o = 150 and
α = 0.0024.

     Craig (1971) confirmed that smoothed distribution patterns of
ΣCO_2, alkalinity, ^{14}C, ^{13}C and dissolved oxygen in the deep sea were
in agreement with a model based on in situ production or consumption.
However, an additional flux of particulate matter was required
together with the oxidation of dissolved organic carbon at specific
depths to account for observed rates of oxygen depletion. Vertical
differences in dissolved oxygen and dissolved organic carbon in the
Pacific were described by a vertical diffusion-advection model when a

consumption term, consistent with Riley's estimates of oxygen uptake, was included. Although this did not provide an estimate of the particulate matter supply or detailed description of fine-scale features of dissolved oxygen profiles, a box model calculation for the gradient of dissolved oxygen concentration between high latitude surface water in the Greenland Sea and the North Atlantic deep water (with residence time derived from ^{14}C data) gave an average value (1 km to the bottom) of 1 μl $O_2 \cdot liter^{-1} \cdot y^{-1}$ for oxygen consumption. Similar calculations for differences in dissolved organic carbon amounted to 36% of this figure with an identical ratio calculated from profiles of dissolved oxygen and organic carbon in the deep Pacific. Craig (1971) concluded that approximately one-third of the oxygen consumed in the deep ocean was used to oxidize dissolved organic carbon, while the remainder (0.67 μl $O_2 \cdot liter^{-1} \cdot y^{-1}$ integrated between 1000 and 4000 m) would oxidize particulate matter. This amounts to 5.3 g $C \cdot m^{-2} \cdot y^{-1}$ consumed below 1000 m.

A similar approach, which also involves mass balance calculations of depositional loss of organic matter, has been based on oxygen flux in the New York Bight (Garside and Malone 1978). It was assumed that net change in dissolved oxygen over time as a result of photosynthesis and respiration in the well-mixed water column with little horizontal transport was proportional to changes in particulate organic carbon. Exchange of dissolved oxygen across the sea surface (F_s) and sediment-water interface (F_b) were assumed to be equal. The distribution of surface temperature, salinity, and dissolved oxygen was used to calculate surface oxygen flux. This was then correlated with suspended particulate organic carbon sampled throughout the year. The intercept of each regression was used as an estimate of oxygen flux to the sediment. Oxygen production by photosynthesis (P) was calculated from ^{14}C-uptake measurements and respiration (R) in the water column was estimated by balancing the steady-state equation

$$F_s + P = R + F_b \qquad\qquad\qquad (4)$$

Monthly estimates of all terms in the equation are presented by Garside and Malone (1978). Daily values for monthly means of P and R were converted to carbon, and the difference between these numbers (as organic carbon produced but not respired, thus assumed to be sedimented) are tabulated in Table 2. Since respiration in the water column was calculated by summation and organic carbon available for deposition was determined by difference, these estimates of depositional flux must be viewed cautiously. However, sedimentation rates in other productive coastal waters (discussed below) can have similar values. Also, Garside and Malone list sources (from sewage, river discharge and sludge dumping) which contribute organic carbon to the Bight in addition to that supplied by phytoplankton. If the organic carbon supply is much greater than that due to phytoplankton production alone, these calculations imply that more carbon is deposited (to be respired or buried in the sediments) than is respired in the water column in this coastal environment.

Table 2. Organic carbon sedimentation during the year in New York Bight calculated from an oxygen budget where mean surface flux F_s, benthic uptake F_b, and photosynthetic production (P) of oxygen were estimated (data from Garside and Malone 1978). Water column respiration (R) was calculated as the difference between production and benthic uptake. Sedimentation (P-R) is assumed to be organic carbon produced but not respired in the water column. Dashes indicate that calculations could not be made either because production was less than respiration or because negative values implied no uptake of oxygen.

Month	P	R	P-R
	$g\ C \cdot m^{-2} \cdot d^{-1}$		
1973 S	1.22	0.48	0.75
O	0.55	0.18	0.37
N	0.90	0.63	0.27
D	0.26	0.30	-
1974 J	0.26	-0.27	-
F	1.70	0.34	1.36
A(10-11)	1.08	0.71	0.37
A(27-29)	1.54	0.57	0.97
M	0.76	-0.05	-
J	3.01	1.93	1.08
J	2.33	1.04	1.29
A	1.99	0.70	1.29
Annual $\overline{X}$	1.39	0.57	0.82
$g\ C \cdot m^{-2} \cdot y^{-1}$	507	208	299

Dissolved nutrients, as well as oxygen, have also been used in mass balance calculations to provide an estimate of the flux of particulate organic carbon sinking in the deep ocean (Eppley and Peterson 1979). If it is assumed that phytoplankton production is regulated by the vertical mixing or transport of dissolved nitrate into the euphotic zone, then at steady-state the downward settling of nitrogen in organic form must be balanced by upward mixing diffusion or biological transport of inorganic nitrogen from deep water. This assumes that nitrification is unimportant in the euphotic zone (Fig. 1). In a closed steady-state system production would depend entirely on the rate of nitrogen regeneration. If losses occur and a steady-state still exists, it must be maintained by the vertical flux of nitrate from deep water.

Eppley et al. (1979) discussed measurements (using ^{15}N labelled substrates) which allow the distinction to be made between phytoplankton ammonium assimilation (production based on regenerated nitrogen) and nitrate assimilation (new production). Estimates of particulate nitrogen export are assumed to be equivalent to measures of nitrate assimilation by phytoplankton. It has also been noted (Dugdale and Goering 1967) that maximum observed rates of primary production are related to the ratio of new production (expressed as nitrate assimilation) and total production (as carbon fixation). Recent studies show the relationship to be non-linear and variable in form (compare Eppley et al. 1979 and Eppley and Peterson 1979). When

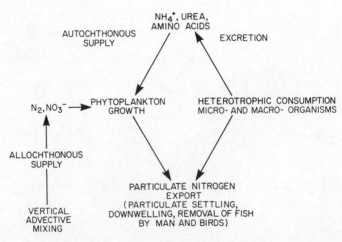

Figure 1. Phytoplankton growth is considered to be due to supply of nitrogen from allochthonous (external, 'new') and autochthonous (internal, 'regenerated') sources. At steady-state, particulate nitrogen export rates must equal the upward flux of dissolved inorganic nitrogen which supports 'new production' (redrawn from Eppley and Peterson 1979).

new production (converted from nitrate uptake to carbon assimilation by use of the Redfield ratio) expressed as a percentage of total production is plotted against total production, data from various areas (oligotrophic to upwelling) form an asymptotic curve. Eppley and Peterson (1979) used the initial linear slope of this curve to predict new production from estimates of total primary production (Table 3). Under the assumptions of steady-state discussed above, organic carbon (nitrogen) sedimentation is considered to be equivalent to this new production.

While these mass balance calculations have limited spatial and temporal resolution, this could be provided by measurements to consider factors which control assimilative fluxes of inorganic nitrogen (Harrison 1978). Eppley and Peterson (1979) noted that new production in shallow inshore waters (< 200 m) was often lower than would have been predicted from estimates of total production, possibly due to the increased supply of ammonia regenerated from sediments. Benthic nitrogen regeneration may supply all of the estimated requirements for phytoplankton production in certain coastal environments (Rowe et al. 1975), although sufficient vertical mixing must occur to deliver ammonia released at the sediment surface to the euphotic zone. Different organisms in the pelagic food web (the relative importance of micro- and macro-heterotrophs, for example) may also affect rates of nitrogen remineralization in the euphotic zone relative to upward transport of dissolved nitrate. The calculated export flux of from 7 to 27 g $C \cdot m^{-2} \cdot y^{-1}$ from the euphotic zone in central Atlantic and Pacific Oceans (Table 3), however, is consistent with rates of particulate organic deposition measured in traps at 200 m (18.5 g $C \cdot m^{-2} \cdot y^{-1}$, Table 1).

A further example of material balances applied to estimate depositional fluxes of particulate matter in the ocean involves the use of particle-reactive radionuclides where a relatively constant supply has occurred over the time-scale set by the half-life of the nuclide. These studies are rare because few areas have a complete inventory for a radionuclide in one area. These requirements are satisfied by ^{210}Pb (measured as excess over ^{226}Ra and products of the ^{238}U decay series).

Benninger (1978) measured and tabulated an inventory of excess ^{210}Pb in Long Island Sound and concluded that the total annual input (mostly atmospheric) was equivalent to the flux required to maintain levels measured in sediment cores from the central part of the Sound. Annual deposition of excess ^{210}Pb (the input needed to maintain concentrations at steady-state against loss by radioactive decay) of 1.5 $DPM \cdot cm^{-2} \cdot y^{-1}$ almost equalled the estimated atmospheric deposition (1.67 $DPM \cdot cm^{-2} \cdot y^{-1}$) expressed per unit of muddy bottom in Long Island Sound. Benninger hypothesized that a rapid removal of soluble ^{210}Pb occurs by absorption onto suspended particles which are trapped in the basin by bottom residual flow. On an annual basis, 90% of the

Table 3.   Estimates of total phytoplankton production (total carbon
           fixation) and new production expressed as a percent of
           total production derived from measures of nitrate assimi-
           lation (and applying the Redfield ratio to convert to
           carbon fixation) in various oceanic water types and areas
           calculated by Eppley and Peterson (1979). Absolute rates
           of new production are assumed to balance the export of
           organic matter (due primarily to sinking) at steady
           state. New production as percent of total production is
           thus equivalent to the percentage of total primary
           production lost by export. Total primary production by
           phytoplankton is estimated from measures of ^{14}C uptake.
           It must be emphasized that many values are extrapolated
           from relatively few measurements over an annual period and
           there is uncertainty that assimilation provides an
           adequate measure of either net or gross primary production
           (see Peterson 1980).

	Total Primary Production $(g\ C \cdot m^{-2} \cdot y^{-1})$	New Production (% of total)	Organic Carbon Export from Euphotic Zone $(g\ C \cdot m^{-2} \cdot y^{-1})$
Oligotrophic waters (central subtropical areas)	26	6	2
Transitional waters between subtropical and subpolar zones	51	13	7
Equatorial divergences and oceanic subpolar zones	73	18	13
Inshore waters	124	30	37
Neritic waters	365	46	168
Antarctic Ocean	325	45	146
Atlantic Ocean	102	26	27
Indian Ocean	84	21	18
Pacific Ocean	55	14	8
Arctic Ocean	1	0	0

210_{Pb} entering the Sound and not lost by radioactive decay, is accumulated in the fine-grained sediments. This estimated deposition rate is similar to that measured in Puget Sound (1.2 $DPM \cdot cm^{-2} \cdot y^{-1}$) from material settled in traps and accumulated in bottom sediments (Lorenzen et al. 1981).

Vertical Concentration Profiles

Rapid decreases in concentration of suspended particulate number, mass, and organic content occur within the upper hundred meters of the ocean with relatively little change at deeper depths. While a heterogeneity of particulate concentrations occurs in the deep sea (Wangersky 1976), which is superimposed on this general decrease in concentration, Sheldon et al. (1972), McCave (1975), Lal and Lerman (1975) and Lerman et al. (1977) have demonstrated decreases in particle number, volume and size with increasing depth. Gordon (1977) confirmed that concentrations of suspended organic carbon and nitrogen were not constant in deep water by demonstrating a continuous decrease with increasing depth. Simple exponential or logarithmic functions can be used to describe these distributions (Table 1).

Except for particles of high lipid content, the specific gravity of particulate matter is greater than that of seawater, and thus gravitational settling should occur. However, eddy diffusion and mixing by other sources of kinetic energy will contribute to the vertical movement of particles in a water column. The relative importance of gravitational settling and the vertical transport of micron-size particles has been discussed by Lerman et al. (1974). In comparison to the micron-size particles, larger and more dense aggregates of particles, like fecal pellets or inorganic mineral grains, have a greater density difference from seawater and gravitational settling predominates in controlling the sinking rate (Komar et al. 1981).

Profiles of the mass, number or volume of suspended particles can be used to estimate vertical flux by applying Stoke's Law to a known size distribution of particles of known density if a settling velocity is measured directly or assumed (Smayda 1970; Riley 1970; Komar et al. 1981). However, as mentioned above, particulate matter is transported by eddy turbulence as well as gravity. Thus as Riley (1970) indicated, if particle concentrations exist at steady-state, the net transport (F) is the difference between the two vertical fluxes where

$$F = SC - A(\Delta C / \Delta Z) \tag{5}$$

The flux due to settling is derived from the sinking rate (S) times concentration (C). Flux due to eddy diffusion is calculated as the product of the coefficient (A) times the mean vertical concentration gradient (ΔC) over a depth interval (ΔZ).

Riley (1970) calculated values of F for the transition zone (100-900 m) in the Sargasso Sea using generally accepted values of A (1 and 0.5 $cm^{-2} \cdot s^{-1}$) and observed values for C and S. He observed sinking velocities for small particles of 1.0 and 0.25 $m \cdot d^{-1}$ in laboratory settling experiments. Much higher rates (10-1000 $m \cdot d^{-1}$) have been observed for larger particles in the laboratory (Smayda 1970; Komar et al. 1981) and organic aggregates (Shanks and Trent 1980) and fecal pellets from gelatinous zooplankton (Bruland and Silver 1981) under natural conditions. However, these values observed by Riley are similar to those estimated for small detrital (0.4 $m \cdot d^{-1}$) and larger calcareous particles (2.2 $m \cdot d^{-1}$) calculated from ^{14}C and ^{55}Fe transport from surface layers to deep (2500 m) water in the Pacific Ocean (Lal and Somayajulu 1977). Estimates of small (1-6 µm) particle settling velocity derived from empirically determined particle volume-size distributions and Stokes Law gave values of 0.1 $m \cdot d^{-1}$ in deep North Atlantic water (> 200 m) with a mean velocity for larger particles (6-15 µm) in the surface layer of 1.3 $m \cdot d^{-1}$ (Lerman et al. 1977). These values may be compared to the calculations by Riley et al. (1949) of a sinking rate of 3 $m \cdot d^{-1}$ for phytoplankton which was necessary to provide a realistic description in a model predicting the vertical distribution of phytoplankton.

The vertical flux rates which Riley (1970) calculated for particulate matter in the Sargasso Sea, using average particulate organic carbon concentrations of 100 and 30 µg $C \cdot liter^{-1}$ at 100 m and 900 m, respectively, correspond to rates summarized for similar depths in Table 1 (40 and 3.5 g $C \cdot m^{-2} \cdot y^{-1}$). Estimates of settling and eddy transport obviously depend on the particulate carbon concentrations used in the calculation. It is useful to consider the magnitude of this dependence since Gordon (1977) reported that long-term (between year) variation in suspended particulate organic carbon may occur in the North Atlantic Ocean. For example, mean concentrations between 0 and 100 m decreased three-fold between 1967-1968 (44.7 µg $C \cdot liter^{-1}$, n = 15) and 1971-1973 (16.0 µg $C \cdot liter^{-1}$, n = 75) for stations south of the Gulf Stream. If the lower value is used in the calculation, vertical particulate flux at 100 m becomes 6.3 g $C \cdot m^{-2} \cdot y^{-1}$. Gordon's (1977) estimate of concentrations at deeper (500-1000 m) depths (5.3 µg $C \cdot liter^{-1}$) is also 16% lower than Riley's value, and flux at 900 m is reduced by a similar amount. Vertical flux calculations by these computations are also dependent upon the gradient in concentration over depth as well as the absolute concentration. Riley (1970) pointed out that the low contribution of eddy flux to total flux of particulate organic carbon in his estimates (8%) was due to the lack of a large gradient over the transition zone.

Nakajima and Nishizawa (1972) found steep gradients in concentrations of particulate carbon within the surface 60-90 m of the Bering Sea which were described by an exponential equation of the form

$$C = C_0 e^{-k(Z-Z_0)} \tag{6}$$

where C_0 and C are concentrations of particulate carbon at the surface (Z_0) and and a subsurface depth (Z), respectively, and k is the coefficient of depth- dependent decrease in concentration. The formulation implies that a constant fraction of suspended particulate matter is lost per unit volume over the depth interval (equivalent to kv where v is the settling velocity) assuming a steady-state in particulate concentration and a constant sinking rate throughout the depth interval. Nakajima and Nishizawa (1972) assumed a sinking rate of 1 $m \cdot d^{-1}$ and from the mean value of k (0.032, range 0.019-0.029) for the Bering Sea estimated that an average of 3.2% (1.9-3.9%) of the suspended particulate carbon would be removed from the surface 60-90 m layer per day. This loss, equivalent to a mean of 88 and a range of 24-169 g $C \cdot m^{-2} \cdot y^{-1}$, amounted to approximately 50% of observed rates of phytoplankton production estimated by ^{14}C uptake (mean 177 and range of 124-230 g $C \cdot m^{-2} \cdot y^{-1}$). If half of the annual primary production is consumed within the upper 60-90 m of the water column, the carbon remaining must either be solubilized and not re-introduced to particulate form or deposited. Average concentrations of particulate carbon at all the stations sampled with the same assumed sinking rate give a mean settling flux of 73 g $C \cdot m^{-2} \cdot y^{-1}$ (41% of the estimated primary production) from the surface layer.

Knowledge of the size distribution of suspended particles provides a more accurate estimate of settling flux for application of the Stokes Law than consideration of total particulate matter concentrations alone. Lal and Lerman (1975) and McCave (1975) reviewed observations which showed that over specified ranges the size distribution of suspended material in seawater may be described by an inverse power function

$$N = a \, d^{-m} \tag{7}$$

where N is the cumulative number of particles greater than size d and a and m are constants calculated for each distribution.

Data presented by McCave (1975) illustrate that two or more curves may be required to describe the relationship between particle number and size. Since the slope coefficient (m) reflects the shape of the size distribution the size range over which the observations are made is crucial. McCave (1975) summarized data from various sources to conclude that values between 2 and 4 were typical for marine suspensions with particles in the 1 to 100 μm size interval. Lerman et al. (1977) obtained slope values of 4 $\pm$ 0.3 at different depths in the North Atlantic Ocean for particles in the size range of 1-6 μm. Values of m > 3 reflect the relative absence of coarser particles in the samples (greater volume in smaller size grades) while values < 3 indicate that more volume exists in larger particle sizes.

McCave (1975) and Lerman et al. (1977) used particle number
distributions to calculate volume distributions in different size
grades and then assigned density values to calculate mass distributions.   The combination of these values with a Stokes settling
velocity was used to compute vertical particle flux as described
above.   The upper limit of particle size distribution is a critical
variable for flux calculations since settling velocity depends on
values of d^2.

McCave (1975) estimated mass from particle size distributions.
Depending on the shape of the distribution, one or two segments of
the curve were fitted for different values of m over the size range
of 1 to 512 μm.   He observed that although concentrations of large
particles in the open ocean are low, their high settling velocity
(calculated to be 105 $m \cdot d^{-1}$ for 400 μm particles) accounts for the
bulk of the vertical flux.   Substitution of values of m of 2.4 and
3.2, typical of shallow (above 100 m) and deep (below 2000 m) water
particle size distributions with a = 2000 and d = 1 to 512 μm in
equation 7, allowed McCave (1975) to calculate that 56.5 and 1.0 g
$C \cdot m^{-2} \cdot y^{-1}$ would be deposited at 100 m and 2000 m, respectively --
rates consistent with those derived from sediment trap observations
(Table 1).   However, 47 to 89% of the calculated flux was attributable to the largest size grades of particles (128 to 512 μm) sinking
at rapid rates (36-105 $m \cdot d^{-1}$).   The calculations show that while
particles smaller than 32 μm constitute almost all of the suspended
mass concentration, these contribute minimally to the settling flux
with sinking rates one to two orders of magnitude lower than large
particles.

Large particles are not adequately sampled by conventional water
bottles because of their low concentration (Sheldon et al. 1972).
However, fecal pellets and large aggregates (marine snow) can be
observed in situ, collected in traps or filtered from large volume
water samples.   Settling rates of fecal pellets from salps and pteropods (2700 and 1800 $m \cdot d^{-1}$) which Bruland and Silver (1981) measured
at sea are the highest recorded to date.   Shanks and Trent (1980)
observed average sinking rates of macroscopic aggregates in situ and
in settling chambers of 68 $m \cdot d^{-1}$.   Settling of aggregates in the
upper 10 to 15 m in Monteray Bay, California, was estimated to remove
from 3 to 5% of the suspended particulate organic carbon (and 4 to 22%
of the particulate nitrogen) per day.   This amounted to an average
deposition rate of 132 g $C \cdot m^{-2} \cdot y^{-1}$ and 21.5 g $N \cdot m^{-2} \cdot y^{-1}$.   Sediment
trap collections at 50 m during the same year in this region provided
an estimate of deposition of 158 g $C \cdot m^{-2} \cdot y^{-1}$ (20.8 g $N \cdot m^{-2} \cdot y^{-1}$)
(Knauer et al. 1979).

Bishop et al. (1977, 1978 and 1980) used a large volume (1-20 m^3)
in situ filtration system to collect particulate matter from various
depths to 400 m in the equatorial and southeast Atlantic Ocean and
1500 m in the Panama Basin.   Cumulative size distributions of numbers

of different particle types (foraminifera, diatoms, fecal matter,
fecal pellets) plotted in the form of equation 7 were calculated
from microscopic observations of particles trapped on filter pads.
Particle size distributions and bulk chemical analysis of size
fractions were used to calculate mass and chemical fluxes from two
settling models based on Stoke's Law. Numbers of large particles
were smaller than would have been predicted by extrapolating from
Coulter Counter measurements for 1 to 100 µm particles, as used by
McCave (1975). In spite of this, at 388 m in the equatorial
Atlantic Ocean, fecal matter and pellets (753 µm) were responsible
for over 98% of the total vertical mass flux (11.3 g $C \cdot m^{-2} \cdot y^{-1}$) with
particles between 1 µm and 53 µm estimated to contribute less than
2% of this amount (based on a calculated sinking velocity of less
than 0.2 $m \cdot d^{-1}$). After comparison of the vertical organic carbon
flux with estimates of ^{14}C primary production, Bishop et al.
concluded that 87% of the carbon, 91% of the nitrogen and 94% of the
phosphorus fixed above 388 m did not settle below this depth.

Bishop et al. (1977) also analyzed material collected on
filters for major ions (Na, K, Mg, Ca and Sr), iron, carbonate,
silica and various radioisotopes. Sinking material was composed of
two major constituents besides organic matter.   Carbonate (109
mmoles $CaCO_3 \cdot m^{-2} \cdot y^{-1}$) and opal (121 mmoles $Si \cdot m^{-2} \cdot y^{-1}$) were transported in the form of coccoliths and diatom fragments. Calculated
flux of ^{210}Pb (1.8 $DPM \cdot cm^{-2} \cdot y^{-1}$) at 388 m exceeded the rate of
atmospheric supply to the sea surface (0.6 $DPM \cdot cm^{-2} \cdot y^{-1}$) estimated
for the tropical North Atlantic (cited from Bason 1975) but the flux
rate is similar to values observed in Long Island Sound and Puget
Sound mentioned above.

Similar observations and flux calculations at 400 m off the
coast of Africa in the southeast Atlantic Ocean at stations exhibiting various levels of primary production (Bishop et al. 1978) yielded
flux rates for > 53 µm particles from 11.1 g $C \cdot m^{-2} \cdot y^{-1}$ at the most
productive site (highest suspended particulate concentrations) to
0.5 g $C \cdot m^{-2} \cdot y^{-1}$ in a region of minimum suspended matter. Over 90% of
the organic matter produced in the euphotic zone was estimated to be
recycled above 400 m in areas of high production (> 180 g $C \cdot m^{-2} \cdot y^{-1}$)
while recycling efficiency was greater than 99% in areas of low
productivity (< 40 g $C \cdot m^{-2} \cdot y^{-1}$). The observations clearly show that
rates of vertical flux depend on the supply of products of primary
production from surface waters.

The possible importance of horizontal variability in primary
production in surface waters in determining flux rates measured
vertically at various depths was also indicated by observations in
the Panama Basin (Bishop et al. 1980). Flux measurements calculated
as described above decreased from maximum rates at 40 m (58.5 g
$C \cdot m^{-2} \cdot y^{-1}$) to minimum rates (0.05-1.8 g $C \cdot m^{-2} \cdot y^{-1}$) below 1100 m.
However, sediment traps deployed 100 m above the bottom (at 2570 m)

at the time of collection of large volume filtration samples, yielded
a mass flux rate 40 times higher than that calcuated for particles
collected by filtration. Particles settled into traps may have been
derived from nutrient-rich water upwelled along the Peruvian coast
while in situ filtration of particles was carried out at a time when
tropical water with low rates of phytoplankton production was present
at the surface in the sampling area. Differences in the ratio of
organic carbon: $CaCO_3$ in trapped (0.6) and filtered (5) material
(although expected due to oxidation of organic matter settling
between 1500 m and 2570 m) might also indicate that material settled
in traps was not entirely derived from vertically sinking particles.
Resuspension of previously deposited material could enhance apparent
deposition in traps suspended close to the bottom.

A final example of the use of analyses of suspended matter to
estimate vertical flux of particulate matter involves measurements
of pairs of nuclides in a radioactive decay series where the parent
nuclide is soluble in seawater while the daughter nuclide is not.
Changes in the relative concentrations of a parent-daughter pair are
believed to arise from the removal of the insoluble decay product
through deposition. Broecker et al. (1973) and Minagawa and
Tsunogai (1980) summarized studies which have used measurements of
$^{234}Th - {}^{238}U$, $^{230}Th - {}^{234}U$, and $^{210}Pb - {}^{226}Ra$ to estimate the
residence time of particle reactive substances in surface ocean
waters. If it is assumed that ^{234}Th is uniformly distributed within
suspended particulate matter, the residence time for thorium in
surface waters (100-200 m) varies between 0.12 and 0.7 years.
Unlike nitrogen and phosphorus, which are extensively recycled
within the euphotic zone before removal by deposition, particle
reactive substances such as thorium are not regenerated but are
continuously transported downwards by particle sedimentation.

Geochronology

Sedimentation can be studied after the fact by the presence of
buried marker substances in undisturbed sediment cores. These
substances provide an estimate of sedimentation if the chronology is
known (i.e., if horizons can be dated). McCave (1975) has summarized
numerous observations of localized distributions of wind-blown
quartz, clay minerals and specific biogenic compounds. The presence
of organic compounds beneath areas of high surface production
provides evidence for the rapid transport of particulate matter
through the water column and incorporation in sediments. The
presence of annual lamina (varves) in coastal sedimentary basins
(Soutar et al. 1977) demonstrates that the history of depositional
events may be preserved in certain sedimentary environments.

The preservation of organic matter in sediments reflects the
relative rates of supply and loss after deposition by aerobic and
anaerobic oxidation. Müller and Suess (1979) compared rates of

organic carbon accumulation in various oceanic and coastal waters
from cores dated by isotope stratigraphy, ^{14}C dating or biostrati-
graphy, and observed that the fraction of organic carbon derived
from primary production (y, in per cent) was related to the rate of
sediment accumulation (x, in cm·(1000 y)$^{-1}$) by the equation

$$y = 0.03 \ x^{1.30} \tag{8}$$

Approximately 0.01% of organic carbon produced by phytoplankton is
accumulated annually in slowly accumulating pelagic sediments (Central Pacific) with < 1% organic carbon content. In areas of higher
production where sediments accrete rapidly (a few cm·(1000 y)$^{-1}$),
burial of organic carbon is equivalent to 10 to 20% of the annual
phytoplankton carbon production. Depositional supply must exceed
rates of utilization to maintain high levels of particulate organic
matter in the sediments even though anaerobic decomposition may
occur, primarily through reduced inorganic sulfur pathways (Howarth
and Teal 1980).

Williams et al. (1978) estimated the residence time of pools of
organic carbon in surface and deep water and abyssal red clay sediments of the Central and Northeast Pacific Ocean by measuring [14]C
activity. Dissolved organic carbon (0.45 µm) in the water column,
has apparent ages of 740 y BP at the surface and 2500 to 3500 y BP
at 2000 m. The ^{14}C activity of suspended particulate organic carbon
at 2000 m, however, was only 10% less than that at the surface. A
rapid sinking of some fraction of suspended particulate matter would
explain this distribution and account for the high ^{14}C activity (and
low apparent age) associated with the upper 4 to 8 cm of sediment.
The penetration of carbon of relatively recent origin below the
surface mixed layer of sediment shows that some organic carbon (7-13%
of the total) was mixed into the bottom independently from the clay
minerals and the associated organic matter. The residence time and
loss of organic carbon before burial could have been calculated if
measures of ^{14}C activity in material sedimenting to the sediment
surface had been made.

Isotopes such as [230]Th, ^{210}Pb, and [239,240]Pu also accumulate in
bottom deposits and these particle-reactive isotopes may be used to
trace transport and storage of fine-grained sediments where inven-
tories can be established. For example, Benninger and Krishnaswami
(1981) measured ^{210}Pb and Pu-isotopes in sediments of the inner New
York Bight. Since the inventory exceeded amounts supportable by
atmospheric deposition, they assumed that lateral redistribution must
have occurred. Studies of the flux of particle reactive substances
such as radionuclides which become associated with particulate matter
and are not recycled provide a direct assessment of transfer of
material from the water column to the sediment and permit an estimate
of residence time in the water column.

Sedimentation

The use of various types of traps or collectors on fixed moorings to directly measure the downward flux of particulate matter in freshwater and marine environments has recently been summarized by Bloesch and Burns (1980), Reynolds et al. (1980) and Blomqvist and Håkanson (1981). Straight-walled cylinders or funnels exposed in small stratified lakes, where horizontal water motion was minimal, collected downward settling particles in direct proportion to their mouth opening area (Kirchner 1975). Rates of horizontal advective transport exceed vertical settling rates by three-to-four orders of magnitude in marine waters. Here the design of collectors has an important influence on the amount and particle size of material entering and retained in traps (Hargrave and Burns 1979; Gardner 1980a, b; Blomqvist and Kofoed 1981).

Gardner's laboratory flume tank experiments demonstrated that water exchange rather than vertical settling carries particles into traps exposed to horizontal currents of a few centimeters per second or more. When cylinders with appropriate dimensions defined by the aspect (height:mouth opening area) ratio are used, a non-turbulent zone forms above the bottom into which particles settle and accumulate. Bloesch and Burns (1980) concluded that values of at least five are required for calm waters and greater than ten for turbulent conditions in large lakes and the ocean. Other attempts to minimize turbulent eddies around and within traps have consisted of placing baffles across the mouth opening (Soutar et al. 1977) or by attaching collectors to a free floating mooring (Staresinic et al. 1978).

Recent studies have compared rates of particle flux into traps with independent measures of particulate matter deposition to provide a field assessment for trap collection efficiency. Minagawa and Tsunogai's (1980) observations of differential decay in the nuclide pair ^{234}Th and ^{238}U in suspended matter in the coastal waters of Funka Bay were compared with settled material collected in traps. Evidence that suspended particles remove thorium from seawater was provided by the similar specific activity of ^{234}Th associated with settled and suspended particles. However, deposition rates of material collected over short periods in traps was only 8 to 58% of that calculated on the basis of thorium removal. Poor trapping efficiency might have occurred because cylinders (aspect ratio of 4:1) with funnels placed in the mouth opening were used to collect settling material. The actual collection area of such a trap in non-stagnant water cannot be determined (Hargrave and Burns 1979).

Lorenzen et al. (1981) also compared the amounts of unsupported ^{210}Pb deposited in cylindrical traps (aspect ratios of 3:1) in Dabob Bay, Puget Sound, with the steady-state accumulation rate of this isotope in underlying sediments. An efficiency of 118 + 13% with an 8 to 10% sampling precision indicated that the traps collected a

representative sample of deposited material in this environment.
Knauer et al. (1979) also observed average particulate ^{210}Pb flux into
traps under upwelling and other conditions (2.0 and 0.2 DPM·cm^{-2}·y^{-1}
respectively) which are comparable to cited estimates of the average
atmospheric import to the northeast Pacific (0.5 - 1.0 DPM·cm^{-2}·y^{-1}).

Bloesch and Burns (1980) reviewed other, non-design related,
problems associated with the use of traps to measure vertical flux of
particulate matter. Length of exposure and decomposition of material
during the collection period are recognized to have significant
effects. Short-term variations in depositional events over days or
weeks which affect quantity and quality of settled material, for
example, cannot be resolved if traps are exposed over month-long
intervals. In addition, Iturriaga (1979) observed that decomposition
of settled material resulted in a significant loss of organic matter
(up to 35% per day for phytoplankton at 20°C) which was temperature
dependent. Heterotrophic consumption is also directly related to the
organic content of settled material (Hargrave 1978; Iseki et al.
1980). Either short-interval exposure periods or the addition of
preservatives such as sodium azide to traps (Honjo 1980) can be used
to minimize or avoid changes in settled material due to decomposition.

Collection of material in traps offers an advantage over other
methods for estimation of vertical flux in that deposited material
is collected. While fragile particle aggregates and fecal pellets
may disintegrate on collection, those that do remain intact, along
with cellular or skeletal remains of plankton, can be enumerated.
Wiebe et al. (1976), Knauer et al. (1979), Rowe and Gardner (1979),
Honjo (1980), and Iseki (1981a), for example, used size fractionation
or direct counts to estimate the proportion of the organic carbon
flux at their oceanic study sites attributable to deposition of fecal
pellets and fecal matter (> 63 μm). Numbers of pellets deposited
(10^2 - 10^3·m^{-2}·d^{-1}) were lower than those observed in a coastal bay
(10^4 - 10^6·m^{-2}·d^{-1}) (Hargrave et al. 1976). Salps may also be
responsible for producing a large fraction of the material deposited
in certain areas (Wiebe et al. 1979; Iseki 1981b).

Skeletal remains of planktonic organisms or body parts such as
vertebrae, scales or teeth of larger animals in settled material may
be counted and analyzed to provide indirect information about production processes in overlying or adjacent water masses (Soutar et al.
1977). Deuser et al. (1981b) counted separate species of planktonic
foraminifera recovered from sediment traps exposed throughout the
year in the deep water near Bermuda to show that species number,
size, weight, and oxygen and carbon isotope ratios reflected seasonal
changes in hydrographic conditions above the thermocline. The rapid
movement (within a few weeks) of foraminifera and other biogenic
particles from the surface to 3200 m (equivalent to sinking rates on
the order of 10^2 m·d^{-1}) was demonstrated by the synchrony between
changes in surface production and deposition rates over two years

(Deuser et al. 1981a). The flux of small particles and inorganic material was also highest during the early months of each year in phase with increased phytoplankton production. This confirms that biologically produced aggregates which settle rapidly account for much of the vertical flux of particulate matter.

Isotopic and elemental fluxes have been determined directly by analysis of trap contents. For example, Brewer et al. (1980) used elemental ratios, normalized to aluminum concentration, to calculate flux of elements assumed to be derived from atmospheric input (K, Ti, Al, La, V, Co and ^{234}Th) and those associated with biogenic particles (Ca, Sr, Mg, Si, Ba, ^{226}Ra, U and I). The ratio of reactive elements such as Mn, Cu, Fe, Sc and ^{230,234}Th relative to alumnium indicative of scavenging from the water column increased with depth in the Sargasso Sea and off Barbados.

These and other recent studies have demonstrated that the depth at which traps are suspended (relative to the surface and bottom) and phytoplankton production rates are critical factors which determine observed rates of particle deposition. Resuspension of sediments can enhance deposition rate in traps suspended near the bottom. Also, the degree of stratification, as measured by the depth of the mixed layer above a thermocline, is important in determining the nature and production of planktonic communities near the surface. Hargrave (1975) used limited data from different studies available at the time to derive an empirical relationship

$$C_f = 108.9 + 0.27 \ C_p - 6.24 \ Z_m \quad (MR^2 = 0.79) \quad (n = 8) \qquad (9)$$

between annual carbon sedimentation (C_f, g $C \cdot m^{-2} \cdot y^{-1}$), phytoplankton production (C_p, g $C \cdot m^{-2} \cdot y^{-1}$) and mixed-layer depth (Z_m, m) during periods of stratification. Only eight studies over a limited depth range were compared but most of the variance in the regression was associated with differences in annual phytoplankton production ($MR^2 = 0.55$). The same data plotted with C_f as a function of the ratio of $C_p:Z_m$ gave the equation

$$C_f = 4.9 + 3.9 \ (C_p/Z_m) \quad (r^2 = 0.92) \tag{10}$$

A similar comparison of observations to 20 m depth in a coastal embayment where monthly rates of carbon sedimentation and phytoplankton production per mixing depth were determined yielded an equation with a low correlation coefficient ($r^2 = 0.22$) (Hargrave 1980). Advective exchange through tidal transport may remove a significant fraction of phytoplankton production from the embayment where these observations were made. Platt and Conover (1971) calculated that 58% of daily phytoplankton production was exported over a 24 h period. A more significant correlation ($r^2 = 0.53$) existed when sedimentation measured as nitrogen flux and phytoplankton production rates were compared.

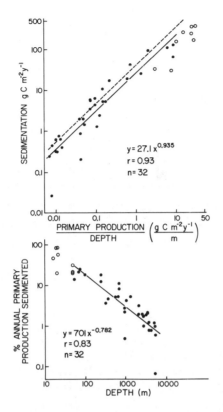

Figure 2. Upper panel: Annual sedimentation of particulate organic carbon compared with phytoplankton production and water column depth in various studies summarized by Suess (1980). Measurements of sedimentation in areas where depth was < 50 m (plotted as open circles) were not included in the regression calculation (solid line). Dotted line indicates the regression line described by equation 11 derived by Suess (1980). Lower panel: Annual sedimentation rates as above expressed as a percentage of annual carbon production plotted against depth.

Suess (1980) reviewed published observations of carbon sedimentation in traps suspended in different marine locations. He expressed organic carbon flux (C_f) at any depth below the euphotic zone as a function of annual phytoplankton production (C_p), both as g $C \cdot m^{-2} \cdot y^{-1}$, and depth (Z) (m) by the regression

$$C_f = C_p (0.0238Z - 0.212)^{-1} \quad (r^2 = 0.79) \quad (n = 33) \quad (11)$$

Data used to derive this relationship, including observations at depths shallower than 50 m, were used above (equation 2) to evaluate the dependence of organic carbon sedimentation on depth alone. A power curve ($r^2 = 0.74$, n = 40) adequately described the data (Table 1).

Since primary production and total water column depth interact to determine the supply of organic carbon available for deposition at any depth, data reviewed by Suess (1980) were replotted to express sedimentation as a function of the ratio of production:depth (Fig. 2). The correlation coefficient was increased by logarithmic transformation. The same data were recalculated to express sedimentation as a percentage of that produced plotted against depth. The relationship (an inverse power curve) quantifies the conventional view that carbon sedimentation below 1000 m amounts to a few percent or less of that produced at the surface.

Suess (1980) analyzed the data used to derive Figure 2 further by calculating the difference in carbon deposition at various depths from that produced at the surface. By assuming a settling rate of particulate matter of 100 $m \cdot d^{-1}$, he transformed depth into time and calculated a second-order rate constant for carbon loss through _in situ_ consumption during settling. Oxygen uptake per volume was calculated by assuming a Redfield ratio to convert from carbon loss. Rates were similar to those derived by Riley (summarized in Table 1) and from the circulation models discussed above. Vertical organic carbon fluxes determined by sediment trap studies are thus consistent with potential supplies from surface phytoplankton production and with previous estimates of depth-dependent consumption.

ACKNOWLEDGMENTS

I thank Drs. D. C. Gordon, Jr., G. C. Harding, G. Harrison and G. A. Riley, for reading and providing comments on the manuscript.

REFERENCES

Benninger, L. K. 1978. ^{210}Pb balance in Long Island Sound. Geochim. Cosmochim. Acta 42: 1165-1174.

Benninger, L. K., and S. Krishnaswami. 1981. Sedimentary process in the inner New York Bight: evidence from excess ^{210}Pb and ^{239,240}Pu. Earth Planet. Sci. Lett. 53: 158-174.

Bishop, J. K., J. M. Edmond, D. R. Ketten, M. P. Bacon, and W. B. Silkers. 1977. The chemistry, biology, and vertical flux of particulate matter from the upper 400 m of the equatorial Atlantic Ocean. Deep-Sea Res. 24: 511-548.

Bishop, J. K., D. R. Ketten, and J. M. Edmond. 1978. The chemistry, biology and vertical flux of particulate matter from the upper 400 m of the Cape Basin in the southeast Atlantic Ocean. Deep-Sea Res. 25: 1121-1161.

Bishop, J. K. B., R. W. Collier, D. R. Kettens, and J. M. Edmonds. 1980. The chemistry, biology and vertical flux of particulate matter from the upper 1500 m of the Panama Basin. Deep-Sea Res. 27A: 615–640.

Bloesch, J., and N. M. Burns. 1980. A critical review of sedimentation trap technique. Schweiz. Z. Hydrol. 42: 15–55.

Blomqvist, S., and L. Håkanson. 1981. A review on sediment traps in aquatic environments. Arch. Hydrobiol. 91: 101–132.

Blomqvist, S., and C. Kofoed. 1981. Sediment trapping – a sub-aquatic experiment. Limnol. Oceanogr. 26: 585–590.

Brewer, P. G., Y. Nozaki, D. W. Spencer, and A. P. Fleer. 1980. Sediment trap experiments in the deep North Atlantic: isotopic and elemental fluxes. J. Mar. Res. 38: 703–741.

Broecker, W. S., A. Kaufman, and R. M. Trier. 1973. The residence time of thorium in surface seawater and its implications regarding the rate of reactive pollutants. Earth Planet. Sci. Lett. 20: 35–44.

Bruland, K. W., and M. W. Silver. 1981. Sinking rates of fecal pellets from gelatinous zooplankton (salps, pteropods, doliolids). Mar. Biol. 63: 295–300.

Craig, H. 1971. The deep metabolism in oxygen consumption in abyssal ocean water. J. Geophys. Res. 76: 5078–5086.

Deuser, W. G., E. H. Ross, and R. F. Anderson. 1982a. Seasonality in the supply of sediment to the deep Sargasso Sea and implications for the rapid transfer of matter to the deep ocean. Deep-Sea Res. 28: 495–505.

Deuser, W. G., E. H. Ross, C. Hembleben, and M. Spindler. 1981b. Seasonal changes in species composition, number, mass, size, and isotopic composition of planktonic foraminifera settling into the deep Sargasso Sea. Palaeogeogr. Palaeoclimatol. Palaeoecol. 33: 103–127.

Dugdale, R. C., and J. J. Goering. 1967. Uptake of new and regenerated forms of nitrogen in primary productivity. Limnol. Oceanogr. 12: 196–206.

Eppley, R. W., and B. J. Peterson. 1979. Particulate organic matter flux and planktonic new production in the deep sea. Nature 282: 677–680.

Eppley, R. W., E. H. Renger, and W. G. Harrison. 1979. Nitrate and phytoplankton production in southern California coastal waters. Limnol. Oceanogr. 24: 483–494.

Gardner, W. D. 1980a. Sediment trap dynamics and calibration: a laboratory evaluation. J. Mar. Res. 38: 17–39.

Gardner, W. D. 1980b. Field assessment of sediment traps. J. Mar. Res. 38: 41–52.

Garside, C., and T. C. Malone. 1978. Monthly oxygen and carbon budgets of the New York Bight Apex. Estuarine Coastal Mar. Sci. 6: 93–104.

Gordon, D. C., Jr. 1977. Variability of particulate organic carbon and nitrogen along the Halifax-Bermuda section. Deep-Sea Res. 24: 257–270.

Gordon, D. C., Jr., P. J. Wangersky, and R. W. Sheldon. 1979.
    Detailed observations on the distribution and composition of
    particulate organic material at two stations in the Sargasso
    Sea. Deep-Sea Res. 26: 1083-1092.
Hargrave, B. T. 1975. The importance of total and mixed-layer depth
    in the supply of organic material to bottom communities. Symp.
    Biol. Hung. 15: 157-165.
Hargrave, B. T. 1978. Seasonal changes in oxygen uptake by settled
    particulate matter and sediments in a marine bay. J. Fish.
    Res. Board Can. 35: 1621-1628.
Hargrave, B. T. 1980. Factors affecting the flux of organic matter
    to sediments in a marine bay, pp. 243-263. In: K. R. Tenore
    and B. C. Coull [eds.], Marine Benthic Dynamics. Univ. S.
    Carolina Press, Columbia.
Hargrave, B. T., and N. M. Burns. 1979. Assessment of sediment
    trap collection efficiency. Limnol. Oceanogr. 24: 1124-1136.
Hargrave, B. T., G. A. Phillips, and S. Taguchi. 1976. Sedimenta-
    tion measurements in Bedford Basin, 1973-1974. Fish. Mar. Ser.
    Rept. 608.
Harrison, W. G. 1978. Experimental measurements of nitrogen
    remineralization in coastal waters. Limnol. Oceanogr. 23:
    684-694.
Honjo, S. 1980. Material fluxes and modes of sedimentation in the
    mesopelagic and bathypelagic zones. J. Mar. Res. 38: 53-97.
Howarth, R. W., and J. M. Teal. 1980. Energy flow in a salt marsh
    ecosystem: the role of reduced inorganic sulfur compounds. Am.
    Nat. 116: 862-872.
Hutchinson, G. E. 1938. On the relation between the oxygen deficit
    and the productivity and typology of lakes. Int. Rev.
    Hydrobiol. 36: 336-355.
Iseki, K. 1981a. Vertical transport of particulate organic matter
    in the deep Bering Sea and Gulf of Alaska. J. Oceanogr. Soc.
    Jpn. 37: 101-108.
Iseki, K. 1981b. Particulate organic matter transport to the deep
    sea by salp fecal pellets. Mar. Ecol. Prog. Ser. 5: 55-60.
Iseki, K., F. Whitney, and C. S. Wong. 1980. Biochemical changes
    of sedimented matter in sediment trap in shallow coastal
    waters. Bull. Plankton Soc. Jpn. 27: 27-36.
Iturriaga, R. 1979. Bacterial activity related to sedimenting
    particulate matter. Mar. Biol. 55: 157-169.
Kirchner, H. B. 1975. An evaluation of sediment trap methodology.
    Limnol. Oceanogr. 20: 657-661.
Knauer, G. A., J. H. Martin, and K. W. Bruland. 1979. Fluxes of
    particulate carbon, nitrogen, and phosphorus in the upper water
    column of the northeast Pacific. Deep-Sea Res. 26: 97-108.
Komar, P. D., A. P. Morse, L. F. Small, and S. W. Fowler. 1981. An
    analysis of sinking rates of natural copepod and euphausiid
    pellets. Limnol. Oceanogr. 26: 172-180.
Lal, D., and A. Lerman. 1975. Size spectra of biogenic particles
    in ocean water and sediments. J. Geophys. Res. 80: 423-430.

Lal, D., and B. L. K. Somayajulu. 1977. Particulate transport of radionuclides ^{14}C and ^{55}Fe to deep waters in the Pacific Ocean. Limnol. Oceanogr. 22: 55-59.

Lerman, A., K. L. Carder, and P. R. Betzer. 1977. Elimination of five suspensoids in the oceanic water column. Earth Planet. Sci. Lett. 37: 61-70.

Lerman, A., D. Lal, and M. F. Dacey. 1974. Stokes settling and chemical reactivity of suspended particles in natural waters, pp. 17-47. In: R. J. Gibbs [ed.], Suspended Solids in Water. Plenum Press.

Lorenzen, C. J., F. R. Shuman, and J. T. Bennett. 1981. In situ calibration of a sediment trap. Limnol. Oceanogr. 26: 580-585.

McCave, I. N. 1975. Vertical flux of particles in the ocean. Deep-Sea Res. 22: 491-502.

Minagawa, M., and S. Tsunogai. 1980. Removal of ^{234}Th from a coastal sea: Funka Bay, Japan. Earth Planet. Sci. Lett. 47: 51-64.

Müller, P. J., and E. Suess. 1979. Productivity, sedimentation rate, and sedimentary organic matter in the oceans. I. Organic carbon preservation. Deep-Sea Res. 26: 1347-1362.

Nakajima, K., and S. Nishizawa. 1972. Exponential decrease in particulate carbon concentration in a limited depth interval in the surface layer of the Bering Sea, pp. 495-505. In: A. Y. Takenouti et al. [eds.], Biological Oceanography of the Northern North Pacific Ocean. Idemitsu Shoten, Toxyo, Japan.

Peterson, B. J. 1980. Aquatic primary productivity and the $^{14}C-CO_2$ method: a history of the productivity problem. Ann. Rev. Ecol. Syst. 11: 359-385.

Platt, T., and R. J. Conover. 1971. Variability and its effect on the 24 h chlorophyll budget of a small marine basin. Mar. Biol. 1: 52-65.

Reynolds, C. S., S. W. Wiseman, and W. D. Gardner. 1980. Aquatic sediment traps and trapping methods. Freshw. Biol. Assoc. Occas. Publ. 11.

Riley, G. A., H. Stommel, and D. F. Bumpus. 1949. Quantitative ecology of the plankton of the western North Atlantic. Bull. Bingham Oceanogr. Collect. Yale Univ. 12: 1-169.

Riley, G. A. 1951. Oxygen, phosphate and nitrate in the Atlantic Ocean. Bull. Bingham Oceanogr. Collect. Yale Univ. 13: 1-126.

Riley, G. A. 1970. Particulate organic matter in sea water. Adv. Mar. Biol 8: 1-118.

Rowe, G. T., C. H. Clifford, K. L. Smith, Jr., and P. C. Hamilton. 1975. Benthic nutrient regeneration and its coupling to primary productivity in coastal waters. Nature 255: 215-217.

Rowe, G. T., and W. D. Gardner. 1979. Sedimentation rates in the slope water of the northwest Atlantic Ocean measured directly with sediment traps. J. Mar. Res. 37: 581-668.

Shanks, A. L., and J. D. Trent. 1980. Marine snow: sinking rates and potential role in vertical flux. Deep-Sea Res. 27: 137-143.

Sheldon, R. W., A. Prakash, and W. H. Sutcliffe. 1972. The size distribution of particles in the ocean. Limnol. Oceanogr. 17: 327-340.

Smayda, T. 1970. The suspension and sinking of phytoplankton in the sea. Oceanogr. Mar. Biol. Ann. Rev. 8: 353-414.

Soutar, A., S. A. Kling, P. A. Crill, E. Duffrin, and K. W. Bruland. 1977. Monitoring the marine environment through sedimentation. Nature 266: 136-139.

Staresinic, N., G. T. Rowe, D. Shaughnessey, and A. J. Williams. 1978. Measurement of the vertical flux of particulate matter with a free-drifting sediment trap. Limnol. Oceanogr. 23: 559-563.

Suess, E. 1980. Particulate organic carbon flux in the oceans - surface productivity and oxygen utilization. Nature 288: 260-263.

Wangersky, P. J. 1976. Particulate organic carbon in the Atlantic and Pacific Oceans. Deep-Sea Res. 23: 457-465.

Wiebe, P. H., S. H. Boyd, and C. Winget. 1976. Particulate matter sinking to the deep-sea floor at 2000 m in the Tongue of the Ocean, Bahamas, with a description of a new sedimentation trap. J. Mar. Res. 34: 341-354.

Wiebe, P. H., L. P. Madin, L. R. Haury, G. R. Harbison, and L. M. Philbin. 1979. Diel vertical migration by Salpa aspera and its potential for large-scale particulate organic matter transport to the deep-sea. Mar. Biol. 53: 249-255.

Williams, P. M., M. C. Stenhouse, E. M. Druffel, and M. Koide. 1978. Organic ^{14}C activity in an abyssal marine sediment. Nature 276: 698-701.

Wyrtki, K. 1962. The oxygen minima in relation to ocean circulation. Deep-Sea Res. 9: 11-23.

MEASUREMENT OF BACTERIOPLANKTON GROWTH IN THE SEA AND ITS REGULATION

BY ENVIRONMENTAL CONDITIONS

Farooq Azam and Jed A. Fuhrman*

Institute of Marine Resources, A-018
Scripps Institution of Oceanography
University of California, San Diego
La Jolla, California 92093
*Marine Sciences Research Center
State University of New York
Stony Brook, L. I., New York 11794

INTRODUCTION

The last few years have witnessed something of a revolution in our view of the role of bacteria in the marine environment. It is now being recognized that bacteria play a quantitatively significant role in the flow of energy and matter in marine ecosystems. The central theme of this "new view" is based on two lines of evidence: (1) bacterial biomass is a significant part of the total biomass in the oceans (Hobbie et al. 1977), and (2) the bacterioplankton is, metabolically, a highly active component of marine biota (Pomeroy 1974; Williams 1981; Azam and Hodson 1977; Hagström et al. 1979; Fuhrman and Azam 1980, 1982). This thinking is diametrically opposed to the conventional wisdom which portrays bacterioplankton as quantitatively trivial and, even today, insists that dormancy due to insufficient nutrients is the dominant physiological state of most bacteria.

Microscopical evidence for the existence of a large bacterial biomass (Francisco et al. 1973; Zimmerman and Meyer-Reil 1974; Hobbie et al. 1977) is definitive; it is already accepted as a fact by marine microbiologists and other marine scientists alike. However there has been considerable skepticism in regard to the quantitative significance of bacterioplankton production in the energetics and trophic-dynamics of marine foodwebs. This skepticism is justified. The methodology for quantification of bacterioplankton growth is quite recent and still not demonstrably precise. It is remarkable,

179

however, that in the last five years no less than six independent
methods have been proposed (as opposed to the widespread acceptance
of essentially one single method, the "^{14}C-method", for phytoplankton
production). It is also worth noticing that all methods indicate
high production rates by bacterioplankton.

The purpose of this communication is to evaluate the conceptual
bases of the current methods for measuring bacterioplankton growth,
and also to discuss some recent estimates of bacterial production
rates in the context of the role of bacteria in the flow of matter
and energy in coastal marine ecosystems, and factors affecting it.

Definition of Growth

The definition of growth is a significant conceptual problem,
and there is no simple solution even when considering the growth of
pure cultures. Bacterial growth may be variously defined as increase
in cell numbers, cell biomass, protein, DNA, or RNA. These measurements can only yield identical results during balanced growth wherein, by definition, all cell constituents increase at the same rate.
The doubling time of any cell constituent would then be the cell
doubling time.

In batch cultures, and probably even more so in nature, growth
does not remain balanced for long. Cellular composition changes to
adjust to changing environmental conditions such as nutrient limitation, accumulation of inhibitory metabolites, change in pH, etc.
Cell size also changes with varying growth conditions. Bacteria
remain in balanced growth only if their environment is unchanged (an
artificial condition created experimentally in the chemostat). We
know now that there are measurable diel variations in the concentrations of some bacterial nutrients (Burney et al. 1981, 1982) and
that there are concomitant variations in assemblage frequency of
dividing cells (FDC) and hence growth rate (Hagström and Larsson,
this volume). Furthermore, Hanson and Lowery (1983) have presented
evidence for unbalanced growth in Antarctic microplankton. Balanced
growth may thus be too elusive in assemblages of marine bacteria to
be used as an a priori basis for growth rate measurements.

If the growth is unbalanced then its definition in terms of
macromolecular synthesis will be imprecise. This is because the
relationships between macromolecular syntheses and growth rate vary
with the growth rate itself. In an assemblage of bacteria, different
species (and even different individuals of a given species) might
grow at different growth rates (some dormant; Stevenson 1978). Thus,
a single relationship may not describe the macromolecular synthesis
in terms of growth of the assemblage. The models of growth should
also take into account that bacteriovores may keep the bacterial
density in check (Azam et al. 1983) and maintain a quasi-steady-state
between growth and death.

Since no single model for bacterial growth accurately describes the state of growth of assemblages of bacteria in the sea, it would appear we have to accept that it is not possible to define growth precisely at the present time. Various biochemical manifestations of growth (such as DNA, RNA, protein and biomass syntheses or increase in cell numbers, etc.) may be employed as indicators of growth and interpreted according to the ecological scenario at hand. The appropriateness of a method for measuring growth may depend on the reason for measuring bacterial growth.

Reasons for Measuring Bacterial Growth

Studies to quantify the flux of material and energy through bacteria in marine foodwebs have become possible due to the recent development of methods to estimate secondary production. Knowing the rate of bacterial secondary production and assimilation efficiency one can calculate the rate of utilization of organic matter. These measurements are central to a quantification of the role of bacteria in marine foodweb energetics and trophodynamics. Another application for growth rate measurement is in studies to elucidate control mechanisms of bacterial growth in the sea. Growth rate measurement may also be employed in determining how ecosystem perturbation (e.g., pollution) might alter bacterial growth and hence the role of bacteria in the perturbed ecosystem (Azam et al., in press).

DISCUSSION OF METHODS

Prerequisites of a Suitable Method

Table 1 considers a number of methods and lists the necessary (but not sufficient) conditions for any suitable approach for measuring bacterial production. These conditions are described below.

1. Specificity for bacteria. Because bacterial production is generally measured in samples of seawater which also contain many other organisms, it is essential that the method be specific for bacteria. This is particularly important in methods involving the use of radiotracer probes to label cell DNA, RNA, protein, etc., where the label might be taken up by non-bacterial organisms as well, or where measurements of intracellular constituents, occurring in organisms besides bacteria, are an integral part of the growth rate measurement.

2. The method should be applicable in a variety of growth states, not just balanced growth. Since we do not know whether bacteria in a sample are in balanced growth, the method should not depend on a relationship between the rate of synthesis of a macromolecule and the growth rate if such relationship is growth-rate-dependent. This problem applies, in principle, to all methods based

on the synthesis of DNA, RNA, and protein. However, the variation of
DNA and protein per cell with growth rate in cultures is very small
compared with that of RNA (Maaloe and Kjeldgaard 1966; Fuhrman and
Azam 1980) and therefore, the rate of synthesis of DNA or protein may
be preferable bases for growth rate measurement.

  3. The experimental manipulation should not significantly
change the growth rate. It is difficult to know whether this
condition is met in a given experimental protocol. Prefiltration or
long incubations in vitro ("bottle effect") are likely to change the
growth rate; enrichments with nutrients may have similar effect.

Table 1. Comparison of the suitability of some current methods
          for measuring bacterial growth rates in seawater. Only
          those methods which attempt to measure growth rates are
          listed. The conditions listed here are necessary but not
          sufficient for the validity of the method (cf. text).

Criterion of suitability [a]	1	2	3	4
FDC	++[b]	+	++	++
^{35}S-SO_4 assimilation	--	+	+	+
^3H-adenine incorporation into RNA	--	--	+	++
^3H-thymidine incorporation into RNA	++	+	+	++
Increase in cell number	++	++	-	-
Increase in cell ATP	-	+	-	+

[a] Criteria:
   1. Method should be specific for bacteria
   2. Conversion factor should not be growth-rate dependent
   3. Experimental manipulation should not change growth rate
   4. Sensitive enough to allow short incubations (i.e., min-hr)

[b] ++ Meets criterion
   + Probably meets criterion
   - Probably does not meet criterion
   -- Does not meet criterion

Short incubations (radioactive tracer techniques) or, ideally, no incubation (e.g., FDC method) are therefore preferable.

4. The method should have demonstrable reliability, precision, and sensitivity compatible with the problem at hand. The precision and sensitivity is greatest for the radioisotopic methods; the reliability has to be judged against a standard and a clear definition of the aspect of growth being measured. This problem of calibration is discussed later.

Suitable Methods

1. Frequency of dividing cells (FDC). In this method the percentage of total cells in the state of division (i.e., septated but not separated) is related to an assemblage growth rate (Hagström et al. 1979). This relationship is highly temperature-dependent; thus the sample temperature must be known (Hagström et al. 1979). Conceptually, this is probably the most elegant of the existing methods. Microscopic examination of preserved samples is the only experimental manipulation and no incubation is required. The method is specific for the organism examined; in fact it could be made specific for a given species of bacteria if species of interest could be marked for microscopic identification (e.g., nitrifiers can be selectively stained with fluorescent antibodies; Ward and Perry 1980). The microscopy can be undertaken days to weeks after the sample is taken. The precision and other aspects of the method are discussed in Hagström and Larsson (this volume). The conversion factors for calculation of growth rates from measured FDC have been determined from chemostat cultures of bacteria (Hagström et al. 1979). Newell and Christian (1981) found that the relation between growth-rate (μ) and FDC is best determined as the regression of the natural logarithm of μ on FDC. These authors, and Newell and Fallon (1982) also point out the need to assume steady-state growth. Additionally, the assumption that all cells are equally active is usually made. The method is somewhat tedious, but potentially it can be automated.

2. [^{35}S]-Sulfate assimilation. The seawater samples are incubated with [^{35}S]-sulfate and the rate of incorporation of radiosulfur into the protein fraction is detemined (Cuhel et al. 1982). Since the percent sulfur in bacterial cell proteins is known with fair accuracy in cultures (Cuhel et al. 1982) the rate of [^{35}S] incorporation can be used to calculate a protein doubling time. The method has three main limitations: (1) since sulfate is incorporated by both algae and bacteria, this method can be used only where the contribution of algae is insignificant. Size-fractionation to separate bacteria from the algae may alleviate the problem in some instances. (2) Seawater contains high concentrations of sulfate, which reduces greatly the specific activity of the added radiosulfur and thus the sensitivity of the method. The method is more successful in lakes

and other bodies of water with low sulfate concentrations. (3)
Uncertainty is introduced because it is difficult to assess the
natural contribution of organic sulfur compunds to bacterial needs.

3. ^{3}H Adenine incorporation into DNA and RNA. This method was
originally for RNA synthesis alone (Karl 1979), but has since been
modified for DNA synthesis as well (Karl 1981). The samples are
incubated with [2-^{3}H] adenine to label the cellular pools, including
cellular ATP and dATP, which are precursors of RNA and DNA, respectively. Two measurements are necessary to determine the rates of RNA
and DNA synthesis: (1) the [^{3}H] specific activity of the ATP and dATP
pools, and (2) the rate of incorporation of ^{3}H into the RNA and DNA
fractions. It is then possible to calculate the rates of RNA and DNA
synthesis.

Since all organisms possess ATP and since ^{3}H-adenine labels the
ATP pools of both bacteria and algae (Karl et al. 1981), the method
is not specific for bacteria. The authors propose this method for
growth rates of entire microbial assemblages, not for bacteria alone.
The accuracy of the method is dependent on the as-yet untested
assumption that all microorganisms in a given water sample (bacteria,
phytoplankton and heterotrophic protozoa) have similar intracellular
^{3}H-ATP specific activities after incubation with ^{3}H-adenine. Size-
fractionation might be tried as a means of separating bacteria from
other organisms, but cell ATP pools can change due to filtration
stress (Karl and Holm-Hansen 1978). Further details of this method
are given in the chapter by Karl and Winn (this volume).

4. ^{3}H Thymidine incorporation into DNA. Seawater samples are
incubated with [methyl-^{3}H]-thymidine to label the DNA synthesized
during the incubation. A set of assumptions is used to convert the
rate of thymidine incorporation into the rate of DNA synthesis and
then to rate of production of new cells (Fuhrman and Azam 1982). At
present, it is not possible to use the [^{3}H] specific activity of
deoxythymidine triphosphate (dTTP) which is an immediate precursor
of the DNA. Although dTTP can be assayed sensitively (Lindberg and
Skoog 1970) it is difficult to separate bacteria from other organisms
to measure dTTP in bacterial metabolic pools. Moriarty and Pollard
(1981) have suggested an isotope dilution method for determining the
extent of isotope dilution, which is discussed in the chapter by
Moriarty (this volume). This problem has been examined indirectly
(Fuhrman and Azam 1982) and the uncertainty in the estimates of DNA
synthesis has been reduced, allowing conservative estimates of the
rate of DNA synthesis to be made. The method is specific for heterotrophic bacteria, since negligible incorporation of ^{3}H-thymidine
(added at 5 nM) occurred in algae or cyanobacteria in autoradiograms;
autoradiography also showed that virtually all the active bacteria are
capable of utilizing nanomolar concentrations of thymidine (Fuhrman
and Azam 1982). The method is sensitive enough to measure significant
incorporation within minutes to hours and is operationally simple.

5. Increase in cell number. In this method, bacteria are separated substantially from bacteriovores by filtering the seawater sample through 3 m Nuclepore filter. Increase in bacterial abundance is then followed microscopically with time (Fuhrman and Azam 1980). Bacterial abundance (and preferably also average cell volume) is measured to compute the rate of increase of bacterial biomass. This method is specific for bacteria, and requires no conversion factors for calculating rates of bacterial production. It assumes however that the filtration does not change the growth rate. It also assumes that removal of > 3 μm particles and organisms does not change the growth rate. These assumptions are largely untested at the present time. The method is less sensitive than the radioisotope methods because its precision depends on the precision of enumerating bacteria. Its simplicity however is an appealing aspect; the only manipulations prior to cell enumeration are filtration and periodic fixation of the samples. Obviously, this method ignores the growth on > 3 μm particles, and also the effect of grazing by < 3 μm flagellates on growth rate measurements. The effect of micropredators may be greatly reduced by measuring growth rates in < 1 μm fraction, although we have observed inhibition of growth after filtration through 1 μm pore size filters (unpublished).

6. Increase in ATP. In this method, bacteria are substantially separated from bacteriovores by differential filtration, as above. Instead of enumerating bacteria microscopically, however, the rate of increase in cellular ATP is followed as a measure of biomass production (Sieburth et al. 1977). The method is not specific for bacteria in samples where small algae might pass the filter used. Also, the growth rate might change due to sample manipulation as in method #5 above, and filtration stress might change cell ATP level. Where applicable, this method is very sensitive because it is based on ATP determination.

CALIBRATION OF METHODS

Since different methods are based on different definitions of growth one would not expect them to yield identical results. It is therefore necessary to develop a valid calibration system so that various growth related parameters can be compared and put in perspective. So far, methods have been calibrated with continuous cultures of marine bacteria (Hagström et al. 1979) and with bacterial assemblages freshly collected from the sea and either filtered through 3-μm filters to remove most grazers (Fuhrman and Azam 1980, 1982) or diluted into filter-sterilized seawater (Fuhrman and Azam 1980; Kirchman et al. 1982).

A calibration system should allow (1) simultaneous measurement of various growth related parameters in a physicochemical regime similar to that in the sea; (2) determination of the biochemical

characteristics of bacteria in the system; and (3) experimental
variation of growth rate, so that growth-rate dependent changes in
cells' biochemical characteristics can be determined.

A promising and logical calibration system is a seawater culture,
either batch (Ammerman and Azam 1982) or continuous (Hagström et al.,
in preparation). To prepare such cultures, particle-free (0.2 μm-filtered) seawater is innoculated with a small volume of 0.6 μm-filtrate
(which contains bacteria). In batch mode, bacteria grow with about
10 h doubling time; in a chemostat it is possible to achieve a range
of growth rates similar to those measured for natural assemblages.
Whereas in chemostat culture the growth rate is known from the
dilution rate, growth rates in the batch mode can be followed either
by microscopic enumeration of bacteria or by measuring cellular ATP.
Measurement of ATP may be easier and more sensitive than microscopic
enumeration. A seawater culture satisfies the criteria for a
suitable calibration system set out above. However, this method
cannot be applied to the measurement of bacteria growth on particles.

DISCUSSION OF RESULTS

Estimates of Bacterial Production Rates

Recent reviews have compiled the available information on
bacterial production in marine environments (Williams 1981; Van Es
and Meyer-Reil 1982; Ducklow 1983). It is apparent from these reviews
that there is still only limited information available on the patterns
of bacterial production in the marine environment. Moreover, the data
have been obtained by a variety of different techniques (including
those discussed in this paper). This makes it difficult to know
whether the observed differences in bacterial production are due to
the use of different methods or to real differences in production.
It is of urgent importance at this point in the development of marine
microbial ecology to evaluate the various methods, intercalibrate
them, and decide upon a method (or a set of methods) best suited for
measuring bacterial production. Recent work in this direction has
been done by Newell and Fallon (1982) and Fallon et al. (1983).

Table 2 lists some estimates of bacterial production rates by
those methods which are specific for bacteria. These data are taken
from the reviews by Van Es and Meyer-Reil (1982) and Williams (1984),
and some additions have been made. Since different techniques were
used, one can only say, at this time, that in coastal temperate waters
the rate of bacterial production is generally on the order of 1-100 μg
$C \cdot 1^{-1} \cdot d^{-1}$. These measurements are mostly from the euphotic zone of
nearshore locations. Fuhrman et al. (1980), working in the Southern
California Bight, found pronounced onshore (high) offshore (low)
gradients in bacterial production rate. In coastal waters off the
eastern United States, where the continental shelf is much broader

Table 2.   Estimates of bacterial biomass production (expressed as
            cell carbon).   Some of the data are abstracted from Van
            Es and Meyer-Reil (1982).

Method	Area and Season	Production Rate ($\mu g\ C \cdot l^{-1} \cdot d^{-1}$)	Reference
Increase in cell volume	Sea of Japan, Coastal Summer, Fall	18–160 (av. 55)	Vyshkvartsev 1980
Increase in cell volume	California, Coastal Summer	10–34	Fuhrman and Azam 1980
DNA synthesis	California, Coastal Summer	6–69[a]	Fuhrman and Azam 1980
DNA synthesis	Saanich Inlet, Br. Columbia, Summer	55–93[a]	Furhman and Azam 1980
DNA synthesis	Antarctica, 9 Coastal Locations, Summer	0.0004–2.9	Furhman and Azam 1980
DNA synthesis	Hudson River Plume USA, Spring	7–10	Ducklow and Kirchman 1983
DNA synthesis	Southeastern USA, Coastal, Summer	4–102[b]	Newell and Fallon 1982
FDC[c]	Southeastern USA, Coastal, Summer	28–181[b]	Newell and Fallon 1982
FDC	Southeastern USA, Coastal, Spring	14–422	Newell and Christian 1981
FDC	Baltic Sea, Coastal Year-round	~ 0–32	Larsson and Hagström 1982

[a] Recalculated from data in Fuhrman and Azam (1980), using the
  conversion factor in Fuhrman and Azam (1982).
[b] Assumes 121×10^{-15} g $C \cdot \mu m^{-3}$ cell volume.
[c] Frequency of dividing cells.

than off California, such distinct onshore-offshore gradients of
bacterial growth rates may not occur close to shore (Newell and
Fallon 1982, Ducklow and Kirchman 1983). If one assumes that all
bacteria in the samples studied to date were growing, the average
doubling times for warm-water nearshore samples were about 0.3-1.5 d
and the generation times for the offshore locations and colder-water
coastal areas were on the order of a few to several days.

Significance in the Food Web

In the nearshore and offshore waters studied so far (listed in
Table 2), bacterioplankton growth accounted for the consumption of
approximately 10-50 percent of the photosynthetically fixed carbon
(Hagström et al. 1979; Fuhrman and Azam 1980, 1982; Newell and
Christian 1981; Larsson and Hagström 1982; Newell and Fallon 1982;
Ducklow and Kirchman 1983). Therefore, although the measurements
are few in number and need refinement, even conservative estimates
of bacterial secondary production show that bacterial utilization of
organic matter is a major channel for material and energy flow in
pelagic food webs. This pathway is probably comparable in magnitude
to the more extensively studied direct link between phytoplankton
and macrozooplankton.

Factors Affecting Bacterial Growth in the Sea

Although there is some uncertainty in measurement of the
absolute rates of bacterial production, the relative rates can be
determined with greater confidence. This allows observation of the
patterns of bacterial production in large areas of the oceans to see
what environmental factors are important determinants of bacterial
growth. The hypotheses formulated on the basis of the field work may
then be tested in laboratory experimental systems to gain insights
into the physiology and biochemistry of bacterial growth in the sea.

Of particular interest is the relationship between the supply of
nutrients and bacterial growth (although other physico-chemical
factors may also be of considerable importance; Ducklow 1983). Few
detailed studies on this subject have been reported where bacterial
growth was the focus (Fuhrman et al. 1980; Eppley et al. 1981;
Hagström and Larsson 1982; Ducklow and Kirchman 1983). Fuhrman et
al. 1980), in an extensive study in the Southern California Bight
found that the per-cell rate of thymidine incorporation (a measure
of the specific growth rate, μ) was strongly correlated with the
abundance of chlorophyll a in seawater. This observation fits with
the generally agreed upon notion that bacteria derive their nutrition
in the open sea and many coastal waters mainly from phytoplankton.
In some coastal zones there may be other sources of dissolved organic
matter (DOM) in addition to phytoplankton; bacteria may utilize
organic nutrients derived from estuaries and other allochthonous
sources, as has been observed in the Hudson River plume by Ducklow
and Kirchman (1983).

The mechanisms of supply of phytoplanktonic DOM to the bacterioplankton are important in the regulation of matter and energy flux
from algae to bacteria. It is generally agreed that the dominant
mechanisms are: (1) exudation by healthy algae; (2) algal pool
spillage during herbivore "sloppy-feeding"; (3) feces and animal
excretions; and (4) decay of detritus. The relative significance of
these mechanisms for supplying the directly utilizable DOM (UDOM)
components (e.g., amino acids, sugars) for bacterial growth is not
known. This, however, is a central issue in elucidating the nature
of coupling between the primary producers and microheterotrophic
foodweb via the bacterioplankton, and is presently receiving
considerable attention.

The true magnitude of algal exudation has been a controversial
issue for over two decades, and considerable literature exists on
this subject (e.g., Hellebust 1965; Sharp 1977; Larsson and Hagström
1982). It appears that the observed rate of exudation is a highly
variable fraction of the photosynthetic rate. In recent publications
on the role of bacterioplankton as a link in the marine food web it
has generally been assumed that perhaps one-third of the photosynthetically fixed carbon is exuded (e.g. Williams 1981; Azam et al.
1983). The actual rate of exudation has eluded direct quantification
because of difficulties and uncertainties in the methodology. An
unresolved problem is to what extent bacteria can rapidly utilize
significant fractions of the exudates such that the experimentally
measured rate of exudation is an underestimate of the real rate
(Lancelot 1979; Williams 1981).

Relationship Between Growth and Nutrient Regime

In order to consider regulation of growth we must know the
nutrient regime in bacterial microenvironments. Bacteria growing on
particles may be in a quite different nutrient regime than those
growing in the free-living state. Since bacteria take up only low
molecular weight solutes (polymeric nutrients must first be
hydrolyzed) growth on particles requires hydrolysis of particulate
organic matter (POM); the concentration of DOM thus produced must be
quite high within the microenvironments. Seawater concentrations of
dissolved solutes on the other hand are extremely low (picomolar to
nanomolar; Mopper et al. 1981; Ammerman and Azam 1981). Thus, it
seems plausible that particles would support rapid growth, and DOM
would support slow or no growth.

Recent studies show, however, that: (1) 80-90% of bacterial
growth is due to free-living bacteria (Williams 1970; Azam and Hodson
1977; Fuhrman and Azam 1980, 1982); (2) rapid bacterial growth (8-12 h
doubling time) occurs in particle free (0.22 μm-filtered) unenriched
seawater (Fuhrman and Azam 1980; Ammerman and Azam 1982). Such
seawater cultures could be grown in a chemostat for 30 generations
(Hagström et al., in preparation); (3) thymidine microautoradiography

showed that at least 50% of bacteria were growing, although only a minute fraction was attached to particles (Fuhrman and Azam 1982). Thus, many, if not most, free-living bacteria grow. This counter-intuitive conclusion raises the question of bacterial adaptations for growth in the free-living state.

Distribution and Growth of Bacteria in Structured Nutrient Fields

In response to microenvironmental nutrient gradients, one might visualize a spatial structuring of bacterial assemblages, and clus-tering of some bacteria in the vicinity of exuding algae or colonized detritus (Azam and Ammerman 1982). The "cluster hypothesis" provides a framework for studying the relationship between nutrient supply and bacterial growth. It also provides a unifying theme for bacteria-organic matter interaction. In regard to regulation of bacterial growth some predictions can be made about environmental conditions likely to cause metabolic shift-up or shift-down (Table 3).

One may speculate that the day-night transitions in algal photo-synthesis are probably the most dramatic modulators of the coupling between UDOM production and utilization. Algal exudation is most likely a major mode of UDOM producton during daytime; particle and polymer hydrolysis might be the dominant route at night. "Sloppy-

Table 3.   Possible changes in the environment of marine bacteria in the euphotic zone.

Relatively Inactive Condition[a]	Relatively Active Condition[a]	Time-Frame
Dispersed	"Clustered"	Sec-min
Free-living	Attached; animal microflora	Sec-hr
Night	Day	Hr
Low primary production	High primary production	Seasonal
Low temperature	High temperature	Seasonal

[a] Change from left column to right presents shift-up, from right to left is shift-down.

feeding" may increase at night (Eppley et al. 1981). If exudation
stops or is greatly reduced at night then bacterial clusters may
disperse; bacteria may experience reduced nutrient concentrations and
shift-down (unless compensatory mechanisms such as multiphasic transport systems are operative; Azam and Hodson 1981). Clustering around
decomposing particles may be sustained or may become more intense.
Mechanisms controlling bacterial growth may thus vary though the day.

Hagström and Larsson (1983) and Fuhrman et al. (1982 and unpublished) observed that bacterial growth rate is maximal in the afternoon or evening followed by only a 30-50% decrease during the night.
This suggests a "buffering" due to polymer and particle accumulation
during the day (Burney et al. 1981, 1982). Therefore, bacteria may
thus avoid drastic shift-down or shift-up by being able to respond to
changing patterns of UDOM production. They may in fact be able to
regulate the part of the UDOM production attributed to particle or
polymer breakdown.

Growth Versus Dormancy

It is often thought that a large part of the bacterial assemblage in seawater is dormant or inactive because of starvation.
Stevenson (1978) suggests that a significant fraction of the bacterial assemblage is in a state of reversible or "exogenous" dormancy,
wherein phenotypic development is interrupted due to unfavorable
environmental conditions. It is pertinent to inquire what changes
in a bacterium's environment could shift it between activity and
dormancy. It is also instructive to note the time-frame for various
environmental changes (Table 3), and ask whether these time-frames
are likely to necessitate dormancy as an adaptive strategy.

Short-term changes in nutrient conditions in the photic zone
take place in seconds (movement of bacteria within, and variations
in structured nutrient fields) to hours (diel variations in rate and
mode of UDOM producion). Such changes on time scales of seconds to
hours are likely to simply shift the pattern of UDOM fluxes and
perhaps cause metabolic shift-up or shift-down within the time-frame
of the change. Experimental observation shows that growth rates
remain constant in short-term incubations (Fuhrman and Azam 1980).

Seasonal changes (e.g., in temperature, light, or plant nutrients) may affect bacterial growth directly or by changing the rate of
algal photosynthesis. Here, one might imagine a shift of bacteria to
a different metabolic state (e.g., dormancy) commensurate with the
expected long-term physiocochemical properties of the environment.
For example, Larsson and Hagström (1982) found great reduction in
bacterial growth in the Baltic during winter. Nevertheless, dormancy
as a dominant physiological state of aquatic bacteria (Stevenson
1978) does not appear to be a tenable concept in view of the recent
work discussed here showing that bacteria, including free-living

ones, are eminently adapted for growth in the sea.

CONCLUSIONS

    1. Several methods for measuring bacterial production have been
proposed recently. Two methods (FDC and thymidine incorporation)
have already yielded useful estimates of bacterial production in seawater. Although further refinement and calibration of methods is
needed, the existing methods are already providing insights into (1)
the magnitude of bacterial secondary production, and (2) control
mechanisms of the growth of bacteria. Proposed methods are adequate
for eutrophic as well as oligotrophic environments, and thymidine
incorporation has also been applied (with modification) to marine
sediments (Moriarty and Pollard 1981; Fallon et al. 1983).

    2. In oligotrophic to moderately eutrophic coastal and offshore
waters the average doubling times are on the order of a half a day to
a few days. In several studied regions, bacterial growth accounted
for the consumption of 10-50% of the photosynthetically fixed carbon.
Thus bacterial utilization of organic matter is a major pathway for
material and energy flow in pelagic marine foodwebs.

    3. It now appears that most bacterial production is due to
free-living bacteria which, contrary to previous belief, are
metabolically active rather than dormant.

    4. In the bacterial microenvironment, nutrients are introduced
non-uniformly; gradients of nutrients presumably exist. Bacteria
are hypothesized to respond to such structured nutrient fields to
position themselves for optimal uptake and growth. Control mechanisms in bacterial growth may vary from day to night; exuding algae
may principally determine the pattern of UDOM distribution during
daytime; at night particle and polymer hydrolysis may be dominant.
Bacterial growth rates have been observed maximal in the afternoon
or evening and decrease only 3-50% at night. This suggests a
"buffering" due to polymer and particle accmulation during the day.
Thus bacteria may avoid drastic shift-up or shift-down by being able
to respond to and in fact partly control changing patterns of UDOM
production.

REFERENCES

Ammerman, J. W., and F. Azam. 1981. Dissolved cyclic adenosine
    monohosphate (cAMP) in the sea and uptake of cAMP by marine
    bacteria. Mar. Ecol. Prog. Ser. 5: 85-89.
Ammerman, J. W., and F. Azam. 1982. Uptake of cyclic AMP by natural
    populations of marine bacteria. Appl. Environ. Microbiol. 43:
    869-876.

Azam, F., and J. W. Ammerman. 1982. Growth of free-living marine
    bacteria around sources of dissolved organic matter. EOS 63: 54.
Azam, F., J. W. Ammerman, J. A. Fuhrman, and Å. Hagström. 1983.
    Role of bacteria in polluted marine systems. To be published
    in Proc. of the Workshop on Meaningful Measures of Marine
    Pollution Effects. NOAA.
Azam, F., T. Fenchel, J. G. Field, J. S. Gray, L.-A. Meyer-Reil,
    and F. Thingstad. 1983. The ecological role of water column
    microbes in the sea. Mar. Ecol. Prog. Ser. 10: 257-263.
Azam, F., and R. E. Hodson. 1977. Size distribution and activity
    of marine microheterotrophs. Limnol. Oceaongr. 22: 492-501.
Azam, F., and R. E. Hodson. 1982. Multiphasic kinetics for
    D-glucose uptake by assemblages of natural marine bacteria.
    Mar. Ecol. Prog. Ser. 6: 213-222.
Burney, C. M., K. M. Johnson, and J. McN. Sieburth. 1981. Diel
    flux of dissolved carbohydrate in a salt marsh and a simulated
    estuarine ecosystem. Mar. Biol. 63: 175-187.
Burney, C. M., P. K. Davis, K. M. Johnson and J. McN. Sieburth.
    1982. Diel relationships of microbial trophic groups and in
    situ dissolved carbohydrate dynamics in the Caribbean Sea.
    Mar. Biol. 67: 311-322.
Cuhel, R. L., C. D. Taylor, and H. W. Jannasch. 1982. Assimilatory
    sulfur metabolism in marine microorganisms: Considerations for
    the application of sulfate incorporation into protein as a
    measurement of natural population protein synthesis. Appl.
    Environ. Microbiol. 43: 160-168.
Ducklow, H. W. 1983. Production and fate of bacteria in the oceans.
    Bioscience 33: 494-501.
Ducklow, H. W., and D. L. Kirchman. 1983. Bacterial dynamics and
    distribution during a spring diatom bloom in the Hudson River
    plume, USA. J. Plankton Res. 5: 333-355.
Eppley, R. W., S. G. Horrigan, J. A. Fuhrman, E. R. Brooks, C. C.
    Price, and K. Sellner. 1981. Origins of dissolved organic
    matter in Southern California coastal waters: Experiments on
    the role of zooplankton. Mar. Ecol. Prog. Ser. 6: 149-159.
Fallon, R. D., S. Y. Newell, and C. S. Hopkinson. 1983. Bacterial
    production in marine sediments: will cell-specific measures
    agree with whole-system metabolism? Mar. Ecol. Prog. Ser. 11:
    119-127.
Francisco, D. E., R. A. Mah, and A. C. Rabin. 1973. Acridine-
    orange-epifluorescence technique for counting bacteria in
    natural waters. Trans. Am. Microscop. Soc. 92: 416-421.
Fuhrman, J. A., J. W. Ammerman, and F. Azam. 1980. Bacterioplankton
    in the coastal euphotic zone: distribution, activity, and
    possible relationships with phytoplankton. Mar. Biol. 60:
    201-207.
Fuhrman, J. A., and F. Azam. 1980. Bacterioplankton secondary
    production estimates for coastal waters of British Columbia,
    Antarctica, and California. Appl. Environ. Microbiol. 39:
    1085-1095.

Fuhrman, J. A., and F. Azam. 1982. Thymidine incorporation as a
    measure of heterotrophic bacterioplankton production in marine
    surface waters: Evaluation and field results. Mar. Biol. 66:
    109-120.

Fuhrman, J. A., F. Azam, R. W. Eppley, and Å. Hagström. 1982. Diel
    variations in phytoplankton, bacterioplankton, and related
    parameters in the Southern California Bight. EOS 63: 946
    (Abstract).

Hagström, Å., J. W. Ammerman, S. Henrichs, and F. Azam. Bacterio-
    plankton growth in seawater: II. Organic matter utilization
    during steady-state growth. Submitted to Mar. Ecol. Prog. Ser.

Hagström, Å., and U. Larsson. Diel and seasonal variation in growth
    rates of pelagic bacteria. In: J. E. Hobbie and P. J. leB.
    Williams [eds.], Heterotrophic Activity in the Sea. Plenum
    Press.

Hagström, Å., U. Larsson, P. Horstedt, and S. Normark. 1979.
    Frequency of dividing cells, a new approach to the determina-
    tion of bacterial growth rates in aquatic environments. Appl.
    Environ. Microbiol. 37: 805-812.

Hanson, R. B., and H. K. Lowery. 1983. Nucleic acid synthesis in
    oceanic microplankton from the Drake Passage, Antarctica:
    evaluation of steady-state growth. Mar. Biol. 73: 79-89.

Hellebust, J. A. 1965. Excretion of some organic coumpounds by
    marine phytoplankton. Limnol. Oceanogr. 10: 192-206.

Hobbie, J. E., R. J. Daley, and S. Jasper. 1977. Use of Nuclepore
    filters for counting bacteria by fluorescence microscopy.
    Appl. Environ. Microbiol. 33: 1225-1228.

Karl, D. M. 1979. Measurement of microbial activity and growth in
    the ocean by rates of stable ribonucleic acid synthesis. Appl.
    Environ. Microbiol. 38: 850-860.

Karl, D. M. 1981. Simultaneous rates of ribonnucleic acid and
    deoxyribonucleic acid syntheses for estimating growth and cell
    division of aquatic microbial communities. Appl. Environ.
    Microbiol. 42: 802-810.

Karl, D. M., and O. Holm-Hansen. 1978. Methodology and measurement
    of adenylate energy charge ratios in environmental samples.
    Mar. Biol. 48: 185-197.

Karl, D. M., and C. D. Winn. Adenine metabolism and nucleic acid
    synthesis: Applications to microbiological oceanography. In:
    J. E. Hobbie and P. J. leB. Williams [eds.], Heterotrophic
    Activity in the Sea. Plenum Press.

Karl, D. M., C. D. Winn, and D. C. L. Wong. 1981. RNA synthesis as
    a measure of microbial growth in aquatic environments. I.
    Evaluation, verification, and optimization of methods. Mar.
    Biol. 64: 1-12.

Kirchman, D., H. Ducklow, and R. Mitchell. 1982. Estimates of
    bacterial growth from changes in uptake rates and biomass.
    Appl. Environ. Microbiol. 44: 1296-1307.

Larsson, U., and Å. Hagström. 1982. Fractionated phytoplankton
    primary production, exudate release and bacterial production in

a Baltic eutrophication gradient. Mar. Biol. 67: 57-70.

Lindberg, U., and L. Skoog. 1970. A method for the determination of dATP and dTTP in picomole amounts. Anal. Biochem. 34: 152-160.

Maaloe, O., and N. O. Kjeldgaard. 1966. Control of Macromolecular Synthesis: A Study of DNA, RNA, and Protein Synthesis in Bacteria. Benjamin-Cummings Publishing Co., New York.

Mopper, K., and P. Lindroth. 1982. Diel and depth variations in dissolved free amino acids and ammonium in the Baltic Sea determined by shipboard analysis. Limnol. Oceanogr. 27: 336-347.

Moriarty, D. J. W. Measurements of bacterial growth rates in some marine systems using the incorporation of tritiated thymidine into DNA. In: J. E. Hobbie and P. J. leB. Williams [eds.], Heterotrophic Activity in the Sea. Plenum Press.

Moriarty, D. J. W., and P. C. Pollard. 1981. DNA synthesis as a measure of bacterial productivity in seagrass sediments. Mar. Ecol. Prog. Ser. 5: 151-156.

Newell, S. Y., and R. R. Christian. 1981. Frequency of dividing cells as an estimator of bacterial productivity. Appl. Environ. Microbiol. 42: 23-31.

Newell, S. Y., and R. D. Fallon. 1982. Bacterial productivity in the water column and sediments of the Georgia (USA) coastal zone estimated via direct counting and parallel measurements of thymidine incorporation. Microb. Ecol. 8: 33-46.

Pomeroy, L. R. 1974. The ocean's food web, a changing paradigm. Bioscience 24: 499-504.

Sharp, J. H. 1977. Excretion or organic matter by marine phytoplankton: do healthy cells do it? Limnol. Oceanogr. 22: 381-399.

Sieburth, J. McN., K. M. Johnson, C. M. Burney, and D. M. Lavoie. 1977. Estimation of in situ rates of heterotrophy using diurnal changes in organic matter and growth rates of picoplankton in diffusion culture. Helgol. Wiss. Meeresunters. 30: 565-574.

Stevenson, L. H. 1978. A case for bacterial dormancy in aquatic systems. Microb. Ecol. 4: 127-133.

Van Es, F. B., and L.-A. Meyer-Reil. 1982. Biomass and metabolic activity of heterotrophic marine bacteria, pp. 111-179. In: K. C. Marshal [ed.], Advances in Microbial Ecology. Plenum Press.

Vyshkvartsec, D. I. 1980. Bacterioplankton in shallow inlets of Posyeta Bay. Microbiology 48: 603-609.

Ward, B. B., and M. J. Perry. 1980. Immunofluorescent assay for the marine ammonium-oxidizing bacterium Nitrosococcus oceanus. Appl. Environ. Microbiol. 39: 913-918.

Williams, P. J. leB. 1970. Heterotrophic utilization of dissolved organic compounds in the sea. I. Size distribution of population and relationship between respiration and incorporation of growth substrates. J. Mar. Biol. Assoc. UK 50: 859-870.

Williams, P. J. leB. 1981. Incorporation of microheterotrohpic processes into the classical paradigm of the planktonic food web. Kiel. Meersforsch. Sonderh. 5: 1-28.

Williams, P. J. leB.   1984.   A review of measurements of respiration
    rates of marine plankton populations.   In: J. E. Hobbie and P.
    J. leB. Williams [eds.], Heterotrophic Activity in the Sea.
    Plenum Press.
Zimmerman, R., and L.-A. Meyer-Reil.   1973.   A new method for
    fluorescence staining of bacterial populations on membrane
    filters.   Kiel. Meersforsch. 30: 24-27.

ADENINE METABOLISM AND NUCLEIC ACID SYNTHESIS: APPLICATIONS TO

MICROBIOLOGICAL OCEANOGRAPHY

David M. Karl and Christopher D. Winn

Department of Oceanography
University of Hawaii
Honolulu, Hawaii   96822

INTRODUCTION

Marine microbial ecology is currently one of the least developed areas of microbiological research. This situation has been, in part, due to the limited availability of methods for evaluating the in situ rates of metabolism and growth of naturally occurring microbial populations. In fact, it may be fair to state that our present understanding of the integrated functioning of marine ecosystems is methods limited. A period of rapid advance in our understanding of microbiological oceanographic processes following the successful development and application of each new experimental approach, is evidence of this limitation.

Classical microbiological methods (e.g., plate count, light-dark bottle) have proven unsatisfactory for enumerating the sparse populations and for assessing the relatively low rates of production which occur in most marine environments. During the past 20 years a significant effort has been invested in the development and evaluation of new experimental approaches, and a variety of methods (some as yet untested) are now available to evaluate the distribution, abundance and metabolic activities of microorganisms in nature. Many of these techniques make use of our expanding fundamental knowledge of biochemistry, physiology and cell biology. One area where significant progress has been made is in the study of cellular nucleotides and nucleic acid metabolism. The well-established and obligatory role of nucleotides in cellular energetics, biosynthesis and metabolic regulation has provided the motivation for initiating laboratory and field studies. During the past few years, a coordinated study of cellular nucleotide concentrations, their intermediary

metabolism, turnover rates and rates of polynucleotide biosynthesis
has been undertaken in marine and freshwater microbial assemblages.
This approach, referred to as "environmental nucleotide fingerprinting" (Karl 1980), provides non-selective, corroborative, and redun-
dant information on total microbial biomass, physiological potential,
growth, and cell division rates. It should be emphasized from the
start that the methods described herein are aimed at evaluating the
combined activities of all microorganisms in the sea. Previous
attempts to separate bacteria from unicellular algae, protozoa, and
other microorganisms, whether by physical or chemical means, have
generally been unsuccessful due to overlapping size spectra and to
the non-specificity of antibiotics and inhibitors. Although certain
critics have focused their attention on the inability of our methods
to measure strictly bacterial processes, we feel that the non-selec-
tivity of nucleotide fingerprinting is one of the most unique and
important aspects of this approach. Obviously, investigators must
look to different methods for information on how the total biomass
carbon and metabolic activity are partitioned among the individual
microbial taxa in nature. In this regard, each methodological
approach should be viewed as a separate "experiment" that yields
novel information on the ecosystem as a whole. The most obvious
conclusion which can be drawn from the extensive literature on
marine microbiological methods is that there is no single approach
which is universally acceptable. A second conclusion is that
despite the numerous methods related publications on microbial
growth in the sea, it is nearly impossible to calibrate any single
measurement in situ with any degree of accuracy due to uncertainties
regarding the validity of the assumptions which accompany each
approach. Intercalibrations are equally difficult since each method
measures a slightly different aspect of metabolism, growth or cell
division. Hence there is no a priori reason for different methods
to yield comparable ecological information. The need for conducting
multiple "experiments" to define the growth state of natural
microbial assemblages cannot be overemphasized.

One of the most useful indices in our environmental nucleotide
fingerprinting approach is the measurement of nucleic acid, particularly DNA, synthesis. In general, two distinct approaches have been
used to measure microbial DNA synthesis in aquatic environments.
One approach utilizes [methyl-^{3}H] thymidine incorporation as a unique
measure of bacterial DNA synthesis (Brock 1967; Tobin and Anthony
1978; Fuhrman and Azam 1980, 1982; Moriarty and Pollard 1981, 1982;
see also Moriarty this volume). The assumptions implicit in this
approach have been discussed in the above references, and by Karl
(1982). The second approach uses [2-^{3}H] adenine as a precursor to
nucleic acid synthesis (Karl 1979, 1981; Karl et al. 1981a) and
differs significantly from the ^{3}H-thymidine method both in theory
and practice. The most noteworthy differences are: (1) ^{3}H-adenine
is used to simultaneously measure rates of both RNA and DNA
synthesis, ^{3}H-thymidine measures only DNA synthesis, (2) ^{3}H-adenine

is assimilated by both bacteria and eucaryotic microalgae so that
experiments with natural microbial assemblages yield nucleic acid
synthesis rates for the entire microbial community, and (3) a
sensitive assay for the immediate precursor to adenine incorporation
into nucleic acid exists so that its specific activity can be
measured directly.

The present paper is intended to summarize the recent work from
our laboratory on adenine metabolism and rates of nucleic acid
synthesis and demonstrate how these methods might be used to address
specific hypotheses in microbiological oceanography.

DISCUSSION

Metabolic Considerations and Assumptions

Background. RNA and DNA synthesis in microorganisms is required
for growth and cell division and the rates of synthesis are closely
coupled to these processes. The cell content of RNA in bacteria and
microalgae (expressed as mg RNA per mg dry weight) is positively
correlated with growth rate and is regulated with great precision
(see reviews by Maaloe and Kjeldgaard 1966; Kjeldgaard 1967; Nierlich
1978). By comparison, DNA concentrations (expressed as mg DNA per
mg dry weight) are relatively insensitive to changes in growth rate,
chemical composition of the growth medium or to the nature of the
growth limiting nutrient. Correlations between rates of RNA and DNA
synthesis and cell growth appear to be universally applicable to
microorganisms and lend themselves well to the study of the growth
of mixed populations encountered in marine ecosystems.

Most microorganisms possess the ability to utilize exogenous
supplies of adenine, and certain other nucleic acid bases, as a
supplement to de novo synthesis. The mechanisms and implications of
these so-called "salvage pathways" have been reviewed elsewhere
(Hochstadt 1974). When adenine is transported into a cell, it is
rapidly incorporated into a number of derivatives including ATP, dATP
(deoxy-adenosine triphosphate), ADP and AMP, and eventually into the
metabolically stable nucleic acids. It has been demonstrated that
adenine uptake is directly coupled to nucleic acid synthesis such
that the rate of entry of adenine into the cell (k_1, Fig. 1) does
not exceed the biosynthetic requirements for ATP and dATP (sum of k_2
plus k_3). However, the actual quantity of adenine taken up by
microbial cells is much less than the value k_2 plus k_3 due to
variable but substantial contributions of adenine from de novo
synthesis and internal recycling of nucleic acid bases. This is
true not only for adenine but for all nucleic acid precursors.

When radioactive adenine is added to a culture or seawater
sample, it is diluted to an unknown extent by existing pools of

I. NTP REDUCTASE PATHWAY (ATP → dATP)

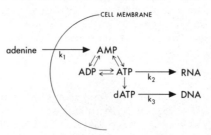

II. NDP REDUCTASE PATHWAY (ADP → dADP → dATP)

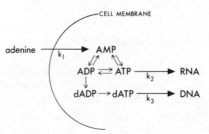

Figure 1. Known pathways for the flow of exogenous adenine into RNA and DNA of microorganisms (From Karl 1981).

adenine and adenine-containing compounds (e.g., ATP, ADP, AMP, adenosine). This process decreases the specific radioactivity of the introduced radiotracer prior to transport into the cells. Since the salvage pathways generally cannot supply the total amount of adenine required by the cells, this process must be supplemented by de novo synthesis of adenine. This further decreases the specific activity of the introduced radiotracer after transport into the cells but before incorporation into nucleic acids. Consequently, before an absolute rate of RNA or DNA synthesis can be obtained, it is necessary to determine the specific radioactivity of the immediate nucleotide triphosphate precursor, as well as the total amount of radioactivity incorporated into the individual nucleic acid pools. Without knowledge of both precursor specific activity (ATP in the case of ^{3}H-adenine assimilation) and the extent of end product labeling, rates of nucleic acid synthesis in seawater samples cannot be measured. This concern, of course, is not unique to ^{3}H-adenine experiments but is true for the use of all stable or radioisotopic tracers.

The two alternative schemes presented in Figure 1 indicate the known pathways by which adenine nucleotides are reduced to the corresponding deoxyadenosine derivatives prior to incorporation into DNA (Thelander and Reichard 1979). It has been demonstrated experimentally that ATP and ADP achieve isotopic equilibrium (i.e., internal specific activity of ATP pool = internal specific activity of ADP pool) over time periods of less than 1 min (Karl 1981). Consequently, we assume that the specific radioactivity of the cellular dATP is at all times equal to that which we measure for ATP, regardless of whether the reduction occurs at the diphosphate or triphosphate level (see Fig. 1). This assumption is consistent with all existing theoretical models and experimental data.

The detailed stepwise procedures and rationale for measuring rates of microbial RNA and DNA synthesis in environmental samples have been presented elsewhere (Karl 1979, 1981, 1982; Karl et al. 1981a, b). Briefly, a seawater sample is incubated with ^{3}H-adenine for a predetermined time period. For reasons which will be discussed later, a time course is preferred to single end-point measurements. Following the incubation, a subsample is extracted for isolation of nucleic acids. By thin layer chromatography, the ATP is purified and radioassayed. The mean ATP pool specific radioactivity is then calculated from the chemical and radiochemical ATP data. RNA and DNA are also separated and purified, and their total radioactivities measured. From the incorporation and precursor specific radio-activity data, the rates of RNA and DNA synthesis can be calculated and expressed as mass of adenine (or deoxyadenine) incorporated into RNA (or DNA) per unit volume per unit time. Biomass-specific rates of RNA and DNA synthesis can also be derived by dividing the total measured rates by biomass (ATP). Alternatively, relative RNA-to-DNA synthesis rate ratios can be determined directly from the incorpora-tion data, thereby eliminating the necessity for ATP pool specific radioactivity measurements. If time course data are available, an independent calculation of the rates of RNA and DNA synthesis can be performed using the total radioactivity accumulated in RNA and DNA and the integral of the ATP specific activity over the length of the time course experiment (Yamazaki and Leung 1981; Winn and Karl 1983).

Assumption. Several testable assumptions are implicit in the application of this method to the analysis of natural microbial assemblages, the most critical of which are:
   1. All (or in practice, most) microorganisms in the ocean assimilate exogenous adenine by a common pathway and exhibit a similar "salvage" response under in situ conditions.
   2. Addition of radioactive adenine does not affect the ATP cell quota, ATP turnover rate, or the rate of microbial RNA and DNA synthesis.
   3. There is no intracellular compartmentalization of microbial ATP pools, or if there is, such compartmentalization does not affect the accuracy of the measured rates.

Experimental Verification of Assumption

The ability of E. coli to utilize exogenous adenine supplies in

preference to de novo synthesis has been extensively investigated
(Hochstadt 1974). In 49 out of 50 strains of marine and freshwater
bacteria examined, nanomolar concentrations of exogenous adenine were
utilized by a similar mechanism (Karl et al. 1981a). This is not
altogether surprising since the biosynthesis of adenine is highly
endergonic. The existence of these salvage pathways enables microorganisms to conserve energy that would otherwise be required for
purine biosynthesis (Karl 1979).

More recently, we have investigated the occurrence and extent
of adenine salvage in marine alage. Twenty-two unialgal, but not
axenic, cultures of marine phytoplankton representing most major
taxa (see Table 1) were grown to early logarithmic phase under
constant light on f/2 medium (Eppley et al. 1967), at which time
they were incubated with 1 μCi ^{3}H-adenine·ml^{-1} (specific radioactivity = 15 Ci·mmol^{-1}) for a 6 h period. ATP, ^{3}H-ATP, ^{3}H-RNA and
^{3}H-DNA were measured as described previously (Karl et al. 1981a).
Since the cultures were not axenic and since bacteria are known to
assimilate ^{3}H-adenine, we endeavored to set an upper limit on the
contribution of bacteria to the total uptake and assimilate of
^{3}H-adenine. A subsample of each culture was fixed in glutaraldehyde,
stained with acridine orange, and the bacteria enumerated with UV
fluorescence microscopy (Hobbie et al. 1977). The majority of the
cells were large rods (1.0 - 1.5 μm x 0.5 μm). Biovolume was
calculated by assuming that all cells conformed to the largest
dimensions (1.5 x 0.5 μm) which biased the estimate in favor of
bacterial biovolume. This value was converted to biomass (Ferguson
and Rublee 1976) and finally to ATP (Karl 1980). This upper bound
on the contribution of bacterial ATP was compared to the total ATP
concentration measured for each culture. Bacterial ATP ranged from
0.14 - 27% of the total ATP and only those cultures where bacterial
ATP was   1% of the total are reported in this study (14 out of 22
cultures). As an independent confirmation of algal uptake, subsamples of each culture were also prepared for microautoradiography
(Meyer-Reil 1978). Labeled phytoplankton were observed in all 22
cultures. Table 1 summarizes the results of the uptake experiments.
The ATP pool's specific radioactivity data, by themselves, provide
convincing indirect evidence for the uptake of ^{3}H-adenine. If the
metabolism of ^{3}H-adenine was restricted to contaminating bacteria
(< 1% of total ATP in the culture) it would have been impossible to
attain the specific activities presented, since bacterial specific
activities rarely exceed values of 1.0 nCi·pmol^{-1} (Karl et al. 1981a;
see also Fig. 5 for waters > 150 m), even under saturating conditions
of ^{3}H-adenine with a specific activity of 15 nCi·pmol^{-1} (Karl et al.
1981a). Finally, the uptake and incorporation of ^{3}H-adenine into
ATP and nucleic acids in environmental samples proceeds without a
detectable lag period indicating that microbial communities in
nature are probably utilizing these salvage pathways under in situ
conditions.

Table 1. Evidence for ^{3}H-adenine uptake and assimilation by marine phytoplankton.

Culture[1]	^{3}H-RNA Produced (nCi/ml)	^{3}H-DNA Produced (nCi/ml)	DNA:RNA	ATP Pool Specific Radioactivity (nCi/pmol)
CHRYSOPHYTA				
Bacillariophyceae				
Chaetoceros pelagicus	79.2	13.7	0.17	0.70
Cyclotella nana	149.5	27.0	0.18	0.68
Lauderia borealis	163.4	31.3	0.19	0.73
Nitzchia sp.	75.7	17.9	0.24	1.79
Rhizosolenia robusta	43.6	7.1	0.16	2.14
Chrysophyceae				
Cricosphaera carterae	245.7	24.4	0.10	0.61
Emiliania huxleyi	218.1	42.2	0.19	1.50
Monochrysis lutheri	247.9	26.9	0.11	0.61
Phaeocystis sp.	35.9	6.0	0.17	1.13
PYRROPHYTA				
Desmophyceae				
Exuviella sp.	60.9	7.4	0.12	1.28
Dinophyceae				
Gymnodinium simplex	35.0	6.1	0.17	0.81
Gymnodinium sp.	39.3	6.5	0.17	1.94
Peridinium foliaceus	23.1	2.5	0.11	0.25
UNIDENTIFIED				
Microcell MC-1	17.3	9.7	0.56	0.66

[1] Phytoplankton cultures courtesy of J. Jordan, Food Chain Research Group, Scripps Institution of Oceanography.

The second assumption is that the addition of ^{3}H-adenine to environmental samples neither alters the steady-state concentration and turnover rate of cellular ATP, nor affects the rate of nucleic acid synthesis. We have evaluated this assumption by: (a) measuring the total ATP pool levels in natural microbial assemblages with time following the addition of ^{3}H-adenine (Karl et al. 1981a), (b) measuring the rate at which ^{3}H-ATP is "chased" out of the cellular ATP pools upon the addition of saturating concentrations of non-radio-active adenine to a microbial community previously labeled with tracer levels of ^{3}H-adenine, (Karl et al. 1981a), and (c) measuring the effects of ^{3}H-adenine isotope dilution on measured rates of RNA and DNA synthesis (Karl 1982). From these experimental data we conclude that there is no luxury uptake of adenine, even at saturating concentrations, and no apparent expansion of ATP cell quotas or stimulation of RNA or DNA synthesis upon the addition of ^{3}H-adenine to the sample. These results support our previous model indicating that the uptake of adenine by the cells (k_1, Fig. 1) does not exceed the net removal of ATP and dATP required for macromolecular synthesis.

The final assumption concerns intracellular compartmentalization of the adenine nucleotide pool which might be expected to arise through either barrier (membrane) or kinetic processes (see discussion in Karl 1979). We have now performed several hundred time course experiments on the rates of RNA and DNA syntheses in cultures and environmental samples and have observed what we interpret to be kinetic compartmentalization of the ATP precursor pool during the initial time periods of labeling (< 30 s for cultures and $<$ 1 h for most environmental samples; see Karl et al. 1981a, b; Karl 1981). For this reason we have stressed the utility of performing time course experiments. The effects of compartmentalization decrease with increasing incubation time suggesting that it might arise, in part, from diffusion-controlled processes. This is supported by the fact that we have observed effects of near equal magnitude in cultures of bacteria and eucaryotic algae. If the effect were due to the existence of organelles (barrier compartmentalization processes), one might argue that it would not be observed in bacteria. Intracellular compartmentalization and time-dependent changes in precursor specific radioactivity are problems inherent to all tracer experiments and certainly are not unique to the use of ^{3}H-adenine. This technique simply provides a convenient means of assessing the problem.

For ecological applications, the assumption regarding intracellular compartmentalization is inevitably further complicated by the heterogeneity of natural populations, all containing ATP pools. As currently employed, our method assumes that all of the individual cellular ATP pools are in isotopic equilibrium. This assumption is also implicit, although not often stated, in all radiotracer experiments with mixed populations. It is very difficult to evaluate this assumption directly but it is encouraging to note that laboratory cultures of algae yield ATP pool specific activities which are comparable to those measured in bacteria grown under identical laboratory conditions. This implies that the salvage pathways in both bacteria and unicellular algae supply similar percentages of the required cell quota of adenine. In fact, this appears to be true under field conditions as well, since there is no substantial difference in the magnitude of the ATP pool specific activities between different size fractions of seawater samples analyzed to date (Karl et al. 1981b) which might be expected if only a portion (e.g., bacteria) of the community was actively assimilating

^{3}H-adenine. We have performed "nutrient addition" perturbation experiments to selectively stimulate the rates of nucleic acid synthesis of photoautotrophs (by the addition of inorganic nutrients) or chemoheterotrophs (by the addition of organic nutrients), and have not been able to detect any substantial changes in the mean specific activity of the total microbial ATP following 12 h incubations (relative to control samples), despite biomass increases of up to 60% (Table 2). Finally, we have recently performed a series of ^{32}PO$_4$ and ^{3}H-adenine double label experiments to investigate further the assumption regarding uniform isotope equilibrium among the various microbial taxa. Figure 2 summarizes data resulting from these double label experiments, which demonstrate: (1) a significant ($p > 0.05$) and positive correlation between the ^{32}P rate ratios and those derived from ^{3}H-adenine, (2) a numerical correspondence between the two data sets (i.e., slope of the regression line is not significantly different from 1.0). If we tacitly assume that all micro-

Table 2. Effects of nutrient additions on the measured ATP concentrations and ATP pool specific radioactivities following incubation with ^{3}H-adenine.

Treatment[1]	ATP Concentration[2] (ng/1)	ATP Pool Specific Radioactivity[3] (nCi/pmol)
Control	44.5	2.14 ± 0.19
	50.5	
+NH$_4$	40.3	1.85 ± 0.51
	42.3	
+PO$_4$	30.3	1.42 ± 0.71
	40.3	
+Glucose	73.2	1.84 ± 0.11
	63.8	

[1] Nutrients were added to yield final concentrations of 5 µM.
[2] ATP concentration at end of 12 h incubation period. ATP concentration at start of the experiment was 43.6 ± 0.16 ng/1.
[3] Measured at end of 12 h incubation period. Values represent mean ± 1 standard deviation (n = 4).

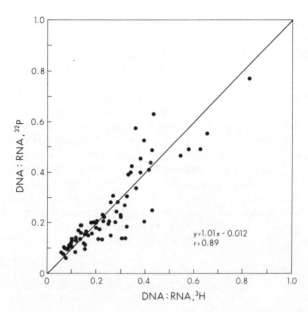

Figure 2.   DNA:RNA incorporation rate ratios determined using $^{32}PO_4$ and ^{3}H-adenine in double-label tracer experiments. All data are from 4 h incubations with surface seawaters (0–90 m) collected at the VERTEX I station (36°48'N, 121°46'W) (From Orrett and Karl, in prep.).

organisms are capable of assimilating $^{32}PO_4$, then we must conclude that all, or at least an identical subset of microorganisms is also capable of assimilating ^{3}H-adenine. Otherwise it would be extremely difficult to explain the correspondence between the two data sets. This excellent agreement between the two independent methods also supports our assumption that adenine and deoxyadenine comprise 25 mol percent of the nucleotide bases in RNA and DNA, respectively, if we acknowledge that the α-P specific activity of all nucleotide bases is equivalent. All of the above observations and data support the assumption that marine microorganisms assimilate adenine by a common and uniform pathway under in situ conditions.

Laboratory Calibration of the Method

We have recently completed a series of laboratory experiments to evaluate the accuracy of the ^{3}H-adenine method for estimating rates of RNA and DNA synthesis in algal and bacterial cultures. The experiments were designed to compare the nucleic acid synthesis rates using ^{3}H-adenine with the actual rates determined from direct measurements of RNA and DNA concentrations and culture doubling times. A complete discussion of the experiments, results, and interpretations will be presented elsewhere (Winn and Karl,

submitted for publication); however, a selected summary of the
results is presented in Figures 3 and 4. The most important result
is the fact that our radiochemical technique provides an accurate
estimate of the actual rate of RNA and DNA synthesis in both
bacterial and algal cultures. The overestimate during the initial
period of labeling observed with the algal cells (Fig. 4) is consistent with our previous data on intracellular compartmentalization
(Karl et al. 1981a) and with our current model of the assimilation
and metabolism of adenine by marine microorganisms. However, it
should be emphasized that accurate rates of both RNA and DNA
synthesis are achieved after incubation periods as short as 10-20%
of the doubling time.

Selected Ecological Applications

     Several of our publications present and discuss the rates of
seawater RNA and DNA synthesis (Karl 1979, 1981, 1982; Karl et al.
1981a, b; Fellows et al. 1981). Table 3 presents a summary of ATP
biomass, rates of RNA and DNA syntheses and biomass-specific rates
of nucleic acid production for several additional marine environ-

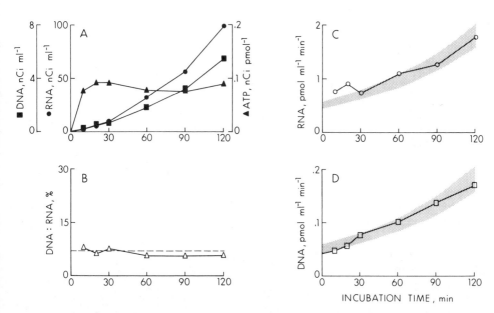

Figure 3. ^{3}H-adenine calibration experiment performed with the
marine bacterium Serratia marinorubra. The dashed line in frame B
indicates the mean DNA: RNA rate ratio as determined by independent
chemical determinations and the stippled areas in frames C and D
indicate the known rates of RNA and DNA synthesis (mean $\pm$ 1 standard
deviation). Initial ATP concentration of culture was $\overline{600}$ ng ATP·1^{-1}
and the doubling time was 60 min (From Winn and Karl, submitted).

Table 3. Summary of representative euphotic zone measurements of microbial biomass and rates of RNA and DNA synthesis.

Sample Location	ATP-Biomass (ng/l)	TOTAL RATE RNA Synthesis (pmol A/l/h)	TOTAL RATE DNA Synthesis (pmol dA/l/H)	DNA:RNA (x 100%)	RNA Synthesis (pmol A/ng ATP/h)	TOTAL RATE DNA Synthesis (pmol dA/ng ATP/h)
Kaneohe Bay, Hawaii (21°26'N, 157°47'W)						
1 m	418	1,440	144	10.0	3.44	0.34
Ke-ahole Point, Hawaii (19°55'N, 156°10'W)						
1 m	53.1	150	*	--	2.90	--
North Pacific Ocean near Point Sur, California (36°07'N, 122°47'W)						
1 m	205	822	--	--	4.00	--
Equatorial Pacific Ocean near Galapagos Islands (0°48'N, 86°09'W)						
1 m	63	240	--	--	3.81	--
Central Pacific Ocean near Hawaii (18°43'N, 156°50'W)						
1 m	27.2	35	12.8	36.6	1.29	0.47
50 m	32.3	28	11.3	40.4	0.87	0.35
130 m	19	61	17.5	28.7	3.21	0.92
North Pacific Ocean, VERTEX I Station (36°48'N, 121°46'W)						
1 m	175	190	42	22.1	1.09	0.24
20 m	210	305	69	22.6	1.45	0.33
50 m	228	270	60	22.2	1.18	0.26

*DNA synthesis data not available.

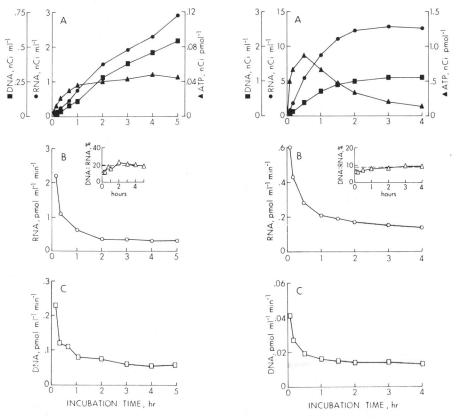

Figure 4. ^{3}H-adenine calibration experiments performed with _Chorella cordata_ (left) and _Cylindrotheca_ sp. (right). Details as per Figure 3. ATP concentrations and culture doubling times were: 10 ng·ml^{-1}, 20.5 h; and 4 ng·ml^{-1}, 17 h, for _C. cordata_ and _Cylindrotheca_ respectively (From Winn and Karl, submitted).

ments. It is interesting to note that despite substantial variations in microbial biomass (ATP) and total rates of nucleic acid synthesis (variations greater than 10 to 50-fold), the biomass-specific rates of RNA and DNA synthesis display only minor variation among the different ecosystems.

Figure 5 presents a detailed depth profile of ATP and rates of RNA and DNA synthesis for water samples collected during the Vertical Transport and Exchange ("VERTEX" I) experiment off Point Sur, California. As might be expected, the total rates of RNA and DNA synthesis decrease with increasing water depth and closely follow the depth distribution of ATP. However, the biomass-specific

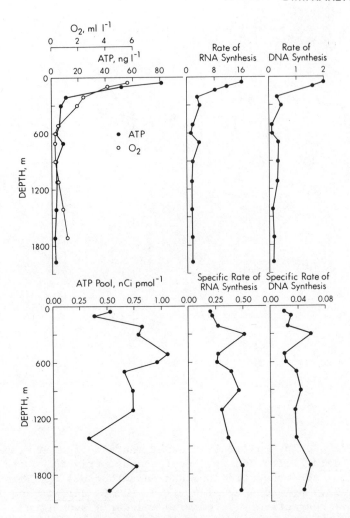

Figure 5.   ATP, O$_2$, rates of RNA and DNA synthesis, specific activity
of the ATP pool, and specific rates of RNA and DNA synthesis for the
water column at the VERTEX I station (36°48'N, 121°46'W).   All water
samples were incubated <u>in situ</u> for 10 h.   Rates of RNA and DNA
synthesis are expressed as pmol A (or dA) incorporated into RNA (or
DNA)·l^{-1}·h^{-1} and specific rates are expressed as pmol A (or dA)
incorporated·ng ATP^{-1}·h^{-1} (From Karl and Knauer, in press).

rates of RNA and DNA synthesis are nearly uniform throughout the
entire water column, suggesting that the microbial community growth
rate is uniform with depth from 0-2000 m.   This was contrary to our
expectations of the vertical distribution of microbial growth in the
ocean; however, we have recently confirmed these observations with
additional data from the tropical North Pacific Ocean (Winn and Karl

1983) and eastern tropical Pacific Ocean (Karl and Knauer, in preparation). We believe this relationship could well be a common feature of the mesopelagic zone of all oceanic environments.

Extrapolation to Growth Rate and Carbon Production Estimates

The majority of investigations concerned with the production, flux, and turnover of organic matter in natural ecosystems have relied upon carbon as the basic unit of cell mass and growth. If the assumptions inherent in the application of nucleic acid precursors to ecological studies were determined to be valid under in situ conditions the ^{3}H-adenine method as described would yield reliable estimates of the total rates of microbial RNA and DNA synthesis under natural environmental conditions. While we would argue that nucleic acid synthesis rate data, by themselves, provide useful information regarding the growth of microorganisms in nature, it is often necessary to extrapolate the DNA synthesis data to estimates of total microbial production (μg C produced per unit volume per unit time) for comparison with other experimental data. Of course, the precision and accuracy of the estimates are adversely affected by the common practice of extrapolation, since the "conversion factors" are, for the most part, highly variable. Our calculations to date have assumed a mean DNA:C ratio of 50 which is based upon laboratory experiments (Holm-Hansen 1969; Mandelstam and McQuillen 1976) and direct field measurements of total DNA and organic carbon. The latter approach is acceptable only where the amount of non-living particulate DNA and carbon are expected to be negligible, which may severely limit the use of an empirically determined extrapolation factor. In a recent intercomparison of methods conducted in Hawaiian coastal waters (PRPOOS experiment; Eppley et al., in prep.) the DNA extrapolated microbial carbon production estimates were, on the average, 160% of the values obtained by the standard ^{14}C primary production technique. This is consistent with the presumed importance of microheterotrophic production (i.e., POC produced from detrital DOC and/or POC) since the DNA synthesis technique measures both autotrophic and heterotrophic processes. Finally, measurements of RNA and DNA synthesis, when used in conjunction with steady-state RNA and DNA concentration data, may be used to estimate the mean community turnover time and specific growth rate (μ). This must be considered a minimum value for the growth rate since the presence of detrital (i.e., non-living) RNA or DNA would tend to overestimate the value for microbial standing stock and underestimate the growth rate.

Table 4 presents a summary of ATP concentrations, rates of DNA synthesis and carbon production (extrapolated from rates of DNA synthesis) and specific growth rates (measured by two independent procedures). It is evident that the specific growth rates calculated from the DNA-biomass data are substantially lower than those based on the ATP-biomass data (assuming a DNA:ATP ratio of 5).

Table 4. Carbon production and community specific growth rate estimates extrapolated from measurements of DNA synthesis at VERTEX II station (154°30'N, 107°30'W).

Depth (m)	ATP Concentration (ng/l)	ATP Pool Specific Radioactivity (nCi/pmol)	Rate of DNA Synthesis (pmol/l/hr)	Total Microbial Production[1] (µg C/l/hr)	Specific Growth Rate[2] (day⁻¹)	DNA Concentration (µg/l)	Specific Growth Rate[3] (day⁻¹)
5	57	0.76	12.7	0.84	0.98	3.0	0.09
10	59	1.32	15.0	0.99	1.16	2.5	0.13
15	68	1.66	29.7	1.97	1.92	4.0	0.16
20	60	1.28	24.1	1.60	1.77	2.6	0.21
25	69	2.34	53.0	3.51	3.38	2.1	0.55
35	84	2.51	48.6	3.22	2.54	2.1	0.51
45	71	1.40	14.2	0.94	0.88	4.6	0.07
55	71	1.27	24.5	1.62	1.51	2.7	0.20
65	63	0.70	14.3	0.95	1.00	1.9	0.16

[1] Carbon production extrapolated from DNA synthesis measurements by converting pmol DNA to pg DNA and by assuming that DNA is 2% of the cell carbon.

[2] Calculated from DNA synthesis and ATP-biomass data as, μ = (doublings day⁻¹) x 0.693, assuming a DNA:ATP ratio of 5.

[3] Calculated from DNA synthesis and DNA-biomass data as, μ = (doublings day⁻¹) x 0.693.

We interpret this to be due to the presence of "detrital" (i.e., non-living) DNA (Holm-Hansen et al. 1968; Sutcliffe et al. 1970) in the marine environment.

The development of new methods to distinguish between "living" and "non-living" nucleic acids in environmental samples would undoubtedly be useful in this regard and might also provide more accurate direct estimates of the mean C:DNA ratios of living microorganisms. Finally, it should be emphasized that extrapolation of nucleic acid synthesis data to estimates of carbon production assumes that marine microorganisms in nature exhibit "balanced growth", with all cellular components increasing at the same rate. As Eppley (1981) points out, the concept of balanced growth is fundamental (but generally not explicitly stated) to the use of nutrient assimilation rates as measures of rate constants for growth. Karl (1981) has already demonstrated that cell growth (as measured by RNA synthesis) and cell division (as measured by DNA synthesis) may be uncoupled in mixed assemblages of asynchronously growing marine microorganisms over incubation periods as short as 1 h.

CONCLUSION

This paper has attempted to summarize the theoretical principles involved in the application of ^{3}H-adenine as a measure of microbial RNA and DNA synthesis in the marine environment. We have not discussed methods employing other nucleic acid precursors, such as ^{3}H-thymidine or ^{3}H-uridine. In our view nucleic acid synthesis measurements employing ^{3}H-adenine and ^{3}H-thymidine are not mutually exclusive methods in microbiological oceanography, since each approach yields different information regarding microbial growth in the oceans. Unless evidence is presented to suggest otherwise, they should be considered complimentary techniques in experimental micro-biological oceanography. Finally, the potential significance of ecological information to be derived from estimates of in situ rates of RNA and DNA synthesis more than justifies the level of research effort that is required to ensure a proper interpretation and accu-rate extrapolation to estimates of microbial growth and production.

ACKNOWLEDGMENTS

We thank C. Andrews, D. Burns, D. Craven, and K. Orrett for their helpful comments and criticisms of an earlier draft of this manuscropt, K. Orrett for providing the unpublished data presented in Figure 2, and the scientists and crew members of the R/V CAYUSE (VERTEX I: G. A. Knauer, chief scientist) and R/V WECOMA (VERTEX II: L. Small, chief scientist) for their assistance in sample collection and analyses. This research was supported in part by National Science Foundation grants OCE78-25446 and OCE80-05180.

REFERENCES

Brock, T. D.   1967.   Bacterial growth rates in the sea: Direct
    analysis by thymidine autoradiography.   Science 155: 81-83.
Eppley, R. W.   1981.   Relations between nutrient assimilation and
    growth in phytoplankton with a brief review of estimates of
    growth rate in the ocean, pp. 251-263.   In: T. Platt [ed.],
    Physiological bases of phytoplankton ecology.   Can. Bull. Fish.
    Aquat. Sci. 210: 346 p.
Eppley, R. W., R. W. Holmes, and J. D. H. Strickland.   1967.   Sink-
    ing rates of marine phytoplankton measured with a fluorometer.
    J. Exp. Mar. Biol. Ecol. 1: 191-208.
Fellows, D., D. M. Karl, and G. Knauer.   1981.   Vertical distribu-
    tion, production and sedimentation of adenosine triphosphate in
    the upper 1550 meters of the Northeast Pacific Ocean.   Deep-Sea
    Res. 28A: 921-936.
Ferguson, R. L., and P. Rublee.   1976.   Contribution of bacteria to
    standing crop of coastal plankton.   Limnol. Oceanogr. 21:
    141-145.
Fuhrman, J. A., and F. Azam.   1980.   Bacterioplankton secondary
    production estimates for coastal waters of British Columbia,
    Antarctica, and California.   Appl. Environ. Microbiol. 39:
    1085-1095.
Fuhrman, J. A., and F. Azam.   1982.   Thymidine incorporation as a
    measure of heterotrophic bacterioplankton production in marine
    surface waters: evaluation and field results.   Mar. Biol. 66:
    109-120.
Hobbie, J. E., R. J. Daley, and S. Jasper.   1977.   Use of Nuclepore
    filters for counting bcteria by fluorescence microscopy.   Appl.
    Environ. Microbiol. 33: 1225-1228.
Hochstadt, J.   1974.   The role of the membrane in the utilization of
    nucleic acid precursors.   CRC Crit. Rev. Biochem. 2: 259-310.
Holm-Hansen, O.   1969.   Algae: Amounts of DNA and organic carbon in
    single cells.   Science 163: 87-88.
Holm-Hansen, O., W. H. Sutcliffe, Jr., and J. Sharp.   1968.   Measure-
    ment of deoxyribonucleic acid in the ocean and its ecological
    significance.   Limnol. Oceanogr. 13: 507-514.
Karl, D. M.   1979.   Measurement of microbial activity and growth in
    the ocean by rates of stable ribonucleic acid synthesis.   Appl.
    Environ. Microbiol. 38: 850-860.
Karl, D. M.   1980.   Cellular nucleotide measurements and applications
    in microbial ecology.   Microbiol. Rev. 44: 739-796.
Karl, D. M.   1981.   Simultaneous rates of ribonucleic acid and
    deoxyribonucleic acid syntheses for estimating growth and cell
    division of aquatic microbial communities.   Appl. Environ.
    Microbiol. 42: 802-810.
Karl, D. M.   1982.   Selected nucleic acid precursors in studies of
    aquatic microbial ecology.   Appl. Environ. Microbiol. 44:
    891-902.
Karl, D. M., and G. A. Knauer.   1983.   Vertical distribution,

transport and exchange of carbon in the Northeast Pacific
Ocean: Evidence for multiple zones of bacterial activity.
Deep-Sea Res. In press.

Karl, D. M., C. D. Winn, and D. C. L. Wong. 1981a. RNA synthesis
as a measure of microbial growth in aquatic environments. I.
Evaluation, verification and optimization of methods. Mar.
Biol. 64: 1-12.

Karl, D. M., C. D. Winn, and D. C. L. Wong. 1981b. RNA synthesis
as a measure of microbial growth in aquatic environments. II.
Field applications. Mar. Biol. 64: 13-21.

Kjeldgaard, N. O. 1967. Regulation of nucleic acid and protein
synthesis in bacteria. Adv. Microb. Physiol. 1: 39-95.

Maaloe, O., and N. O. Kjeldgaard. 1966. Control of Macromolecular
Synthesis. W. A. Benjamin Inc., New York.

Mandalstam, J., and K. McQuillen. 1976. Biochemistry of Bacterial
Growth. John Wiley and Sons, New York.

Meyer-Reil, L.-A. 1978. Autoradiography and epifluorescence
microscopy combined for the determination of number and
spectrum of actively metabolizing bacteria in natural waters.
Appl. Environ. Microbiol. 36: 506-512.

Moriarty, D. J. W., and P. C. Pollard. 1981. DNA snythesis as a
measure of bacteria productivity in seagrass sediments. Mar.
Ecol. Prog. Ser. 5: 151-156.

Moriarty, D. J. W., and P. C. Pollard. 1982. Diel variations in
bacteria productivity in seagrass (Zostera capricorni) beds
measured by rate of thymidine incorporation into DNA. Mar.
Biol. 72: 165-173.

Nierlich, D. P. 1978. Regulation of bacterial growth, RNA and
protein synthesis. Ann. Rev. Microbiol. 32: 393-432.

Sutcliffe, W. J., Jr., R. W. Sheldon, and A. Prakash. 1970.
Certain aspects of production and standing stock of particulate
matter in the surface waters of the Northwest Atlantic Ocean.
J. Fish. Res. Board Can. 27: 1917-1926.

Thelander, L., and P. Reichard. 1979. Reduction of ribonucleotides.
Ann. Rev. Biochem. 48: 133-158.

Tobin, R. S., and D. H. J. Anthony. 1978. Tritiated thymidine
incorporation as a measure of microbial activity in lake
sediments. Limnol. Oceanogr. 23: 161-165.

Winn, C. D., and D. M. Karl. 1983. Microbial productivity and
growth rate estimates in the tropical North Pacific Ocean.
Biol. Oceanogr. in press.

Yamazaki, H., and K. Leung. 1981. Determination of the total rates
of synthesis and degradation of RNA in bacterial cultures.
Can. J. Microbiol. 27: 168-174.

MEASUREMENTS OF BACTERIAL GROWTH RATES IN SOME MARINE SYSTEMS USING

THE INCORPORATION OF TRITIATED THYMIDINE INTO DNA

D. J. W. Moriarty

Division of Fisheries Research
CSIRO Marine Laboratories, P. O. Box 120
Cleveland, Qld 4163, Australia

INTRODUCTION

Heterotrophic microorganisms, especially bacteria, play an important part in decomposition processes, nutrient cycling and food chains in aquatic systems. In quantifying their role, measurement of the growth rate of the whole bacterial population is necessary, but has proved difficult. A number of different methods have been proposed for measuring microbial growth rates, but many are not specific for bacteria or do not include the whole population.

Some techniques have involved culture and enumeration of bacteria. The best approach for this type of technique is to alter natural conditions as little as possible and to use direct microscopy to count bacteria (Meyer-Reil 1977). Another technique based on direct microscopy uses the frequency of dividing cells to calculate growth rate of bacteria in seawater (Hagström et al. 1979). Besides the problem of determining whether a pair of cells is dividing, there are difficulties with this technique, particularly in the relation-ship between the frequency of dividing cells and growth rate (Newell and Christian 1981). This technique is not applicable to sediments, where many cells remain attached to each other and are bound in large aggregations of slime and particulate matter.

Azam, in this volume, has discussed in some detail the require-ment for measuring the growth rate of natural microbial populations. It is evident that the ideal method should involve minimal handling of the bacterial population and be applied quickly enough so as not to alter natural or in situ growth rates or to be influence by bacterial grazers. The use of radioactive nucleic acid precursors, especially thymidine, to measure the rate of DNA synthesis, has many

217

of the prerequisites of the ideal method. As with all other techniques, there are disadvantages as well as advantages in using
measurements of nucleic acid synthesis. This paper considers the
measurement of growth rates calculated from the rate of tritiated
thymidine incorporation into DNA. A more extensive review of
methods based on the synthesis of RNA and DNA is published elsewhere
(Moriarty, in press).

RELATIONSHIPS BETWEEN DNA AND RNA SYNTHESIS AND GROWTH

Growth in microorganisms is a complex process involving
synthesis of protein, RNA, and DNA, usually culminating in cell
division. DNA synthesis is directly proportional to division rate
in bacteria. The regulation of DNA synthesis occurs primarily at
initiation (Lark 1969). Once initiated, DNA synthesis proceeds to
completion, and the termination of replication triggers a cycle of
division. Initiation of replication is affected by growth conditions. Under conditions of rapid growth, more than one replication
fork may proceed along the chromosome at one time, but the rate of
travel is not affected by growth rate (Lark 1969). This close relationship between growth and DNA synthesis means that measurement of
the rate of DNA synthesis is a good measure of bacterial growth
rates. Unlike bacteria, eukaryotes do not synthesize DNA continuously throughout a growth cell cycle, but only at one stage. DNA
synthesis in eukaryotes is dependent on continued protein synthesis,
and replication stops if protein synthesis stops (Lark 1969). Thus
growth in eukaryotes is more complex. The method chosen to measure
DNA synthesis in a natural environment should, therefore, be
reasonably specific to bacteria for ease of interpretation.

The relationship of RNA synthesis to growth is much more
complex than that of DNA synthesis. Cells that are growing rapidly
(generation time of 1-2 h) do show a direct correlation between RNA
synthesis and growth rate. In cells that are growing more slowly,
however, there is some relationship between growth rate and RNA
synthesis, but it is not a simple linear function (Nierlich 1974).
This is further complicated by the different rates at which the main
types of RNA are synthesized. About 97% of total RNA is ribosomal
and transfer RNA, but these account for only about 50% of RNA
synthesis in rapidly growing cells. Messenger RNA, which is
unstable, comprises the rest (Nierlich 1974). In slowly growing
cells, net synthesis of stable RNA may cease completely when an
excess of these RNA forms is present; however, as they do turn over,
a slow rate of synthesis still occurs (Nierlich 1978).

In order to calculate division rates of cells in a natural
population, accurate information is needed on the rate of synthesis
of each form of RNA, the amount of RNA present and the growth state
of the cells. The amount of RNA per cell is variable and depends on

the growth state (Maaloe and Kjeldgaard 1966). Cells that are growing slowly generally have an excess of stable RNA. As most cells in
natural populations are not likely to be in a state of rapid growth,
the lack of a clear relationship between growth state and RNA content
as well as synthesis rates means the at measurement of the total
amount of RNA synthesis is not possible, as discussed by Fuhrman and
Azam (1980). Several reviews on growth and RNA synthesis in bacteria
discuss the complexities of the processes and it is only selected
strains of bacteria growing under defined conditions and at particular growth rates, that show precise relationships (Edin and Broda
1968; Nierlich 1974, 1978; Maaloe and Kjeldgaard 1966). Thus there
are severe problems associated with the use of RNA synthesis for
estimating microbial growth rates (Karl 1979, 1981).

The utilization or potential for utilization of exogenous
adenine for nucleic acid synthesis by algae as well as heterotrophic
bacteria, is a real disadvantage in trying to estimate bacterial
growth rates. As explained below, the great advantage of thymidine
is that may be used to measure heterotrophic bacterial growth rates
in the presence of microalgae. In order to study the growth and
activity of particular members of the complex microbial community,
we need techniques that are specific rather than general in their
application.

Enzymology of Thymidine Incorporation into DNA

Thymidine (thymine-2-deoxyribose; Tdr) is unique among nucleosides because the only function of its nucleotides in cells is
participation in the synthesis of DNA (O'Donovan and Neuhard 1970).
Thymidine is readily incorporated into DNA via a salvage pathway,
but in some bacteria the incorporation stops after a short time due
to breakdown of thymidine (O'Donovan and Neuhard 1970). De novo
synthesis proceeds via dUMP directly to dTMP (Fig. 1). Catabolism
of thymidine starts with conversion to thymine and ribose-1-phosphate
by the action of an inducible phosphorylase. The best radioactive
label is [methyl-^{3}H] because subsequent conversion to uracil removes
the label. The tritiated methyl group can be transferred to a wide
variety of compounds, but DNA is not labelled, as demonstrated in
microorganisms that lack thymidine kinase (Fink and Fink 1962).
[2-^{14}C] Thymidine, on the other hand, does label DNA after catabolism, because the label is retained in the resulting uracil (Grivell
and Jackson 1968). The absence of tritium incorporation into DNA in
some eukaryotic microorganisms led Grivell and Jackson (1968) to
show that these organisms lacked thymidine kinase. As Kornberg
(1980) pointed out, thymidine meets reasonable well the criteria for
pulse labelling. These are that the precursor should be rapidly and
efficiently taken up by bacteria, be stable during uptake, be
converted rapidly into the nucleotides and specifically label DNA
with little dilution by intracellular pools. He also outlined
pitfalls in its use, of which some are particularly relevant to

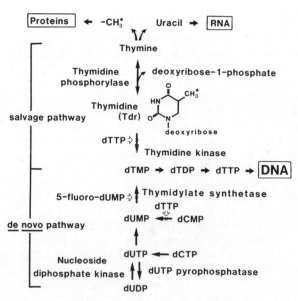

Figure 1. Some pathways of thymidine metabolism. The asterisk
shows the position of tritium labelling in thymidine. Sites of
feedback inhibition of dTTP and inhibition by 5-fluoro-dUMP are
indicated (adapted from Kornberg 1980).

environmental studies and are discussed in detail below.

     Thymidine is converted to dTMP by thmidine kinase (Fig. 1).
This enzyme must be present for labelling of DNA to occur to a
significant extent. Thymidine kinase was thought to occur in most
organisms (Kornberg 1980), but some groups of microorganisms are now
known not to contain it. These include fungi (Neurospora crassa,
Aspergillus nidulans and Saccharomyces cerevisiae) and Euglena
gracilis (Grivell and Jackson 1968), and a number of cyanobacteria
(blue green algae) (Glaser et al. 1973). It is also absent from the
nuclei of various eukaryotic algae, but may be present in chloroplasts although the amount of label incorporated from tritiated
thymidine into chloroplast DNA was slight and required hours or days
of incubation to be shown by autoradiography (Stocking and Gifford
1959; Sagan 1965; Steffensen and Sheridan 1965; Swinton and Hanawatt
1972).

     We have been unable to obtain significant incorporation of
[methyl-^{3}H] Tdr into DNA of four species of marine microalgae
(Thalassiosira, Isochrysis, Platymonas and Synechococcus), which
suggests that they lack thymidine kinase (Pollard and Moriarty, in
preparation). As there are no reports of the presence of thymidine
kinase in the nuclei of small eukaryotic algae, fungi or

cyanobacteria, it seems reasonable to generalize, and conclude the this salvage pathway is lacking in all members of these groups of microorganisms. The lack of thymidine kinase in blue green algae and many eukaryotic microorganisms is a considerable advantage for studies on heterotrophic bacterial production in the marine environment. Protozoa probably do contain the enzyme (Plant and Sagan 1958; Stone and Prescott 1964), but as explained below their contribution to label DNA in short term experiments is probably small. Thus the use of thymidine provides specific information about the growth of heterotrophic bacteria which has not been available previously.

Most bacteria that lack thymidine kinase are mutants specially selected for biochemical studies. Two wild type strains of Pseudomonas have been reported not to incorporate thymidine into DNA (Ramsay 1974). The technique used to demonstrate this was autoradiography, which is insensitive compared to liquid scintillation counting of purified DNA. Ramsay's results could mean that these bacteria lacked thymidine kinase, or that they had a deficient membrane transport mechanism. A few species of Pseudomonas have been found not to utilize thymidine, due to a deficient cell membrane transport system (Pollard and Moriarty, in preparation). Fuhrman and Azam (1980) have found good agreement between bacterial growth rates in seawater measured by the incorporation of thymidine and by counting the increase in cell number. The results of an autoradiographic study on bacteria in sea water support the view that most aerobic marine heterotrophic bacteria can utilize thymidine (Fuhrman and Azam 1982). Anaerobic bacteria with strict and limited nutrient requirements may not be able to utilize thymidine, particularly if they can transport only a limited range of metabolites. Desulfovibrio, for example, does not appear to be able to utilize exogenous thymidine (G. W. Skyring, personal communication).

The possibility that some bacteria in seawater may be unable to incorporate thymidine into DNA means that estimates of bacterial productivity may be too low, but this disadvantage is considerably outweighed by the advantages of using thymidine to measure DNA synthesis over other techniques for estimating growth rates of bacteria in natural populations.

Kinetics

Bacteria, with their active transport systems, take up organic molecules much more rapidly than do algae or protozoa, and can utilize nanomolar concentrations of organic molecules in their environment more effectively than algae or protozoa (Wright and Hobbie 1966; Fuhrman and Azam 1980). Thus in a short time period (e.g., 10 min at 25°C, 20 min at 15°C) tritiated thymidine should be taken up preferentially by bacteria in a mixed community. We have conducted autoradiography on surface sediment and epiphytic populations of microorganisms and have found that bacteria were

heavily labelled, whereas very few silver grains were formed over
diatoms and cyanobacteria after 15 min (Moriarty and Pollard 1982).
Autoradiography of the same populations with tritiated adenine,
however, showed more label in microalgae than with thymidine;
bacteria also were more heavily labelled.

Uptake of labelled thymidine by organisms should not be confused
with incorporation into DNA, although in bacteria the latter may be
the main fate of labelled thymidine assimilated intact (Hollibaugh et
al. 1980; Fuhrman and Azam 1980). As mentioned above, thymidine is
readily incorporated into DNA in bacteria, but thymidine phosphorylase soon converts thymidine to thymine and deoxyribose-1-phosphate
(Fig. 1). Labelled thymidine concentration within cells may be
rapidly depleted, so it is important to measure the rate of label
incorporation into DNA and not simply label uptake. In our work with
sediments we have found that label incorporation into DNA proceeds
linearly for 5-8 min at high temperatures (27°-31°C) and 20-30 min at
lower temperatures (15°-18°) in sediments, and over 1 h in seawater
(Fig. 2). We had interpreted the change from the linear rate to be
due to adsorption of thymidine by clay in sediments, but although
this undoubtedly is a factor, degradation by thymidine phosphorylase
may also have occurred. Experimental studies to determine growth
rates must be carried out in the initial linear period of incorporation of label into DNA. Uptake of thymidine by cells and incorporation into TCA-insoluble fractions are different processes, probably
with different kinetics which may be uninterpretable in a mixed
population.

The kinetic studies (Fig. 2) show that thymidine is very rapidly
taken up and is incorporated into DNA in less than 1 minute. We
presumed that this was indicative of bacterial activity (Moriarty and
Pollard 1981), because protozoa, the other main group of microorganisms with thymidine kinase, are generally particulate feeders and
probably would not have membrane transport mechanisms that are as
efficient as those of bacteria. We are not aware of any literature
on this topic and so are conducting studies with cycloheximide, an
inhibitor of DNA synthesis in eukaryotes (Cooney and Bradley 1962;
Venkatesan 1977). Preliminary results indicate that cycloheximide
has no effect on ^{3}H-thymidine incorporation into DNA in sediments
(Moriarty and Pollard 1982). In other words, although bacteria are
not the only organisms that utilize thymidine, they are the only
ones that do so significantly in a short experimental period.

Purification of DNA

In those organisms that do not contain thymidine kinase,
degradation is the only fate of thymidine. The amount of tritium
appearing in DNA is negligible, but other compounds are labelled and
thus DNA must be purified when working with mixed populations,
particularly if bacteria are minor components. In seawater, Fuhrman

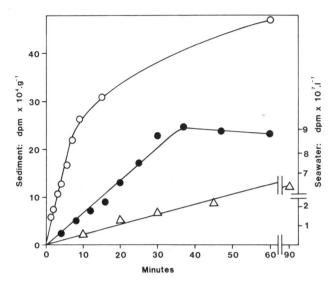

Figure 2.   Time course of incorporation of tritiated thymidine into
DNA.   △ = seawater; ● = sediment (17°C); ○ = sediment (31°C) (from
Moriarty and Pollard 1981, 1982).

and Azam (1980) have found that 90% of the total uptake of tritiated
thymidine was by microorganisms that passed through a 1 μm filter
and about 80% of that was incorporated into DNA. The proportion
incorporated into DNA will vary with different samples, so DNA
should be purified, or the proportion of tritium in DNA should be
checked for each environment.

   We based our methods on the Schmidt-Thannhauser procedure
(reviewed by Munro and Fleck 1966) but modified them to obtain
maximum recovery of DNA from sediment with low background radioactive
contamination (Moriarty and Pollard 1981). Recovery of standard
amounts of DNA varied from about 40% to 70%; it was higher when humic
acid was present. One problem with the technique is that DNA has to
be precipitated from solution by ice-cold strong acid, and we have
found that losses occurred, probably due to hydrolysis. Recoveries
of DNA were generally in the range of 55% ± 5%. In order to
calculate growth rates of bacteria, it is essential, therefore, to
measure the actual recovery of DNA from environmental samples.

   The problem of measuring background values due to adsorption of
radioactivity is an important one. It is important to ensure that
control experiments, in which bacterial DNA synthesis is prevented,
contain the same amount of labelled thymidine during extraction of
DNA. We achieved this in sediments by mixing control sediments in
NaOH first, then adding the same amount of thymidine as in the
experimental samples. Background values could also be due to

adsorption of or exchange of tritiated water or other products of
thymidine metabolism in active cells onto compounds such as DNA,
protein and humic acids. Such processes would not occur in zero-
time or formalin-killed controls. Thus true background values may
be higher than those actually measured.

Isotope Dilution

    The specific radioactivity of exogenous thymidine is diluted
during incorporation into DNA, primarily by de novo synthesis of
dTMP (Fig. 1). A technique for measuring the dilution of label from
an exogenous precursor during synthesis of a macromolecule is to add
different quantities of unlabelled precursor as well, and to measure
the effect on the amount of label actually incorporated into the
macromolecule (Forsdyke 1968, Hunter and Francke 1974). We showed
that this technique worked well with bacterial populations in
sediments (Moriarty and Pollard 1981). A plot of the reciprocal of
isotope incorporated into DNA against total amount of thymidine
added is extrapolated to give the amount of dilution of isotope in
DNA itself (Fig. 3). We referred to the negative intercept on the
abscissa rather loosely as pools of thymidine and other endogenous
precursors that dilute the added labelled thymidine (Moriarty and
Pollard 1981). In fact, true pool sizes can be calculated only if
the actual amount of label taken up by the cells is known. We used
the negative intercept on the abscisssa to determine the specific
activity of tritiated thymidine actually incorporated into DNA.

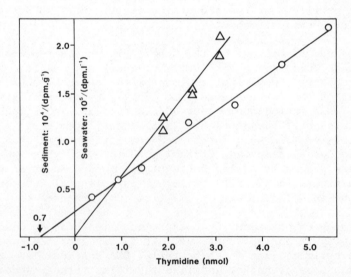

Figure 3.   Isotope dilution plots for incorporation of tritiated
thymidine into DNA. Δ = seawater; 0 = sediment (from Moriarty and
Pollard 1981).

This technique measures the dilution of labelled thymidine in dTTP (the final precursor for DNA synthesis) by all other precursors of dTTP, including de novo synthesis. Provided DNA is purified before counting radioactivity that is incorporated, it doesn't matter if only a small proportion of the thymidine that is taken up by cells is used for DNA synthesis, because only the dilution of isotope in thymidine pools that are actually being used for DNA synthesis in growing cells is measured. Thus this method is not subject to errors inherent in trying to extract and quantify nucleotides from cells.

The amount of dilution of labelled thymidine incorporated into DNA in various marine systems varies considerably. Fuhrman and Azam (1980) initially assumed that there was little dilution in seawater samples that they analyzed. We have used our technique on seawater from a number of environments, and have found that there was no significant isotope dilution in planktonic bacteria when a high concentration of thymidine (16 nM) was used (Fig. 3; Moriarty 1983; Moriarty and Pollard 1982).

Subsequently, Fuhrman and Azam (1982) reported that isotope dilution may be as high as 4 to 7-fold in seawater bacteria. Their reasons for suggesting this were based on the relative difference in incorporation of labelled thymidine and ^{32}P into DNA in samples in the presence and absence of the inhibitor mitomycin C. The difference between the rates of incorporation of the labelled substrates, however, may have been due to the synthesis of RNA, and not to dilution of thymine in DNA. This is because tritiated thymidine is incorporated only into DNA and not RNA (any results to the contrary would be due to imperfect extraction and separation of the macromolecules), whereas ^{32}P will label both nucleic acids, and furthermore, mitomycin C may block RNA synthesis as well as DNA synthesis. A better procedure to test for isotope dilution is the one that we have used (Moriarty and Pollard 1982). These problems with measuring the amount of isotope dilution by de novo synthesis of dTMP can be avoided by using a high enough concentration of thymidine to supply all the thymine required for DNA synthesis, in which case de novo synthesis is switched off by feedback inhibition (for more detail see Moriarty, in press). About 10-20 nM thymidine is generally sufficient to do this (unpublished observations). Once this concentration is determined for a particular environment, only one incubation is needed with thymidine to estimate a growth rate, which considerably simplifies ecological work.

In sediments there appears to be a correlation between growth rate of bacteria and dilution of isotope in DNA. Dilution is greater in systems with fast growth rates than in systems with slowly growing bacteria (Table 1). These results suggest that in slowly growing bacteria the exogenous thymidine is sufficient for DNA synthesis, and de novo synthesis of dTTP is inhibited. In rapidly growing bacteria, however, the supply of exogenous thymidine presumably is insufficient

Table 1.   Degree of participation (DP) of exogenous thymidine in DNA
            synthesis by bacteria in a seagrass sediment (Moriarty and
            Pollard 1982).

DP %	Production rate No. cells produced$\cdot$h$^{-1}\cdot$g^{-1}
16	2×10^7
18	1×10^7
25	7×10^6
33	6×10^6
57	2×10^6
80	8×10^5

to maintain the level of dTTP required, and thus de novo synthesis
proceeds as well. The degree of participation of exogenous thymidine
in DNA synthesis was generally in the range of 20% to 70% in sediment
populations. Similar values were found by Rosenbaum-Oliver and
Zamenhof (1972) for E. coli DNA synthesis.

     Small pools of thymidine inside and outside cells may also exist,
but are probably unimportant. Since publishing our original studies,
in which we reported evidence suggesting that substantial pools of
thymidine occurred in three sediment samples (Moriarty and Pollard
1981), we have analyzed many different environmental samples and
have found no further evidence for substantial dilution of tritiated
thymidine (see e.g., Moriarty and Pollard 1982). Those experiments
which showed substantial dilution were carried out with comparatively
large amounts of sediment. Probably most of the thymidine was
adsorbed to sediment, as the actual concentration available to
bacteria was very small. The kinetics of transport into the cell
may be different at very low substrate concentrations. With less
sediment dilution is low, and no biphasic curves are obtained
(Moriarty and Pollard 1982).

Growth Rate Calculation

     To calculate growth rate from the rate of synthesis of a
macromolecule measured by incorporation of a precursor, the total
amount of the macromolecule per cell and the amount of precursor in
the macromolecule must be known. As pointed out by Fuhrman and Azam
(1980), the RNA content of bacterial cells is variable, and depends
on growth rate, whereas DNA content of all bacteria varies little
over a wide range of growth rates. Thus to estimate growth rate of
bacteria DNA synthesis is more suitable than RNA synthesis.
We assumed that thymine constituted an average of 25% of the
bases in DNA, and the average genome size was 2.5×10^9 daltons

(range 1 x 10^9 to 3.6 x 10^9) (Moriarty and Pollard 1981); these values are similar to those used by Fuhrman and Azam (1980).

Another way to determine a conversion factor is to estimate it directly, by measuring the growth rate of bacteria in seawater culture by direct microscopy and measuring the rate of incorporation of thymidine into DNA. This has been done by Kirchman et al. (1982) and has one particular advantage. The conversion factor calculated with their procedure takes into account any bacteria that cannot utilize thymidine for DNA synthesis. There are two conditions that need to be met in both the seawater culture and the natural marine environment, viz., (1) the time period for assay with tritiated thymidine must be short enough to ensure that only DNA is labelled, and (2) isotope dilution should be measured, or better still, prevented by using a high thymidine concentration. If these conditions are not met the conversion factors in culture would be different from those in the sea.

BACTERIAL PRODUCTIVITY IN WATER AND SEDIMENTS

Some examples of the range of bacterial growth rates and productivities that we have measured in seawater are shown in Table 2. Diurnal variation of growth rates in the water column occurs in some bodies of water. In the surface sediment of seagrass beds there is a marked diurnal cycle in bacterial growth rates (Fig. 4).

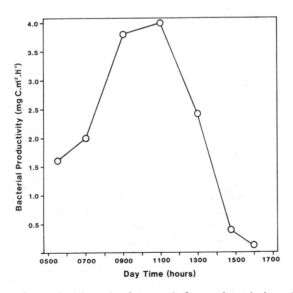

Figure 4. Diurnal variation in bacterial productivity in surface sediment of seagrass beds (adapted from Moriarty and Pollard 1982).

Table 2.   Bacterial productivity in seawater.

Environment	Productivity $\mu g\ C \cdot l^{-1} \cdot h^{-1}$	Reference
Moreton Bay, Queensland	1 - 3	Moriarty and Pollard 1982
Great Barrier Reef, lagoon off Lizard Is.	0.2 - 0.6	Moriarty and Pollard, in preparation
Hamelin Pool, Shark Bay, W.A. (hypersaline)	0.1 - 0.3	Moriarty 1983

Although not shown here, we have carried out studies in a number of environments and over a full diel period. The large increase in growth rates (often 5-10 fold) occurred only in sediments closely associated with seagrass during daylight hours (Moriarty and Pollard 1982). Bacterial productivities in the seagrass beds in autumn were about 80 mg $C \cdot m^{-2} \cdot d$ in the surface sediment (0-3 mm depth) and about 50 mg $C \cdot m^{-3} \cdot d$ in the water column (average depth 1 m). If we assume the bacterial conversion efficiency is 50% then a total of 260 mg $C \cdot m^{-2} \cdot d^{-1}$ would be required from the primary producers to support this production. Further organic C would be required for slime production by the sediment bacteria and to support the bacterial population below the sediment surface. As there are technical difficulties in measuring bacterial growth rates accurately in anaerobic cores, we have not yet obtained many results for bacterial productivity around seagrass roots. Preliminary results suggest that the productivity is less than in the surface layers.

A substantial proportion of primary production probably is utilized by bacteria within the seagrass beds and through them by animals. The primary production utilized by bacteria includes dead roots and leaves as well as dissolved organic matter excreted by roots, rhizomes and leaves and by benthic and epiphytic algae. The primary producers would depend, at least to some extent, on release of nutrients from decomposing organic matter by bacteria. We are trying to quantify some of these interrelationships.

Measurement of bacterial growth by DNA synthesis rates indicates how many new cells are produced. It does not tell us how much total organic matter, especially extracellular products such as slime, is produced by bacteria. Bacteria that are not dividing and synthesizing DNA may also produce extracellular products. Such products are probably trophically important to deposit-feeders, and accurate measurement of their production in natural environments is a challenge awaiting investigation.

ACKNOWLEDGMENTS

I am grateful to Peter C. Pollard for preparing the figures.

REFERENCES

Azam, F. 1983. Growth of bacteria in the oceans. In: J. E. Hobbie and P. J. LeB. Williams [eds.], Heterotrophic Activity in the Sea. Plenum Press, New York.

Cooney, W. J., and S. G. Bradley. 1962. Action of cycloheximide on animal cells, pp. 237-244. In: M. Finland and G. M. Savage [eds.], Antimicrobial Agents and Chemotherapy - 1961. American Soc. Microbiol., Michigan.

Edin, G., and P. Broda. 1968. Physiology and genetics of "ribonucleic acid control" locus in Escherichia coli. Bacteriol. Rev. 32: 206-226.

Fink, R. M., and K. Fink. 1962. Relative retention of H^3 and C^{14} labels of nucleosides incorporated into nucleic acids of Neurospora. J. Biol. Chem. 237: 2889-2891.

Forsdyke, D. R. 1968. Studies of the incorporation of [5-^{3}H] uridine during activation and transformation of lymphocytes induced by phytohemagglutinin. Biochem. J. 107: 197-205.

Fuhrman, J. A., and F. Azam. 1980. Bacterioplankton secondary production estimates for coastal waters of British Columbia, Antarctica, and California. Appl. Environ. Microbiol. 39: 1085-1095.

Fuhrman, J. A., and F. Azam. 1982. Thymidine incorporation as a measure of heterotrophic bacterioplankton production in marine surface waters: evaluation and field results. Mar. Biol. 66: 109-120.

Glaser, V. M., M. A. Al-Nui, V. V. Groshev, and S. V. Shestakov. 1973. The labelling of nucleic acids by radioactive precursors in the blue-green algae. Arch. Mikrobiol. 92: 217-226.

Grivell, A. R., and J. J. Jackson. 1968. Thymidine kinase: evidence for its absence from Neurospora crassa and some other microorganisms, and the relevance of this to the specific labelling of deoxyribonucleic acid. J. Gen. Microbiol. 54: 307-317.

Hagström, Å., U. Larsson, P. Horstedt, and S. Normark. 1979. Frequency of dividing cells, a new approach to the determination of bacterial growth rates in aquatic environments. Appl. Environ. Microbiol. 37: 805-812.

Hollibaugh, J. T., J. A. Fuhrman, and F. Azam. 1980. Radioactively
    labeling of natural assemblages of bacterioplankton for use in
    trophic studies. Limnol. Oceanogr. 25: 172-181.
Hunter, T., and B. Francke. 1974. In vitro polyoma DNA synthesis
    characterization of a system from infected 3T3 cells. J.
    Virol. 13: 125-139.
Karl, D. M. 1979. Measurement of microbial activity and growth in
    the ocean by rates of stable ribonucleic acid synthesis. Appl.
    Environ. Microbiol. 38: 859-860.
Karl, D. M. 1981. Simultaneous rates of RNA and DNA syntheses for
    estimating growth and cell division of aquatic microbial
    communities. Appl. Environ. Microbiol. 42: 802-810.
Kirchman, D., H. W. Duckow and R. Mitchell. 1982. Estimates of
    bacterial growth from changes in uptake rates and biomass.
    Appl. Environ. Microbiol. 44: 1296-1307.
Kornberg, A. 1980. DNA Replication. W. H. Freeman, San Francisco.
Lark, K. G. 1969. Initiation and control of DNA synthesis. Ann.
    Rev. Biochem. 38: 569-604.
Maaloe, O., and N. O. Kjeldgaard. 1966. Control of macromolecular
    synthesis. A study of DNA, RNA, and protein synthesis in
    bacteria. In: Microbial and Molecular Biology Series. W. A.
    Benjamin, Inc., New York, Amsterdam.
Meyer-Reil, L.-A. 1977. Bacterial growth rates and biomass produc-
    tion, pp. 223-236. In: G. Rheinheimer [ed.], Microbial Ecology
    of a Brackish Water Environment. Springer-Verlag, Berlin,
    Heidelberg, New York.
Moriarty, D. J. W. 1983. Bacterial biomass and productivity in
    sediments, stromatolites and water of Hamelin Pool, Shark Bay,
    W. A. Geomicrobiology. In press.
Moriarty, D. J. W. 1983. Measurement of bacterial growth rates in
    aquatic systems using rates of nucleic acid synthesis. Adv.
    Aquat. Microbiol. In press.
Moriarty, D. J. W., and P. C. Pollard. 1982. Diel variation of
    bacterial productivity in seagrass (Zostera capricorni) beds
    measured by rate of thymidine incorporation into DNA. Mar.
    Biol. 72: 165-173.
Moriarty, D. J. W., and P. C. Pollard. 1981. DNA synthesis as a
    measure of bacterial productivity in seagrass seidments. Mar.
    Ecol. Prog. Ser. 5: 151-156.
Munro, H. N., and A. Fleck. 1966. The determination of nucleic
    acids, pp. 113-176. In: D. Glick [ed.], Methods of Biochemical
    Analysis, Vol. 14. Interscience Publishers, John Wiley and
    Sons, New York, London, Sydney.
Newell, S. Y., and R. R. Christian. 1981. Frequency of dividing
    cells as an estimator of bacterial productivity. Appl.
    Environ. Microbiol. 42: 23-31.
Nierlich, D. P. 1974. Regulation of bacterial growth. Science
    184: 1043-1050.
Nierlich, D. P. 1978. Regulation of bacterial growth, RNA, and
    protein synthesis. Ann. Rev. Microbiol. 32: 393-432.

O'Donovan, G. A., and J. Neuhard. 1970. Pyrimidine metabolism in microorganisms. Bacteriol. Rev. 34: 278-343.

Plant, W., and A. Sagan. 1958. Incorporation of thymidine in the cytoplasm of Amoeba proteus. J. Biophys. Biochem. Cytol. 4: 843-847.

Ramsay, A. J. 1974. The use of autoradiography to determine the proportion of bacteria metabolizing in an aquatic habitat. J. Gen Microbiol. 80: 363-373.

Rosenbaum-Oliver, D., and S. Zamenhof. 1972. Degree of participation of exogenous thymidine in the overall deoxyribonucleic acid synthesis in Escherichia coli. J. Bacteriol. 110: 585-591.

Sagan, L. 1965. An unusual pattern of tritiated thymidine incorporation in Euglena. J. Protozool. 12: 105-109

Steffensen, D. M., and W. F. Sheridan. 1965. Incorporation of 3H-thymidine into chloroplast DNA of marine algae. J. Cell Biol. 25: 619-626.

Stocking, C. R., and E. M. Gifford, Jr. 1959. Incorporation of thymidine into chloroplasts of Spirogyra. Biochem. Biophys. Res. Commun. 1: 159-164.

Stone, G. E., and D. M. Prescott. 1964. Cell division and DNA synthesis in Tetrahymena pyriformis deprived of essential amino acids. J. Cell Biol. 21: 275-281.

Swinton, D. C., and P. C. Hanawalt. 1972. In vivo specific labeling of Chlamydomonas chloroplast DNA. J. Cell Biol. 54: 592-597.

Venkatesan, N. 1977. Mechanism of inhibition of DNA synthesis by cycloheximide in Balb/3T3 cells. Biophys. Acta 478: 437-453.

Wright, R. T., and J. E. Hobbie. 1966. Use of glucose and acetate by bacteria and algae in aquatic ecosystems. Ecology 47: 447-464.

BACTERIAL GROWTH IN RELATION TO PHYTOPLANKTON PRIMARY

PRODUCTION AND EXTRACELLULAR RELEASE OF ORGANIC CARBON

Bo Riemann and Morten Søndergaard*

Freshwater Biological Laboratory
University of Copenhagen,
Helsingørsgade 51, DK-3400 Hillerød, Denmark
*Botanical Institute
University of Aarhus
Nordlandsvej 68, DK-8240 Risskov, Denmark

INTRODUCTION

Carbon flow in pelagic waters is dominated by phytoplankton photosynthesis and the subsequent transport and decomposition by heterotrophic organisms of the produced organic matter. Present understanding of the interactions between autotrophic and heterotrophic organisms suffers from limited knowledge of the secondary production. One major problem is the part played by bacteria in returning dissolved organic carbon to organic particles which can be utilized by higher trophic levels (Pomeroy 1974). In this context it is especially important to elucidate the dynamic relationship between phytoplankton, bacteria and zooplankton and the extent to which such interactions are controlled by extracellular organic carbon released by the phytoplankton.

Recent evidence has suggested that extracellular organic carbon (EOC) released by phytoplankton populations is of importance to bacterial secondary production and thus to the carbon and energy flow in the aquatic environment (Nalewajko 1977; Larsson and Hagström 1979; Coveney 1982). The term "extracellular organic carbon" only covers release from healthy algae (Nalewajko 1977). Hobbie and Rublee (1977) demonstrated a close relationship between phytoplankton primary production and the potential bacterial uptake of organic substrates. Several workers have shown that EOC released from phytoplankton can be utilized by bacteria (Bell and Mitchell 1972; Herbland 1975; Nalewajko et al. 1976; Iturriaga and Hoppe 1977; Bell and Sakshaug 1980; Iturriaga 1981).

233

So far most studies have been qualitative in approach or have
compared the heterotrophic utilization of EOC with primary production.
Few have treated the quantitative role of EOC products in relation to
bacterial growth. Smith et al. (1977) concluded that the metabolism
of bacteria prevented EOC accumulation in upwelling regions and that
EOC was probably of minor importance to the bacterial carbon budget.
Independent measurements of bacterial production were, however, not
presented. Using dark CO_2 uptake as an estimate of bacterial production and a differential filtration technique to measure EOC transport
from algae to bacteria, Coveney (1982) reported that bacterial assimilation of the EOC products accounted for from 30 to 90% of bacterial
production. In the northern Baltic proper, Larsson and Hagström
(1982) found that extracellular products constituted on average 50%
of the carbon source for the bacterial production observed.

The present article reviews recent studies on bacterial growth
in relation to phytoplankton primary production, release of newly
formed photosynthetic products, and zooplankton grazing. Diel
variations of bacterial secondary production measured by several
independent methods are presented.

DISCUSSION

Bacterial growth in relation to phytoplankton production, EOC and
zooplankton grazing during a spring diatom bloom

To understand the flux of carbon between natural planktonic
assemblages, it is necessary to measure biomass as well as rates of
carbon transport through the phytoplankton, bacteria and zooplankton.
In 1980, a study was carried out during a spring diatom bloom in the
eutrophic Lake Mossø in order to characterize carbon pool sizes and
important routes of carbon flow (Riemann et al. 1982b).

During the increase of the diatoms, release of EOC and bacterial
uptake of EOC remained low. Total release averaged 3% of the total
^{14}C fixation (Fig. 1). After the peak of the bloom, total release
of EOC increased to a maximum of 12%. Between 30 and 50% of the
extracellular products were utilized by bacteria during the 5 hr
incubation. Transport of EOC to bacteria was measured as ^{14}C
activity in small free-living bacteria (< 1.0-μm > 0.2-μm) corrected
for glucose uptake in different size fractions (Derenbach and
Williams 1974). These values are similar in magnitude to those
reported by Bell and Sakshaug (1980), Berman and Gerber (1980),
Mague et al. (1980) and Larsson and Hagström (1979, 1982) and
support the conclusions (1) that total release of EOC represents a
minor proportion of the carbon fixed by the algae, (2) that
exponentially growing algae release less EOC than stressed algae
(Sharp 1977; Mague et al. 1980) and (3) that the products released
are labile to heterotrophic assimilation (Bell and Sakshaug 1980;

Iturriaga 1981). In this context it must be emphasized that most
data published on release and assimilation of EOC are net values.
An underestimation of true release and uptake is inherent in current
in situ usage of the ^{14}C technique. This is further supported by
the high respiration rates of EOC found by Iturriaga (1981).

During the diatom bloom, phytoplankton biomass (measured as
chlorophyll a, ATP, TANC and by direct counting) had a similar time
pattern to the carbon fixation, with a distinct biomass maximum from
April 10-14. By contrast, bacterial numbers showed small variations
throughout the bloom (total numbers varied between 11 and 18 x 10^6
ml^{-1}). Distribution of the bacteria in two size fractions also
varied little, with about 50% found in the phytoplankton fraction
(> 1.0-μm) and 50% in a small particle fraction (< 1.0-μm > 0.2-μm)
(Fig. 2). The numbers of glucose-active bacteria (measured by microautoradiography) from both size fractions increased throughout the
diatom bloom (Fig. 2). The percentage of active bacteria (percentage
of total numbers) was correlated with the glucose assimilation rate
(r = 0.827, p < 5%) and with total EOC release (r = 0.809, p < 5%).

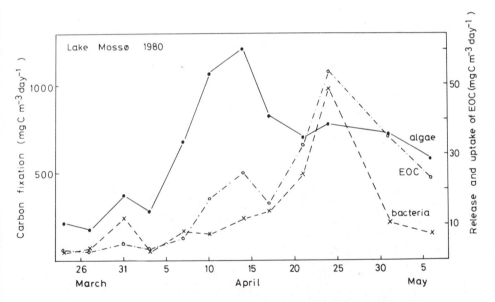

Figure 1. Changes in daily carbon fixation into algae (size
fraction > 1.0-μm, corrected for bacterial uptake by means of the
distribution of ^{3}H-glucose uptake in various size fractions, left
axis), into bacteria (size fraction < 1.0-μm > 0.2-μm, corrected for
bacteria found in size fraction > 1.0-μm, right axis), and dissolved
extracellular organic carbon (EOC, size fraction < 0.2-μm, right
axis) during the spring diatom bloom in Lake Mossø. Redrawn from
Riemann et al. (1982b).

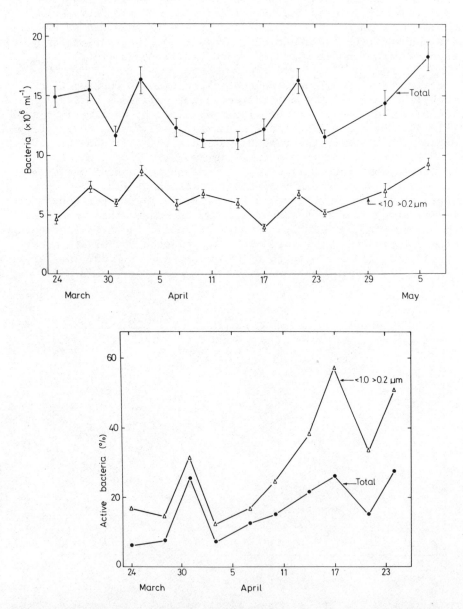

Figure 2. Changes in number of bacteria (top) and percentage glucose active bacteria (% of bacterial number) using micro-autoradiography in two size fractions (bottom) during the spring diatom bloom in Lake Mossø. Bars represent range of two samples. Redrawn from Riemann et al. (1982b).

The flow of carbon through the phytoplankton, bacteria and zooplankton is given for three selected dates to demonstrate important
carbon flow characteristics during various stages of growth of the
phytoplankton (Fig. 3). During the increase of the phytoplankton
bloom (April 3) about 2% of the fixed carbon was released as EOC, and
the bacteria assimilated about half the EOC released during 10 hours.
Assuming zero or reduced release of EOC from the phytoplankton at
night, the bacteria could theoretically assimilate most of the EOC
products throughout 24 hours and thus prevent any significant accumulation. A similar conclusion can be reached for the turnover of EOC
material during the peak of the bloom (April 10) and after its
culmination (April 21).

An estimate of bacterial production was made by means of dark
$^{14}CO_2$ uptake. These estimates are tentative (Overbeck 1979; Jordan
and Likens 1980), however, assuming that true bacterial secondary
production lies between EOC uptake and the production calculated
from the dark $^{14}CO_2$ fixation, a subsequent increase in the bacterial
biomass should be expected, since zooplankton grazing as a maximum
would ingest a small fraction of the bacteria (Fig. 3). No such
increase in the bacterial biomass could be verified by cell counting.
As bacterial cells do not disappear by sedimentation unless attached
to larger and heavier particles (Jassby 1973, cited in Jordan and
Likens 1980), and no such attachment was observed, we have proposed
that lysis of the bacterial cells was an important factor in
regulating bacterial biomass.

Most of the recent literature reports rather high transport
rates of EOC to bacteria, both from freshwater systems (Nalewajko
and Schindler 1976; Coveney 1982) and from marine environments
(Derenbach and Williams 1974; Berman 1975; Herbland 1975; Smith et
al. 1977; Larsson and Hagström 1982). This supports the conclusion
that EOC in most circumstances neither accumulates on a seasonal
basis nor on a diel basis. In fact, Mague et al. (1980) and
Søndergaard and Schierup (1982) demonstrated that most of the EOC
material consists of small molecules < 700 dalton. Compounds with
molecular weights above 1,500 dalton have however, also been
reported (Nalewajko and Schindler 1976). Iturriaga (1981) found
that the products released by phytoplankton were characterized by a
low molecular size fraction, which appeared after 1 hr of photoassimilation, and by a higher molecular weight fraction which
appeared after 4 hrs of photoassimilation. He also reported higher
rates of uptake and respiration of low molecular weight compounds,
whereas high molecular weight products (< 100,000 > 1,000 dalton)
were incorporated and respired at lower rates (8-16% day^{-1}). A
similar conclusion was made by Chrost (1981) from studies in the
Rhode River estuary.

Bell and Sakshaug (1980), on the other hand, found long EOC
turnover times, between 160 and 440 h, and an accumulation of EOC

which increased from 20 µg C·l⁻¹ at the onset of a marine phytoplankton bloom to a steady state concentration of 160-200 µg C·l⁻¹. Their
technique included addition of labelled EOC substrates produced from
cultures of the same types of phytoplankton present in situ. These
EOC products were diluted with stored seawater and added at several
different concentrations to natural populations of bacteria during a
diatom bloom in Trondheim Fjord. The long turnover times of EOC may
reflect species-specific EOC products (Hellebust 1974), some of which
may be quickly assimilated by bacteria while others may be more
refractory. Cahet and Martin (1979), Martin (1980), and Martin and
Bianchi (1980) found that both the quantity and molecular composition
of EOC were factors regulating bacterial production. Changes in the
composition of EOC rapidly induced new bacterial enzyme systems; after
this induction period no substantial accumulation of EOC products
should be expected. To our knowledge, no investigations have shown
major proportions of the EOC to be refractory to bacterial attack in
natural plankton populations. High molecular weight EOC products
were reported by Søndergaard and Schierup (1982) from Lake Mossø, and
these coincided with the scenescent phase of the bloom and with the
occurrence of large (up to several hundred microns long) filamentous
bacteria. By means of microautoradiography, the filamentous bacteria
were seen to be very active (Riemann et al. 1982b). Whether or not
these high molecular weight products in fact favored the growth of
the filamentous bacteria is still a moot question.

Measurements of EOC transport from algae to bacteria using the
¹⁴C method have certain disadvantages, which must be carefully
considered before the results are interpreted and compared with
estimates of bacterial production. All results are net values
calculated on the assumption of rapidly attained isotopic equilibrium
in the products released. Any delay in achieving equilibrium leads
to underestimation of release (Storch and Saunders 1975). Grazing
on bacteria would also lead to an underestimate, while grazing by
zooplankton on algae can cause cell rupture and overestimation of
true release. Algal lysis during incubation would also produce an
overestimate.

In the past few years, there have been a number of estimates of
importance of EOC compunds in relation to total bacterial carbon
assimilation. Most estimates are that EOC products may constitute
from 10 to 50% of the total bacterial carbon assimilation. The
entire field is too new to know whether variations reflect different
methods used, the different importance of EOC in different types of
water, or changes caused by the plankton populations present at
different times of the year. The physiological state of the phytoplankton and species composition are also important factors (Chrost
1981; Søndergaard and Schierup 1982). The suggested lysis of bacteria as a control of the biomass may be important even in short-time
experiments (hours) and could mask the interactions between release
and assimilation of EOC.

Diel patterns in bacterial growth in relation to phytoplankton EOC release

Assuming that EOC release from natural populations of phyto-plankton is characterized by changes in the molecular weight distribution dependent on light and incubation periods (Iturriaga 1981; Søndergaard and Schierup 1982), diel changes should also be expected in the bacterial secondary production. Such changes may reflect changes in phytoplankton photosynthesis via release of low molecular weight products at the start of photosynthesis (sunrise) and a probable decrease during the day (and night?) when high molecular weight compounds with low uptake rates are released (Iturriaga 1981).

Examples of diel variations in the bacterial secondary produc-tion in a eutrophic lake are given in Figures 4 and 5 as rates of ^{3}H-thymidine incorporation into DNA (Riemann et al. 1982a). During November 17-18, ^{3}H-thymidine incorporation increased during the morning and decreased slowly until 8 p.m. (sunset 4 p.m.). At midnight a new increase was seen, followed by a decrease before sunrise on November 18, which initiated a new increase. On the other hand ^{3}H-glucose uptake was low during the morning of November 17, increased at sunset, and continued to increase slowly during the night.

On April 6-7, exactly the same variations were found in ^{3}H-thymidine incorporation as was seen during November 17-18. These diel variations suggest that sunrise may initiate conditions which increase the activity of the bacteria. Phytoplankton EOC products may be possible candidates for explaining the variations observed. During the first hours of sunlight, low molecular weight EOC compounds may initiate high bacterial activity. Later on, when these substrates have been taken up by the bacteria, phytoplankton EOC compounds with higher molecular weight are released. These compounds are probably taken up by the bacteria at lower rates (Iturriaga 1981), leading to reduced bacterial activity. It was interesting to note that after a slow decrease in the thymidine incorporation into DNA during the day, a marked increase was observed 4 hours after sunset on April 6-7 and 8 hours after sunset on November 17-18. Whether these changes are directly related to quality and quantity of EOC products is difficult to evaluate, since very little is known about the composition and amount of EOC products released during night.

The patterns observed demonstrate an interesting scenario for growth of bacteria in natural populations in relation to the possible changes in the EOC pool released by phytoplankton. However, the absolute amounts of carbon produced by the bacteria per unit volume during the 24 hours were very low compared to the phytoplankton primary production and EOC release (Riemann et al. 1982a).

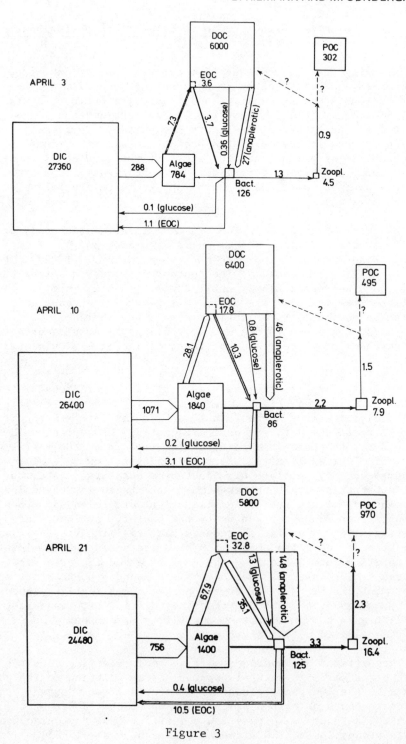

Figure 3

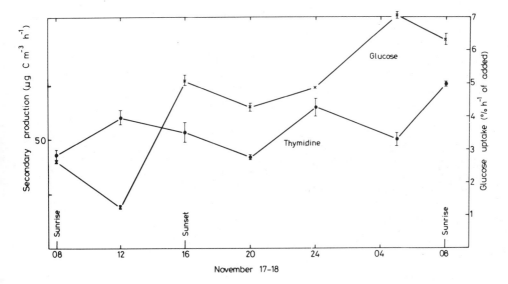

Figure 4. Diel changes in ^{3}H-thymidine incorporation into DNA and ^{3}H-D-glucose uptake during November 17-18 in Frederiksborg Slotssø. Bars represent SE, n = 3. From Riemann et al. (1982a).

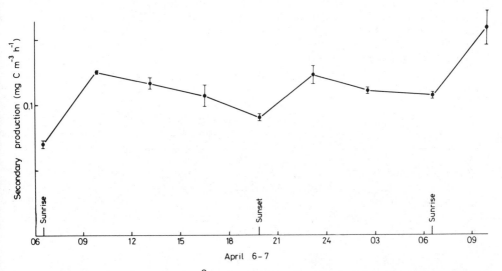

Figure 5. Diel changes in ^{3}H-thymidine incorporation into DNA during April 6-7 in Frederiksborg Slotssø. Bars represent SE, n = 3. From Riemann et al. (1982a).

In an attempt to evaluate the results obtained from thymidine incorporation into DNA, we measured bacterial secondary production by the methods of $^{35}S-SO_4$ dark uptake (Jordan and Likens 1980), frequency of dividing cells (Hagström et al. 1979; Newell and Christian 1981), uptake of EOC products and ^3H-thymidine incorporation during two 24 hour periods in two eutrophic Danish lakes: Frederiksborg Slotssø and Lake Mossø (Table 1).

In both lakes, the phytoplankton was dominated by blue-green algae (Microcystis in Frederiksborg Slotssø; Aphanizomenon and Oscillatoria in Mossø), and low values of EOC release were observed. In situ uptake of $^{35}S-SO_4$ and ^3H-thymidine was performed every 3 hours from subsamples taken from a 5 litre bottle.

^3H-thymidine incorporation into DNA revealed very low values, even lower than those obtained from EOC uptake of the bacteria. The calculations of total release and bacterial EOC assimilation are based on the maximum values observed during 24 hours incubation with ^{14}C. In Frederiksborg Slotssø, the maximum was found shortly after sunset. In Lake Mossø the activity of the bacterial size fraction increased slowly during the entire period of incubation. In contrast to thymidine incorporation, both the frequency of dividing cells (FDC) and $^{35}S-SO_4$ uptake were markedly higher. Dark CO_2 uptake could not be measured from Lake Mossø since uptake was not linear with time.

Comparing bacterial production estimates, it is essential to separate the results into those where the biomass of the bacteria is used to calculate the production and those which do not directly include the biomass value. Dark CO_2 fixation, EOC uptake and $^{35}S-SO_4$ uptake do not need the biomass in calculating the production. ^3H-thymidine incorporation and FDC production require the biomass value in the calculation, since the first results obtained from the two techniques are given as cells produced per unit volume of water and time.

Considering the results in the first group, dark CO_2 uptake measurements especially have been discussed in the literature. Major problems concern the conversion factor from the amount taken up to total carbon assimilated. This factor may vary from 0.4% to 30% (Overbeck 1979), though most workers have used 6% (Jordan and Likens 1980; Coveney 1982). Further, a substantial part of the uptake may be due to ^{14}C labelled organic substances released by the phytoplankton (Riemann et al. 1982b). Such data should always be treated with caution (Jordan and Likens 1980).

EOC uptake by bacteria includes only a fraction of the bacterial production; this fraction may vary from a few percent up to more than 50% (Larsson and Hagström 1982).

Table 1. Primary production, total release of extracellular organic
          carbon (EOC), and bacterial production estimates obtained
          from various methods during two diel studies: Frederiksborg
          Slotssø, September 28-29, and Lake Mossø, October 6-7,
          1981.*

	Frederiksborg Slotssø	Lake Mossø
	$mg\ C \cdot m^{-3} \cdot 24\ h^{-1}$	
Primary production	855	530
Total release of EOC	60	57
Bacterial production estimates:		
$^{14}CO_2$-dark uptake	83	?
EOC uptake	45	40
^{35}S-SO_4 dark uptake	31-78	36-62
Frequency of dividing cells	72	57
^3H-thymidine incorporation into DNA	5	5

* All experiments were carried out in situ from subsamples taken
  from 5 litre bottles. Primary production is the sum of all size
  fractions. Total release of EOC is the sum of carbon fixation in
  size fractions < 1.0-μm > 0.2-μm + < 0.2-μm, corrected for
  bacterial uptake in size fraction > 1.0-μm by means of glucose
  uptake distribution (Derenbach and Williams 1974). The values of
  bacterial uptake of $^{14}CO_2$ in the dark, EOC, and ^{35}S-SO_4 are all
  corrected for attached bacteria and bacteria > 1.0-μm by the
  glucose uptake data. ^{35}S-SO_4 uptake was further corrected for 40%
  uptake in other pools than in protein (Cuhel et al. 1982; Riemann,
  unpublished data). A C:S ratio of 50 was used (Jordan and Likens
  1980). ^3H-thymidine incorporation into DNA followed Fuhrman and
  Azam (1980), except that DNA was purified (Fuhrman and Azam
  1982). Frequency of dividing cells followed Hagström et al.
  (1979) slightly modified in Newell and Christian (1981).

^{35}S-SO$_4$ uptake has previously been discussed by Jordan and
Peterson (1978), Monheimer (1978) and Jordan and Likens (1980).
Major drawbacks of the method include the fact that phytoplankton may
take up a substantial fraction, even in dark experiments (Monheimer
1978). This will require a complete size separation between algae
and bacteria. Such conditions are seldom fulfilled and can easily
lead to overestimation of the bacterial production. In seawater,
the technique is not useful due to very high pools of SO$_4^{-2}$, and
even in lake water the SO$_4^{-2}$ concentration often exceeds 10 mg·l^{-1}.
This implies that a very small uptake of ^{35}S (% uptake of added) has
to be multiplied by a large pool factor to give total uptake. Otherwise, a large amount of tracer has to be added or the incubation time
must be extended. Further, ^{35}S-SO$_4$ may be present in the cell in
other pools than the proteins (Cuhel et al. 1982). The results in
Table 1 are corrected for 40% of the radioactivity found in other
cellular pools (Cuhel et al. 1982; Riemann, unpublished data).

The results of the two methods which both need the biomass value
of the bacteria to calculate production (thymidine incorporation into
DNA and FDC), are directly comparable with respect to turnover of the
bacterial biomass. Whatever the biomass of the bacterial population
might be, a comparable turnover is given by the two methods, since
the biomass value is only a factor to convert cells produced into
carbon produced.

The application of the thymidine incorporation technique to
natural populations of bacteria was presented by Fuhrman and Azam
(1980). Their technique included additions of ^{3}H-thymidine in
quantities large enough to overcome isotope dilution from external
pools of unlabelled thymidine. They discussed de novo synthesis of
thymidine and concluded that a true evaluation of the importance of
de novo synthesis would require measurements of the precursor to
DNA, thymidine-triphosphate. Recently, Moriarty and Pollard (1981)
suggested that one could overcome this problem by adding increasing
amounts of unlabelled thymidine to a fixed amount of ^{3}H-thymidine.
They plotted unlabelled thymidine against the reciprocal activity in
DNA and used the intercept with the x-axis to calculate total isotope
dilution from both external and internal sources. Half the results
presented by Moriarty and Pollard (1981) showed that more than one
line existed. However, they calculated that the results were nearly
the same (variation up to a factor of 2), no matter what line the
calculations were based on. A comparison between results obtained
from the two techniques was presented by Riemann et al. (1982a). In
all experiments, more cells appeared to be produced by means of the
Moriarty and Pollard dilution technique than with results calculated
according to Fuhrman and Azam (1980, 1982). This suggests that de
novo synthesis of thymidine may be of ecological importance in
measuring bacterial secondary production in freshwater and in
sediments (Moriarty and Pollard 1981). Moriarty (1983) mentioned
that isotope dilution experiments performed in seawater indicated de

novo synthesis of thymidine to be of minor importance. More data
are needed to evaluate whether these differences between the two
methods reflect varying growth conditions of the bacteria. Such
experiments are currently being carried out in our laboratories.

The FDC technique was introduced by Hagström et al. (1979) as
an approach to measure bacterial secondary production in seawater.
The method was recently evaluated by Newell and Christian (1981),
and Larsson and Hagström (1982) compared the results obtained from
FDC measurements with phytoplankton production and EOC products in
coastal environments. Major drawbacks of the method are the
time-consuming cell counting procedure, difficulties in seeing
whether small cells are dividing or not, and lack of knowledge about
FDC in relation to diurnal/diel variations in cell numbers and cell
volume. Krambeck et al. (1981) gave two examples on diel changes in
cell volume, numbers of cells, FDC and relation to EOC products.
Unfortunately, no information was presented on these experiments
with respect to whether sampling was made in bottles or directly in
the lake. Nevertheless, Krambeck et al. (1981) demonstrated only
small diel changes in FDC. More pronounced changes were seen in
cell volume, probably an effect of the nutrient regime, regulated by
EOC products. Other diel patterns in FDC in relation to cell
numbers suggest an inverse relationship between cell numbers and FDC
in eutrophic lakes (Riemann, unpublished data).

The estimates of bacterial production obtained from the various
methods (Table 1) did not give a clear picture of the importance of
phytoplankton EOC products to the bacteria. The diel changes in
thymidine incorporation into DNA, nevertheless, suggested that EOC
products were important to the bacteria, but total bacterial carbon
assimilation, calculated from the thymidine incorporation experiments, was probably too low. Certainly more comparable data are
needed before any of these methods may be recommended for use in
studies of carbon metabolism in eutrophic pelagic waters. Although
this is an unexciting conclusion, the results presented in Table 1
nevertheless support the validity of this statement.

At present, thymidine incorporation into DNA in particular
looks like an interesting method to the aquatic ecologist. The
method fulfills most of the requirements needed in a good ecological
tool: specific to one group of organisms (bacteria), reproducible
with CV values generally less than 10%, and very short incubation
periods required. The validity of the results is still based on a
considerable number of assumptions, but work in progress supports an
optimistic attitude.

ACKNOWLEDGMENTS

One of the authors (BR) is very grateful to Dr. J. E. Hobbie

who arranged a late invitation to this meeting. We wish to thank D.
Clayre for linguistic suggestions and H. Jessen Hansen for help with
the sulfate uptake experiments. Technical assistance was performed
by W. Martinsen, S. Sørensen and B. Pihlkjaer. This work was partly
supported by the Danish Natural Science Research Council (J. nr.
11-1816 and 11-2588).

REFERENCES

Bell, W. H., and R. Mitchell. 1972. Chemotactic and growth
    responses of marine bacteria to algal extracellular products.
    Biol. Bull. 143: 265-276.
Bell, W. H., and E. Sakshaug. 1980. Bacterial utilization of algal
    extracellular products. 2. A kinetic study of natural
    populations. Limnol. Oceanogr. 25: 1021-1033.
Berman, T. 1975. Size fractionation of natural aquatic populations
    associated with autotrophic and heterotrophic carbon uptake.
    Mar. Biol. 33: 215-220.
Berman, T., and C. Gerber. 1980. Differential filtration studies
    of carbon flux from living algae to microheterotrophs, micro-
    plankton size distribution and respiration in Lake Kinneret.
    Microb. Ecol. 6: 189-198.
Cahet, G., and Y. Martin. 1979. Production primaire et activité
    bactérienne en eutrophisation expérimentale (lagune du Brusc.
    Automne 1977). Publ. Sci. Tech. CNEXO. Actes Collq. 7:
    351-366.
Chrost, R. J. 1981. The composition and bacterial utilization of
    DOC released by phytoplankton. Kiel. Meeresforsch. Sonderh. 5:
    325-332.
Coveney, M. F. 1982. Bacterial uptake of photosynthetic carbon from
    freshwater phytoplankton. Oikos 38: 8-20.
Cuhel, R. L., C. D. Taylor, and H. W. Jannasch. 1982. Assimilatory
    sulfur metabolism in marine microorganisms: sulfur metabolism,
    protein synthesis, and growth of Alteromonas luteo-violaceus
    and Pseudomonas halodurans during perturbed batch growth.
    Appl. Environ. Microbiol. 43: 151-159.
Derenbach, J. R., and P. J. leB. Williams. 1974. Autotrophic and
    bacterial production: fractionation of plankton populations by
    differential filtration of samples from the English Channel.
    Mar. Biol. 25: 263-269.
Fuhrman, J., and F. Azam. 1980. Bacterioplankton secondary produc-
    tion estimates for coastal waters of British Columbia, Antarc-
    tica, and California. Appl. Environ. Microbiol. 39: 1085-1095.
Fuhrmann, J., and F. Azam. 1982. Thymidine incorporation as a
    measure of heterotrophic bacterioplankton production in marine
    surface waters: evaluation and field results. Mar. Biol. 66:
    109-120.
Hagström, Å., U. Larsson, P. Horstedt, and S. Normark. 1979.
    Frequency of dividing cells, a new approach to the determination

of bacterial growth rates in aquatic environments. Appl.
Environ. Microbiol. 37: 805–812.

Hellebust, J. A. 1974. Extracellular products, p. 838–863. In:
W. D. P. Stewart [ed.], Algal Physiology and Biochemistry.
Blackwell Scientific Publ., Oxford.

Herbland, A. 1975. Utilization par la flore hétérotrophe de la
matiere organique naturelle dans l'eau de mer. J. Exp. Mar.
Biol. Ecol. 19: 19–31.

Hobbie, J., and P. Rublee. 1977. Radioisotope studies of hetero-
trophic bacteria in aquatic ecosystems, pp. 441–476. In: J.
Cairns, Jr. [ed.], Aquatic Microbial Communities. Garland
Publ., New York.

Iturriaga, R. 1981. Phytoplankton photoassimilated extracellular
products; heterotrophic utilization in marine environments.
Kieler Meeresforsch. Sonderh. 5: 318–324.

Iturriaga, R., and H.-G. Hoppe. 1977. Observations of heterotrophic
activity on photoassimilated organic matter. Mar. Biol. 40:
101–108.

Jassby, A. D. 1973. The ecology of bacteria in the hypolimnion of
Castle Lake, California. Ph.D. thesis, University of
California, Davis.

Jordan, M. J., and G. E. Likens. 1980. Measurments of planktonic
bacterial production in an oligotrophic lake. Limnol. Oceanogr.
25: 719–732.

Jordan, M. J., and B. J. Peterson. 1978. Sulfate uptake as a
measure of bacterial production. Limnol. Oceanogr. 23: 146–150.

Krambeck, C., H.-J. Krambeck, and J. Overbeck. 1981. Microcomputer-
assisted biomass determination of plankton bacteria on scanning
electron micrographs. Appl. Environ. Microbiol. 42: 142–149.

Larsson, U., and Å. Hagström. 1979. Phytoplankton exudate release
as an energy source for the growth of pelagic bacteria. Mar.
Biol. 52: 199–206.

Larsson, U., and Å. Hagström. 1982. Fractionated phytoplankton
primary production, exudate release, and bacterial production
in a Baltic eutrophication gradient. Mar. Biol 67: 57–70.

Mague, T. H., E. Friberg, D. L. Hughes, and I. Morris. 1980. Extracellular release of carbon in marine phytoplankton; a
physiological approach. Limnol. Oceanogr. 25: 262–279.

Martin, Y. P. 1980. Succéssion écologique de communautés bactériennes au cours de l'evolution d'un écosystème phytoplanctionique
marin expérimental. Oceanol. Acta 3: 293–300.

Martin, Y. P., and M. A. Bianchi. 1980. Structure, diversity, and
catabolic potentialities of aerobic heterotrophic bacterial
populations associated with continuous cultures of natural
marine phytoplankton. Microbiol. Ecol. 5: 265–279.

Monheimer, R. H. 1978. Difficulties in interpretation of microbial
heterotrophy from sulfate uptake data: laboratory studies.
Limnol. Oceanogr. 23: 150–154.

Moriarty, D. J. W. 1983. Measurement of bacterial growth rates in
marine systems using nucleic acid precursors. In: J. E. Hobbie

and P. J. leB. Williams [eds.], Heterotrohic Activity in the Sea. Plenum Press, New York.

Moriarty, D. J. W., and P. C. Pollard. 1981. DNA synthesis as a measure of bacterial productivity in seagrass sediments. Mar. Ecol. Prog. Ser. 5: 151-156.

Nalewajko, C. 1977. Extracellular release in freshwater algae and bacteria: extracellular products of algae as a source of carbon for heterotrophs, pp. 589-624. In: J. Cairns, Jr. [ed.], Aquatic Microbial Communities. Garland Publ., New York.

Nalewajko, C., T. G. Dunstall, and H. Shear. 1976. Kinetics of extracellular release in axenic algae and in mixed algal-bacterial cultures: significance in estimation of total (gross) phytoplankton excretion rates. J. Phycol. 12: 1-5.

Nalewajko, C., and D. W. Schindler. 1976. Primary production, extracellular release, and heterotrophy in two lakes in the ELA, Northwestern Ontario. J. Fish. Res. Board Can. 33: 219-226.

Newell, S. Y., and R. R. Christian. 1981. Frequency of dividing cells as an estimator of bacterial productivity. Appl. Environ. Microbiol. 42: 23-31.

Overbeck, J. 1979. Dark CO_2 uptake - biochemical background and its relevance to in situ bacterial production. Arch. Hydrobiol. Beigh. Ergebn. Limnol. 12: 38-47.

Pomeroy, L. R. 1974. The ocean's food web, a changing paradigm. Bioscience 24: 499-504.

Riemann, B., J. Fuhrman, and F. Azam. 1982a. Bacterial secondary production in freshwater measured by ^{3}H-thymidine method. Microb. Ecol. 8: 101-114.

Riemann, B., M. Søndergaard, H.-H. Schierup, S. Bosselmann, G. Christensen, J. B. Hansen, and B. Nielsen. 1982b. Carbon metabolism during a spring diatom bloom in the eutrophic Lake Mossø. Int. Rev. Ges. Hydrobiol. 67: 145-185.

Sharp, J. H. 1977. Excretion of organic matter by marine phytoplankton: Do healthy cells do it? Limnol. Oceanogr. 22: 381-399.

Smith, W. O., Jr., R. T. Barber, and S. A. Huntsman. 1977. Primary production off the coast of northwest Africa: excretion of dissolved organic matter and its heterotrophic uptake. Deep-Sea Res. 24: 35-47.

Søndergaard, M., and H.-H. Schierup. 1982. Release of extracellular organic carbon during a diatom bloom in Lake Mossø: molecular weight fractionation. Freshwat. Biol. 12: 313-320.

Storch, T. A., and G. W. Saunders. 1975. Estimating daily rates of extracellular dissolved organic carbon released by phytoplankton populations. Ver. Int. Verein Limnol. 19: 952-958.

DIEL AND SEASONAL VARIATION IN GROWTH RATES OF PELAGIC BACTERIA

Åke Hagström and Ulf Larsson*

Department of Microbiology
University of Umeå, S-901 87 Umeå, Sweden
*Askö Laboratory
Institute of Marine Ecology, University of Stockholm
Box 6801, S-113 86 Stockholm, Sweden

INTRODUCTION

With the recognition of bacteria as a trophic element in the aquatic food web, comparable to the algae and not just mediators of degradative processes, growth rates of natural bacteria have become essential information (Pomeroy 1974; Williams 1981). During the last few years several methods to measure growth of bacteria have been developed (Christian et al. 1982; Fuhrman and Azam 1982; Hagström et al. 1979; Moriarty and Pollard 1981; Sieburth et al. 1977), but their advantages and disadvantages are still being discussed. For details see chapter by Azam (this volume). The resulting data on bacterial growth have come from a wide variety of geographic locations and techniques, hence generalizations about the variability of bacterial growth rate in the pelagic ecosystem are still not possible (Azam, this volume; Williams 1981). Circumstantial evidence such as diel variations of bacterial cell volumes and of frequency of dividing cells (Krambeck et al. 1981), however, indicate short term variability. Diel patterns in bacterial heterotrophic activity in the pelagic ecosystem, recorded using tracer techniques (Ammerman and Azam 1981; Spencer 1979), also seem to agree well with the variations in amounts of various dissolved substances seen by direct analyses (Burney et al. 1981; Mopper and Lindroth 1982). As suggested by many workers, e.g., Wiebe and Smith (1977) and Larsson and Hagström (1979, 1982), one of the major sources of organic carbon to bacterial growth could be phytoplankton extracellular release of exudates.

This possibility makes short term shifts in rate of growth likely to occur. Furthermore, such a close relationship between algal and bacterial growth implies seasonal variations of bacterial production following the annual course of primary production. In a temperate sea such as the Baltic, such pronounced variations have been shown to occur (Larsson and Hagström 1982).

The frequency of dividing cells (FDC) of natural bacterial populations was used as a direct measure of growth rate by Hagström et al. (1979) and later by Newell and Christian (1981). In this technique, a relationship between the FDC of mixed bacterial isolates and growth rate at various temperatures is established using continuous cultures. FDC values determined from natural samples are translated into growth rate using this empirical relationship. The FDC method involves no incubation, and since binary fission is the mode of growth found in all procaryotes this parameter is believed to reflect growth with a minimum of bias (Hagström et al. 1979; Christian et al. 1982).

The purpose of this paper is to examine the variation of growth rate over different time scales in order to reveal possible regulatory mechanisms governing the bacterial growth rate in the sea. The effect on bacterial growth of the daily peak in algal photosynthesis and the temperature dependence of bacterial growth are discussed in relation to in situ measurements of bacterial growth rate.

MATERIALS AND METHODS

Investigation Areas

Three stations were sampled during the course of one full year. Two stations, "1" and "3", are situated in the southern archipelago of Stockholm, Sweden. Station "3", at the head of the fjord Himmerfjärden, is affected by a wastewater treatment plant (the equivalent of 150,000 persons), while Station "1", known as "Asko Bojen", is exposed to the open Baltic. The Station "N1" is situated in the Northern Bothnian Sea in the Norrbyn archipelago 40 km south of the city of Umeå (Sweden). As a rule, the water at all three stations is stratified during the summer and early autumn. The thermocline is situated between 15 and 20 m depth at Station "1" and somewhat shallower, between 10 and 15 m depth at Stations "3" and "N1". The trophogenic layer extends approximately to the depth of the thermocline.

Typical surface water salinities were 3-5°/°° at "N1", 6.4-7.4°/°° at "1", 5.3-6.6°/°° at "3", and 28-28.5°/°° at "TJ". Station TJ was sampled during one diel experiment and is situated inside the island Tjärnö on the Swedish west coast, close to the city of Strömstad.

In situ Carbon-14 Uptake Measurements

Carbon-14 uptake was determined during two diel cycle experiments. Water samples from eight depths (0, 1, 2, 4, 6, 8, 10, 14 m) at Station "N1" and seven depths (0, 1, 2, 3, 5, 7, 10 m) at Station "TJ" were incubated with the addition of 10 μCi carrier-free $NaH^{14}CO_3$. Radioactivity in intact seawater and in filtrates (through 0.2 μm pore-size Nuclepore filters) was determined in 10 ml samples transferred to scintillation vials and acidified (pH of 2) with HCl. The $^{14}CO_2$ remaining in the samples was stripped with air for 15 minutes and 10 ml of Instagel (Packard Instruments) was added. Samples were counted in a Nuclear Chicago Mark II liquid scintillation counter. Uptake was measured in both light and dark bottles and dark values were subtracted.

Determination of FDC

Direct counts of bacteria were performed on 0.2 μm filters using fluorescent staining. Water samples were taken at 10 a.m. $\pm$ 20 min with a Ruttner sampler. At stations "1" and "3", 25 ml samples were taken every fifth meter and pooled (Station "1": 0, 5, 10, 15, 20 m; Station "3": 0, 5, 10 m). The water column (0-14 m) at Station "N1" was sampled at every second meter and pooed. Four subsamples of 5 ml were drawn from the pooled samples and added to tubes containing 0.4 ml of filtered (0.22 μm) buffered (hexamethylenediamine, 20 g/100 ml, pH 7.2) formaldehyde (20% wt/wt) containing acridine orange (0.125 g/100 ml). To ensure an even filtration, subsamples (0.1 to 0.6 ml) were mixed with 5 ml of particle-free water in a 13 mm stainless steel funnel (Millipore Corp.). The filters were rinsed three times with 5 ml of particle-free water and mounted on glass slides in cinnamaladehyde and eugenol (1:2). The cover slips were sealed with clear nail varnish. The "N1" samples and samples from the diel experiments were stained using ethidium bromide (8 mg/ 100 ml) instead of acridine orange; rinsing was found unnecessary and filters were mounted in pure eugenol. The filters were counted with epifluorescence using Zeiss WG optics (Planapochromat, 63/1.40, kpl 12.5 W, magnification: x 984). Ten fields were counted until a total of 300 cells were recorded. Dividing bacteria were counted on a minimum of ten fields. If less than 30 dividing cells were recorded a maximum of 20 additional fields were examined. Bacteria showing an invagination, but not a clear zone between cells, were considered as one dividing cell.

Careful preparation of the samples (bright bacteria, low background stain, and flat filter surface) is important to ensure precision of the FDC counts. As a test of the counting-procedure four replicate samples from 23 sampling occasions were examined. Of these, 10 had a CV (coefficient of variation) of less than 10% and 19 had less than 20%. No CV above 29% occurred. The temperature-dependent relationship between frequency of dividing cells (FDC) and

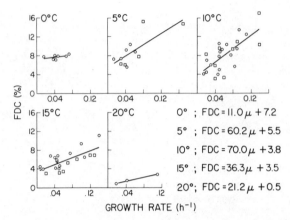

GROWTH RATE (h⁻¹)

Figure 1.   Relationship between FDC and growth rate at five different temperatures.   Batch (□) and continuous (○) cultures isolated
from the Baltic Sea were grown to steady state before sampling.

growth rate (μ) used to translate FDC of field samples to growth
rate is given in Fig. 1.   New data points have been added to the
data presented in Hagström et al. (1979) and Larsson and Hagström
(1982) and the relations recalculated.

Size Determination of Bacteria

Mean cell length was determined from epifluorescence micrographs
using a vernier caliper.

RESULTS

Bacterial growth rate at three stations with different levels of
primary production but with similar temperature regimes are shown in
Fig. 2.   In a previous paper we have reported bacterial production
in relation to primary production in a eutrophication gradient of
the northern Baltic (Larsson and Hagström 1982).   The FDC data from
two stations "1" and "3" within that study together with data from a
third station in the northern Bothnian Sea, "N1", allow us to
compare rate of growth of pelagic bacteria at different levels of
carbon supply.   The annual primary production at stations "3", "1"
(1978) and "N1" (1980) were 280, 160, and 80 g $C \cdot m^2$, respectively.
The general patterns of growth at the three stations are very
similar (Fig. 2a).   During the winter months no bacterial growth was
recorded and both onset and end of the growth periods coincide at
all stations.   The threefold difference in the primary production is
reflected in higher maximum growth rates of the bacterial
communities.

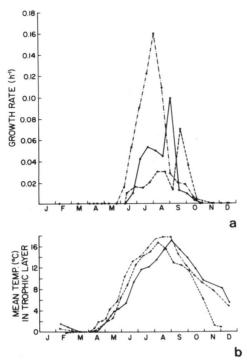

Figure 2a. Top: Growth rates (h^{-1}) of pelagic bacteria at three
stations within the Baltic Sea with different annual primary produc-
tion: Station "N1" 80 g $C \cdot m^{-2} \cdot yr^{-1}$ (---), Station "1" 160 g $C \cdot m^{-2} \cdot yr$
(--), Station "3" 280 g $C \cdot m^{-2} \cdot yr^{-1}$ (-·-). Figure 2b. Bottom: Mean
temperatures of the trophogenic layer at Stations "N1", "1" and "3".

By comparing the curves of growth rate and temperature it
becomes evident that significant measurable rates do not occur when
the water temperature is below 7-10°C. This is true both for the
start of growth during early summer and at the end of the growth
period. Indications of slow growth at low temperatures can, however,
be found in early spring at all three stations where there were
peaks in FDC of 6-7% (data not shown). The same value of FDC is
obtained from the calibration curve at 0°C when extrapolated back
to zero growth rate (Fig. 1). These high FDC values occur at the
beginning of the spring bloom, while the lowest FDC values of the
year coincide with the spring bloom peak (Larsson and Hagström
1982). After the spring bloom has declined at Station "1", there is
a conspicuous peak in bacterial biomass (Fig. 3), although measured
FDC values indicate zero growth rate. The bacterial biomasses at
Stations "3" and "N1" also increase at a low rate during this period
without FDC indicating growth. This increase in biomass at Station
"1" corresponds to a doubling time of about 14 days if grazing on
the bacteria is assumed to be negligible.

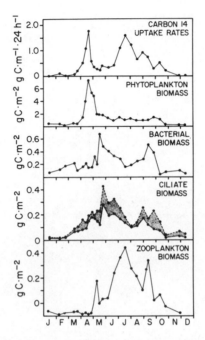

Figure 3. Dynamics of four major trophic components in the pelagic
ecosystem as reflected by standing crops and phytoplankton primary
production. Samples were collected at Station "1", situated in the
northern Baltic proper. The carbon-14 uptake rates and the bacteria
biomass are from Larsson and Hagström (1982), the phytoplankton and
ciliate masses are courtesy of R. Hobro (shaded area is the biomass
of the autotrophic ciliate Mesodinium rubrum), and the net zooplankton (>90 μm) are courtesy of Larsson and Johansson (unpublished).

     The finding that bacterial growth probably is slow during
spring can be supported from data on the annual dynamics of the
major functional groups in the pelagic ecosystem as seen at Station
"1". The four groups depicted in Fig. 3 include phytoplankton as
the primary producers, bacteria as consumers of dissolved organic
matter, ciliates as grazers on algae and bacteria, and zooplankton
as possible feeders on all of the above three groups. Detectable
growth rates of bacteria coincide with the increase in primary
productivity in June, which in turn coincides with the peak in net
zooplankton (>90 μm) biomass. The phytoplankton biomass in this
figure shows an almost constant level during the summer but with
picoplanktonic forms (<2 μm) added it has a peak parallel to that of
primary production and net zooplankton (data not shown) (Larsson
and Hagström 1982 and unpublished data). There is a negative
correlation among net zooplankton, ciliates, and bacteria. The
data presented in Fig. 3 thus suggest complex interactions within
the pelagic ecosystem leading to the control of bacterial annual

dynamics. Bacterial growth is stimulated by exudate release from
phytoplankton, which mainly occurs during summer, and coincides with
high water temperatures, a prerequisite for high growth rates. On
the other hand, bacteria are controlled by ciliates and net zooplankton; the latter also controls the ciliates during summer. During
spring, net zooplankton biomass is very low, and it is reasonable to
assume that the grazing pressure on the ciliates during this period
is low. From the biomass increase, it was estimated that the ciliate
population showed a doubling time of about 30 days during spring.
This is most probably an underestimate, as is the estimate presented
for bacteria, but this argument serves to show that both bacterial
and ciliate growth rates could be low during spring, and that ciliate
predation probably has not biased the estimate of bacterial growth
rate based on biomass.

Temperature affects both the growth rate and size of bacteria.
As temperature increases a given FDC value gives an increased growth
rate (Fig. 1). To further test this a temperature shift experiment
was carried out. By lowering the temperature of a mixed culture kept
at constant growth rate in a chemostat, the effect found in Fig. 1
can be reproduced. Table 1 gives the results from three such experiments where the media and flow rates, and thus growth rates, were
constant. Consecutive temperature downshifts every 7 days resulted
in fewer and larger cells with higher FDC.

Each morning, the algal primary production can be expected to
present a shift up of available food for the bacterial community. In
order to investigate this we monitored growth of bacteria and algal
release of exudates during one full day. The first experiment was
conducted on 19 August 1980 at Station "N1" (Fig. 4a). The mean
temperature of the water column (0-14 m) stayed within the range of
15.0 to 15.4°C with the high value during the afternoon. The course
of the primary production, measured as ^{14}C uptake in unfractionated
sea water, shows the expected diurnal curve. The pool of dissolved
labelled extracellular algal material, defined as the fraction
passing a 0.2 μm filter, is also given. It must be noted that the
amounts of extracellular material shown do not represent total
release since bacterial consumption during the incubation will
remove produced material through assimilation and respiration. The
release of exudates appears to synchronize the growth of bacteria.
During the first part of the day, the growth rate of bacteria rises
and increased numbers of bacteria are found in the early afternoon.
At this time the bacterial mean cell length has been reduced, most
likely due to cell division. The mean generation time during this
experiment was 30 hours.

In Fig. 4b a second diurnal experiment is depicted. The station
chosen is situated on the Swedish West Coast and represents a truly
marine environment with salinities of about 30°/∘∘ as compared to the
Baltic Sea with 5-10°/∘∘. The temperature in the layer 0-10 m during

Table 1. Temperature dependence of FDC in mixed continuous cultures
          isolated from the Baltic Sea. Baltic Sea water supple-
          mented with 2 mg·liter^{-1} yeast extract (Difco) and mineral
          salts PO_4-P, NH_4-N, and NO_3-N.

Temp °C	Growth rate (h^{-1})	FDC[a] (% $\pm$ SD)	Number (10^6 ml^{-1})	Length/width[b]	Volume[c]
20	0.05	5.3 $\pm$ 0.1	6.7	1.16/0.48	0.18
15	0.05	8.3 $\pm$ 0.8	3.7	1.56/0.55	0.35
10	0.05	8.8 $\pm$ 0.7	3.2	1.58/0.57	0.35
5	0.05	10.3 $\pm$ 0.9	2.8	1.63/0.63	0.44
15	0.035	6.7 $\pm$ 0.4	6.1	–	–
10	0.035	6.0 $\pm$ 0.2	6.1	–	–
5	0.035	7.3 $\pm$ 0.5	3.1	–	–
15	0.015	4.5 $\pm$ 0.3	12.0	–	–
10	0.015	4.1 $\pm$ 0.2	7.0	–	–

a) Four subsamples were filtered and counted separately.
b) Length and widths were determined from microphotographs.
c) Volume was calculated according to $\pi \frac{w^2 L}{4} - \frac{w^3}{12}$.

the experiment ranged from 5.0 to 5.7°C with the high value in the
afternoon. The estimated growth rates show a drastic increase during
the day, still the lowest generation time was 27 hours and the mean
value is about 80 hours. The bacterial numbers do not increase as a
result of the peak in growth rate. In this study integrated (0–10 m)
and discrete (3 m) samples were treated separately but no significant
difference in FDC could be seen.

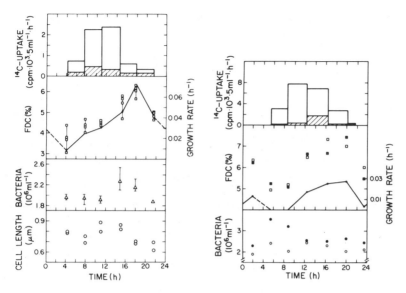

Figure 4.   A. (left) Observations at Station "Nl" on 19 August 1980
of phytoplankton primary production and bacterial growth parameters.
Water temperature was 15.0-15.4°C.   The top panel shows total
^{14}C uptake in photosynthesis (cpm) plus the cpm released in algal
exudates.   The next panel shows the FDC as open squares and the
growth rate as a solid line.   Bacteria numbers and cell length are
given in the two lower panels.   B. (right) Diel experiment performed
during spring conditions at Tjärnö situated in the northern part of
the Swedish west coast.   Measurements reported were made in the same
way as those of Station "Nl".   The mean temperature of the water
column was 5.0 to 5.7°C.   Filled circles and squares represent
integrated (0-19 m) and open circles and squares discrete (3 m)
water samples.

DISCUSSION

        We have shown that growth of bacteria in the pelagic ecosystem
occurs, and that changes in the rate of growth can occur within a few
hours.   Furthermore, we conclude that a strong seasonal effect on
bacterial growth due to the changing temperature exists in temperate
sea areas.   The FDC when used as a growth parameter depends on a
temperature calibrated relationship between FDC and growth rate of
mixed bacterial cultures.   This empirical relationship is based on
earlier observations that in dividing bacteria the septum separation
is virtually constant when the growth rate of the bacteria is changed
(Grover et al. 1977; Sargent 1975).   Consequently, the frequency of
dividing cells will increase as the growth rate of the bacteria
increases.   With changing temperature of growth this relationship
changes.   Bacteria, when kept growing at constant rate therefore

show high FDC values at low rather than at high temperatures. This point has been demonstrated both in the original FDC growth rate curves (Fig. 1) and in the temperature shift experiment. In a similar shift experiment Herbert and Bell (1977) demonstrated identical shifts in cell size and numbers due to decreasing temperature. Although they did not record dividing cells, their work does agree with the data presented in this paper.

The FDC method does not indicate bacterial growth until the beginning of June. As is obvious from the increase in bacterial biomass during spring, this is not the case. If however, the argument of low grazing pressure during spring is valid, bacterial growth rates as estimated from the increase in bacterial biomass are indeed low during this period. Still this uncertainty raises the fundamental question of whether or not a given FDC value of a natural population can be correctly translated to growth rate via mixed-culture data. The mixed population in the chemostat does maintain morphological diversity (Hagström et al., unpublished) but changes in number of physiological types can not be excluded. Thus the calibration curves may misrepresent the field data as discussed by Newell and Christian (1981). One additional aspect of the observed diel variation is that the time of sampling becomes important. In our case all samples for the curves showing seasonal variation were taken at 10 a.m. although the highest values in the diel experiments were found during the early afternoon. During the summer this is a good estimate of the mean growth rate. However, the recorded zero growth rate may be an artifact caused by sampling early in the day. Such an interpretation would be in accordance with the measurements of an increase in bacterial biomass during spring. The precision of the FDC method as well as other measurements of bacterial growth rate in situ can not be determined since the true rates are unknown. With regard to the technical aspects of the counting procedure, Newell and Christian (1981) reported a deviation of less than 0.7% between two operators counting FDC, and Larsson and Hagström (1982) found a coefficient of variation typically below 20% for replicate samples. When this uncertainty is considered in relation to the equations used, the relative error in the growth rate estimate is likely to increase with decreasing FDC. Especially in the equation for 0°C, a small change in FDC gives rise to a large difference in growth rate. However, the low growth rates found during periods of low water temperature are too pronounced to be explained only by the uncertainties of the method or an unrepresentative sampling time.

Bacteria growing at low temperatures (0-5°C) are generally considered to fall into two categories: 1) psychrophiles growing at temperatures of 0° or lower and up to 20°C; and (2) psychrotrophic bacteria which grow at 0°C but have maximum growth temperatures exceeding 25°C (Morita 1975). As shown by competition experiments, the psychrophiles are likely to outcompete the psychrotrophic bacteria at 0°C (Harder and Veldkamp 1971). Evidence of such temperature-induced succession was found by Sieburth (1967) who demonstrated seasonal shifts in temperature tolerance of bacteria from the sea. The range of maximal growth rates found for cultured bacteria at 0°C is around 0.04-0.15 hr^{-1} (Baross and Morita 1978; Herbert and Bell 1977; Mohr and Krawiec 1980) with psychrophilic bacteria showing the highest rates. Although such rates would be ecologically significant if achievable in nature, this is probably not the case. The isolates cited above when grown at 10°C commonly show a two to threefold increase in growth rate over growth at 0°C (Herbert and Bell 1977). Psychrotrophic bacteria usually show even larger differences in growth rate at the same temperature interval (Baross and Morita 1978). In the natural environment this means inconceivable high rates (~ 0.5 hr^{-1}) that are not confirmed by present in situ growth estimates (Azam 1983; Williams 1981; Larsson and Hagström 1982). Using the in situ growth rates found at 10°C (this paper) of up to 0.03 h^{-1}, a transfer to 0°C would yield generation times of about 100 hours provided the substrate for growth remains unchanged.

The primary production in the Northern Baltic with the exception of a diatom spring bloom, seen at Stations "1" and "3", is almost zero during the period of low temperature, hence the rate of supply of substrate for bacterial growth must be low (Larsson and Hagström 1982). In temperate areas such as the Baltic the majority of the metabolic activity occurs during summer. Selection pressure can therefore be expected towards a bacterial population growing rapidly at summer water temperatures and then surviving at a low rate of growth at low temperature. Surviving at low temperature however does not mean dormant inactive bacteria, as shown by changes in the FDC in the 5°C diel experiment.

The close relationship between bacterial growth and substrate supply is evident in the diel experiments reported here and in other experments (Brown et al. 1981). Yet the generally held view that bacteria only degrade particulate organic matter has produced the image of bacteria slowly consuming a huge pile of debris while releasing inorganic minerals (Steele 1974). On the contrary, reports of diel cycles in carbohydrates, amino acids, and nucleotides both as macromolecules and monomers clearly point to immediate consumption of released dissolved material as a major route of heterotrophic growth (Ammerman and Azam 1981; Burney et al. 1981; Fuhrman et al. 1980; Mopper and Lindroth 1982).

The data presented in this paper suggest a close covariation between bacterial and ciliate biomass, furthermore both these groups are negatively correlated to net zooplankton biomass. Decreasing bacterial biomass during summer while high growth rates prevail illustrates the existence of an effective control of bacterial biomass. To what extent the result of grazing by one group of predators, or the combined effect of several groups of organisms including flagellates, is not known. Anyhow intense mineralization can be anticipated during this period, and hence the phytoplankton primary production reaches its summer maximum at the same time as the biomass of the net zooplankton. This could serve to illustrate a point made some 17 years ago by Johannes (1965) and later by Williams (1981) that bacteria are less efficent mineralizers than zooplankton. Using N and P rich organic substrates, bacteria contribute to mineralization not by releasing inorganic nutrients, but as food for higher trophic levels where metabolic waste products are excreted. Indeed, bacteria may compete with phytoplankton for inorganic nutrients (Rhee 1972; Krempin et al. 1981). If nutrient supply controls bacterial growth, mineralization by zooplankton may be the rate limiting step during periods with low nutrient availability, and thus affect both diel and annual variations in bacterial growth rates.

ACKNOWLEDGMENTS

    This work has been supported by the National (Swedish) Environmental Protection Board grants, SNV-7-537 and 7-80, and the Swedish Natural Science Research Council Grant, B 4452-100. We want to thank Bengt-Owe Jansson, Ragnar Elmgren and Sture Hansson for reading the manuscript and Mrs. Christina Tjärner for excellent technical assistance.

REFERENCES

Ammerman, J. W., and F. Azam. 1981. Dissolved cyclic adenosine monophosphate (cAMP) in the sea and uptake of cAMP by marine bacteria. Mar. Ecol. Prog. Ser. 5: 85-89.

Azam, F. 1983. Growth of bacteria, and its relationship to the production of organic matter in the sea. In: J. E. Hobbie and P. J. leB. Williams [eds.], Heterotrophic Activity in the Sea. Plenum Press, New York.

Baross, J. S., and R. T. Morita. 1978. Microbial life at low temperature: Ecological aspects, pp. 9-71. In: D. J. Kushner [ed.], Microbial Life in Extreme Environments. Academic Press, New York.

Brown, E. J., D. K. Button, and D. S. Lang. 1981. Competition between heterotrophic and autotrophic microplankton for dissolved nutrients. Microb. Ecol. 7: 199-206.

Burney, C. M., K. M. Johnson, and J. McN. Sieburth. 1981. Diel flux of dissolved carbohydrate in a salt marsh and a simulated estuarine ecosystem. Mar. Biol. 63: 175-187.

Christian, R. R., R. B. Hanson, and S. Y. Newell. 1982. Comparison of methods for measurements of bacterial growth rates in mixed batch culture. Appl. Environ. Microbiol. 43: 1160-1165.

Fuhrman, J. A., and F. Azam. 1982. Thymidine incorporation as a

measure of heterotrophic bacterioplankton production in marine
surface waters: Evaluation and field results. Mar. Biol. 66:
109-120.

Fuhrman, J. A., J. W. Ammerman, and F. Azam. 1980. Bacterioplankton
in the coastal euphotic zone: Distribution activity and possible
relationships with phytoplankton. Mar. Biol. 60: 210-207.

Grover, N. B., C. L. Woldringh, A. Zaritsky, and R. F. Rosenberger.
1977. Elongation of rodshaped bacteria. J. Theor. Biol. 67:
181-193.

Hagström, Å., U. Larsson, P. Hörstedt, and S. Normark. 1979.
Frequency of dividing cells, a new approach to the determi-
nation of bacterial growth rates in aquatic environments.
Appl. Environ. Microbiol. 37: 805-812.

Harder, W., and H. Veldkamp. 1971. Competition of marine psychrophilic bacteria at low temperatures. Antonie van Leewenhock J.
Microbiol. Serol. 37: 51-63.

Herbert, R. A., and C. R. Bell. 1977. Growth characteristics of an
obligately psychrophilic Vibrio sp. Arch. Microbiol. 113:
215-220.

Johannes, R. E. 1965. Influence of marine protozoa on nutrient
regeneration. Limnol. Oceanogr. 10: 434-442.

Krambeck, C., H.-J. Krambeck, and J. Overbeck. 1981. Microcomputer-
assisted biomass determination of plankton bacteria on scanning
electron micrographs. Appl. Environ. Microbiol. 42: 142-149.

Krempin, D. W., S. M. McGrath, J. Beeler Soohoo, and C. W. Sullivan.
1981. Orthophosphate uptake by phytoplankton and bacterioplankton from the Los Angeles Harbor and Southern California
coastal waters. Mar. Biol. 64: 23-33.

Larsson, U., and Å. Hagström. 1979. Phytoplankton exudate release
as an energy source for the growth of pelagic bacteria. Mar.
Biol. 52: 199-206.

Larsson, U., and Å. Hagström. 1982. Fractionated phytoplankton
primary production, exudate release, and bacterial production
in a Baltic eutrophication gradient. Mar. Biol. 67: 57-70.

Mohr, P. W., and S. Krawiec. 1980. Temperature characteristics and
Arrhenius plots for nominal psychrophiles, mesophiles and
thermophiles. J. Gen. Microbiol. 121: 311-317.

Mopper, K., and P. Lindroth. 1982. Diel and depth variations in
dissolved free amino acids and ammonium in the Baltic Sea
determined by shipboard HPLC analysis. Limnol. Oceanogr. 27:
336-347.

Moriarty, D. J. W., and P. C. Pollard. 1981. DNA synthesis as a
measure of bacterial productivity in seagrass sediments. Mar.
Ecol. Prog. Ser. 5: 151-156.

Morita, R. Y. 1975. Psychrophilic bacteria. Bacteriol. Rev. 39:
144-167.

Newell, S. Y., and R. R. Christian. 1981. Frequency of dividing
cells as an estimator of bacterial productivity. Appl.
Environ. Microbiol. 42: 23-31.

Pomeroy, L. R. 1974. The ocean's food web, a changing paradigm.
Bioscience 24: 499-504.

Rhee, G-Yill. 1972. Competition between an alga and an aquatic
    bacterium for phosphate. Limnol. Oceanogr. 17: 505-514.
Sargent, M. G. 1975. Control of cell length in Bacillus subtilis.
    J. Bacteriol. 123: 7-13.
Sieburth, J. McN. 1967. Seasonal selection of estuarine bacteria
    by water temperature. J. Exp. Mar. Biol. Ecol. 1: 98-121.
Sieburth, T. McN., K. M. Johnson, C. M. Burney, and D. M. Lavoie.
    1977. Estimated in situ rates of heterotrophy using diurnal
    changes in dissolved organic matter and growth rates of pico-
    plankton in diffusion cultures. Helgol. Wiss. Meeresunters.
    30: 565-574.
Spencer, J. J. 1979. Light-dark discrepancy of heterotrophic
    bacterial substrate uptake. FEMS Microbiol. Lett. 5: 343-347.
Steele, J. H. 1974. The Structure of Marine Ecosystems. Harvard
    University Press, Cambridge, MA. 128 pp.
Wiebe, W. T., and D. F. Smith. 1977. Direct measurement of
    dissolved organic carbon release by phytoplankton and
    incorporation by microheterotrophs. Mar. Biol. 42: 213-223.
Williams, P. J. leB. 1981. Incorporation of microheterotrophic
    processes into the classical paradigm of the planktonic food
    web. Kiel. Meeresforsch. 5: 1-28.

ORGANIC PARTICLES AND BACTERIA IN THE OCEAN

Peter J. Wangersky

Department of Oceanography
Dalhousie University
Halifax, Nova Scotia, B3H 4J1, Canada

INTRODUCTION

The bacterial decomposition of particulate matter has always been considered one of the major pathways in the recycling of nutrients in the oceans. It is only recently, however, that any attempt has been made to partition nutrient regeneration between the various trophic levels. The purpose of this review is to examine the suitability of naturally occurring organic particles as substrates for bacterial growth and as sites for nutrient regeneration. Because of the dual meaning of the word, substrate will be used only to refer to the substances used for microbial growth.

It is clear that all particles are not created equal in the sight of the bacteria. Particles in the ocean have different sources, distributions, and pathways of degradation. Particle dynamics has probably received the most attention; there have been several reviews over the past decade (Riley 1970; Conover 1975; Wangersky 1978a, Cauwet 1978). Accordingly, in this review I will slight dynamics and emphasize particle source, particles as attachment sites and as substrates, problems of sampling particles, and nutrient regeneration from particles.

SOURCES OF PARTICLES

In earlier days the particulate matter in the ocean was divided into organic or inorganic materials. The inorganic materials were thought to be principally clays, delivered to the ocean in the form of very fine particles. In this form they were presumed to remain suspended for very long periods, even to hundreds of years (Kuenen

1950), before settling to the ocean floor.   The organic matter was considered to be the remains of planktonic organisms, falling toward the ocean floor in a constant slow rain.

A major part of the inorganic material coming down the rivers is deposited in the estuaries as a result of flocculation into larger units under the joint influence of organic materials and increases in salinity.   These effects have been demonstrated in the laboratory (Kranck and Milligan 1980) and in nature (Sakamoto 1972; Sholkovitz 1976; Kranck 1979, Pierce and Siegel 1979; Yamamoto 1979; Sholkovitz and Price 1980).   For this reason, most inorganic particles from land do not reach the pelagic ocean and have to be considered separately as a part of estuarine ecology.   In the same way, it was long thought that river run-off introduced amounts of particulate organic matter large enough to affect the biology of the coastal zones.   Recent work, however, suggests that most of the organic particles are also deposited either in the estuaries or immediately in the coastal zone (Biggs and Flemer 1972; Martin et al. 1977; Sholkovitz et al. 1978; Pocklington and Leonard 1979; Tan and Strain 1979).   We must deal with two separate universes, that of the estuaries and that of the open ocean.

Particles in Estuaries

In many of the rivers studied, the burden of suspended inorganic material carried by the river decreases as the water becomes saline (Sholkovitz and Price 1980).   Some of the dissolved trace metals are also removed (Sholkovitz 1976, 1978; Sholkovitz and Copeland 1981), perhaps with the precipitation of humic acids.   Some rivers, such as the St. Lawrence, display a turbidity maximum near the head of the estuary; this maximum is maintained in part by resuspension of bottom sediments by tidal turbulence and the breaking of internal waves (Kranck 1979, Silverberg and Sundby 1979).   The inorganic material suspended in this fashion sediments relatively quickly, most of it in the estuary and the remainder in the immediate coastal area (Kranck 1975; Yamamoto 1979).

The organic materials in suspension follow two slightly differ-ent courses in a river.   The terrestrial particulate material behaves much as the inorganic particles (Sholkovitz et al. 1978; Pocklington and Leonard 1979; Tan and Strain 1979) and decreases with increased salinity.   Materials from an aquatic source derive mostly from phyto-plankton (Biggs and Flemer 1972; Martin et al. 1977; de Souza Lima and Williams 1978).   This particulate matter often increases in the seaward parts of the estuary where phytoplankton respond to inorganic nutrients carried by the river water or regenerated from organic materials at the river mouth (Stanley and Hobbie 1981).

The behaviors of both the organic and inorganic particles are in fact closely linked.   One mechanism of their removal is the

change in surface charge on suspended particles; this results from
the acquisition of a surface coating of organic material, derived
from the natural surfactants present in sea water (Neihof and Loeb
1972; Hunter and Liss 1979). As the surface charges on the particles
approach zero with increasing salinity, the particles begin to coagu-
late into large flocs and drop out of suspension. This production
of "macroflocs" has been studied both in the estuaries and in the
laboratory (Sakamoto 1972; Kranck and Milligan 1980); SEM studies
show that these particles consist of aggregations of mineral grains,
organic matter, biogenic debris, phytoplankton, and bacteria (Pierce
and Siegel 1979).

Material deposited on the shelf may ultimately enter the deep
ocean through a process of resuspension during periods of increased
turbulence (Eggiman et al. 1980; Pak et al. 1980b). Such particles
are most likely to be found as components of turbid layers at intermediate depths or as nepheloid layers just above the sea floor.
Material of this kind will be quite different from particles which
have reached the deep ocean by direct fall-out from the surface
layers. The particles from shelf sediments would presumably have
been stripped of much of their organic content during their sojourn
in shallower water, and may thus be less attractive as sites for
bacterial colonization.

Particles in Open Ocean

The particulate material of the open ocean is heterogeneous and
not easy to sample. Small samplers, such as the Niskin bottle, did
a reasonable job of collecting the slowly sedimenting materials but
missed the larger, faster-falling particles almost entirely. The
larger particles could be collected only with a submersible pump
(Bishop and Edmond 1976) or with sedimentation traps (Rowe and
Staresinic 1977/1979, Honjo and Roman 1978; Fellows et al. 1981).
These newer collection methods have demonstrated that the particulate
load of normal sea water is much greater than could be calculated
from bottle collections or from examination of bottom deposits. One
reason for the discrepancy is biological degradation occurring on
the particles and at the water-sediment interface. For example,
Eadie and Jeffrey (1973) have suggested from stable carbon isotope
ratios that less than 5% of the particulate matter present in the
surface waters is ever incorporated into the deep-sea sediments.

One type of inorganic particles, for example, the red clay in
deep-sea deposits, originate from terrestrial soil and dust carried
by winds (Delany et al. 1967; Duce et al. 1980). When this material
enters the ocean, it penetrates the surface film composed of hydrophobic and surface-active materials (Wangersky 1976a). This film is
chemically and physically suited to the formation of organic coatings
on particles. Yet, the surface film is not the only source of
organics for Tanoue and Handa (1979) concluded that the concentration

of organic materials on silt and clay sediment was determined by the
amounts of surfactants present in the water column. Indeed,
inorganic particles can adsorb fatty acids and hydrocarbons (Meyers
and Quinn 1971a,b, 1973; Morris and Calvert 1975; Barcelona and
Atwood 1979), fatty-acid methyl esters (Ehrhardt et al. 1980), and
organic pollutants (Pfister et al. 1969; Oloffs et al. 1973; Miller
and Zepp 1979). Because only a small amount of inorganic material
enters the ocean from the atmosphere, there seems to be enough
surface-active organic matter in the euphotic zone of the ocean to
alter the surface properties of the material. It has even been
suggested that the aerosol particles over the oceans are themselves
covered by a layer of organic materials (Goetz 1965; Ruppersberg and
Schellhase 1980). It may well be that until bacterial degradation
has removed any such layer of organic materials, the inorganic
materials are not available for reaction with sea water.

Living organisms larger than bacteria make up a class of organic
particles but they have been little studied as habitats for microbes.
One exception to this is the community associated with the Sargasso
weed, which thrives even in the Sargasso Sea (Hanson 1977); here,
nitrogen fixation occurred at rates great enough to supply some 40%
of the weed's requirement and some 50-70% of the released DOM was
used by epiphytic and planktonic organisms associated with the
weed. Measurable phosphate was found in the water within the weed
clump but was absent in the surrounding water. Sargassum clumps are
very rich communities indeed, comprised of many trophic levels; they
maintain themselves in a region apparently devoid of inorganic
nutrients. If further investigation confirms the high rates of
nitrogen fixation, the limiting factor on community metabolism may
be the availability of phosphorus, both organic and inorganic, in
the surface layers. The limiting rate may be that of phosphate
remineralization by microbes.

Other biologic sources of large particulate matter are the
mucus webs and houses constructed by some large zooplankters. These
webs, used for feeding, are discarded either in flight from danger
or after they become clogged with unwanted particulate materials
(Alldredge 1972, 1976; Gilmer 1972; Barham 1979). The webs are not
rare in the oceans; Barham (1979) has seen as many as 1 m^{-3} in deep
dives. However, they break apart in sample bottles. Little is
known of the ability of the webs to capture marine bacteria and even
less of the suitability of the material as a site for bacterial
attachment. They must be easily degraded, and probably serve as
point sources of nutrient regeneration.

Other large organisms obviously must make contributions to the
particulate load of oceanic waters. Lasker (1964) has calculated
that the moults of one crustacean species alone, if untouched by
bacteria, could account for the entire suspended load of the oceans.
Since the organic content of deep sea sediments is low and since

there are few recognizable organic remnants in this sediment, it is obvious that these moults must be partially or wholly decomposed before reaching the sediment. Bacteria capable of such degradation have been described by Seki (1965). Particles arising from larger organisms, particularly those retaining enough of their mass to fall rapidly, are important to benthic organisms (Rowe and Staresinic 1977/1979) and to the bacteria of the sea floor but contribute little to regeneration of inorganic nutrients in the water column. Their primary importance may be as a mechanism for the transport of bacteria from the surface layers to the sea floor.

The fecal material produced by marine invertebrates is a rich source of organic material for bacteria and even, through the bacteria, for other invertebrates (Johannes and Satomi 1966). Most of the studies have examined copepod fecal pellets; their sinking rates are relatively rapid, between 100 and 900 $m \cdot day^{-1}$, far faster than the unaggregated bits and pieces of which the pellets are composed (Fowler and Small 1972). Furthermore, the protective organic skin covering the pellets serves to preserve the silicate and calcium carbonate tests of food organisms from solution in sea water. The organic skins carry their own collections of gut fauna, which break up the skin while the pellet is falling through the water column. Once the skin has been consumed, the organic matter in the body of the pellet is available to colonization and the carbonates and silicate to solution. The rate of degradation of the skin is temperature dependent, but it has been estimated that the pellets begin to break up below 1000 m. There is also some evidence that pellets may stay together even after the outer membranes have been consumed (Honjo and Roman 1978; Small et al. 1979).

The discovery of this method of transport for silicates and carbonates has helped to clear up a major problem in marine sedimentation. In those deep-sea sediment cores rich in carbonate it is not uncommon to find some dissolution of the relatively thick-walled foraminifera while the delicate coccoliths are still perfectly preserved. If coccoliths were transported to the bottom in protected fecal pellets, at least some could reach the sea floor without exposure to sea water. The factors affecting the degradation of the membranes of fecal pellets on the ocean floor are unknown, but it is known that intact fecal pellets have been found in sediments (Wangersky and Joensuu 1967). However, very gentle methods of fractionation have to be employed.

In addition to copepods, many other marine organisms, benthic and pelagic, produce fecal matter which serves as a habitat for bacteria. Kofoed (1975a,b) has reported on the fecal matter produced by a deposit-feeding prosobranch, Hydrobia ventrosa, while Pomeroy and Deibel (1980) have studied three genera of pelagic tunicates. The tunicates produce feces which closely resemble the flocculent organic aggregates from the suface waters of productive areas of the

oceans. These feces, when fresh, are contained in an amorphous
gelatinous matrix which serves as an excellent culture material for
bacteria. Similarly, Iseki (1981) has shown that salps, which often
appear in huge swarms, produce fecal pellets indistinguishable, after
bacterial degradation, from particulates produced by aggregation of
smaller particles. It may not be possible to distinguish fecal
material from non-pellet producing animals from the loosely structured organic aggregates of surface waters once bacterial growth has
become well established. As a result, the only estimates we may
ever be able to make of the importance of these materials in the
regenerative process will have to be based on extrapolations from
laboratory experiments.

Another type of particle in the open ocean is one modified by
the adsorption of various pollutants (Wangersky 1978a). Chlorinated
hydrocarbons, polycyclic aromatic hydrocarbons, and the simpler
hydrocarbons have all been found to adsorb on particulate matter and
to prefer organic materials to clays (Herbies 1977). The resulting
particle may have physical, chemical, and biological properties
quite different from those of either the original particle or the
pollutant. For example, the new particle may be considerably larger
and may therefore sediment faster or slower than the original
particle (Kajihara and Matsuoka 1978). It is also true that the
chemical reactivity of the pollutant may be changed; Miller and Zepp
(1979) found that model compounds adsorbed in particles were
photo-oxidized slightly faster than when in solution.

The pollutant itself can become a particle and a site for
bacterial colonization. For example, Seki et al. (1974) found
bacterial colonies on oil globules floating in the Philippine Sea.
In the presence of surfactants, many hydrocarbons will both adsorb
onto the surfaces of micelles and dissolve in the centers (Mukerjee
and Cardinal 1978). While little work has been done on the influence
of micelles on bacterial growth, and vice versa, it is certainly to
be expected that the presence of these micro-environments rich in
organic matter will be noted by any passing bacteria. In that
portion of the surface waters regularly traversed by bubbles injected
by breaking waves, both micelles and bacteria will be collected on
the surfaces of the bubbles.

Collection of pollutants on particle surfaces may have a
variety of effects. If the pollutant covers much of the surface of
the particle and is itself resistant to microbial degradation, the
particle may be protected from bacterial action until after sedimentation. This could result in a lower rate of nurient regeneration
in the surface waters, and therefore a lower average primary productivity. If the pollutant can be metabolized by the bacteria, then
adsorption may simply concentrate the material and make it more
available to the microbes. Oloffs et al. (1973) investigated
chlorinated hydrocarbons adsorbed on sediments and found that

lindane was metabolized but DDT and its derivatives were not. The
worst ecological effect would occur when a material toxic to the
bacteria coats the particles. Bacteria would not attack the
particles, even on the sea floor, and the result would be a major
removal of nutrients from the water column.

The bacteria may themselves form large particles; they may
cement together material which might otherwise pass through the
filter used to define the particulate state or they may form
colonies large enough to be caught by the filter. Barber (1966) and
Sheldon et al. (1967) have reported on the formation of particles in
sea water by bacteria. Some mechanisms involved in this formation,
such as the cementing together of smaller particles by bacterial and
fungal filaments, have been demonstrated by Paerl (1973, 1974, 1977).
The clumping together of bacteria to form particles has been shown
by Seki (1970, 1971), who found large clumps after the spring phytoplankton bloom. In Narragansett Bay, a carbon-rich area, Sieburth
(1968) found that pseudomonads produced organic materials with high
molecular weights which clumped other species of bacteria.

The various mechanisms for the production of particulate matter
in sea water are generally investigated in the laboratory, where
variables are minimized, to allow the study of one mechanism at a
time. In nature, particles often result from the agglomeration of
the smaller particles formed by different processes. The early work
on the composition of naturally occurring particles has been reviewed
by Riley (1970). More recently, Melnikov (1975) has described
particulate matter collected in various areas of the southeast
Pacific. He found higher proportions of undefined organic detritus
in particles collected in oligotrophic waters (92-97%) than in
eutrophic areas (32-65%) where much of the particulate matter was in
the form of phytoplankton.

Much of our information on the composition of particulates comes
from the examination of material which has already been subjected to
an averaging process by the mechanism used for collection. For
example, passage of the more fragile particles through the narrow
ports of our sampling gear must result in disintegration into smaller
parts. If we wish to study the particles as they exist in the
oceans, we must find better sampling methods. If we are concerned
only with particles in the surface waters, it may be possible to
collect at least the larger particles by hand. For example, the
particles collected by Silver et al. (1978) and Trent et al. (1978)
from the surface waters of Monterey Bay were little universes, each
with its own flora and fauna. Most of the POC in those waters was
in the form of the macroscopic aggregates, with only minor
quantities present in the regions between.

Our usual methods of sampling the deeper water tell us only how
much organic particulate material is present in the water parcel

collected by our sampling gear; they do not tell us whether this
material is distributed evenly, as small particles throughout the
water sampled, or as one large clump which has been broken up by the
act of sampling. The survival in deep water of the large aggregates
found in the surface layers can only be investigated with some form
of sampling gear which can scan large volumes of water. A promising
technique for this is an in situ camera which takes a picture of
particles in a cubic meter of water. The resolution is about 0.5 mm
(Honjo et al. 1983).

COMPOSITION OF PARTICLES

     In the ocean, the organic particulates are normally no more
than half the total particulate load. The exceptions are the bloom
periods in temperate and boreal waters and the upwelling regions,
where the organic materials of phytoplankton make up almost the
whole of the particulate load.

     Not surprisingly, the content of carbohydrates and proteins in
the organic aggregates is not much different from that in living
phytoplankton (Artemiev 1973; Smetacek and Hendrikson 1979). Large
seasonal variations in total protein, phosphorus, and carbohydrate
are found, as well as seasonal differences in the proportions of the
individual amino acids and carbohydrate monomers making up the larger
compounds (Matsuda et al. 1975; Daumas 1976; Maita and Yanada 1978).
These differences probably reflect both differences in species
composition and differences in nutritional state of the phytoplankton
community. Alldredge (1979) compared the chemical compositions of
the macroscopic aggregates to those of the total particulate matter
in the Santa Barbara Channel and the Gulf of California and found
that the larger aggregates had significantly less organic material
and a lower C:N ratio than the total particulate matter. This would
suggest that bacterial degradation was already in progress on the
larger particles, particularly on those showing evidence of
zooplankton origin.

     With depth, the total amount of particulate organic material
decreases sharply while the components change in a non-uniform
manner. The water-soluble carbohydrates and the proteins disappear
first, until at greater depths most of the organic matter is composed
of water-insoluble carbohydrates (Handa 1970; Artemiev 1974). These
changes are reflected in the common observation of changes with depth
of the C:N ratio of particulate matter reviewed by Riley (1970).
While some of the early workers considered the changes in POC with
depth to be within the analytical error of the methods of determination and felt that essentially no change in either content or
concentration occurred below perhaps 400-500 m (Menzel and Ryther
1968, 1970; Menzel 1970), more recent data clearly demonstrates an
exponential decrease in POC content with depth in all oceans and to

all depths (Wangersky 1976a). Bacterial utilization is also strongly
suggested by changes in stable carbon isotope content with depth
(Eadie et al. 1978).

Most of the measurements of changes in composition with depth
have been made on particles caught by water bottle samplers; thus,
these data refer largely to the small particles which remain in
suspension for some considerable period. It is only in the last few
years that sediment traps have again become popular, giving us access
to the larger particles which fall too swiftly to be caught in our
usual samplers. While the organic chemistry of these particles is
only beginning to be studied, preliminary results (Wakeham et al.
1980) suggest that lipids become progressively impoverished in the
particulate matter with depth, decreasing even faster than the total
organic content. In spite of the relatively quick onset of bacterial
degradation of these materials, evident in samples from even the
shallowest trap at 389 m, material labile to bacterial decomposition
was found in the deepest trap, at 5068 m, and could be expected to
reach the sea floor. While there are still many problems to be
solved in the design and deployment of sediment traps, it is only
with their use that we will begin to understand the dynamics of
organic particulate materials in the oceans.

DISTRIBUTION OF PARTICULATE ORGANIC CARBON

The literature on the distribution of POC in the oceans refers
almost entirely to those particles caught by the usual water
samplers; data from large-volume samplers, such as the in situ pump
employed by Bishop and Edmond (1976), and from the many kinds of
sediment traps are still too scanty for any quantitative regional
estimates. Even the POC data we do have on regional distributions
is often marred by the lack of standardization of sampling methods.
While any one laboratory may display commendable consistency over
long periods of time, and while even single samples may be
considered reasonable representations of the universe sampled
(Wangersky 1974), small changes in the method of sampling, such as a
change in filter type, may result in quite different estimates of
the material being measured (Wangersky and Hincks 1980). However,
when one system of sampling and one analytical technique is used
throughout, the values and distributions of POC are much the same in
all of the oceanic regions (Wangersky 1976b).

The general outlines of the POC distribution are now fairly well
accepted. The highest values are to be found near the surface, with
a considerable fall-off in values in the top 50-100 m. The region
of high particulate content may extend to deeper water in areas with
particularly strong mixing, such as the Gulf Stream. Below this
area, the POC content falls off exponentially with depth (Nishizawa
and Tsunogai 1974; Wangersky 1976b; Gordon 1977; Gordon et al.

1979). Values near the surface vary regionally and seasonally, depending upon the local productivity of the waters; values as high as 0.3 mg POC·liter^{-1} have been reported, and I myself have seen 0.2 mg C·liter^{-1} at the height of the spring plankton bloom on the Scotian Shelf. More typical numbers would be 30–50 µg C·liter^{-1} in the surface waters, and 1–10 µg C·liter^{-1} in deep water. The POC values given above were measured on samples collected with either Selas Flotronic® 0.45 µm silver filters or Whatman GF/C® glass fibre filters. However the absolute value of POC reported is dependent upon sampling technique, and there are values in the literature from usually reliable laboratories that are several times higher than these. Most laboratories now use either the Flotronic or one of the glass fibre filters and find values much like those above.

Our sampling techniques for POC are all discontinuous and usually involve single samples taken at considerable vertical and horizontal intervals. In our analyses we then interpolate smoothly between these sampling points and write a paper on the organic aggregates of tropical and subtropical waters of the North Atlantic based on the sampling points (Riley et al. 1964). This kind of thinking is permissible only in the early development of a field; there certainly are important discontinuities in both space and time in the distributions of POC. Particle-rich layers have long been observed from research submersibles; like UFO's and poltergeists they have been fully accepted only by those who have seen them, although there is a good description in the literature (Costin 1970). No one would dispute the existence of regions of high or low POC associated with particular water masses; many have reported patches of phyto-plankton in surface waters. Wangersky (1967b) for the Antarctic Intermediate Water in the South Pacific, and Pak et al. (1980a) for the oxygen minimum layer off Peru have reported changes in POC levels in layers at various depths. These layers are broad enough to be found by our usual methods of sampling, although the Antarctic Intermediate Water had to be sought deliberately; the method of placing bottles at standard depths would have missed the layer in about half of the casts. It would be difficult indeed to attempt to sample a layer perhaps less than a meter in thickness even if its location were known precisely. The proper sampling of any such thin layer awaits the invention of sampling gear which can first find the layer.

Seasonal changes in POC were always expected in the surface waters because of seasonal changes in phytoplankton growth. Seasonal changes in the whole water column were reported in the North Pacific by Nishizawa and Tsunogai (1974) and off Bermuda by Gordon (1977). More recently, the changes have been confirmed by sediment trap work in the Sargasso Sea (Deuser et al. 1981). Many workers did not accept seasonality in POC in the deeper water because of our calculations of the sinking speed of the small particles caught in water samplers. These particles are close to

the density of sea water, and should take many years to sink to the
bottom. We should not be able to see any seasonality of particles in
deeper waters as a result. It is much easier to accept a seasonality
in the supply of larger particles to the deep water; the rapid sinking of the particles should ensure a response time measured in days.
However, a large proportion of these larger particles are fecal
pellets, the remains of appendicularian houses, and other structures
which are readily attacked by bacteria. Indeed, all of the sediment
trap work has shown the rapid breakdown of such materials. Therefore, the size distribution of the larger particles should change
during their descent, shifting toward the smaller sizes, and pieces
of the larger structures should break off to become either part of
the POC or part of the dissolved material. The seasonality of the
small particles thus becomes understandable as a necessary consequence of the seasonality of large particles, and it is now understandable that the large particles are a source for the occasional
"cloud" of POC to be found in the deeper water (Wangersky 1974,
1976b, 1978b).

In our calculations of organic carbon fluxes for particulate
materials, we should also consider that for many of these particles
there will be no single sinking rate; as the particle disintegrates,
the sinking rate will decrease. Many of the smaller particles which
break off will have sinking rates close to zero. If estimates of
utilization are based entirely on changes in material caught in
sediment traps set at different depths, then the rates calculated
will be too high by the amount lost to the suspended and dissolved
organic fractions. A true estimate of the rate of change of organic
material, or of any class of organic compounds, must include measurements of large particles, suspended small particles, and dissolved
organic matter. However, changes in the dissolved carbon would be
very small (DOC pool is 7-10 times bigger than POC pool) and
impossible to measure with current methods.

PARTICLES AND MICROBIAL ACTIVITY

Bacteria and Particle Formation

The formation of organic particles by bubbling may be affected
by bacteria but it is difficult to know what is cause and what is
effect. Barber (1966) considered that organic aggregates were
formed primarily by bacterial activity, and Menzel (1966) claimed
that bacteria-free sea water would not form particles when bubbled.
Yet, recent work has demonstrated particle formation both by bubble
bursting and bubble dissolution (Johnson 1976; Johnson and Cooke
1980). In addition to forming particles, bubbles also collect and
concentrate bacteria on their surfaces as they rise (Blanchard and
Syzdek 1970, 1972, 1974; Bezdek and Carlucci 1972; Blanchard 1978).
This collection and subsequent bursting could be responsible for the

long-range aerial transport of both bacteria and viruses (Blanchard
and Syzdek 1974; Baylor et al. 1977). These mechanisms would
certainly also result in an accumulation of bacteria on the particles
resulting from bubble dissolution.

Particles, as well as bubbles, serve to concentrate microbes and
the addition of particles to sea water has been used to collect both
viruses and bacteria (Dixon and Zielyk 1969; Mitchell and Jannasch
1969; Bitton and Mitchell 1974; Gerba and Schaiberger 1975; Bitton
et al. 1979). The removal of terrestrial bacteria, such as E. coli,
from the water column soon after mixing with sea water has long been
known, and is often attributed to some antibacterial substance
present in the sea water. The simpler explanation of attachment to
particles and subsequent sedimentation has not been thoroughly
investigated, although the accumulation of such bacteria by particle-
feeders, such as clams and oysters, has resulted in the closing of
these fisheries in the vicinity of most centers of population even
in Nova Scotia.

The complementary process of particle formation by bacterial
activity is also well documented. The binding together of flocculant
material by bacterially produced carbohydrate fibrils was shown by
Friedman et al. (1969), and the mechanism of attachment was investigated by Marshall et al. (1971). They found a two-step process, the
first being an attraction resulting from the difference in surface
charge between bacteria and non-living particles. This step is
reversible and the strength of the attachment is relatively small.
Bacteria best attached to hydrophobic materials of low charge, next
best to hydrophilic materials of positive charge, and least to
materials of negative charge (Fletcher and Loeb 1979). The second
step is the extrusion of the polysaccharide fibrils by the bacteria,
and is irreversible. The process should result in a skewing of
particle size distributions towards the larger sizes, the point of
equilibrium being determined by the balance between particle
cementation and utilization (Paerl 1974). The activity of fecal
bacteria in the formation of flocculent aggregates in the ocean has
been shown by Ogawa (1977a,b).

Particles and Decomposition

Experiments have shown that the bacteria associated with
particles are quite capable of decomposing the materials by themselves. Miyoshi (1976), in his investigation of the decomposition
of plankton in the laboratory, found a rapid breakdown into soluble
materials during the first days followed by a gradual decrease in
rate. The rates were temperature dependent, and decreased sharply if
the system was allowed to go anaerobic. In the natural environment,
conclusions must be drawn from comparisons of materials in suspension
and in sediment traps (Hendrikson 1976), from comparisons of samples
taken at different depths (Yanada and Maita 1978), or from

comparisons of suspended material with surface sediments (Iturriaga 1979). In each of these cases, changes in the chemical composition of the particulates suggested that bacterial decomposition was taking place at a rate which was temperature dependent, very much faster in the summer than in the winter.

Biochemical assays of microbial mass and activity on natural (pine needles) and artificial surfaces incubated in an estuary showed that a surface which is a food source as well as a site for colonization supports a microbial population with much greater enzyme activity than an inert surface (Bobbie et al. 1978). In future studies, care will have to be taken to ensure that the inorganic particles used do not acquire an organic coating upon immersion in the sea water (Neihof and Loeb 1972). This could probably best be accomplished by treating the sea water with successive additions of clean particles, until all of the surface-active organics are removed. The usual method of removing organics, oxidation by high-intensity UV light, results in the formation of peroxides which may prove injurious to the organisms.

The question then raised is whether the bacteria on particles are important in the decomposition of dissolved organic materials. The literature in this area of research is contradictory and confusing, but there can be no doubt that attached bacteria can make use of dissolved organic matter. Aizatulin and Khailov (1972), studying the degradation of proteins and polysaccharides in sea water, concluded that the attached bacteria were transformers, taking in large molecules and excreting smaller ones. Organisms associated with particles are also capable of the uptake of glucose (Iturriaga 1979), arginine, and glutamic acid (Hollibaugh 1976). Satoh and Hanya (1976) showed decomposition of urea by the particulate fraction, but not by the free-living bacteria.

It is, of course, how many microbes are on particles versus how many are free living in a given habitat that determines whether attached or free-living forms dominate the heterotrophic processes. There are also questions about the activity per cell. In the Humber estuary, where the particulate load is high, Goulder (1977) concluded the organic decomposition was carried out mainly by attached bacteria. In estuaries with lower suspended loads the free-living bacteria might be more important. Hanson and Wiebe (1977) studied a saltmarsh estuary and found most of the heterotrophic activity in the particulate fraction greater than 3 μm. In coastal waters, the activity was associated with even larger particles. In the Gulf Stream, however, the activity seemed to be present in the smallest size fraction; Azam and Hodson (1977) found the same in the Pacific Ocean where 90% of the activity passed through a 1 μm filter. Harvey and Young (1980) also found the active bacteria to be positively correlated with particulate matter; they, too, were working with samples from salt marshes.

The field data thus suggest that the increased activity of bacteria on particles may be important in those regions of the ocean where the particle load is usually high, but that the free-living forms may be more important in the open ocean.

Direct count methods, such as the acridine-orange technique of Hobbie et al. (1977), provide no estimate of the metabolic state of the bacteria observed; many workers believe that most of the bacteria might well be present in some inactive form. Thus, Hobbie (1979) found that the activity of bacteria could vary over a range of 10,000-fold, while the numbers varied only by a factor of 10.

Methods for counting the metabolically active cells have usually depended upon the uptake of organic substrates containing radioactive carbon (Faust and Correll 1977) or hydrogen (Hoppe 1978). After the period of incubation, the number of cells taking up the isotope is measured by autoradiography and direct count. The percentage of actively metabolizing bacteria, as measured by these techniques, ranges from 40-85% of the total count.

Particles and Pollutants

While the primary purpose of this paper is to examine the natural relationships between bacteria and particulate matter in the oceans, the association between particles and pollutants makes necessary the examination of the possible effects of pollution on bacterial metabolism. It is not possible simply to equate pollution with deleterious effects; for example, if the water body is turbulent enough that oxidation of organic industrial wastes will not turn the region anaerobic, the addition of such wastes may just increase the overall productivity of the region (Sibert and Brown 1975; Ishida et al. 1977). Even the addition of fairly massive amounts of hydro-carbons seems at most to slow down microbial activity, and even then only temporarily as oil-oxidizing bacteria replace those normally present (Atlas and Bartha 1972; Hughes and McKenzie 1975; Atlas et al. 1976; Walker et al. 1976a; Walker and Colwell 1976, 1977; Walker et al. 1976b; Fujisawa et al. 1977; Larson et al. 1979; Gearing et al. 1980; Prahl et al. 1980; Stahl 1980). In fact, it is true that most of the organic material in petroleum is little different from other naturally occurring organics in that bacteria will oxidize them successfully as long as there is enough oxygen available. The key to successful elimination of all but the heaviest fractions of the petroleum is sufficient turbulence to resupply the necessary oxygen. Lacking this, the very much slower and less complete anaerobic degradation will apply.

In the context of the relationship between bacteria and particles, hydrocarbons are important because some of them adsorb to pre-existing particles, and others can themselves be considered as particles which collect other organic materials as well as bacteria.

Some petroleum fractions, including some unidentified volatiles (Atlas and Bartha 1972) and some photo-oxidation products (Larson et al. 1979), may either retard or prevent bacterial growth; but the most likely result of petroleum additions should be local increases in bacterial growth, perhaps of species unusual in normal sea water, and possible local regions in and around the particles where the Eh becomes temporarily reducing. Where heavy metals have been enriched on the surfaces of particles, perhaps by collection in the microlayer at the air-water interface (Elzerman and Armstrong 1979) or in wind-generated foam (Eisenreich et al. 1978), local regions in which bacterial growth is retarded might be expected. However, oxidation of the organic fraction, either chemical or bacterial, should free these materials into the water column. In this manner the particles may act as transporters of heavy metals down into deeper water (Rohatgi and Chen 1975).

Particles containing or consisting of hydrocarbons may also collect the naturally occurring surfactants from the water column and, more seriously from the point of view of the bacteria, may collect and concentrate the chlorinated hydrocarbons (Duinker and Hillebrand 1979; Osterroht and Smetacek 1980). It has been shown that the chlorinated biphenyls will inhibit heterotrophic bacteria at microgram per liter levels (Blakemore and Carey 1978). These are levels which are seldom reached in the water column; however, they may be relatively common concentrations on the surfaces of particles, or in the hydrophobic centers of micelles. Our measurements of chlorinated hydrocarbon concentrations should probably be made with reference to the particulate load of the water column, rather than on a simple weight-volume basis.

SUMMATION: THE PARTICLE AS A HAPPY HOME FOR BACTERIA

In much of the early work in phytoplankton dynamics it is implicit, although not stated, that regeneration of inorganic nutrients is something which occurs primarily in the deeper water layers, and that surface productivity is maintained, in between mixing episodes, by the movement of nutrients upward from deeper water by eddy diffusivity. More recent work, using stable isotope dilution methods (Harrison 1978), has suggested that regeneration of nutrients occurs mostly in surface waters and is a rapid process with a turnover time measured in days. This remineralization is certainly not due entirely to bacteria; just about every kind of non-photosynthetic organism present in the surface waters has been shown to participate in the breakdown of organic material. However, experimental work on laboratory microcosms has shown that in the absence of all other organisms, bacteria could do the job all by themselves (Skopintsev 1973/76; Melnikov and Volostnykh 1974; Rajagopal 1974; DePinto and Verhoff 1977; Yamada et al. 1979).

The degradation of organic particulate matter by bacteria, with associated regeneration of nutrients, is also well accepted. There is some evidence that such degradation may result in the formation of nutrient micropatches (Shanks and Trent 1979). The participation of bacteria on particles in the large-scale degradation of dissolved organic matter is less well documented, largely because the proper kinds of experiments have yet to be conducted. One such experiment, using sterilized and unsterilized particles, showed a linear increase in nitrification with the addition of suspended particles. Addition of sterilized particles resulted in a lag period before nitrification rates increased (Kholdebarin and Oertli 1977).

The importance of attached bacteria in the regeneration of nutrients in surface waters will depend upon the activity and relative numbers of the free-living bacteria. If we accept the estimates based on uptake of labelled substrates, that somewhere between 40-90% of the free-living bacteria are active and that some 80-90% of the heterotrophic activity is in the smallest size fraction, then the attached bacteria are obviously important only in the degradation of particulate organic materials or in habitats where particles are especially abundant.

The function of the particle in bacterial degradation of dissolved organic matrials may be two-fold: the particle furnishes a surface and a larger mass which ensures that the access of the bacteria to these dissolved materials is not limited by molecular diffusion (Wangersky 1977); and, the particle supplies some usable organic material directly which ensures that the organism is in the proper physiological state to use any dissolved organic molecules it may encounter.

Not all of the particulate matter stays around in the surface waters long enough to be degraded. The recent spate of sediment trap work, along with the in situ high-speed pump studies already mentioned, has shown that the larger, heavier particles, such as fecal pellets, fall out of the surface waters relatively intact. Shanks and Trent (1980) have calculated that, based on measured sinking rates, some 3-5% of the POC and 4-22% of the PON present in the surface waters is removed daily by the sinking of particles. I suspect that this is an overestimate, since it is based on an extrapolation implying no change in size during the journey. Such particles, even if not consumed by any particle-feeders in the upper waters, should be subject to steady decomposition by their own inhabitants. I would expect a rapid initial sinking, with a steady decrease in rate as pieces break off into smaller bits.

A similar picture of disintegration into smaller pieces, with a consequent reduction in sinking rate, should be expected for many of the larger particles, and has been documented for some kinds of fecal materials. Thus, the re-supply of smaller particulate materials to deeper water should be a natural consequence of bacterial degradation of the larger, faster-falling materials, and explains the seasonality in particulate matter observed in the deep ocean.

ACKNOWLEDGEMENT

This work was funded by a grant from the Natural Sciences and Engineering Research Council of Canada.

LITERATURE CITED

Aizatulin, T. A., and K. M. Khailov. 1972. Kinetics of the trans- formation of proteins and polysaccharides dissolved in the sea water through the interaction with detritus. Okeanologia 12: 809-816.

Alldredge, A. L. 1972. Abandoned larvacean houses: a unique food source in the pelagic environment. Science 177: 885-887.

Alldredge, A. L. 1976. Discarded appendicularian houses as sources of food, surface habitats, and particulate organic matter in planktonic environments. Limnol. Oceanogr. 21: 14-23.

Alldredge, A. L. 1979. The chemical composition of macroscopic aggregates in two neritic seas. Limnol. Oceanogr. 24: 855-866.

Artemiev, V. E. 1973. Carbohydrates in the suspended matter of the Pacific Ocean. Okeanologia 13: 809-813.

Artemiev, V. E. 1974. Comparative characteristics of the composi- tion of carbohydrates of phytoplankton, suspended matter and bottom sediments of the ocean. Okeanologia 14: 1012-1016.

Atlas, R. M., and R. Bartha. 1972. Biodegradation of petroleum in seawater at low temperature. Can. J. Microbiol. 18: 1851-1855.

Atlas, R. M., E. A. Schofield, F. A. Morelli, and R. E. Cameron. 1976. Effects of petroleum pollutants on Arctic microbial populations. Environ. Pollut. 10: 35-43.

Azam, F., and R. E. Hodson. 1977. Size distribution and activity of marine microheterotrophs. Limnol. Oceanogr. 22: 492-501.

Barber, R. T. 1966. Interaction of bubbles and bacteria in the formation of organic aggregates in seawater. Nature (London) 211: 257-258.

Barcelona, M. J., and D. K. Atwood. 1979. Gypsum-organic inter- actions in the marine environment: sorption of fatty acids and hydrocarbons. Geochim. Cosmochim. Acta 43: 47-53.

Barham, E. G. 1979. Giant larvacean houses: observations from deep submersibles. Science 205: 1129-1131.

Baylor, E. R., M. B. Baylor, D. C. Blanchard, L. D. Syzdek, and C. Appel. 1977. Virus transfer from surf to wind. Science 198: 575-580.

Bezdek, H. F., and A. F. Carlucci. 1972. Surface concentration of marine bacteria. Limnol. Oceanogr. 17: 566-569.

Biggs, R. B., and D. A. Flemer. 1972. The flux of particulate carbon in an estuary. Mar. Biol. 12: 11-17.

Bishop, J. K. B., and J. M. Edmond. 1976. A new large volume fil-
    tration system for the sampling of oceanic particulate matter.
    J. Mar. Res. 34: 181-198.

Bitton, G., B. N. Feldberg and S. R. Farrah. 1979. Concentration
    of enteroviruses from seawater and tapwater by organic floccu-
    lation using non-fat dry milk and casein. Water Air Soil
    Pollut. 12: 187-195.

Bitton, G., and R. Mitchell. 1974. Effect of colloids on the
    survival of bacteriophages in seawater. Water Res. 8: 227-229.

Blakemore, R. P., and A. E. Carey. 1978. Effects of polychlorinated
    biphenyls on growth and respiration of heterotrophic marine
    bacteria. Appl. Environ. Microbiol. 35: 323-328.

Blanchard, D. 1978. Jet drop enrichment of bacteria, virus, and
    dissolved organic material. Pure Appl. Geophys. 116: 302-308.

Blanchard, D., and L. Syzdek. 1970. Mechanism for the water-to-air
    transfer and concentration of bacteria. Science 170: 626-628.

Blanchard, D., and L. D. Syzdek. 1972. Concentration of bacteria
    in jet drops from bursting bubbles. J. Geophys. Res. 77:
    5087-5099.

Blanchard, D., and L. D. Syzdek. 1974. Importance of bubble
    scavenging in the water-to-air transfer of organic matter and
    bacteria. J. Rech. Atmos. 13: 529-540.

Bobbie, R. J., S. J. Morrison, and D. C. White. 1978. Effects of
    substrate biodegradability on the mass and activity of the
    associated estuarine microbiota. Appl. Environ. Microbiol. 35:
    179-184.

Cauwet, G. 1978. Organic chemistry of sea water particulates:
    concepts and developments. Oceanol. Acta 1: 99-105.

Conover, R. J. 1975. Transformation of organic matter, pp. 221-
    499. In: O. Kinne [ed.], Marine Ecology, Vol. IV, Dynamics.
    John Wiley and Sons, New York.

Costin, J. M. 1970. Visual observations of suspended-particle
    distribution at three sites in the Caribbean Sea. J. Geophys.
    Res. 75: 4144-4150.

Daumas, R. A. 1976. Variations of particulate proteins and
    dissolved amino acids in coastal seawater. Mar. Chem. 4:
    225-242.

Delany, A. C., A. C. Delany, D. W. Parkin, J. J. Griffin, E. D.
    Goldberg, and B. E. F. Reimann. 1967. Airborne dust collected
    at Barbados. Geochim. Cosmochim. Acta 31: 885-909.

DePinto, J. V., and F. H. Verhoff. 1977. Nutrient regeneration
    from aerobic decomposition of green algae. Environ. Sci.
    Technol. 11: 371-377.

de Souza Lima, H., and P. J. leB. Williams. 1978. Oxygen consump-
    tion by the planktonic population of an estuary - Southampton
    Water. Estuarine Coastal Mar. Sci. 6: 515-521.

Deuser, W. G., E. H. Ross, and R. F. Anderson. 1981. Seasonality in
    the supply of sediment to the deep Sargasso Sea and implications
    for the rapid transfer of matter to the deep ocean. Deep-Sea
    Res. 28: 495-505.

Dixon, J. K., and M. W. Zielyk. 1969. Control of the bacterial
    content of water with synthetic polymeric flocculants. Environ.
    Sci. Technol. 3: 551-558.
Duce, R. A., C. K. Unni, B. J. Ray, J. M. Propero, and J. T. Merrill.
    1980. Long-range atmospheric tranpsort of soil dust from Asia
    to the tropical North Pacific: Temporal variability. Science
    209: 1522-1524.
Duinker, J. C., and M. T. J. Hillebrand. 1979. Behaviour of PCB,
    pentachlorobenzene, hexachlorobenzene, α-HCH, γ-HCH, β-HCH,
    Dieldrin, endrin and p,p'-DDD in the Rine-Meuse estuary and the
    adjacent coastal area. Neth. J. Sea Res. 13: 256-282.
Eadie, B. J., and L. M. Jeffrey. 1973. δ13-C analyses of oceanic
    particulate organic matter. Mar. Chem. 1: 199-209.
Eadie, B. J., L. M. Jeffrey and W. M. Sackett. 1978. Some observa-
    tions on the stable carbon isotope composition of dissolved and
    particulate organic carbon in the marine environment. Geochim.
    Cosmochim. Acta 42: 1265-1269.
Eggiman, D. W., P. R. Betzer and K. L. Carder. 1980. Particle
    transport from the West African shelves of Liberia and Sierra
    Leone to the deep sea: a chemical approach. Mar. Chem. 9:
    283-306.
Ehrhardt, M., C. Osterroht, and G. Petrick. 1980. Fatty-acid methyl
    esters dissolved in seawater and associated with suspended
    particulate matter. Mar. Chem. 10: 67-76.
Eisenreich, S. J., A. W. Elzerman, and D. E. Armstrong. 1978.
    Enrichment of micronutrients, heavy metals, and chlorinated
    hydrocarbons in wind-generated lake foam. Environ. Sci.
    Technol. 12: 413-417.
Elzerman, A. W., and D. E. Armstrong. 1979. Enrichment of Zn, Cd,
    Pb, and Cu in the surface microlayer of Lakes Michigan,
    Ontario, and Mendota. Limnol. Oceanogr. 24: 133-144.
Faust, M. A., and D. L. Correll. 1977. Autoradiographic study to
    detect metabolically active phytoplankton and bacteria in the
    Rhode River estuary. Mar. Biol. 41: 293-303.
Fellows, D. A., D. M. Karl, and G. A. Knauer. 1981. Large particle
    fluxes and the vertical transport of living carbon in the upper
    1500 m of the northeast Pacific Ocean. Deep-Sea Res. 28:
    921-936.
Fletcher, M., and G. I. Loeb. 1979. Influence of substratum
    characteristics on the attachment of a marine pseudomonad to
    solid surfaces. Appl. Environ. Microbiol. 37: 67-72.
Fowler, S. W., and L. F. Small. 1972. Sinking rates of euphausiid
    fecal pellets. Limnol. Oceanogr. 17: 293-296.
Friedman, B.A., P. R. Dugan, R. M. Pfister, and C. C. Remsen. 1969.
    Structure of exocellular polymers and their relationship to
    bacterial flocculation. J. Bacteriol. 98: 1328-1334.
Fujisawa, H., M. Murakami, and T. Manabe. 1977. Ecological studies
    on hydrocarbon-oxidizing bacteria in Japanese coastal waters.
    I. Some methods of enumeration of hydrocarbon-oxidizing

bacteria. Bull. Jpn. Soc. Sci. Fish. 43: 659-668.

Gearing, P. J., J. N. Gearing, R. J. Pruell, T. L. Wade, and J. G.
    Quinn. 1980. Partitioning of No. 2 fuel oil in controlled
    estuarine ecosystems. Sediments and suspended particulate
    matter. Environ. Sci. Technol. 14: 1129-1136.

Gerba, C. P., and G. E. Schaiberger. 1975. Aggregation as a factor
    in loss of viral titer in seawater. Water Res. 9: 567-571.

Gilmer, R. W. 1972. Free-floating mucus webs: a novel feeding
    adaptation for the open ocean. Science 176: 1239-1240.

Goetz, A. 1965. Parameters for biocolloidal matter in the
    atmosphere, pp. 79-97. Proc. Atmos. Biol. Conf.

Gordon, D. C., Jr. 1977. Variability of particulate organic carbon
    and nitrogen along the Halifax-Bermuda section. Deep-Sea Res.
    24: 257-270.

Gordon, D. C., Jr., P. J. Wangersky and R. W. Sheldon. 1979.
    Detailed observations on the distribution and composition of
    particulate organic material at two stations in the Sargasso
    Sea. Deep-Sea Res. 26: 1083-1092.

Goulder, R. 1977. Attached and free bacteria in an estuary with
    abundant suspended solids. J. Appl. Bacteriol. 43: 399-405.

Gundersen, K., C. W. Mountain, D. Taylor, R. Ohye and J. Shen. 1972.
    Some chemical and microbiological observations in the Pacific
    Ocean off the Hawaiian Islands. Limnol. Oceanogr. 17: 524-531.

Handa, N. 1970. Organic components of particulate matter in sea
    water in the Kuroshio and Oyashio areas. Proc. 2 CSK Symp.,
    Tokyo 1970: 207-211.

Hanson, R. B. 1977. Pelagic Sargassum community metabolism: carbon
    and nitrogen. J. Exp. Mar. Biol. Ecol. 29: 107-118.

Hanson, R. B., and W. J. Wiebe. 1977. Heterotrophic activity assoc-
    iated with particulate size fractions in a Spartina alterniflora
    salt-marsh estuary, Sapelo Island, Georgia, U.S.A., and the
    continental shelf waters. Mar. Biol. 42: 321-330.

Harrison, W. G. 1978. Experimental measurements of nitrogen remin-
    eralization in coastal waters. Limnol. Oceanogr. 23: 684-694.

Harvey, R. W., and L. Y. Young. 1980. Enrichment and association
    of bacteria and particulates in salt marsh surface water.
    Appl. Environ. Microbiol. 39: 894-899.

Hendrikson, P. 1976. Abbauraten von organischem Kohlenstoff im
    Seston und in Sinkstoffen der Kieler Bucht. Kiel. Meeresforsch.
    3: 103-119.

Herbies, S. E. 1977. Partitioning of polycyclic aromatic hydro-
    carbons between dissolved and particulate phases in natural
    waters. Water Res. 11: 493-496.

Hobbie, J. E., R. J. Daley, and S. Jasper. 1977. Use of Nuclepore
    filters for counting bcteria by fluorescence microscopy. Appl.
    Environ. Microbiol. 33: 1225-1228.

Hobbie, J. E. 1979. Activity and bacterial biomass. Arch.
    Hydrobiol. Beih. Ergebn. Limnol. 12: 59-63.

Hollibaugh, J. T. 1976. The biological degradation of arginine and
    glutamic acid in seawater in relation to the growth of phyto-

plankton. Mar. Biol. 36: 303-312.

Honjo, S., K. Doherty, Y. Agrawal, and V. Asper. 1983. Direct
    optical assessment of large amorphous aggregates in the deep
    ocean. Deep-Sea Res. In press.

Honjo, S., and M. R. Roman. 1978. Marine copepod fecal pellets:
    production, preservation and sedimentation. J. Mar. Res. 36:
    45-57.

Hoppe, H.-G. 1978. Relations between active bacteria and
    heterotrophic potential in the sea. Neth. J. Sea Res. 12:
    78-98.

Hughes, D. E., and P. McKenzie. 1975. The mirobial degradation of
    oil in the sea. Proc. R. Soc. Lond. (B) 189: 375-390.

Hunter, K. A., and P. S. Liss. 1979. The surface charge of
    suspended particles in estuarine and coastal waters. Nature
    (London) 282: 823-825.

Iseki, K. 1981. Particulate organic matter tranpsort to the deep
    sea by salp fecal pellets. Mar. Ecol. Prog. Ser. 5: 55-60.

Ishida, Y., A. Uchida, and H. Kadota. 1977. Ecological studies on
    bacteria in the sea and lake waters polluted with organic
    substances. IV. Determination of bacterial degradable organic
    matter in aquatic environments. Bull. Jpn. Soc. Sci. Fish. 43:
    885-892.

Iturriaga, R. 1979. Bacterial activity related to sedimenting
    particulate matter. Mar. Biol. 55: 157-169.

Johannes, R. E., and M. Satomi. 1966. Composition and nutritive
    value of fecal pellets of a marine crustacean. Limnol.
    Oceanogr. 11: 191-197.

Johnson, B. D. 1976. Nonliving organic particle formation from
    bubble dissolution. Limnol. Oceanogr. 21: 444-446.

Johnson, B. D., and R. C. Cooke. 1980. Organic particle and
    aggregate formation resulting from the dissolution of bubbles
    in seawater. Limnol. Oceanogr. 25: 653-661.

Kajihara, M., and M. Matsuoka. 1978. Settling of oil particles
    adsorbed on suspended matter. Bull. Fac. Fish. Hokkaido Univ.
    29: 259-269.

Kholdebarin, B., and J. J. Oertli. 1977. Effect of suspended
    particles and their sizes in the nitrification in surface
    water. J. Water Pollut. Control Fed. 49: 1693-1697.

Kofoed, L. H. 1975a. The feeding biology of Hydrobia ventrosa
    (Montagu). I. The assimilation of different components of the
    food. J. Exp. Mar. Biol. Ecol. 19: 233-241.

Kofoed, L. H. 1975b. The feeding biology of Hydrobia ventrosa
    (Montagu). II. Allocation of the components of the
    carbon-budget and the significance of the secretion of
    dissolved organic matter. J. Exp. Mar. Biol. Ecol. 19: 243-256.

Kranck, K. 1975. Sediment deposition from flocculated suspensions.
    Sedimentology 22: 111-123.

Kranck, K. 1979. Dynamics and distribution of suspended particulate
    matter in the St. Lawrence estuary. Nat. Can. 106: 163-173.

Kranck, K., and T. Millligan. 1980. Macroflocs: production of marine snow in the laboratory. Mar. Ecol. Prog. Ser. 3: 19-24.

Kuenen, P. H.   1950. Marine Geology. John Wiley & Sons, Inc., New York.

Larson, R. A., T. L. Bott, L. L. Hunt, and K. Rogenmuser. 1979. Photooxidation products of a fuel oil and their antimicrobial activity. Environ. Sci. Technol. 13: 965-969.

Lasker, R. 1964. Moulting frequency of a deep-sea crustacean Euphausia pacifica. Nature (London) 203: 96.

Maita, Y., and M. Yanada. 1978. Particulate protein in coastal waters, with special reference to seasonal variation. Mar. Biol. 44: 329-336.

Marshall, K. C., R. Stout, and R. Mitchell. 1971. Mechanism of the initial events in the sorption of marine bacteria to surfaces. J. Gen. Microbiol. 68: 337-348.

Martin, A. G., C. Raiux, and J. R. Graull. 1977. Distribution de la matiére organique particulaire dans l'estuairie de la Penzé (Nord-Finistére). J. Rech. Oceanogr. 2: 13-19.

Matsuda, O., T. Endo and H. Koyama. 1975. On the balance and seasonal variation of dissolved and particulate phosphorus in an eutrophicated coastal environment. J. Fac. Fish. Anim. Husb. Hiroshima Univ. 14: 217-240.

Melnikov, I. A. 1975. Microplankton and organic detritus in the water of the south-east Pacific. Okeanologia 15: 146-157.

Melnikov, J. A., and B. V. Volostnykh. 1974. Some data about the destruction of organic matter and the regeneration of the mineral forms of phosphorus and nitrogen in the Antarctic waters of the Atlantic Ocean. Tr. Inst. Okeanol. Akad. Nauk SSSR 98: 261-269.

Menzel, D. W. 1966. Bubbling of sea water and the production of organic particles: a re-evaluation. Deep-Sea Res. 13: 963-966.

Menzel, D. W. 1970. The role of in situ decomposition of organic matter on the concentration of non-conservative properties in the sea. Deep-Sea Res. 17: 751-764.

Menzel, D. W., and J. H. Ryther. 1968. Organic carbon and the oxygen minimum in the South Atlantic Ocean. Deep-Sea Res. 15: 327-337.

Menzel, D. W., and J. H. Ryther. 1970. Distribution and cycling of organic matter in the ocean, pp. 31-54. In: D. W. Hood [ed.], Symposium on Organic Matter in Natural Waters. Inst. Marine Sci. Occas. Publ. 1.

Meyers, P. A., and J. G. Quinn. 1971a. Interaction between fatty acids and calcite in sea water. Limnol. Oceanogr. 16: 992-997.

Meyers, P. A., and J. G. Quinn. 1971b. Fatty acid-clay mineral association in artificial and natural sea water solutions. Geochim. Cosmochim. Acta 35: 628-632.

Meyers, P. A., and J. G. Quinn. 1973. Factors affecting the association of fatty acids with mineral particles in sea water. Geochim. Cosmochim. Acta 37: 1745-1759.

Miller, G. C., and R. G. Zepp. 1979. Photoreactivity of aquatic pollutants sorbed on suspended sediments. Environ. Sci. Technol. 13: 860-863.

Mitchell, R.,and H. W. Jannasch. 1969. Processes controlling virus inactivation in seawater. Environ. Sci. Technol. 3: 941–943.

Miyoshi, H. 1976. Decomposition of marine plankton under laboratory conditions. Bull. Jpn. Soc. Sci. Fish. 42: 1205–1211.

Morris, R. J., and S. E. Calvert. 1975. Fatty acid uptake by marine sediment particles. Geochim. Cosmochim. Acta 39: 377–381.

Mukerjee, P., and J. R. Cardinal. 1978. Benzene derivatives and naphthalene solubilized in micelles. Polarity of microenvironment, location and distribution in micelles, and correlation with surface activity in hydrocarbon-water systems. J. Phys. Chem. 82: 1620–1627.

Neihof, R., and G. Loeb. 1972. The surface charge of particulate matter in seawater. Limnol. Oceanogr. 17: 7–16.

Nishizawa, S., and S. Tsunogai. 1974. Dynamics of particulate material in the ocean. I. Production and decomposition of particulate organic carbon in the northern North Pacific Ocean and Bering Sea, pp. 173–174. In: D. W. Hood and E. J. Kelly [eds.], Oceanography of the Bering Sea. Occas. Publ. Inst. Mar. Sci. Univ. Alaska 2.

Ogawa, K. 1977a. Primary participation of fecal bacteria in the formation of suspended organic matter in the sea. I. Ion adsorption and floc formation by Escherichia coli. Bull. Jpn. Soc. Sci. Fish. 43: 1081–1088.

Ogawa, K. 1977b. Primary participation of fecal bacteria in the formation of suspended organic matter in the sea. II. Floc formation by fecal bacteria isolated from marine animals. Bull. Jpn. Soc. Sci. Fish. 43: 1089–1096.

Oloffs, P. C., L. J. Albright, S. Y. Szeto, and J. Lau. 1973. Factors affecting the behavior of five chlorinated hydrocarbons in two natural waters and their sediments. J. Fish. Res. Board Can. 30: 1619–1623.

Osterroht, C., and V. Smetacek. 1980. Vertical transport of chlorinated hydrocarbons by sedimentation of particulate matter in Kiel Bight. Mar. Ecol. Prog. Ser. 2: 27–34.

Paerl, H. W. 1973. Detritus in Lake Tahoe: structural modification by attached microflora. Science 180: 496–498.

Paerl, H. W. 1974. Bacterial uptake of dissolved organic matter in relation to detrital aggregation in marine freshwater systems. Limnol. Oceanogr. 19: 966–972.

Paerl, H. W. 1977. Bacterial sediment formation in lakes: trophic implications, pp. 40–47. In: H. L. Golterman [ed.], Interactions Between Sediments and Fresh Water. Dr. W. Junk B. V., The Hague.

Pak, H., L. A. Codispotti, and J. R. V. Zaneveld. 1980a. On the intermediate particle maxima associated with oxygen-poor water off western South America. Deep-Sea Res. 27: 783–797.

Pak, H., J. R. V. Zaneveld, and J. Kitchen. 1980b. Intermediate nepheloid layers observed off Oregon and Washington. J. Geophys. Res. 85: 6697–6708.

Pfister, R. M., P. R. Dugan, and J. I. Frea. 1969. Microparti-
culates: Isolation from water and identification of associated
chlorinated pesticides. Science 166: 878–879.
Pierce, J. W., and F. R. Siegel. 1979. Particulate material
suspended in estuarine and oceanic waters. Scanning Electron
Microsc. 1979: 555–562.
Pocklington, R., and J. D. Leonard. 1979. Terrigenous organic
matter in sediments of the St. Lawrence Estuary and the
Saguenay Fjord. J. Fish. Res. Board Can. 36: 1250–1255.
Pomeroy, L. R., and D. Deibel. 1980. Aggregation of organic matter
by pelagic tunicates. Limnol. Oceanogr. 25: 643–652.
Prahl, F. G., J. T. Bennett, and R. Carpenter. 1980. The early
diagenesis of aliphatic hydrocarbons and organic matter in
sedimentary particulates from Dabob Bay, Washington. Geochim.
Cosmochim. Acta 44: 1967–1976.
Rajagopal, M. D. 1974. On assimilation and regeneration of
phosphorus in two different environments. Mahasagar 7: 143–149.
Riley, G. A. 1970. Particulate organic matter in sea water. Adv.
Mar. Biol. 8: 1–118.
Riley, G. A., P. J. Wangersky, and D. Van Hemert. 1964. Organic
aggregates in tropical and subtropical surface waters of the
North Atlantic Ocean. Limnol. Oceanogr. 9: 546–550.
Rohatgi, N., and K. Y. Chen. 1975. Transport of trace metals by
suspended particulates on mixing with seawater. J. Water
Pollut. Control Fed. 47: 2298–2316.
Rowe, G. T., and N. Staresinic. 1977/1979. Sources of organic
matter to the deep-sea benthos. Ambio Spec. Rept. 6: 19–23.
Ruppersberg, G. H., and R. Schellhase. 1980. Warum reagiert das
Aerosol uber dem Atlantik so abnorm auf Anderungen der
relativen Feuchte? Annln. Met. 15: 245–246.
Sakamoto, W. 1972. Study on the process of river suspension from
flocculation to accumulation in estuary. Bull. Ocean Res.
Inst. Univ. Tokyo 5: 1–46.
Satoh, Y., and T. Hanya. 1976. Decomposition of urea by the larger
particulate fraction and the free bacteria fraction in a pond
water. Int. Revue ges. Hydrobiol. 61: 799–806.
Seki, H. 1965. Microbiological studies on the decomposition of
chitin in the marine environment – IX. Rough estimation on
chitin decomposition in the ocean. J. Oceanogr. Soc. Jpn. 21:
17–24.
Seki, H. 1970. Microbial biomass in the euphotic zone of the North
Pacific subarctic water. Pac. Sci. 24: 269–274.
Seki, H. 1971. Microbial clumps in seawater in the euphotic zone
of Saanich Inlet (British Columbia). Mar. Biol. 9: 4–8.
Seki, H., H. Abe, Y. Yamaguchi, and S.-E. Ichimura. 1974. Bacteria
on petroleum globules in the Philippine Sea in January, 1973.
J. Oceanogr. Soc. Jpn. 30: 151–156.
Shanks, A. L., and J. D. Trent. 1979. Marine snow: Microscale
nutrient patches. Limnol. Oceanogr. 24: 850–854.
Shanks, A. L., and J. D. Trent. 1980. Marine snow: sinking rates

and potential role in vertical flux. Deep-Sea Res. 27: 137–143.

Sheldon, R. W., P. T. Evelyn, and T. R. Parsons. 1967. On the occurrence and formation of small particles in seawater. Limnol. Oceanogr. 12: 367–375.

Sholkovitz, E. R. 1976. Flocculation of dissolved organic and inorganic matter during the mixing of river water and seawater. Geochim. Cosmochim. Acta 40: 831–845.

Sholkovitz, E. R. 1978. The flocculation of dissolved Fe, Mn, Al, Cu, Ni, Co and Cd during estuarine mixing. Earth Planet. Sci. Letts. 41: 77–86.

Sholkovitz, E. R., E. A. Boyle, and N. B. Price. 1978. The removal of dissolved humic acids and iron during estuarine mixing. Earth Planet. Sci. Letts. 40: 130–136.

Sholkovitz, E. R., and D. Copeland. 1981. The coagulation, solubility and adsorption properties of Fe, Mn, Cu, Ni, Cd, Co, and humic acids in a river water. Geochim. Cosmochim. Acta 45: 181–189.

Sholkovitz, E. R., and N. B. Price. 1980. The major-element chemistry of suspended matter in the Amazon estuary. Geochim. Cosmochim. Acta 44: 163–171.

Sibert, J., and T. J. Brown. 1975. Characteristics and potential significance of heterotrophic activity in a polluted fjord estuary. J. Exp. Mar. Biol. Ecol. 19: 97–104.

Sieburth, J. McN. 1967. Observations on bacteria planktonic in Narragansett Bay, Rhode Island; a resume. Bull. Misaki Mar. Biol. Inst. Kyoto Univ. 12: 49–64.

Silver, M. W., A. L. Shanks, and J. D. Trent. 1978. Marine snow: microplankton habitat and source of small-scale patchiness in pelagic populations. Science 201: 331–373.

Silverberg, N., and B. Sundby. 1979. Observations in the turbidity maximum of the St. Lawrence Estuary. Can. J. Earth Sci. 16: 939–950.

Skopintsev, B. A. 1973/76. Mineralization regularities of the organic matter of dead phytoplankton. Ambio Spec. Rept. 4: 45–54.

Small, L. F., S. W. Fowler, and M. Y. Unlu. 1979. Sinking rates of natural copepod fecal pellets. Mar. Biol. 51: 233–241.

Smetacek, V., and P. Hendrikson. 1979. Composition of particulate organic matter in Kiel Bight in relation to phytoplankton successison. Oceanol. Acta 2: 287–298.

Stahl, W. J. 1980. Compositional changes and 13C/12C fractionations during the degradation of hydrocarbons by bacteria. Geochim. Cosmochim. Acta 44: 1903–1907.

Stanley, D. W., and J. E. Hobbie. 1981. Nitrogen recycling in a North Carolina coastal river. Limnol. Oceanogr. 26: 30–42.

Tan, F. C., and P. M. Strain. 1979. Organic carbon isotope ratios in recent sediments in the St. Lawrence Estuary and the Gulf of St. Lawrence. Estuarine Coastal Mar. Sci. 8: 213–225.

Tanoue, E., and N. Handa. 1979. Differential sorption of organic matter by various sized sediment particles in recent sediment from the Bering Sea. J. Oceanogr. Soc. Jpn. 35: 199–208.

Trent, J. D., A. L. Shanks, and M. W. Silver. 1978. In situ and laboratory measurments on macroscopic aggregates in Monterey Bay, California. Limnol. Oceanogr. 23: 626–635.

Wakeham, S. G., J. W. Farrington, R. B. Gagosian, C. Lee, H. DeBaar, G. E. Nigrelli, B. W. Tripp, S. O. Smith, and N. M. Frew. 1980. Organic matter fluxes from sediment traps in the equatorial Atlantic Ocean. Nature (London) 286: 798–800.

Walker, J. D., J. J. Calomiris, T. L. Herbert, and R. R. Colwell. 1976a. Petroleum hydrocarbons: degradation and growth potential for Atlantic Ocean sediment bacteria. Mar. Biol. 34: 1–9.

Walker, J. D., and R. R. Colwell. 1976. Measuring the potential activity of hydrocarbon-degrading bacteria. Appl. Environ. Microbiol. 31: 189–197.

Walker, J. D., and R. R. Colwell. 1977. Sampling device for monitoring biodegradation of oil and other pollutants in aquatic environments. Environ. Sci. Technol. 11: 93–95.

Walker, J. D., P. A. Seesman, T. L. Herbert, and R. R. Colwell. 1976. Petroleum hydrocarbons: degradation and growth potential of deep-sea sediment bacteria. Environ. Pollut. 10: 89–99.

Wangersky, P. J. 1974. Particulate organic carbon: sampling variability. Limnol. Oceanogr. 19: 980–984.

Wangersky, P. J. 1976a. The surface film as a physical environment. Ann. Rev. Ecol. Syst. 7: 161–176.

Wangersky, P. J. 1976b. Particulate organic carbon in the Atlantic and Pacific oceans. Deep-Sea Res. 23: 457–465.

Wangersky, P. J. 1977. The role of particulate matter in the productivity of surface waters. Helgol. Wiss. Meeresunters. 30: 546–564.

Wangersky, P. J. 1978a. Production of dissolved organic matter, pp. 115–220. In: O. Kinne [ed.], Marine Ecology, Vol. IV, Dynamics. John Wiley and Sons, New York.

Wangersky, P. J. 1978b. The distribution of particulate organic carbon in the oceans: ecological implications. Int. Revue ges. Hydrobiol. 63: 567–574.

Wangersky, P. J., and A. V. Hincks. 1980. Shipboard intercalibration of filters used in the measurement of particulate organic carbon, pp. 53–62. In: J. Albaiges [ed.], Analytical Techniques in Environmental Chemistry. Pergamon Press, New York.

Wangersky, P. J., and O. I. Joensuu. 1967. The fractionation of carbonate deep-sea cores. J. Geol. 75: 148–177.

Yamada, H., A. Murakami, and M. Kayama. 1979. Mineralization of organic substances in seawater. Bull. Jpn. Soc. Sci. Fish. 45: 1299–1305.

Yamamoto, S. 1979. Size distribution of detrital mineral grains suspended in surface waters of the Yellow Sea and East China Sea. J. Oceanogr. Soc. Jpn. 35: 91–99.

Yanada, M., and Y. Maita. 1978. Production and decomposition of particulate organic matter in Funka Bay, Japan. Estuarine Coastal Mar. Sci. 6: 523–533.

NUTRIENT INTERACTIONS AND MICROBES

Francis A. Richards

Department of the Navy
Office of Naval Research
Branch Office, London, Box 39, FPO New York   09510

MICROBIALLY ALTERED CONSTITUENTS OF GENERAL INTEREST
IN THE MARINE ENVIRONMENT

This paper is concerned with microbial processes that alter the composition of seawater; the alterations represent modifications of the environment that in turn can be expected to influence microbial activity. Such processes can be examined with a view to many of the chemical elements, which may be involved because they are essential to the growth and metabolism of plants and animals, which may be concentrated in or on the bodies of organisms either biochemically or physico-chemically, or which may have special stimulating or toxic effects. As suggested by the organizing committee, most of the presentation will deal with the elements considered in the "Redfield ratios", carbon, nitrogen, phosphorus, and dissolved oxygen. The discussion will, however, be extended to the consideration of oxygen-deficient and anoxic waters, in which the role of sulfur compounds becomes important, and some speculation as to the possible role of phosphate as a free energy source will be included. It will be important to review the assumptions implicit in the Redfield ratios and the departures in nature from the ratios and assumptions. The departures may sometimes be more interesting than the ratios themselves, because they may indicate significant peculiarities in the populations or in the environment.

The vital role of a large number of biologically essential elements is recognized, and it is also recognized that elements such as iron, manganese, molybdenum, etc., may be "limiting" elements, but their cyclic relationship to photosynthesis and respiration has not been demonstrated generally as is the case with phosphorus, nitrogen, carbon, and dissolved oxygen. Other vital elements are

289

present in seawater in such large relative quantities that they
never become limiting, and although they may play an entirely
different role in other aquatic ecosystems, they do not in the
marine environment.

MICROBIALLY MEDIATED CHEMICAL CHANGES IN THE MARINE COMMUNITY

Historical Developments

The dependence of marine animals on the presence of dissolved
oxygen and the availability of oxygen in the marine environment were
primary questions in the classical Forbes-Thompson debate on the
general presence or absence of an azoic zone in the depths of the
ocean.   Investigations intended to settle the matter in the middle
of the 19th century by British chemists and biologists, including W.
L. Carpenter and C. W. Thompson, led to both the discovery of the
Wyville Thompson Ridge, predicted from biological findings, and to
the mounting of the Challenger expedition.   In addition, the
biological studies were accompanied by attempts to observe dissolved
oxygen and its relationship to the occurrence of marine organisms.

Evidence of both photosynthetic production and respiratory
consumption of dissolved oxygen was obtained during the Challenger
expedition, but the subsurface photosynthetic oxygen maximum was so
unexpected that the data were dismissed as inaccurate (E. D. Goldberg, personal communication).   Knowledge of the general nature of
the distribution of dissolved oxygen in the ocean had to await the
application of Winkler's (1888) titrimetric method to the study of
seawater.   This was first done world-wide during the circumnavigation
of the globe by the German ship Planet (Brenneke 1909).

Although W. L. Carpenter (1874) had attempted to predict whether
dredge hauls would be "good" or "bad" from the relationships between
dissolved oxygen and carbon dioxide in the water, the application of
the Winkler method appears to have been the first example of the
routine use of a quick and reliable chemical method to investigate
biological processes in the ocean.   The idea that phosphorus might
be present in seawater in concentrations sufficiently low to limit
biological productivity - Liebig's law of the limits - was early
suggested by Brandt, but there were no suitable analytical methods
for the routine determination of phosphorus compounds at the
concentrations we now know to exist in the ocean, and the biological
relationship of changes in the phosphorus compounds in seawater had
to await the adaptation of a sensitive colorimetric method to the
analysis of seawater by Atkins (1923) of the Marine Biological
Laboratory of the UK, Plymouth.   Atkins' colleague at Plymouth was
H. W. Harvey, and in the next few years the seasonal changes in the
partitioning of phosphorus among inorganic and organic forms - the
latter including phytoplankton, zooplankton, and benthos - were

elucidated. The phosphorus content of the plants and animals could
be determined rather accurately, but the partitioning between
soluble inorganic phosphorus, which we now call reactive phosphorus,
and soluble organic phosphorus had to be inferred because suitable
methods for the estimation of organic phosphorus were not available.

Good analytical methods for nitrogen compounds except for the
nitrite ion were still needed, but the Plymouth group did develop a
tedious colorimetric method for the determination of nitrate (Harvey
1926, 1929; Atkins 1932), and a body of data on nitrate began to
accumulate. The determination of nitrite was easy and reliable,
being an adaptation of the diazotization reaction to the analysis of
seawater (Robinson and Thompson 1948). Ammonia remained difficult to
determine accurately at the levels of concentration occurring in seawater. Many attempts to use the Nessler reaction were frustrating,
tedious, and inaccurate. Other early methods involved concentration
by vacuum distillation and a subsequent oxidation by bromine - a
method that although proposed by a Nobel laureate, August Krogh
(1934), still was inadequate to produce the body of data required to
elucidate the role of ammonia in microbial processes. The introduction of the Berthelot-indophenol method to the analysis of seawater
by Riley (1953) improved the situation as did the cadmium reduction
method for the determination of nitrate, as introduced by Morris and
Riley (1963) and adapted to a column method by Wood, Armstrong and
Richards (1967). Satisfactory methods have now been developed for
the determination of urea (e.g., Emmet 1969; McCarthy 1970) and of
creatine (Whitledge and Dugdale 1972), so the most important
compounds in the nitrogen cycle can now be determined accurately
and, in general, by automated methods.

The inorganic carbon components required careful acid-base
methodology and a knowledge of the basic relationships among the
titration alkalinity, P_{CO_2}, and the concentrations of carbonate,
bicarbonate, and borate. These had been worked out by Buch
(reviewed by Harvey 1955) and refined by later workers, especially
Lyman (1956), as reviewed by Skirrow (1975).

Redfield Ratios

The Woods Hole Oceanographic Institution was formed in 1930, the
research vessel Atlantis was built, and extensive investigations of
the Atlantic were begun, frequently with chemical observations using
analytical methods as outlined above. The observations included
chemical analyses by, among others, Dayton Carritt, Norris Rakestraw,
Richard Seiwell, B. H. Ketchum, and A. C. Redfield. Based on such
observations, Redfield wrote his landmark paper of 1934 in which he
demonstrated linear relationships between the concentrations of phosphate, nitrate, carbon dioxide, and oxygen consumption in large areas
of the Atlantic, Pacific, and Indian oceans. In the areas selected,
the intercepts of the scatter diagrams were generally at the origin

(except for carbon), and Redfield concluded that the regressions represented the simultaneous release of phosphate, nitrate, and carbon dioxide into the water as marine organisms – plants, animals, and bacteria – respired organic matter that had been formed photo-synthetically in the upper layers of the ocean – a conclusion the general features of which remain unchallenged. Redfield's conclusions were strengthened by a review of the elemental composition of plankton organisms by R. H. Fleming (1940), which indicated an average atomic ratio of N:P of 16:1, close to the slope $\Delta N:\Delta P$ of 15:1 found by Redifeld. The carbon ratio was not inconsistent with Redfield's regression slopes, $\Delta C:\Delta P = 106:1$.

Nitrate, Nitrite, and Ammonia

Phosphate was more amenable to the model and was assumed to be microbially regenerated with little concurrent oxygen consumption. Quantitatively this is true, because the bulk of the oxygen consumed during respiratory processes goes to the combustion of fats, proteins, carbohydrates, etc.; less than 1% of photosynthetically produced organic matter is phosphorus, so if its oxidation state is changed during its respiratory release, not much oxygen is involved. Nitrogen is a different matter.

The complications of the nitrogen cycle are reasonably well known, as is the fact that nitrogen exists in seawater in valance states varying from −3 to +5. Gilson (1937) pointed out that, because the most stable combined inorganic form of nitrogen in seawater is nitrate, and that upon decomposition of organic matter nitrogen can be assumed to be released as ammonia or as simple amines and perhaps urea, a significant amount of oxygen will be required for its final nitrification to nitrate. This can be represented by the equation

$$2NH_3 + 4O_2 = 2HNO_3 + 2H_2O$$

Normalizing to one atom of phosphorous requires multiplying the equation by 8

$$16NH_3 + 32O_2 = 16HNO_3 + 16H_2O.$$

Thus, the oxidation of the amino nitrogen associated with one atom of phosphorus requires 64 atoms of oxygen, and the oxygen required to nitrify the ammonia hydrolytically released during the microbial decomposition of nitrogenous matter must be taken into consideration. This was done in the model proposed by Redfield, Ketchum and Richards (1963).

Model Equations

The RKR model is an inorganic chemist's view of the end result of the oxidative (respiratory, microbial) decomposition of typical organic matter ultimately formed by photosynthesis in the upper layers of the ocean. It is based primarily on the average composition of plankton material as reviewed by Fleming (1940), a set of simple assumptions about the chemical (oxidative) state of that material, and a consideration of the end products of its (oxidative) decomposition by microbial respiration. It is represented by the simple and now well-known equations

$$(CH_2O)_{106}(NH_3)_{16}(H_3PO_4) + 106\ O_2 = 106CO_2 + 16NH_3 + H_3PO_4 + 212H_2O.$$

Following Gilson (1937), oxidizing the ammonia

$$16NH_3 + 32\ O_2 = 16HNO_3 + 16H_2O,$$

and adding (complete oxidation) gives

$$(CH_2O)_{106}(NH_3)_{16}(H_3PO_4) + 138\ O_2 = 106CO_2 + 16HNO_3 + H_3PO_4 + 236H_2O,$$

and $\Delta O:\Delta C:\Delta N:\Delta P = -276:106:16:1$, by atoms.

The assumptions of the model should be repeated. They are:

1) The oxidation state of carbon in the model organic matter is that of carbohydrate, CH_2O.

2) Nitrogen is present in the organic matter in the reduced (amino) state. It is released initially to the ambient water in that oxidation state, i.e., NH_3.

3) The oxidation or reduction of phosphorus compounds is quantitatively negligible, but a significant quantity of oxygen is required to oxidize ammonia to nitrate, the stable inorganic nitrogen compound in oceanic waters.

4) The model assumes complete remineralization of the organic matter and the ratios imply that the products are introduced simultaneously into the same volume of water.

Although a good number of examples of the applicability of the model have been published, exceptions are perhaps more abundant than the rule. There are many reasons why exceptions are to be expected; most are because of departures from the assumptions of the model. They include:

1) The elemental ratios of photosynthetically formed organic matter in the marine environment may differ from those of the model. Ketchum (1939a, b, 1947) early demonstrated the response of the

composition of phytoplankton cells to the composition of the medium
in which they were grown (batch culture), and either phosphorus-rich
or phosphorus-poor cells could be grown in nitrogen or phosphorus-
deficient media.   Cells in nature can similarly reflect the nitrogen:
phosphorus ratio of the water.   As is shown by many regression plots
of nitrate vs. phosphate, actual N:P ratios depart widely from the
$\Delta N:\Delta P$ slope of the curves.   Thus, waters rich or poor in one or the
other element may reach the photic zone and phytoplankton produced
there can be expected to have variable ratios of N:P.   In spite of
very low concentrations of nutrients observed in the surface layers
of much of the ocean, there are probably few if any oceanic regimes
where primary productivity is actually nil on account of the absolute
lack of either nitrogen or phosphorus.   Even in the Sargasso Sea,
where the surface waters may contain essentially no phosphate or
combined nitrogen, nutrients are recycled and primary production
continues, as demonstrated by Menzel and Ryther (1960).   In many
Pacific Ocean surface waters nitrogen appears to be "limiting" in
the presence of substantial concentrations of phosphate.   One would
expect the phytoplankton that do grow there to be somewhat
relatively richer in phosphorus than in nitrogen.

2) For the Redfield ratios to hold, inorganic nitrogen,
phosphorus, and carbon compounds should be released to the water and
achieve their ultimate oxidation state simultaneously.   This can
hardly be the case in nature.   It is to be expected that phosphorus,
nitrogen, and carbon are released at different rates during the
decomposition of plankton material, resulting in different rates for
the production of carbon dioxide, phosphate, and nitrate.   The
effects of the time lags may be erased if vertical mixing is
relatively rapid, but the oceanic intermediate maxima in phosphate
and nitrate do not occur at the same depths, which should be the
case if the assumptions of the model were strictly fulfilled.   The
recycling of nitrogen is fairly rapid, at least in upwelling
ecosystems (Whitledge 1981) and as was demonstrated in vitro in
early studies by Von Brand, Rakestraw, and Renn (1937, 1939), but it
would be surprising if phosphorus were recycled at the same rate and
one would expect at least local fractionation of the two elements.

3)   Widely varying ratios of carbon to nitrogen and phosphorus
in marine particulate matter, living and dead, have been reported.
Edmond et al. (1981) reported on the C:P ratios in the two major
components of the suspended particles in the Amazon River plume; the
ratio in the Amazon detritus was ca. 30 (by atoms), in the average
plankton it was ca. 110, and much higher ratios were observed in a
good many of the samples.

Until now little has been known about the chemical composition
and changes therein of particulate matter showering downward through
the water column.   However, sediment traps capable of collecting
relatively large quantities have been built (Honjo et al. 1980) and

Table 1. The atomic ratios in trapped particulate matter from
several locations.

Station	Depth	C	:	N	:	P
Sargasso Sea	976	78	:	15.3	:	1
	3964	109	:	12.3	:	1
Tropical Atlantic	389	185	:	24.4	:	1
	988	279	:	32.1	:	1
	3755	184	:	17.1	:	1
	5068	205	:	24.4	:	1
Central N. Pacific	2778	320	:	30.2	:	1
	4280	318	:	10.3	:	1
Panama Basin	667	134	:	13.5	:	1
	1268	137	:	15.1	:	1
	2869	131	:	14.6	:	1
	3769	98	:	9.8	:	1
	3791	125	:	11.0	:	1
	Sediment	85	:	8.6	:	1

some analyses of material so collected have become available (Honjo
et al. 1982).   Traps were deployed at various depths in the central
Sargasso Sea, the tropical Atlantic, the central North Pacific, and
the north central Panama Basin.   The atomic ratios in the trapped
particulate matter were as shown in Table 1.

The data are too few and, as pointed out by the authors, may be
in some error. Wide departures from the Redfield ratios, with
ratios of about 200:21:1 at mesopelagic depths and up to 300:33:1 at
bathypelagic depths, are evident. The combustible fractions of the
samples had a more consistent depth pattern, increasing with depth
at every station, suggesting continuing microbial decomposition at
all depths.

4) The assumption that the carbon in the organic matter being respired is in the oxidation state of CH_2O implies, upon completion of oxidation, a respiratory quotient of $\Delta O : \Delta CO_2$ of 1:1. (To this must be added the oxygen required for nitrification.) Respiratory quotients are known to vary from this value, but as they are measured they do not generally reflect the complete oxidation of the substrate. The ratios of C:H in planktonic matter vary widely, and differing ratios of carbohydrate to protein to fat in the respired material will result in differing respiratory quotients.

Craig (1969) calculated a $\Delta CO_2 : \Delta O_2$ ratio of 1 for the abyssal waters of the Pacific, whereas the Redfield ratio is 0.77. Kroopnick (1974) commented that the total CO_2 data Craig used were too high, but he also suggested that the higher ratio may be the result of the more rapid liberation of phosphorus and nitrogen than carbon from the sinking organic matter.

Holm-Hansen suggested in a personal communication cited by Kroopnick (1974) that the C:N:P ratios in both particulate and dissolved organic matter do not change with depth, but unless there are quite large errors in the data of Honjo et al. (1982) this seems not to be the case.

It is evident from the above that the hypothetical composition of the organic matter that is respired may not be realized. The remineralization of carbon, nitrogen, and phosphorus probably does not proceed at rates relatively the same as the elemental ratios, and the oxidation of the material is not apt to be completed in the dynamic system of the oceans at the same time and place for all the elements considered. The ocean simply does not act like a beaker or bottle into which planktonic organic matter is introduced and permitted to remineralize. However, in stagnating or weakly circulating systems the beaker or bottle conditions may be approached. Such systems will be discussed later.

Some Applications of the Redfield Ratios

In a later paper, Redfield (1942) related the concentrations of inorganic (reactive) phosphate to the amount of oxygen consumed in the upper and intermediate depths of the North and South Atlantic. In this case, there were marked discrepancies between the concentrations of phosphate actually observed and the amounts predicted from the apparent oxygen utilization (AOU = $O_2' - O_2$, where O_2' is the concentration of oxygen that a seawater sample would have if equilibrated with a standard atmosphere at the sea surface, the temperature and salinity of the sample are as observed, and O_2 is the observed oxygen content). It happens that the areas first examined by Redfield were regions in which photosynthesis drew down phosphorus and nitrate to very low levels in the surface waters, which were just saturated with oxygen. But this is not the case

everywhere in the ocean, and in the region of the Subtropical
Convergence, around 50° South Latitude, waters saturated with
dissolved oxygen but containing significant concentrations of
inorganic nutrients sink and spread northward. These waters are
also laden with photosynthetically produced organic matter. During
the course of the transport, microbes respired this organic matter,
consumed oxygen, and released phosphate and inorganic nitrogen.
Thus, although the $\Delta O_2 : \Delta P$ values might be expected to be constant,
as observed in the North Atlantic, the absolute ratios would differ,
i.e., AOU:P = $\Delta O : \Delta P + P_p$, where P_p is the amount of inorganic phosphate present when the water was just saturated with oxygen. It
should also be pointed out that the convergence is not a point
source but a three-dimensional region, so that the amounts of
inorganic phosphate present in the sinking water vary from place to
place and presumably from time to time, and the value of P_p will be
altered by mixing with other waters with differing initial amounts
of "preformed" phosphates (Redfield's terminology). Redfield used
the concept to calculate the preformed phosphate which, once the
waters were below the influence of photosynthesis, should act as a
conservative property. Redfield's diagrams of the distribution of
preformed phosphate in a north-south section of the Atlantic
indicate a reasonable distribution not unlike the distribution of
the salinity.

     Even a superficial survey of the literature will show the
frequency with which the Redfield ratios are used; often they are
considered norms against which to evaluate observed changes.
Redfield himself (1948) used the concept and the seasonal changes in
phosphate content of the water column in the Gulf of Maine to estimate the changes in the dissolved oxygen content to be expected in
the integrated water column. The calculated oxygen changes were
compared with observations, and the differences were attributed to
exchanges of oxygen across the sea surface. The results indicated
passage of oxygen from the water to the atmosphere during the season
of active photosynthesis and an absorption of oxygen by the water
during the months when respiration predominated. The results were
in reasonable agreement with laboratory determined values of
transfer coefficients, expressed as functions of the difference
between the partial pressure of oxygen in the air and in the water.

     The Redifeld ratios have been used frequently to estimate
biologically mediated changes in one of the components of seawater
from measurements of other components. Fahrbach et al. (1981) used
concurrent decreases in phosphate and increases in dissolved oxygen
with distance offshore in the Peruvian upwelling system as an
indication of upwelling near the shore and of offshore motion in the
upper layers. They found a $\Delta P : \Delta O$ atomic ratio of 1:-295, which they
considered to be consistent with the Redfield, Ketchum and Richards
(RKR) (1963) ratio of 1:-276 and a sign of the reliability of the
nutrient data.

As will be discussed later, Kroopnick (1974) used the Redfield ratio for $\Delta O:\Delta CO_2$ to estimate deep-water excesses of CO_2 in the Pacific originating from the respiratory decomposition of organic matter.   Packard (1969) used the ratios to estimate changes in phosphate and nitrate arising during respiration as estimated from electron transport activity assays, and there are many other examples of the use of the ratios in the literature.

Friederich and Codispoti (1981) studied the mixing of waters of equatorial and of subantarctic origin along surfaces of equal sigma-t in the Peruvian upwelling system. During the austral winter, July to October, 1976, the nitrate contents of the mixtures were close to the concentrations to be expected from the conservative mixing of the waters plus the changes in nitrate as predicted from the changes in dissolved oxygen along the equal-density surface. They found for ΔNO_3^- vs. ΔO_2, m = -4.1 μg-at. $NO_3^- \cdot ml\ O_2^{-1}$ (r = -0.83), close to the RKR relationship, in the units they used of m = -3.9 μg-at. $NO_3^- \cdot ml\ O_2^{-1}$. In winter, March to May, 1977, the equatorial water penetrated with little admixture of subantarctic water to nearly 16°S along the 26.0-sigma-t surface. During the apparent southward progression of the equatorial water, the dissolved oxygen concentration decreased from about 1.6 $ml \cdot liter^{-1}$ near 5°S to nil near 15°S. This would, according to the RKR ratios, predict an increase of about 8.2 μg-at. $NO_3 \cdot liter^{-1}$ -- close to the maximum NO_3^- values of about 7.6 μg-at. $liter^{-1}$ observed between 10 and 15°S.   Otherwise, there was little relationship between the nitrate and oxygen changes, and apparently nitrate reduction and denitrification, which will be discussed later, became significant factors in the nitrogen cycle at the low oxygen concentrations that developed.

ANOXIC MARINE ENVIRONMENTS

General Review

In 1912, the Lake Washington ship canal was opened. The canal joined Lake Union and Lake Washington via a set of locks to Puget Sound. When opened, the levels of the lakes were equalized - the level of Lake Washington dropped some 20 feet and that of Lake Union similarly increased. But when the locks were operated, some seawater from the sound entered the canal-lake system. During the summer season of little fresh water runoff and frequent boat passages through the locks, seawater filled the deeper parts of the lakes and, because of the strong vertical stratification, stagnated. The microbial decomposition of organic matter consumed all of the oxygen, and sulfate reduction and concomitant sulfide production ensued. The noxious and noisome conditions attracted the attention of the public and of Thomas G. Thompson, a young analytical chemist on the faculty of the Chemistry Department of the University of Washington. The public attributed the sulfide problem to the burning of coal,

but Thompson thought perhaps not and began a scientific investigation
of the system. It turned out that the amount of sulfur involved as
sulfide would have required the burning of astronomical amounts of
coal, so the microbial reduction of sulfates had to be invoked to
explain the phenomenon. Although it had been known since the end
of the 19th century that the Black Sea was anoxic and contained
sulfides, Thompson's investigations may have been the first quantitative estimates of the effects of microbial sulfate reduction.

Redfield first suggested that the geomorphology of the Cariaco
Trench, on the Caribbean coast of Venezuela, was such that its deeper
waters might stagnate and lose their dissolved oxygen. Shortly later
a cruise to the area proved that to be the case, and intense chemical
investigations of the system began (Richards and Vaccaro 1955). Such
systems as the Black Sea and the Cariaco Trench exhibit, to varying
degrees, the "beaker or bottle" systems into which organic matter of
photosynthetic origin is introduced and then oxidized by various
processes of microbial respiration. The oceanography of the systems
varies and controls the degree to which they stagnate, from embay-
ments and fjords that stagnate and circulate cyclically (seasonally)
to the extreme case of a few landlocked fjords which, by changes in
sea strand, have been completely cut off from the sea and are
totally stagnant. The Black Sea has a relatively minuscule
connection with the open ocean and is the largest anoxic and
sulfide-bearing marine system.

The Black Sea has a shallow and tenuous connection with the
World Ocean through the narrow and shallow Bosphorus to the Sea of
Marmora, the Dardanelles, and the Mediterranean Sea. The connection
was established around 6000 years ago, when sea level changes
admitted seawater into what was previously a freshwater lake. Large
influxes of fresh water from the numerous rivers draining into the
sea provide a strong halocline and vertical stability, which prevent
extensive vertical mixing and ventilation of the deeper water. The
upper waters are renewed and sustain a rich biota, from which organic
debris rains into the deeper waters to decompose. In the course of
time, microbial respiration has reduced all the dissolved oxygen,
the oxidized nitrogenous ions nitrate and nitrite, and some of the
sulfate, so that at all depths below 150 to 250 m the waters are
anoxic and sulfide-bearing. The minor introductions of "new"
seawater through the Bosphorus are inadequate to replace or flush
the deeper water. The system provides a good laboratory case for
the study of the processes of microbial degradation of marine
organic matter, and, incidentally, provides further tests of the
applicability of the Redfield ratios to extreme cases where electron
acceptors other than free oxygen are utilized by the microbial
populations, whose taxa presumably shift in response to the altered
environmental conditions.

So long as dissolved oxygen is present, the major process

involved in the decomposition of organic matter will be aerobic
respiration by all trophic levels. When such respiration has
reduced oxygen concentrations to very low levels or nil, nitrate
reduction and denitrification follow. After all or nearly all of
the nitrite and nitrate have been reduced, sulfate reducers take
over and sulfides are produced.

In none of the marine systems examined has all the sulfate been
reduced, although that has happened in Powell Lake, a landlocked
fjord in British Columbia, Canada (Williams et al. 1961), but the
sulfate:chlorinity ratio is significantly less in the Black Sea than
in open ocean waters (Redfield et al. 1963). On the other hand, the
sulfate is reduced to nil in many marine sedimentary layers (Martens
and Berner 1974) and apparently carbon dioxide is reduced to methane.
Methane has been determined in the water column in several anoxic
systems (Atkinson and Richards 1967, and others), but there is no
clear evidence for its production in the water column. It seems
more probable that the methane is produced in the sediments and
diffuses upward into the water.

There is no evidence that I know of other electron acceptors
being microbially utilized in the marine environment. However, the
primary productivity in many lakes is adequate to consume all the
oxygen and sulfate, and the production of methane (marsh gas) is
common. It has been suggested that bacteria can then turn to phosphate ions as a free energy source, reducing it to phosphine, PH_3, or
an analog of it. Phosphine so produced might be responsible for the
ignition of the will-of-the-wisp (ignis fatuus) - burning clouds of
marsh gas sometimes observed over swamps and perhaps sometimes
identified as unidentified flying objects.

Different sets of equations and ratios are necessary to express
the chemical changes accompanying the respiratory decomposition of
organic matter under various anoxic conditions.

Nitrate Reduction

$$(CH_2O)_{106}(NH_3)_{16}(H_3PO_4) + 212HNO_3 = 106CO_2 + 212HNO_2 + 16NH_3 + H_3PO_4 + 106H_2O$$

This step may well be best illustrated by observations made in
Darwin Bay in the Galapagos Islands, in which the appearance of one
ion of nitrite was precisely matched by the disappearance of one ion
of nitrate during a period of five months separating two sets of
observations (Richards and Broenkow 1971).

Denitrification

$$3(CH_2O)_{106}(NH_3)_{16}(H_3PO_4) + 424HNO_2 = 318CO_2 + 212N_2 + 48NH_3 + 3H_3PO_4 + 530H_2O$$

Sulfate Reduction

$$(CH_2O)_{106}(NH_3)_{16}(H_3PO_4) + 53SO_4^= = 106CO_2 + 53S^= + 16NH_3 + 106H_2O + H_3PO_4$$

CO_2 Reduction

This cannot be written as an oxidation-reduction reaction like the above, which would result in CO_2 being both the reactant and the product. However, methane production by strict anaerobes is practically universal in anoxic environments. One of the many reaction pathways by which methane may be formed is CO_2 reduction:

$$CO_2 + 2H = HCOOH; \quad HCOOH + 2H = HCOH; \quad HCOH + 2H = CH_3OH;$$
$$\quad \text{formic acid} \qquad \text{formaldehyde} \qquad \qquad \text{methanol}$$
$$CH_3OH + 2H = CH_4$$
$$\quad \text{methane}$$

Phosphate Reduction

The suggested reaction might be

$$2(CH_2O)_{106}(NH_3)_{16}(H_3PO_4) + H_3PO_4 = 212CO_2 + 212H_2O + 32NH_3 + 2H_3PO_4 + PH_3$$

The equation for denitrification has been written to show free N_2 as the product. It is not likely that NH_3 (or other inorganic reduced species) is produced. In the vast marine systems that have been investigated, including the Black Sea, the Cariaco Trench, and the regions of the eastern tropical Pacific in which nitrate reduction and denitrification can be demonstrated (Richards et al. 1973; Cline and Richards 1972; Codispoti and Richards 1976; and others) there is no convincing evidence that the nitrogenous end product of the nitrite reduction is anything but N_2, although N_2O, NH_2OH, and NH_3 could be formed. On the other hand, NO_2^- does accumulate as an intermediate product and substantial concentrations of nitrite are observed just above (or below) sulfide-bearing layers in anoxic fjords such as Saanich Inlet and Lake Nitinat, British Columbia, and in the upper part of the denitrifying zones in the eastern tropical North and South Pacific (Brandhorst 1959; Wooster et al. 1965; Fiadeiro and Strickland 1968; Cline and Richards 1972; among others). Ammonia does accumulate, but I am of the opinion that it arises by hydrolytic release from amino compounds rather than from the reduction of nitrate or nitrite.

The fate of the ammonia hydrolytically released during nitrate reduction and denitrification is uncertain. It might or might not be oxidized at the expense of additional NO_3^- or NO_2^-, possibly by a reaction such as

$$2NH_3 + 3NHO_3 = N_2 + 3HNO_2 + 3H_2O$$

or

$$NH_3 + HNO_2 = N_2 + 2H_2O.$$

In any case, phosphates accumulate during nitrate reduction, denitrification, and sulfate reduction and ammonia accumulates at least during sulfate reduction. Observations in Lake Nitinat, an anoxic marine fjord, demonstrate good linear relationships among sulfide, phosphate, and ammonia concentrations in close to the ratios that would be expected from the reactions for sulfate reduction. Denitrification and sulfate reduction are both processes that serve to fractionate the nutrients, and in the case of denitrification to remove biologically available compounds from the ecosystem. Barlow (1965) pointed out that the mixing of anoxic waters back into the upper layers would tend to enrich the photic zone with nutrients and stimulate primary productivity. However, the ratio of N:P would be altered to the extent that combined nitrogen was lost by denitrification. Barlow's suggestion assumes that the sulfides are completely oxidized and their toxicity destroyed. Bacterial nitrification of ammonia should produce nitrites or nitrates (apparently not free N_2). In any case, nitrogen compounds and phosphates would become available to the phytoplankton, which can utilize any of the three nitrogen compounds.

Tests of the ability of the above stoichiometric model are rather satisfactory in a good many anoxic systems, with demonstrable linear relationships between accumulations of sulfide and ammonia, phosphate and ammonia, etc., but Brewer and Murray (1973) point out "certain deficiencies in the approach; it takes no account of oceanic mixing or of the rate at which the system is driven to steady state; moreover, the reactions may take place in the water column or in the sediment, and the products then transported to the site of the observations, the site of the reaction remaining undefined."

POPULATION CHANGES

It is evident that the differing environments arising as a consequence of the consumption of various electron donors during the continued microbial respiration of organic matter will prove hospitable to differing populations.

It has been observed that various lantern fishes appear to inhabit the extremely oxygen-deficient layers of the eastern tropical Pacific, and various mysids inhabiting such waters have been studied by Childress (1968). However, both nitrate reduction and denitrification are bacterial functions, with nitrate reduction generally a function of facultative organisms.

Hydrogen sulfide is a highly toxic substance, and although some

fish appear able to migrate into and out of the sulfide-bearing
waters of the Cariaco Trench (reviewed by Richards 1975, pp. 53-55),
the presence of sulfides generally excludes all organisms higher
than bacteria. Several bacterial species are known to be sulfate
reducers.

At the interface between oxygenated and sulfide-bearing layers,
mixing can produce solutions containing both H_2S and O_2. Such
mixtures are thermodynamically unstable and can react inorganically
to form thiosuflates, sulfite, and sulfate (Cline and Richards 1969).
Such mixtures should also provide suitable environments for the
chemoautotrophic bacterial genus Beggiotoa, apparently the first
organism shown capable of chemoautotrophism.

The examples of sulfide-bearing environments cited in this
paper are all oxygenated to depths below the photic depth, but there
are small lagoons and other sulfide-bearing environments within the
photic zone. In such environments photosynthetic purple sulfur
bacteria, obligate anaerobes, can carry out photosynthesis according
to the equation

$$H_2S + CO_2 = CH_2O + S.$$

Sulfur compounds other than H_2S can serve as the hydrogen source --
presumably $H_2S_2O_3$, $H_2S_2O_6$, H_2SO_3, etc.

SOME ESTIMATES OF RATES OF INTERACTIONS BETWEEN MICROBES, NUTRIENTS,
AND OXYGEN

Direct measurements of rates of microbial activity in the ocean
are difficult, and most of the older estimates were based on measurements of the end products of the activity rather than the rate of the
activity itself.

Oxygen Consumption Rates

Rates of oxygen consumption have been estimated by methods that
range from the rather simple business of determining oxygen or
nutrient concentrations at two or more times in a water body that
can be assumed not to have been altered during the time by mixing or
by exchange with the atmosphere. Such an approach was used by
Redfield, who followed a water mass in a stagnant pool of water
beneath sill depth in the Gulf of Maine; he estimated the annual
rate of consumption was 0.36 ml·liter^{-1}. E. F. Thompson similarly
estimated an oxygen consumption rate of 2 ml·liter^{-1}·yr^{-1} for the
Red Sea. Barnes and Collias (1958) also followed semi-stagnant
pockets of water in Puget Sound. They found essentially linear
rates of oxygen consumption that averaged:

In Dabob Bay, from   22 Jan to 22 Sept   0.012 ml·liter^{-1}·day^{-1}
"   "   "   "   25 May to 17 June   0.010   "   "
"   "   "   "   9 July to 30 Sept   0.005   "   "
"   Port Susan   23 Sept to 12 Jan   0.017   "   "
"   Lynch Cove   18 Feb to 21 May   0.020   "   "
"   "   "   "   9 March to 9 July   0.018   "   "

The rates of decrease were smooth during the periods shown, and abrupt re-oxygenation occurred with the introduction of "new" seawater.

Other methods of estimating oxygen consumption were used by Sverdrup and Fleming and by Riley (reviewed by Richards 1957). Sverdrup and Fleming used a water mass mixing model; Riley's estimate was based on historic observations of salinity, temperature, and dissolved oxygen. It involved evaluating both advection and diffusion in horizontal and vertical directions from salinity and temperature, with deviations from the physically predicted oxygen content being attributed to biological consumption. Riley's monumental effort was done with a desk calculator and a paucity of data. A modern computerized effort using a much more complete data base might well be a highly enlightening exercise to evaluate the potential of the ocean as, say, a world-wide waste receptacle.

Kroopnick (1974) applied an advection-diffusion model, following the formulation and terminology of Craig (1969):

$$K\partial^2 c/\partial z^2 + J = \partial w\, c/\partial z.$$

Here, c is concentration, z the vertical coordinate (positive upward), K is the vertical eddy diffusion coefficient, w the vertical eddy velocity, J is the in situ production rate and K and w were assumed to be constant. The parameter $z* = K/w$ can be evaluated from profiles of potential temperature, O, vs. salinity, S, by Craig's method. The ratio has the dimension of length, and in the deep water of the Pacific it increases northward from a value of about 0.6 km at 50°S to about 1.2 km at 50°N. For conservative properties such as O and S, J = 0.

The surface waters of both the Atlantic and Pacific contain about 2×10^{-3} moles of total carbon dioxide; Atlantic deep water contains about 10% more, Pacific deep water about 20% more. The excess CO_2 comes from the dissolution of shells, during which 1 mole of $CaCO_3$ yields 2 equivalents of alkalinity, and from the oxidation of organic matter, during which 1 mole of oxygen consumption yields, according to the Redfield ratio, $\Delta CO_2 : \Delta O_2 = 106:138$, or 0.77 moles of CO_2. Thus,

$$RKR = \frac{\Delta CO_2}{\Delta O_2} = 0.77 = (J_{\Sigma CO_2} - \frac{1}{2} J_{ALK})/J_{O_2} = F_{org} J_{CO_2}/J_{O_2}.$$

If F_c is the fraction of carbon added by carbonate dissolution and F_{org} is the fraction from the oxidation of organic matter $(F_{org} = 1-F_c)$, then

$$F_c(\delta_c) + F_{org} \delta_{org} = \delta_J.$$

In this equation, δ_c and δ_{org} were taken as $2°/°°$ and $-23°/°°$, respectively, and δ_J is the mean isotopic composition of the carbon added by the solution and oxidation of descending particulate matter and dissolved organic carbon. The F_c can be calculated directly from the alkalinity and total carbon dioxide data:

$$F_c = \frac{1}{2} J_{alk}/(J_{CO_2} + \frac{1}{2} J_{alk}).$$

Using the above model, Kroopnick calculated values for average rates of oxygen consumption, within the depth interval defined by the linear part of the O-S diagram, ranging from 0.029 to 0.14 micromoles per kilogram per year (= 0.6 to 3 $\mu l \cdot kg^{-1} \cdot yr^{-1}$):

Station	J_{O_2}	$J_{\Sigma CO_2}$	RKR Ratio
	$\mu moles \cdot kg^{-1} \cdot yr^{-1}$		
SCAN 20	−0.058		
YALOC 69-70	−0.029	0.030	0.67
SCAN 30	−0.108		
SCAN 38	−0.092	0.072	0.47
SCAN 43	−0.129		
SCAN 56	−0.080	0.116	0.87
YALOC 69-127	−0.140	0.119	0.76

Kroopnick commented on the fact that the ratio of $\Delta CO_2 : \Delta O_2$ Craig (1969) found was closer to 1 than to the 0.77 predicted by the RKR model. The model departure from 1 arises from the amount of oxygen required in the nitrification of ammonia. Kroopnick, probably correctly, attributed at least part of the discrepancy to apparent errors in CO_2 values arising from production of CO_2 in the stored samples referred to by Craig. Nonetheless, one would expect much of the amino nitrogen to have been released and nitrified in depths of

the ocean less than those considered by Kroopnick; the remaining
organic moieties would not have the same oxygen demand as the RKR
model organic matter, and $\Delta CO_2 : \Delta O_2$ would be between the RKR predicted
value and a value of 1.0 to be expected from the microbial oxidation
of simple carbohydrate. The RKR C:N ratio (by atoms) is 6.625:1;
larger ratios should indicate organic matter that has lost nitrogen
(the obverse, gain of carbon relative to nitrogen, seems unlikely in
deep waters). The set of data reported by Honjo et al. (1982) on the
composition of particulate matter collected in sediment traps from
different depths in the open ocean suggests that the C:N ratios in
deep water are significantly higher than 6.625:1. Thus, the material
being remineralized at depth can be expected to have a smaller oxygen
demand per carbon atom than that initially formed by photosynthesis.

The rates of oxygen consumption discussed above are based on
the distribution of biologically mediated variables, not on actual
rates of biological processes. Packard (1969, 1971) used estimates
of the rates of microbial processes to evaluate oxygen consumption
(= respiratory) rates in the ocean. He related estimates of the
respiratory electron transport system activity (ETSA) to rates of
oxygen consumption under different environmental conditions. In an
intercomparison exercise (Hobbie et al. 1972), Packard's method was
compared with a method based on ATP assays and with estimates arrived
at by following changes in the oxygen content of seawater samples
placed in a respirometer for a least 1 hr. The comparisons were
made in a test at sea in the North Atlantic. The results by the
different methods were reasonably comparable in surface water but
diverged widely in the deep water, where the respiratory rate was
below the sensitivity of the oxygen method. The estimated rates at
one station are shown in in Table 2. The results appear to agree
well with Kroopnick's estimates of oxygen consumption rates in the
deep waters of the Pacific (Kroopnick 1974).

Nitrate Reduction and Denitrification

Like estimates of respiratory oxygen consumption rates, nitrate
reduction and denitrificaion rates have been estimated by following
nitrate and nitrite changes with time in an identified water body.
The earlier cited observations by Richards and Broenkow (1971) of the
simultaneous decrease in nitrate matched by an ion-for-ion increase
in nitrite in Darwin Bay over a two-month period is an example. The
estimated rate was a minimum value, but it is not widely applicable
because the environment was atypical.

Microbial function has also been used to estimate rates of
nitrate reduction in the subsurface waters of the Peru Current by
Packard et al. (1978). In the deeper waters (200 and 250 m), nitrate
reductase activity ranged from 1.1 to 1.4 ng-at. $N \cdot liter^{-1} \cdot hr^{-1}$ and
was associated with the deep nitrite maximum; the authors concluded
the activity represented bacterial respiration. At station C-5 off

Punta San Juan, Peru, the deep nitrate reductase activities were:

Depth (m)	ng-at. $N \cdot liter^{-1} \cdot hr^{-1}$
43	1.10 (the bottom of the primary nitrite maximum)
100	0.29 (the minimum between the two maxima)
200	1.35
225	1.37 (from the secondary nitrite maximum)
250	1.10

Ammonium Nitrification

Ward, Olson and Perry (1982) have determined nitrification rates in the primary nitrite maximum off southern California. Two distinct steps are involved. Bacteria of the genera Nitrosococcus and Nitrosomonas mediate the oxidation of ammonium to nitrite and Nitrobacter-like organisms oxidize nitrite to nitrate. Bacteria were enumerated by immunofluorescent assays to yield bacterial numbers. The average nitrite production rate in the upper 100 m off southern California in July was 4.8×10^{-13} mole$\cdot$cell$^{-1} \cdot$day^{-1} or, for a volume of water, 1.9×10^{-8} mole$\cdot$liter$^{-1} \cdot$day^{-1}. Comparable numbers in November were 5.6×10^{-13} mole$\cdot$cell$^{-1} \cdot$day^{-1} and 1.6×10^{-8} mole$\cdot$liter$^{-1} \cdot$day^{-1}.

Table 2. Estimated respiratory rates of various depths at one station (Hobbie et al. 1972).

Depth (m)	Respiratory Rate (μl $O_2 \cdot liter^{-1} \cdot day^{-1}$)		
	ETS	O_2	ATP
4	43.0	36.0	30.1
10	55.0	27.6	24.6
40	1.3	13.4	3.0
100	0.06	0.3	1.3
200	0.05	1.2	0.4
500		1.5	0.1
700	0.01	0.9	0.1

Chemosynthetic Uptake in a Sulfide-oxygen Mixing Zone
_____

Brewer and Murray (1973) proposed a one-dimensional advection-diffusion model for the oxygen-sulfide interface of the Black Sea. The region represents a mixing interval between the cold winter water above and the warmer, more saline inflow beneath. Their treatment, similar to that of Craig (1969), gave consumption rates in the mixing zone of 1×10^{-6} mole of $CO_2 \cdot kg^{-1} \cdot yr^{-1}$, 0.1×10^{-6} mole of ammonia $\cdot kg^{-1} \cdot yr^{-1}$, and of 0.01×10^{-6} mole of phosphate $\cdot kg^{-1} \cdot yr^{-1}$, which apparently represent chemosynthetic consumption and respiratory production by bacteria in the oxygen-sulfide interface.

CONCLUSION

This discussion cannot be considered a complete review of the subjects touched upon. Rather it is a sampling of some interactions between microbes and nutrients, their consequences, and rates.

Many of the processes described in this paper have been inferred from chemical observations that have been incomplete or of an accuracy inadequate to distinguish between analytical noise and significant trends. The changes that take place when a marine system switches from photosynthesis and aerobic respiration to nitrate reduction to denitrification to sulfate reduction are poorly understood, particularly the formation and fate of various intermediate nitrogen compounds such as N_2O, NH_2OH, NO, etc. Many of the nutrient changes in the water column are clearly the result of microbial activities within the sediments, and these are now the object of intensive studies by a large number of chemists and geochemists.

Perhaps a fundamental kind of research that is needed is the investigation of the metabolic and respiratory functions of marine anaerobes, both obligate and facultative, isolated from anoxic environments.

Man's ever increasing use of and dependence on the oceans requires a fuller understanding of microbial processes to improve our evaluation of the marine environment as source or sink of substances and organisms both harmful and benign.

REFERENCES

Atkins, W. R. G. 1923. The phosphate content of sea water in relationship to the growth of the algal plankton. J. Mar. Biol. Assoc. UK 13: 119-150.
Atkins, W. R. G. 1932. Nitrate in sea water and its estimation by means of diphenylbenzidene. J. Mar. Biol. Assoc. UK 18: 167-192.

Atkinson, L. P., and F. A. Richards. 1967. The occurrence and
    distribution of methane in the marine environment. Deep-Sea
    Res. 14: 673-684.
Barlow, J. P. 1965. Formal discussions of chemical observations in
    some anoxic, sulfide-bearing basins and fjords, pp. 233-234.
    In: Proc. Int. Water Pollution Conf., 2nd, Tokyo. Pergamon
    Press.
Barnes, C. A., and E. E. Collias. 1958. Some considerations of
    oxygen utilization rates in Puget Sound. J. Mar. Res. 17:
    68-80.
Brandhorst, W. 1959. Nitrification and denitrification in the
    eastern tropical Pacific Ocean. J. Cons. Cons. Perm. Int.
    Explor. Mer 25: 3-20.
Brenneke, W. 1909. Ozeanographie: Forschungsreise S. M. S. Planet
    1906/7, 3. K. Siegismund, Berlin.
Brewer, P. G., and J. W. Murray. 1973. Carbon, nitrogen and
    phosphorus in the Black Sea. Deep-Sea Res. 20: 803-818.
Carpenter, W. L. 1874. Summary of the results of the examination
    of samples of sea-water taken at the surface and at various
    depths, pp. 502-511. In: C. W. Thompson [ed.], The Depths of
    the Sea. Macmillan, London.
Childress, J. J. 1968. Oxygen minimum layer vertical distribution
    and respiration of the mysid Gnathophausia ingens. Science
    160: 1242-1243.
Cline, J. D., and F. A. Richards. 1969. Oxygenation of hydrogen
    sulfide in seawater at constant salinity, temperature and pH.
    Environ. Sci. Technol. 3: 838-843.
Cline, J. D., and F. A. Richards. 1972. Oxygen deficient conditions
    and nitrate reduction in the eastern tropical North Pacific
    Ocean. Limnol. Oceanogr. 17: 885-900.
Codispoti, L. A., and F. A. Richards. 1976. An analysis of the
    horizontal regime of denitrification in the eastern tropical
    North Pacific. Limnol. Oceanogr. 21: 379-388.
Craig, H. 1969. Abyssal carbon and radiocarbon in the Pacific. J.
    Geophys. Res. 74: 5491-5506.
Edmond, J. M., E. D. Boyle, B. Grant and R. F. Stallard. 1981. The
    chemical mass balance in the Amazon plume. I. The nutrients.
    Deep-Sea Res. 28: 1339-1374.
Emmet, R. T. 1969. Spectrophotometric determination of urea and
    ammonia in natural waters with hypochlorite and phenol. Anal.
    Chem. 41: 1648-1652.
Farhbach, E., C. Brockmann, N. Lostaunau, and W. Urquizo. 1981. The
    northern Peruvian upwelling system during the ESACAN experiment,
    pp. 134-145. In: F. A. Richards [ed.], Coastal Upwelling.
    American Geophysical Union.
Fiadeiro, M., and J. D. H. Strickland. 1968. Nitrate reduction and
    the occurrence of a deep nitrite maximum in the ocean off the
    west coast of South America. J. Mar. Res. 26: 187-201.
Fleming, R. H. 1940. The composition of plankton and units for
    reporting populations and reproduction. Proceedings of the 6th

Pacific Science Congress, California, 1939 3: 535–540.

Friederich, G. E., and L. A. Codispoti. 1981. The effects of mixing and regeneration on the nutrient content of upwelling waters off Peru, pp. 221–227. In: F. A. Richards [ed.], Coastal Upwelling. American Geophysical Union.

Gilson, H. C. 1937. The nitrogen cycle. John Murray Expedition, 1933–34. Sci. Rep. 2: 21–81.

Harvey, H. W. 1926. Nitrate in the sea. J. Mar. Biol. Assoc. UK 14: 17–88.

Harvey, H. W. 1929. Methods of estimating phosphates and nitrates in sea water. Rapports et proces verbaux, Cons. Int. Perm. Explor. Mer 53: 68.

Harvey, H. W. 1955. The Chemistry and Fertility of Sea Waters. Cambridge University Press.

Hobbie, J. E., O. Holm-Hansen, T. T. Packard, L. R. Pomeroy, R. W. Sheldon, J. P. Thomas, and W. J. Wiebe. 1972. A study of the distributions and activity of microorganisms in ocean water. Limnol. Oceanogr. 17: 544–555.

Honjo, S., J. F. Connell, and P. L. Sachs. 1980. Deep-ocean sediment trap; design and function of PARFLUX Mark II. Deep-Sea Res. 27: 745–754.

Honjo, S., S. J. Manganini, and J. J. Cole. 1982. Sedimentation of biogenic matter in the deep ocean. Deep-Sea Res. 29: 609–625.

Ketchum, B. H. 1939a. The absorption of phosphate and nitrate by illuminated cultures of Nitschia closterium. Am. J. Bot. 26: 399–407.

Ketchum, B. H. 1939b. The development and restoration of deficiencies in the phosphorus and nitrogen composition of unicellular plants. J. Cell. Comp. Physiol. 13: 373–315.

Ketchum, B. H. 1947. The biochemical relations between marine organisms and their environment. Ecol. Monogr. 17: 309–315.

Krogh, A. 1934. A method for the determination of ammonia in water and air. Biol. Bull. 67: 126–131.

Kroopnick. P. 1974. The dissolved O_2–CO_2–^{13}C system in the eastern equatorial Pacific. Deep-Sea Res. 21: 211–277.

Lyman, J. 1956. Buffer mechanism of sea water. Ph.D. thesis, University of California, Los Angeles.

McCarthy, J. J. 1970. A urease method for urea in seawater. Limnol. Oceanogr. 15: 309–312.

Martens, C. S., and R. A. Berner. 1974. Methane production in the interstitial waters of sulfate-depleted sediments. Science 185: 1167–1169.

Menzel, D. W., and J. H. Ryther. 1960. The annual cycle of primary production in the Sargasso Sea off Bermuda. Deep-Sea Res. 6: 351–367.

Morris, A. W., and J. P. Reily. 1963. The determination of nitrate in seawater. Anal. Chim. Acta 29: 272–279.

Packard, T. T. 1969. The estimation of the oxygen utilization rate in sea water from the activity of the respiratory electron transport system in plankton. Ph.D. thesis, University of Washington, Seattle.

Packard, T. T. 1971. The measurement of respiratory electron-
    transport activity in marine plankton. J. Mar. Res. 29:
    235-244.
Packard, T. T., R. C. Dugdale, J. J. Goering and R. T. Barber. 1978.
    Nitrate reductase activity in the subsurface waters of the Peru
    Current. J. Mar. Res. 36: 59-76.
Redfield, A. C. 1934. On the proportions of organic derivatives in
    sea water and their relation to the composition of plankton,
    pp. 176-192. In: James Johnstone Memorial Volume. University
    of Liverpool.
Redfield, A. C. 1942. The processes determining the concentration
    of oxygen, phosphate and other organic derivatives within the
    depths of the Atlantic Ocean. Papers in Physical Oceanography
    and Meteorology 9(2), 22 pp.
Redfield, A. C. 1948. The exchange of oxygen across the sea
    surface. J. Mar. Res. 7: 347-361.
Redfield, A. C., B. H. Ketchum and F. A. Richards. 1963. The
    influence of organisms on the composition of sea-water, pp.
    26-77. In: M. N. Hill [ed.], The Sea, Chapter 2, Vol. 2.
    Interscience Publishers.
Richards, F. A. 1957. Oxygen in the ocean. In: Treatise on Marine
    Ecology and Paleoecology. Geol. Soc. Am. Mem. 67 (1): 185-238.
Richards, F. A. 1975. The Cariaco Basin (Trench). Oceanogr. Mar.
    Biol. Ann. Rev. 13: 10-67.
Richards, F. A., and W. W. Broenkow. 1971. Chemical changes,
    including nitrate reduction, in Darwin Bay, Galapagos
    Archipelago, over a 2-month period, 1969. Limnol. Oceanogr.
    16: 758-765.
Richards, F. A., J. J. Goering, L. A. Codispoti and R. C. Dugdale.
    1973. Nitrogen fixation and denitrification in the ocean:
    biogeochemical budgets, pp. 12-27. In: E. Ingerson [ed.],
    Proceedings of the International Symposium on Hydrogeochemistry
    and Biogeochemistry, Vol. 2, Biogeochemistry. The Clarke Co.,
    Washington, D.C.
Richards, F. A., and R. F. Vaccaro. 1955. The Cariaco Trench, an
    anaerobic basin in the Caribbean Sea. Deep-Sea Res. 3: 214-228.
Riley, J. P. 1953. The spectrophotometric determination of ammonia
    with particular reference to sea-water. Anal. Chim. Acta 9:
    575-589.
Robinson, R. J., and T. G. Thompson. 1948. The determination of
    nitrites in sea water. J. Mar. Res. 7: 42-48.
Skirrow, G. 1975. The dissolved gases - carbon dioxide, pp. 1-192.
    In: J. P. Riley and G. Skirrow [eds.], Chemical Oceanography,
    2nd Edition. Academic Press.
von Brand, T., N. W. Rakestraw and C. E. Renn. 1937. The experi-
    mental decomposition and regeneration of nitrogenous organic
    matter in sea water. Biol. Bull. 72: 165-175.
von Brand, T., N. W. Rakestraw and C. E. Renn. 1939. Further
    experiments on the decomposition and regeneration of
    nitrogenous organic matter in sea water. Biol. Bull. 77:
    285-296.

Ward, B. B., R. J. Olson and M. J. Perry. 1982. Microbial
    nitrification rates in the primary nitrite maximum off southern
    California. Deep-Sea Res. 29: 247-256.
Whitledge, T. E. 1981. Nitrogen recycling and biological popula-
    tions in upwelling ecosystems, pp. 257-273. In: F. A. Richards
    [ed.], Coastal Upwelling. American Geophysical Union.
Whitledge, T. E., and R. C. Dugdale. 1972. Creatine in seawater.
    Limnol. Oceanogr. 17: 309-314.
Williams, P. M., W. H. Mathews and G. L. Pickard. 1961. A lake in
    British Columbia containing old sea-water. Nature 191: 830-832.
Winkler, L. W. 1888. Die Bestimmung des im Wasser gelösten
    Sauerstoffes. Ber. Dtsch. Chem. Ges. 21: 2843-2855.
Wood, E. D., F. A. J. Armstrong and F. A. Richards. 1967.
    Determination of nitrate in seawater by cadmium-copper
    reduction to nitrite. J. Mar. Biol. Assoc. UK 47: 23-31.
Wooster, W. S., T. J. Chow and I. Barrett. 1965. Nitrite
    distribution in Peru Current waters. J. Mar. Res. 23: 210-221.

HETEROTROPHIC UTILIZATION AND REGENERATION OF NITROGEN

Gilles Billen

Laboratoire d'Oceanographie
University of Brussels
Belgium

INTRODUCTION

Because organic nitrogen compounds represent both a nitrogen and an energy source for heterotrophic microorganisms in the sea, a discussion of the processes of their utilization and mineralization can bring insights either into specific aspects of the nitrogen cycle or into general mechanisms of organic matter metabolism in the sea. This paper will be mainly devoted to the latter aspect. There are indeed some technical advantages in focusing on nitrogen instead of on carbon for studying organic matter utilization in the sea, owing to the greater sensitivity of analytical methods for organic nitrogen than for organic carbon. However, parallelism or lack of parallelism between nitrogen and carbon utilization processes will be underlined.

Many examples discussed in this paper originate from data obtained in the Southern Bight of the North Sea and in the English Channel. This area is dominated by a flow of Atlantic water directed to the north east. However, due to the influence of the Scheldt estuary there is a zone of longer residence time of the water masses, of lower salinity, and of higher turbidity just in front of the Belgian coast. This zone receives important nutrient imputs from the land, so that the whole area offers a complete spectrum of situations from highly eutrophic in the Scheldt estuary to almost as oligotrophic as Atlantic water in the central English Channel (Fig. 1).

Existing data and concepts related to heterotrophic utilization and regeneration of nitrogen will be summarized under four main topics:

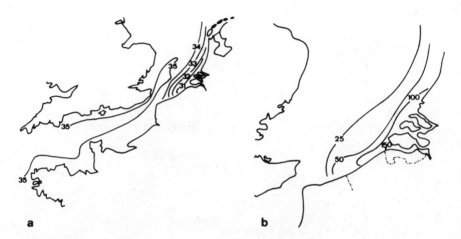

Figure 1.   a. Spring distribution of salinity in the English Channel
and the Southern Bight of the North Sea (Pichot 1980).   b. Total
nitrogen concentrations in the Belgian coastal waters (Nihoul and
Boelen 1976; Hagel et al. 1973).

(1)   Forms and production processes of organic nitrogen compounds
in the sea.
(2)   Hydrolysis of macromolecular organic nitrogen.
(3)   Uptake mechanisms of direct nitrogenous substrates and
competition.
(4)   Nitrogen mineralization by heterotrophic microorganisms.

FORMS AND PRODUCTION PROCESSES OF ORGANIC NITROGEN COMPOUNDS
IN SEA WATER

Dissolved Organic Nitrogen in Sea Water

During the last decade, intensive work has been devoted to the
characterization of dissolved organic compounds in sea water, and to
the geographical and seasonal variations of their concentration.
Although most information available is based on organic carbon
analysis, the same major features also apply to organic nitrogen:
     - Most organic matter in the sea exists in the dissolved rather
     in the particulate phase.
     - An important fraction of dissolved organic matter exists as
     high molecular weight compounds (Degens 1970; Ogura 1974;
     Wheeler 1976; Hama and Handa 1980).
     - A significant part of dissolved organic matter is refractory
     to microbial attack and is defined as humic compounds.
     - Direct organic substrates for microbial growth make up only a

very small part of total dissolved organic matter.

Total Organic and Inorganic Nitrogen

Total nitrogen concentration in sea water reflects the general
level of nitrogen enrichment and varies widely from estuarine and
coastal to open sea environments (see Table 1 and Fig. 1). Obviously, in most situations, total nitrogen concentration does not display
seasonal variations (Fig. 2).

Dissolved organic nitrogen makes up 25 to 80% of total nitrogen
according to the season; its percentage is at a maximum in late
summer and at a minimum in the winter.

Characterization of Dissolved Organic Nitrogen

Numerous authors (Williams 1971, 1975) have pointed out that
only a small fraction (10 to 30%) of the total dissolved organic
carbon in natural waters can be identified as well defined compounds.
As seen in Table 1 and Table 2, the same holds for total dissolved
organic nitrogen. Only a small fraction of the dissolved organic
nitrogen is compounds, such as free amino acids, that are directly
usable by microbes. A much more important part (16-50%, Tuschall
and Brezonik 1980) of organic nitrogen is made up by dissolved
proteins and polypeptides ("combined hydrolyzable amino acids") of
high molecular weight. The remaining fraction is probably made of
humic compounds refractory to microbial attack (Thomas et al. 1971).
Several hypotheses concerning the origin of these compounds have been
suggested including reaction of phenolic compounds with proteinaceous
material (Sieburth and Jensen 1969) or interaction between urea or
amino acids and metallic ions (Degens 1970).

Regional and Seasonal Variations

While total organic nitrogen varies a great deal from estuarine
and coastal to open sea environments, no large variations are found
in the concentration of easily usable organic nitrogen compounds
such as free amino acids. What is observed for regional variations
seems also true for seasonal variations. Distinct seasonal variations in total dissolved organic nitrogen have been reported (Butler
et al. 1979) which show accumulation of organic nitrogen compounds
following the period of intense phytoplanktonic activities. On the
other hand, most authors who have studied the annual cycle of total
free amino acids concentration did not find very clear evidence of
important seasonal change (Andrews and Williams 1971; Riley and
Seagar 1970; Crawford et al. 1974; Billen et al. 1980). Andrews and
Williams (1971) concluded that if such changes did occur, they were
either small or short lived.

Table 1. Typical values reported for total nitrogen and total
          dissolved organic nitrogen in various marine environments
          (in μg-at N liter^{-1}).

Environments	Total Nitrogen	Total Dissolved Organic Nitrogen	Reference
Estuaries			
Rhine	420	105	Van Bennekom 1975
Scheldt	800	200	Van Bennekom 1975
Coastal Areas			
Nearshore North Sea	100	70	Nihoul and Boelen 1976
Offshore North Sea	50	20	Nihoul and Boelen 1976
English Channel		4.6	Banoub and Williams 1973
English Channel	12	6	Butler et al. 1979
Open Ocean (upper layers)			
North Atlantic		6	Holm-Hansen et al. 1966
Pacific		8.2	Thomas et al. 1971
Indian Ocean		7.5	Fraga 1966
Mediterranean		5.2	Banoub and Williams 1972
Open Ocean (deep layers)			
Pacific	35	4	Armstrong et al. 1966
Pacific		3.5	Holm-Hansen et al. 1966
Pacific		6.9	Thomas et al. 1971
Indian Ocean		5	Fraga 1966
Mediterranean		3.3	Banoub and Williams 1972

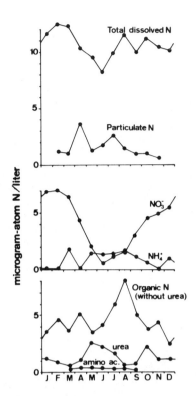

Figure 2.   Seasonal variation of the concentration of various classes of nitrogenous compounds in the water of the Western English Channel (composite data from Butler et al. 1979; Banoub and Williams 1973; Andrews and Williams 1971).

This lack of change is illustrated by data on amino acids in the Southern Bight of the North Sea. Although differences of two orders of magnitude exists at some seasons in the relative rate of heterotrophic amino acid utilization from the Scheldt estuary and the Belgian coastal zone to the English Channel (Fig. 3), no significant differences are found in the pool size of amino acids in these three regions (Table 2).   Thus the difference between eutrophic and oligotrophic waters lies in the concentration of high molecular weight organic nitrogen rather than in the concentration of easily usable substrates like amino acids.

As far as can be judged from the few data available, urea behaves more like an inorganic nutrient than like an organic substrate as it shows important regional (Table 2) and seasonal (Fig. 2) variations.

Table 2. Typical values reported for different classes of organic
          nitrogen compounds dissolved in the water of various marine
          environments (in µg-at N·liter^{-1}).

Environments	Total Amino Acids*	Primary Amines**	Free Amino Acids	Reference
Estuaries				
Scheldt	1.6		0.48	Lancelot, pers. comm.
Scheldt			0.2	Billen et al. 1980
Pamlico River			0.2	Crawford et al. 1974
York River			0.4	Hobbie et al. 1968
Coastal Areas				
North Sea	2.4			Lancelot, pers. comm.
North Sea			0.51	Billen et al. 1980
English Channel	2.6			Lancelot, pers. comm.
English Channel			0.26	Andrews & Williams 1971
English Channel			0.72	Billen et al. 1980
Irish Sea	0.4		0.15	Riley & Segar 1970
Baltic Sea			0.26	Dawson & Gocke 1978
Buzzards Bay	2.2		0.61	Siegel & Degens 1966
Open Ocean (upper layer)				
Atlantic	0.18		0.025	Lee & Bada 1977
Atlantic			0.26	Pocklington 1971
Atlantic		0.67		Liebezeit et al. 1980
Pacific	0.5		0.045	Lee & Bada 1975
Pacific			0.072	Williams et al. 1976
Pacific		0.7		North 1975
Pacific		0.41		Hollibaugh et al. 1980
Open Ocean (deep layer)				
Atlantic	0.1		0.025	Lee & Bada 1977
Atlantic			0.17	Pocklington 1971
Atlantic		0.59		Liebezeit et al. 1980

Environment	Total Nucleic Acid	Reference
Bombay Harbour Bay	0.4	Pillai & Ganguly 1970

Environments	Urea	Creatin	Reference
Estuaries			
Savannah River	6.8		Remsen et al. 1972
Coastal Areas			
English Channel	1.4		Butler et al. 1979
Coastal N. Atlantic	1.29		Remsen 1971
Peru upwelling	3.5		Remsen 1971
Peru upwelling		0.1	Whitledge & Dugdale 1972
Japan Sea	2.8		Mitamura & Saijo 1975
Open Ocean (upper layer)			
Atlantic	0.45		Remsen 1971
Pacific	0.24		McCarthy 1972
Pacific	1.8		Remsen 1971
Open Ocean (lower layer)			
Atlantic	0.5		Remsen 1971

*Total amino acids (free and combined)
** Primary amines (free a.a. and small pept.)

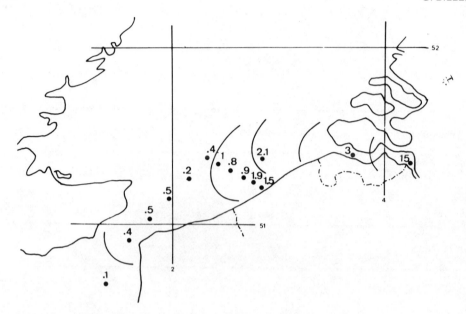

Figure 3. Utilization rate of amino acids (protein hydrolysate) in %·h^{-1} in the Southern North Sea (April 1981).

Supply of Organic Nitrogen in the Sea

Except in local near-shore situations, primary production is by far the most important source of organic matter in the sea. More-over, although rivers can carry large amounts of organic matter, most of it is degraded or sedimented in estuarine systems; thus only a small fraction of the organic matter, mostly made of compounds refractory to bacterial attack, is discharged into coastal zones (see e.g., Wollast and Billen 1981).

The three main processes to be discussed with regard to produc-tion of dissolved organic nitrogen in the sea are:
 - extracellular release of dissolved organic matter by phyto-plankton;
 - spontaneous lysis or zooplankton spillage from phytoplank-tonic cells;
 - excretion by zooplankton.

Excretion of Organic Compounds by Phytoplankton

Studies of extracellular release of organic matter by natural phytoplanktonic populations are numerous [see for instance the reviews by Hellebust (1974), Sharp (1977) and the more recent papers by Larsson and Hagström (1979), and Lancelot (1979)]. Although little agreement exists on the relative importance of this process

with respect to total primary productivity (estimations varying from
a few to 70%), it is certainly true that it represents a significant
source of dissolved organic matter available for bacteria, especially
in oligotrophic waters.

The chemical composition of this excreted material varies from
species to species (Hellebust 1965). In the Southern Bight of the
North Sea, during the spring phytoplankton bloom dominated by the
species Phaeocystis poucheti, Lancelot (1982) demonstrated by ultrafiltration of the ^{14}C-labelled excreted material that organic
compounds of molecular weight higher than 500 d. represent 60-91%
(mean 82%) of the total exudate. Nalewajko et al. (1976) report
somewhat different figures for fresh water phytoplankton: one third
of the excreta was made of compounds with molecular weight higher
than 5000 d. Wiebe and Smith (1977), working in an Australian
estuary, reported that 95% of the excreted material is of low
molecular weight (i.e., lower than 3500 d.).

The high molecular weight fraction of excreted material is
probably mostly polysaccharides but proteins would also be an
important part. In the low molecular weight fraction, on the other
hand, free amino acids are present but they were found by Juttner
and Matuschek (1978) to make up only about 10% of the total. Most
authors also found glycolate, Krebs cycle acids and carbohydrates
(i.e., non-nitrogenous compounds) as dominant constituents of
extracellularly produced organic matter. Only nitrogen fixing
blue-green algae were observed to excrete important amounts of free
amino acids (Stewart 1963; Jones and Stewart 1969). All these data
suggest that organic nitrogen excreted by natural phytoplankton
communities in the sea mainly consists of proteins and peptides,
while only a small amount of free amino acid is directly produced.

Death and Lysis of Phytoplanktonic Cells

Although it was thought in the past that the normal fate of
phytoplankton cells in the sea is to be grazed by zooplankton and
that only few simply die (e.g., Harvey 1945), it becomes more and
more evident that in many aquatic systems, particularly coastal
marine ones, zooplankton grazing cannot explain the variations
observed in phytoplankton biomass and that an important spontaneous
phytoplankton mortality occurs (Jassby and Goldman 1974; Joiris
1977).

When phytoplankton cells are disrupted or lyse, particulate
detritus and dissolved organic matter are produced. Table 3
summarizes some data on detailed nitrogen composition of phytoplankton cells. It can be inferred from these figures that mostly
macromolecular nitrogen compounds are primarily produced on lysis of
phytoplankton. Although data on the pool of free amino acids in
algae is rather scarce, this pool appears to be very small so that

Table 3. Detailed composition of phytoplankton nitrogen (calculated
          from data of Mayzaud and Martin 1975, and Reisner et al.
          1960).

	% dry weight	% of total nitrogen
Total nitrogen	4.3	100
Protein nitrogen	3.65	85
Non-protein nitrogen	0.69	15
Free amino acids	0.23 - 0.35	5 - 8

only minute amounts of free amino acids are released by simple
disruption of algal cells.

Excretion of Organic Nitrogen by Zooplankton

     A controversy exists in the literature concerning the forms of
dissolved nitrogen excreted by zooplankton. Johannes and Webb (1965)
and Webb and Johannes (1967) using high concentration of mixed zooplankton found that considerable amounts of free amino acids were
released, while Corner et al. (1965) and Corber and Newell (1967)
concluded from experiments with much lower zooplankton concentrations
that no significant amounts of nitrogenous substances other than
ammonia were excreted. Jawed (1969) measured the following composition of soluble excretion products of zooplankton: 76-82% ammonia-
nitrogen, 13-18% amino nitrogen, and about 1% urea. Eppley et al.
(1973) found much higher urea excretion (about 50%).

     Soluble nitrogen excretion by fishes was also measured by
Whiteledge and Dugdale (1972) on Peruvian anchoveta and consisted of
about 50% ammonia, 25% creatine, 18% urea.

EXOENZYMATIC HYDROLYSIS OF MACROMOLECULAR ORGANIC NITROGEN

     From the preceding section, it is apparent that an important
part of both the stocks and of the fluxes of organic nitrogen in the
sea is in the form of high molecular weight, polymeric material
(proteins or peptides), either particulate or dissolved. Such
material normally cannot be directly taken up by bacteria and can
only be ultimately absorbed after exoenzymatic hydrolysis (Rogers
1961).

Exoenzymes, and exproteases in particular, can therefore be
inferred to play an important role in aquatic ecosystems. Unfortunately, there is little information on their occurence and activity
in the sea.

Occurence of Exoproteases in Aquatic Environments

Hydrolytic enzymes are present inside and outside the cell.
Exoenzymes, found outside the cell, may be bound to the surfaces,
such as a cell, or may be free (Pollock 1962). It is not easy to
establish whether an enzymatic activity is intracellular or cell-
surface bound, but free exoenzymic activity can be demonstrated
after separation of the cells by gentle filtration. It is often
difficult, however, to decide whether enzymes found free in the
external medium have been liberated by healthy cells or have been
liberated by cell disruption or autolysis.

Use of synthetic protein-dye (Azocoll, Calbiochem) or chitin-
dye (Chitin-Red, Calbiochem) as substrates allowed some authors to
demonstrate the occurence of free proteolytic and chitiniasic exoenzymic activities in lake water, sea water and interstitial waters
of marine sediments (Reichardt et al. 1967; Kim and Zobell 1974).
Another technique is to use aminoacyl derivatives of (Beta)-naphthyl-
amine. These give rise to fluorescent products upon enzymatic
hydrolysis (Roth 1965) and have been found by Somville and Billen
(1983) to be a very convenient and sensitive tool for studying
exopeptidases in estuarine and marine waters. Some results obtained
with this method in the English Channel, the North Sea and the
Scheldt estuary are presented in Table 4. A measurable proteasic
activity is present in all freshly collected unfiltered samples.
In all cases, autoclaved samples lose their activity. Results of
proteasic activity determination with or without prior filtration
of the samples through 0.2 and 0.8 μm membranes show that some
exoproteases may exist as free enzymes in the eutrophic Belgian
coastal zone, while in the more oligotrophic waters of the Channel,
most of them are bound to particles between 0.8 and 0.2 μm. In the
last case, exoenzymes are probably linked to the external surface of
bacterial cells, as demonstrated in some instances by Christison and
Martin (1971).

Control of Exoenyzymes Production by Environmental Factors

Numerous studies of pure bacterial cultures have been concerned
with the physiological regulation of exoenzymes synthesis (see the
review by Glenn 1976). The production of many exoproteases can be
induced by the presence of peptides or proteins in the external
medium; end production inhibition by free amino acids and catabolic
repression (e.g., by glucose) has also been reported for most
exoprotein producing bacteria. Unfortunately, it is not known to
what extent these pure culture observations can be extrapolated to

natural aquatic environments, and at which levels of substrate
concentrations these physiological controls are operative.

   Moreover, besides this "physiological regulation", some kind of
"sociological regulation" (Wuhrmann 1968) must be taken into account
when dealing with natural aquatic communities of microorganisms.
Martin and Bianchi (Martin 1980; Martin and Bianchi 1980; Bianchi
1980), studying the specific composition of bacterial communities in
various aquatic environments, showed that strains which produced
exoenzymes are selected when polymeric detritus is the most important

Table 4. Proteolytic activity of filtered and unfiltered water
          from various marine environments (Billen and Somville,
          unpublished).

Environments	Sample	Proteolytic activity [+] 10^{-8} moles·liter^{-1}·min^{-1})
Belgian coastal zone (West Hinder, June 1981)	unfiltered	0.43
	filtered through 0.2 μm	0.38
	autoclaved	0
English Channel (Off Boulogne, June 1981)	unfiltered	0.66
	filtered through 0.8 μm	0.48
	filtered through 0.2 μm	0.18
	autoclaved	0
Scheldt estuary (Doel, May 1981)	unfiltered	1.82
	filtered through 0.2 μm	0.27
	autoclaved	0
	filtered + autoclaved suspended matter	0.27

[+] Amount of B-Naphthylamide produced from L Leucyl B-Naphthylamide
   per minute of incubation time (Roth 1965).

substrate available such as in senescent phytoplankton cultures, sediments, etc.

Kinetics of Detritus Hydrolysis

Although no direct data are available concerning the kinetics of exoenzymes action in sea water, indirect information can be obtained from experiments on bacterial degradation of dead algal cells or on extracts of algal cultures, since it can be thought that exoenzymic hydrolysis is the first limiting step in the decomposition of these complex materials.

Most authors found or assumed first order kinetics for the degradation of detritus, the value of the kinetic constant depending on the nature of the material being degraded and on the environmental conditions. Table 5 shows the range of values reported for the

Table 5. First order constant of the degradation of various classes of complex organic nitrogen in aquatic environments.

Nature of the Material	$k(day^{-1})$	Reference
Dead phytoplankton cells		
- as a whole	0.04 - 0.1	Golterman 1972
	0.038 - 0.056	Von Brand et al. 1937
- soluble fraction (25% of total N)	1 - 0.2	Otsuki and Hanya 1972
- insoluble labile fraction (42% of total N)	0.056	Otsuki and Hanya 1972
- refractory fraction (33% of total N)	0.005	Otsuki and Hanya 1972
Dissolved macromolecular organic nitrogen		
- labile fraction	0.011	Otsuki and Hanya 1972
- dissolved proteins (in the presence of natural suspended matter)	0.015	Kailov and Finenko 1970

degradation of several classes of particulate or dissolved organic
nitrogen. Otsuki and Hanya (1972) in one of the more complete
studies, suggested that nitrogen compounds in green algae may be
divided into three fractions according to their resistance to
decomposition: a first fraction (about 25% of total N) is made of
soluble, very easily degraded compounds (decomposition constant as
high as 1–0.2 day^{-1}); a second fraction (about 42%) comprises labile
material with a decomposition constant of about 0.06 day^{-1}; a third
fraction (33%) is refractory to degradation (0.05 day^{-1} or less).
In comparison to these studies of detritus, the decomposition of
dissolved proteinaceous matter is much more subject to variations
from one environment to the other. Kailov and Finenko (1970) have
found that detritus enhanced the degradation of dissolved protein.

   The rather low values of the rate constant of degradation of
polymeric nitrogeneous material and the fact that, as shown in the
preceding section, most of the organic nitrogen is supplied as
macromolecules, imply that bacterial acitivity is buffered to a
certain extent against short term variations in phytoplanktonic
production by the large pool of nitrogen polymers. Whether such a
buffer effect also exists for all other classes of organic material
is not clear. Some authors demonstrated a very close coupling
between phytoplankton excretion and bacterial activity, the turnover
time of excreted products being sometimes as short as a few hours
(Nalewajko et al. 1976; Wiebe and Smith 1977; Iturriaga and Hoppe
1977; Larsson and Hagström 1979). As shown above, however, those
very rapidly used excretory products are not likely to contain
nitrogenous compounds, but rather organic acids and monosaccharides.

UPTAKE MECHANISMS OF NITROGENOUS SUBSTRATES AND COMPETITION BETWEEN
MICROBIAL SPECIES FOR THESE SUBSTRATES

Physiological Mechanisms of Nitrogen Uptake

   All heterotrophic microorganisms need nitrogen compounds (either
mineral or organic) for biosynthetic purposes. In addition, most of
them can use organic nitrogen compounds as substrates for energy.
Algae use ammonium or nitrate as nitrogen source, but many species
have been demonstrated to be capable of autotrophic growth with
amino acids or urea as the sole nitrogen source. Finally, ammonium
is an energy source for the chemolithotrophic nitrifying bacteria.

Mineral Nitrogen Assimilation by Heterotrophic Bacteria and Algae

   At low concentration, ammonium is generally assimilated as well
in bacteria as in algae by the glutamine synthetase (GS)/glutamine-
oxoglutarate amino transferase (GOGAT) system, involving ATP dependent synthesis of glutamine from glutamate and NH_3 and NADPH-linked
synthesis of two glutamates from glutamine and oxaglutarate (Tempest

et al. 1970; Brown et al. 1972; Falkowski and Rivkin 1977; Miflin and Lea 1977). This system is much more efficient than glutamate deshydrogenase (GDH), which catalyses the direct glutamate synthesis from NH_3 and oxoglutarate at high ammonium concentrations.

When used as nitrogen source, nitrates are first converted into ammonium via nitrite before being assimilated by the same pathway. In algae, nitrate reductase (and perhaps the nitrate permease system) is generally suppressed by ammonium concentrations higher than about $0.5-1$ µmoles·liter^{-1} unless the cells are severely nitrogen depleted (Eppley and Solorzano 1969; Morris 1974; Conway 1977). This last regulation mechanism has not been demonstrated in marine bacteria (Brown et al. 1972), although it has often been observed that ammonium is preferentially taken up by heterotrophic microorganisms grown on both nitrate and ammonium (Alexander 1977).

In 12 strains of marine pseudomonads, Brown et al. have shown that the GS/GOGAT system is repressed by the presence of amino acids, indicating that amino acids, when available in sufficient amounts, are a preferred source of nitrogen for marine bacteria. When used also as source of carbon, amino acids are generally converted to their respective oxoacids by transamination with oxoglutarate yielding glutamate which is then deaminated by GDH (Brown et al. 1972).

Active Transport Systems

Active transport systems (permeases) for both mineral and organic nitrogen compounds have been demonstrated in bacteria and algae. These system in algae for ammonium and nitrate are very efficient in terms of the half-saturation constant (Table 6). Falkowski (1975) has suggested that nitrate uptake by marine phytoplanktonic algae is related to a NO_3^- activated ATPase associated with the cell membrane. Fewer studies exist on mineral nitrogen permeases in heterotrophic microorganisms, although two distinct ammonium uptake systems have been demonstrated in Saccharomyces cerevisiae (Roon et al. 1975; Dubois and Grenson 1979).

Permeases systems for amino acids have been much more intensively studied. In most microorganisms investigated (e.g., Escherichia coli (Piperno and Oxender 1968; Brown 1971; Rahmanian et al. 1973; Guardiola et al. 1974) Pseudomonas aeruginosa (Kay and Gronlund 1969, 1971) Saccharomyces cerevisiae (Grenson et al. 1966, 1970; Grenson 1966; Gits and Grenson 1967) several amino acid transport systems with high affinity and high substrate specificity were demonstrated, each responsible for the uptake of only single or closely related amino acids. Often, multiple transport systems exist for a single or a group of amino acids.

In some cases, besides these specific very efficient permease systems, an additional system with wider specificity was demonstrated.

Table 6. Transport constant (K_t) for amino acids, urea and mineral
          nitrogen (μmoles/l) of different species and strains of
          heterotrophic bacteria, phytoplanktonic algae and nitri-
          fying bacteria. ([+] Values in brackets refer to a second
          transport system with lower affinity.)

Organism	Transport Constant for			
	Amino Acid	Urea	NH_4^+	NO_3^-
Heterotrophic microorganisms				
Saccharomyces cerevisiae[a]			1-(20)[+]	
Pseudomonas aeruginosa[b]	Phe 0.4			
	Tyr 0.5			
	Try 1.4			
Alteromonas haloplanktis[c]	Ala 4.6			
Heterotrophic strain RP 303[d]	Pro 1.8			
Oligotrophic strain 486[d]	Pro 0.2			
Phytoplanktonic algae				
Skeletonema costatum[e]		4.25		
			0.5-0.4	3.6-0.8
Ditylum brightwellii[f]		0.20	0.6	1.1
Dunaliella tertiolecta[f]			1.4	0.1
Platymonas subcordiformis[g]	Gly 5			
Melosira nummuloides[h]	Arg 7.7			
	Val			
Navicula pavillardi[i]	Glu 18			
Coccolithus huxleyi[f]		0.1	0.1	0.1
Cyclotella nana[f]			0.3, 0.7	0.4
		0.21		
Nitrifying bacteria (Apparent transport constants based on oxidation rates by culture of nitrifying bacteria)				
Mixed culture[j]			500	
Nitrocystis oceanus[k]			150	
Nitrosomonas europaea[l]			86-715	
Enrichment culture from [m] the Scheldt estuary			250	

[a] Dubois and Grenson 1979; [b] Kay and Gronlund 1971; [c] Fein and
MacLeod 1975; [d] Akaji and Tage 1980; [e] Carpenter et al. 1972;
[f] Eppley et al. 1969; [g] North and Stephens 1967; [h] Hellebust 1970;
[i] Lewin and Hellebust 1975; [j] Knowles et al. 1965; [k] Carlucci and
Strickland 1968; [l] Painter 1977; [m] Somville 1980

The affinity of transport systems of different strains and
species of aquatic microorganisms for amino acids, urea, NH_4^+ and
NO_3^- are given in Table 6. From these data, it can be seen that
heterotrophic bacteria have developed amino acid transport systems
with higher affinity for substrates than have phytoplanktonic algae.
The reverse seems to be true for nitrite and ammonium, although few
data are available for mineral nitrogen uptake by bacteria. The K_t
values of phytoplanktonic algae for urea are of the same order of
magnitude as for mineral nitrogen. In contrast, the K_t values of
nitrifying bacteria for ammonium are at least two order of magnitude
higher than those of algae. When comparing these figures however,
it must be remembered that for bacteria, considerable difference can
exist between the value of K_t of an organism in natural water and
the value measured after isolation and cultivation of this organism.
For example, Jannasch (1968) showed that the half saturation constant
of growth of several bacteria increased 2 to 10 fold during repeated
transfers in media enriched in organic substrates.

In situ Determination of Half-Saturation Constant for the Uptake of Nitrogenous Compounds

Following the work of Wright and Hobbie (1966), numerous
measurements of the uptake kinetics of various substrates by intact
natural communities of aquatic microorganisms have been reported.
In most situations, the uptake was found to obey a Michaelis-Menten-
Monod relationship:

$$V = \sum_i \nu_i \simeq V_{max} \, S/(S + K_t) \qquad (1)$$

where V is the total rate of uptake, ν_i is the rate of uptake by
population i in the microbial community, V_{max} is the maximum total
rate of uptake, S is the substrate concentration, and K_t is the
half-saturation constant of uptake.

The validity of this approach when dealing with heterogeneous
microbial communities has been discussed in detail by Williams
(1973). He showed that the validity of relation (1) is better when
the community becomes less diverse (in terms of the values of K_{ti} of
the various populations). Indeed, some cases of departure from
Michaelis-Menten-Monod kinetics have been reported for the uptake of
some substrates by natural population of microorganisms (Vaccaro and
Jannasch 1966; Hamilton and Preslan 1970; Barvenik and Malloy 1979);
they concern generally very oligotrophic environments where it is
known (Martin and Bianchi 1980) that microbial diversity is more
important. In all the other cases the validity of relation (1)
indicates either that one single microbial strain dominates all the
others in the utilization of the substrate, or that all the strains
utilizing the substrate have a very similar value of K_t. Therefore,

the K_t value obtained from this kind of measurement with natural communities characterizes the affinity toward the substrate of the dominant microbial populations.

For substrates used purely for energy, like glucose and acetate, there appears to be a relationship between the K_t value of the microbial community and its natural rate of substrate utilization; the lower the flux of the substrate, the lower the K_t value (Fig. 4 and 5). Since the rate of utilization of a substrate (rather than its concentration) is the best index of the "richness" of an environment in terms of this substrate, this trend can be interpreted as reflecting the competition between microorganisms for their substrates; the lower the availability of a substrate, the higher the selective pressure for developing sophisticated and expensive permease systems with great affinity for this substrate.

The same relationship holds for mineral nitrogen substrates as shown by the data of McIsaac and Dugdale (1969) and Paasche and Kristiansen (1982) from oligotrophic and eutrophic marine areas (Fig. 6). These data show K_t values less than 0.2 μmoles$\cdot$liter^{-1} in oligotrophic waters and higher than 0.5 μmoles$\cdot$liter^{-1} in eutrophic waters.

Such a relationship, however, cannot be demonstrated for amino acids. The K_t values found in a whole range of environmental situations, from oligotrophic ocean area to organic rich sediments, always are very low, ranging from 0.02 to 1.5 μmoles$\cdot$liter^{-1} (Fig. 7). Perhaps the double nature of amino acids as both a carbon and a nitrogen

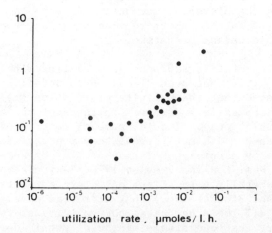

utilization rate . μmoles/ l. h.

Figure 4. Half saturation constant for glucose uptake measured in various aquatic environments plotted against the glucose utilization rate. (Data from Japanese marine and brackish environments, and from Canadian lakes, Seki et al. 1975, 1980a,b).

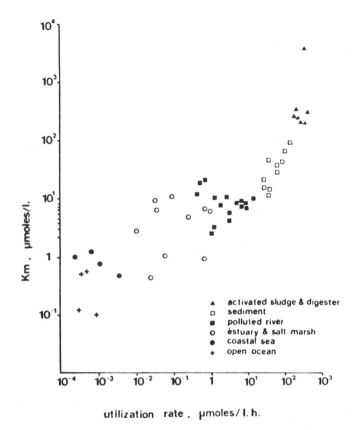

Figure 5. Half-saturation constant for acetate uptake measured in various aquatic environments plotted against the acetate utilization rate. (Data from Seki et al. 1974, 1980; Strayer and Tiedje 1977; Stanley and Staley 1977; Russel and Baldwin 1979; Billen et al. 1980; Billen, unpublished.)

source for microorganisms explains the fact that competition for them is always very severe even when their total flux is important.

Comparing the range of the K_t values found in situ for natural mixed microbiological communities (Fig. 6 and 7) with those found for pure species of microorganisms (Table 6), suggests the following conclusions concerning the competition conditions of the various microorganisms for each class of organic nitrogen compounds.

(i) Amino acids uptake seems to be dominated by bacteria rather than by algae. This conclusion is supported by the work of Williams (1970) and Derenbach and Williams (1974) showing by differential filtration following incubation with labelled amino acids that most of the radioactivity incorporated was associated with particles

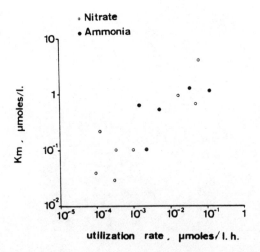

Figure 6. Half-saturation constant for nitrate and ammonia uptake
in oligotrophic and eutrophic marine systems plotted against the
total rate of nitrate or ammonia utilization. (Data from McIssaac
and Dugdale 1969; Paasche and Kristiansen 1982.)

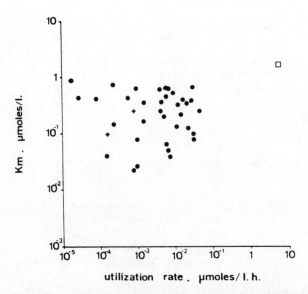

Figure 7. Half-saturation constant for alanine uptake in various
aquatic environments plotted against the alanine utilization rate.
(Data from Seki et al. 1974, 1980; Crawford et al. 1974; Christensen
and Blackburn 1980.)

smaller than 8 m. These works were criticized however by Wheeler
et al. (1977) who emphasized the possibility of artifact due to
postfiltration; using differential filtration prior to incubation
and autoradiography these authors found that phytoplankton were
sometimes responsible for up to 50% of glycine uptake. On the other
hand, Schell (1974) using ^{15}N-labelled glycine and glutamate, showed
that these amino acids only contribute a very small part of the
nitrogen requirements of phytoplankton.

The situation is different for urea which has been shown to
contribute appreciably to nitrogen uptake by phytoplankton, both by
^{15}N studies (McCarthy 1972) and by differential filtration methods
(Remsen et al. 1972; Mitamura and Saijo 1975). The same studies
indicate that the contribution of bacteria to urea decomposition is
minor compared to that of phytoplankton.

(ii) Mineral nitrogen uptake in open water is probably dominated by phytoplankton, although no data are available concerning a
possible ammonium uptake by heterotrophic bacteria. In most situations in the photic layer, autotrophic nitrifying bacteria are unable
to compete with algae for NH_4^+ uptake. These bacteria are only
active in situations where, owing to intense ammonium production or
reduced phytoplankton uptake, ammonia accumulates. Accordingly,
active nitrification is found in sediments (Billen 1976; Henriksen
1980), in the water column below the photic layer (Gundersen and
Mountain 1973), or in polluted estuaries (Billen 1975; Somville
1978) but no convincing evidence of active nitrification in the
photic layer of unpolluted seas has been obtained so far.

Control of Substrate Concentration by Rapidly Growing Microorganisms

The purpose of this section is to investigate the relations
existing in natural waters between the uptake characteristics of
microorganisms discussed in the preceding sections and the concentration of directly usable substrates, such as amino acids, urea or
inorganic nitrogen species.

General Theory

The concentration of a particular substrate results from the
balance between the rate production of this substrate (e.g., by
phytoplanktonic excretion, exoenzymatic hydrolysis, etc.) and the
rate of uptake by the dominant microorganisms population. If this
population is limited by the substrate and is able to grow fast
enough to maintain a steady state between substrate production and
uptake, the concentration of the substrate is independent of its
rate of production and depends only on physiological characteristics
of the microorganisms (Billen et al. 1980).

This can be shown in the following way. The rate of change of substrate concentration (S) can be written:

$$dS/dt = P - V_{max}S/(S + K_t)B \qquad (2)$$

Where P is the rate of production of the substrate, B is the mass of organisms utilizing S, V_{max} is the maximum rate of uptake per organism, and K_t is the transport constant of S by the organism.

On the other hand, the rate of change of the biomass of the organisms can be written:

$$dB/dt = B(YV_{max} \; S)/(S + K_t) - k_dB \qquad (3)$$

where Y is the yield constant, i.e., the mass of organisms formed per unit of substrate taken up and k_d is a first order mortality constant.

At stationary state the solution (2) and (3) is

$$S = K_t/((YV_{max}/k_d)-1) \qquad (4)$$

$$B = (Y/k_d)P \qquad (5)$$

showing that at steady state only the biomass of microorganisms is affected by the rate of production of the substrate. The concentration of the substrate depends only on the transport constant, and on the ratio between the maximum growth rate ($\mu_{max} - YV_{max}$) and the death rate of the organisms.

The question is of course to know how closely a steady state is approached by natural aquatic systems. It has been shown by very simple simulations (Billen et al. 1980) that the time required for reaching a steady state, or for restoring it after a sudden perturbation, is about $1/k_d$, that is, of the order of the generation time of the microorganisms at steady state. Thus bacteria, having short generation times, effectively control the concentration of the substrates they use predominantly. In contrast, algae, having relatively long generation times (and moreover, being limited by other factors like light intensity) are not always able to maintain the concentration of their substrates at a stationary state.

The validity of equation (4) for organic substrates predominantly used by bacteria can be tested with experimental concentration and K_t data for glucose and acetate reported in the literature for various aquatic environments (Fig. 8). Both sets of data agree well for these substrates, showing that these systems are close to being in a steady state, and that the control of substrate concentration by the uptake of microorganisms is effective. The relation obtained in both cases is S $\simeq$ Km/3 which suggests that in all the environments considered the ratio μ_{max}/k_d is about 4.

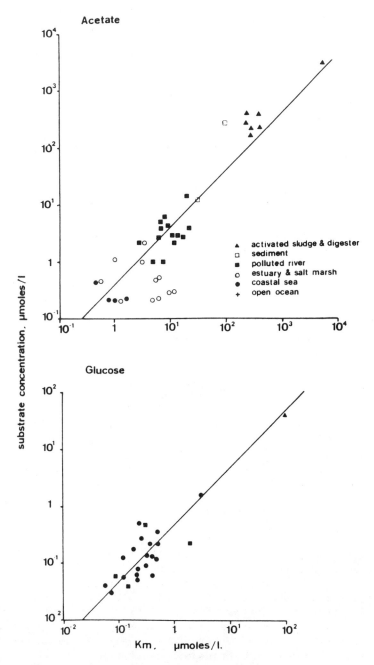

Figure 8.   Relation between the natural concentration of glucose and acetate and the half-saturation constant of uptake in various aquatic environments. (Data from Seki et al. 1974, 1980; Walker and Monk 1971; Russel and Baldwin 1979; Stanley and Staley 1977; Kaspar and Wuhrmann 1978; Billen et al. 1980.)

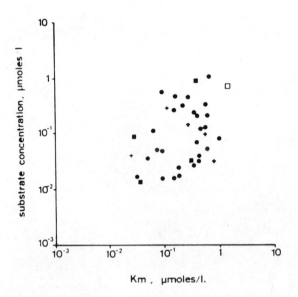

Figure 9.   Relation between the natural concentration of alanine and
its half-saturation constant of uptake in various aquatic environments.   (Data from Seki et al. 1974, 1980; Crawford et al. 1974;
Christensen and Blackburn 1980.)

Application to the Case of Nitrogenous Substrates

Based on the information given above, it appears that amino
acids are predominantly taken up by bacteria while urea and mineral
nitrogen species are mostly used by phytoplankton.   This division is
consistent with the ideas presented above that the rapidly growing
microbes can actively control the concentration of substrates.   Thus,
amino acid concentrations do not change significantly over the
seasons in contrast to the wide fluctuations of urea and inorganic
nitrogen.

Figure 9 shows, as an example, the relation observed between
concentration and the half-saturation for uptake of alanine.   The
theoretical relationship (equation 4) does not fit the data as well
as it did for glucose and acetate (Fig. 8).   This can be explained
by the fact that no microorganism is likely to be limited by a
single individual amino acid.   Nevertheless individual amino acid
concentrations in natural water are well within the range of
experimentally determined transport constants.

NITROGEN MINERALIZATION BY HETEROTROPHIC MICROORGANISMS

In the preceding part of this paper, the gross utilization of organic nitrogen compounds by heterotrophic microorganisms has been discussed without concern for the fate of the compounds. After uptake by the microbes, organic nitrogen compounds can be either incorporated into biomass or excreted as ammonia. In the classical conception, the latter process is thought to be the most important, and microorganisms are viewed as direct mineralizers of organic matter. Some authors however (Rittenberg 1963; Johannes 1968) have stressed the possible importance of the former process, claiming that the bacteria account directly for only a minor fraction of nutrient regeneration and may even compete with algae for mineral nitrogen. This question of organic nitrogen incorporation versus mineralization can now be reexamined in the light of recent physiological and ecological data.

Mineralization of Individual Organic Nitrogen Substrates

Most of the information available on microbiological mineralization of individual organic compounds has been obtained by ^{14}C tracer studies and therefore strictly refers to <u>carbon mineralization</u> even when organic nitrogen compounds are involved. By simultaneously measuring respiration and incorporation of labelled compounds, it is possible to define a growth yield ratio (ratio of incorporation to total uptake) or its complement the C-mineralization ratio (ratio of respiration to total uptake). For amino acids, the latter ratio varies from 0.15 to 0.60. This variability is much lower, however, when considering a single amino acid.

Crawford et al. (1974) and Wright (1974) have shown that amino acids can be divided into three groups: the first (including Glu, Asp, Asn, Pro, Arg) has high ratios for mineralization (often equal or higher than 0.50), the second (including Leu, Isoleu, Lys, Phe, Val) has low values for the mineralization ratio (about 0.15), and the third (Ala, Ser, Tyr) has intermediate values for the mineralization ratio (0.20-0.50). Amino acids of the first group are those which are the most directly metabolized into Krebs' tricarboxylic acid cycle. They are also those for which the maximum rates of uptake are the higher (Wright 1974).

The values just discussed of the C-mineralization ratio are for short term (2-5 hr) experiments. They probably do not reflect the natural steady state metabolism of amino acids by bacteria. In long-term incubation experiments (about 100 hr) Williams and Gray (1970) and Hollibaugh (1978) found much higher values of the C-mineralization ratio, from 0.65 to 0.70.

Because amino acids can be very easily deaminated at early stages of their metabolism in microorganisms, the fate of nitrogen after amino acid uptake is not necessarily similar to that of the carbon skeleton. Using parallel ^{15}N and ^{14}C labelling experiments, Schell (1974) demonstrated that the nitrogen of glycine was

preferentially incorporated with respect to carbon (C uptake/N uptake
ratio = 0.6) while the reverse was true with glutamate (C uptake/N
uptake = 2.6). The mean ratio of ammonium release to total uptake
of amino N was 0.74 in long-term incubation experiments with added
amino acids (Hollibaugh 1978).

Mineralization of Complex Substrates

In natural situations, heterotrophic microbial populations do
not assimilate a single individual compound but rather a mixture of
several substrates. The balance between N-mineralization and
incorporation will then depend on the C:N ratio of the total organic
matter utilized. The amount of ammonia released (ΔNH_4) per unit
carbon taken up (ΔC) is given by the following relation:

$$\Delta NH_4 / \Delta C = 1 / \left(\frac{C}{N}\right)_S - Y / \left(\frac{C}{N}\right)_B \qquad (6)$$

where Y is the growth yield ratio, $(C/N)_S$ is the C:N in the
substrate, and $(C/N)_B$ is the C:N of bacterial biomass.

The experiments of Hollibaugh (1978), Somville (1980) and Billen
(unpublished), who supplemented natural sea water with mixtures of
organic substrates and followed the consumption of the substrates and
the release of ammonia, permit a test of the validity of relation (6)
(Fig. 10). Although the data come from two different environments
and were obtained with different organic substrates, a very good fit
is obtained, with $\frac{Y}{\left(\frac{C}{N}\right)_B}$ = 0.1 gN/gC.

If 4 gC/gN is taken as a reasonable estimate for $(C/N)_B$, Y can
be evaluated as 0.4 in good agreement with the values cited above
for C-assimilation ratio in long-term experiments.

Relation (6) and Fig. 10 also show the lack of parallelism
between the role of bacteria in carbon and nitrogen cycling. When
the (C/N) ratio of the organic matter used by bacteria increases,
ammonium release decreases and, for $(C/N)_S$ higher than 10 gC/gN,
uptake instead of release can even occur during organic matter
degradation. The role of bacteria as nitrogen mineralizers thus does
not necessarily parallel their role as carbon mineralizers. A striking example is provided by data obtained by Joiris et al. (1982) and
Billen (unpublished) in the Belgian coastal zone, the English Channel
and the Scheldt estuary (Table 7). Total heterotrophic activity
throughout the year was estimated by measuring the concentration and
the relative rate of bacterial utilization of the three main classes
of direct organic substrates (free amino acids, monosaccharides, and
glycollate). From these data, ammonium remineralization can be calculated according to equation (4). As seen, the most important

Table 7. Annual means of rate of organic substrate uptake and calculated rate of ammonium release in three marine environments.

Heterotrophic Activity	Eastern Channel (off Boulogne)	Belgian Coastal Zone (off Ostend)	Scheldt Estuary (Hansweert)
Amino acid uptake*	3.1	2.4	6.5
Monosaccharide uptake*	2.6	4.4	16.3
Glycollate uptake*	0.3	1.2	15
Total*	6	8	37.8
C/N of organic matter taken up	6	10	17
NH$_4$ release**	0.4	0.08	-0.115

 * mg C·liter^{-1}·yr^{-1}
 ** mg N·liter^{-1}·yr^{-1}

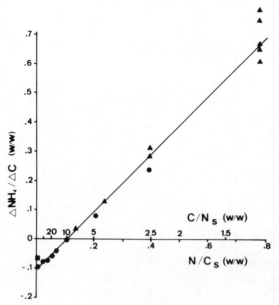

Figure 10. Release or uptake of ammonia per unit carbon taken up by
microbiological communities of marine environments supplemented with
mixed substrates of various C/N ratio. (Data from Hollibaugh 1978
(▲), Somville 1980 (■), Billen, unpublished (●)).

ammonium release occurs in the Western Channel, where heterotrophic
carbon utilization is the lowest. In the Belgian coastal zone, the
ammonium regeneration by bacteria is limited because the most
important substrates for heterotrophic activity are carbohydrates
which in part come from the mucopolysaccharides excreted by
Phaeocystis poucheti (Lancelot 1982). In the heavily polluted
Scheldt estuary there is actually net ammonium uptake due to the
high C/N ratio of the terrigenous organic material being degraded.

Relative Role of Bacteria and Zooplankton in Nitrogen Mineralization in the Sea

There are a few recent studies which indicate the relative role
of bacteria, net plankton, and microplankton in providing nitrogen
for phytoplankton (Table 8). Net zooplankton excretion has been
measured by various authors after concentration by filtration through
180 or 100 μm mesh net. The results suggest a clearcut difference
between oceanic environments, where net zooplankton excretion
accounts for 36-100% of N-requirements of primary producers, and
coastal (and upwelling) environments, where zooplankton excretion
only represents a small fraction (less than 20% and often less than
10%) of phytoplankton nitrogen uptake.

Direct measurements by a ^{15}N isotope dilution technique of

Table 8. Relative role of various mineral nitrogen sources in meeting the requirements of primary producers in different marine environments.

Environments	Period	N requirement for net primary production (mg-at N·m^{-2}·day)	(Net)* zooplankton excretion	(Micro) zooplankton excretion	Bacterial mineralization
			(% of phytoplanktonic N requirements)		
Coastal Areas					
Washington coast[a]	June–July	(4.6)**	3		101
Southern California Bight[b]	March	2.4	7.5		(91)
Saanich Inlet[b] (CEPEX enclosure)	September	16.6	11	(33)	
North Sea Southern Bight					
– Belgian coastal zone[c]	Annual budget	(5.2)	(20)		
– offshore zone[c]	Annual budget	(5.6)	(18)		
Open Ocean Areas					
Pacific off Oregon[a]	June–July	(1.6)	36		
Sub-tropical Pacific central gyre					
– station 1 (0–75 M)[d]	November	1.8	40		
– station 2 (0–75 M)[d]	November	3.0	110		

a Jawed 1973; b Harrison 1978; c Billen 1978; d Eppley 1973

* Separated by filtration through 180 – 100 μm mesh net according to the authors.

** Values in brackets are indirect estimations based on C measurements. The other data result from direct nitrogen flux measurements.

nitrogen remineralization in coastal areas have been reported by
Harrison (1978). By differential filtration, he showed that only
10% of the total ammonium regeneration activity was associated with
particles greater than 35 μm, 40% passed a 1 μm filter, and the
remaining 50% fell between 1-35 μm. Free bacteria are probably
responsible for ammonium production in the < 1 μm fraction and
attached bacteria for a part of the activity in the other fractions.
In the latter, however, other organisms could have a significant
role.

Microzooplankton (either ciliates or flagellates), which graze
on bacteria, are known as efficient mineralizers with high specific
ammonia excretion rates. Fenchel (1980), however, showed that
bacterivorous ciliates cannot play an important role as grazers at
the low bacterial densities normally encountered in marine environments. Available estimates (Harrison 1978) of NH_4 production by
microzooplankton indicates that they produce a maximum of 1/3 of the
total ammonia.

Phytoplankton ammonium excretion is probably of minor importance, although there is some evidence supporting the occurrence of
the process (Prochazkova et al. 1970; Schell 1974).

Most of the planktonic mineralization, at least in coastal
areas, is thus the result of bacterial activity. Direct mineralization of amino acids have been shown by Hollibaugh (1980) to account
for about 60% of the observed flux of ammonia production.

CONCLUSION

According to the preceding discussion, nitrogen cycling in the
photic layer can be viewed as represented in Fig. 11. Primary production requires mainly mineral nitrogen (or urea), part of which
comes from the subphotic layer or the benthos ("new production",
Dugdale and Goering 1967), and the other part comes from remineralization mechanisms in the photic layer ("regenerated production").
There are three fates of the phytoplanktonic material produced. One
fraction sinks to the subphotic layer or the benthos. A second
fraction is grazed by zooplankton, initiating the food chain leading
to fish. The third fraction (comprising all dissolved material
produced and part of the particulates produced) is available for
bacterial utilization. This involves two steps: (i) exoenzymatic
hydrolysis of the macromolecules which form the bulk of the organic
nitrogen; (ii) rapid uptake of the hydrolysis products (mainly amino
acids), the concentration of which are maintained at a constant
level. The organic nitrogen taken up is partly remineralized as
ammonium and partly incorporated into bacterial biomass, according
to the C/N ratio of the organic matter taken up.

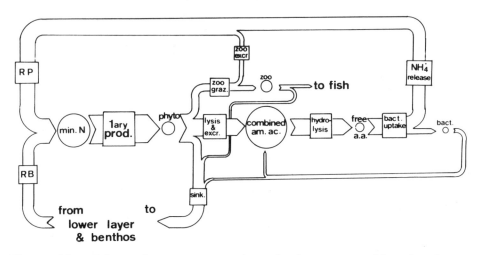

Figure 11.   Schematic representation of nitrogen cycling in the
upper layer of the ocean.   (RP is N regeneration in the photic layer
(corresponding to "regenerated production") RB is regeneration in
the lower layer and the benthos (corresponding to "new production")).

Table 9.   Estimation of planktonic nitrogen regeneration in the
           photic zone by water type of the world ocean.   Data from
           Eppley and Peterson (1979).

	Primary Production $(g\ C \cdot m^{-2} \cdot yr^{-1})$	% New Production (%)	Planktonic N Regeneration in the Photic Layer[+] $(g\ N \cdot m^{-2} \cdot yr^{-1})$
Oligotrophic waters from anticyclonic eddies	25.6	6	4.1
Transitional waters	51	13	7.7
Equatorial divergence and subpolar zones	73	18	10.5
Inshore waters	124	30	15.1
Neritic waters	365	46	34.6

[+]Calculated from the Redfield ratio for primary production.

Bearing in mind this picture of the nitrogen cycle in sea water, it is possible to examine the coherency of the various experimental measurements of nitrogen related microbiological activities in the world ocean. Eppley and Peterson (1979), summarizing the data available on new and regenerated production from ^{15}N measurements, proposed an estimate of global new primary production. From their data, planktonic nitrogen regeneration in the photic zone can be calculated by water types from the world ocean (Table 9).

Table 10. Measurements of the rate of utilization of amino acids by heterotrophic microbial communities in the photic layer of various marine environments.

Environment	Reference	Turnover time (hours)
Pamlico River estuary	Crawford et al. 1974	35
Scheldt estuary	Billen, unpublished	32
Tokyo Bay	Seki et al. 1975	13
Shimoda Bay	Seki et al. 1975	28
Belgian coastal zone	Joiris et al. 1982	66 - 50
Florida Strait	Williams 1975	24 - 50
Eastern English Channel	Billen, unpublished	100
Coastal Eastern Pacific	Williams 1975	60 - 2400
Pacific off California	Williams et al. 1976	40 - 70
Mediterranean	Willaims 1975	96 - 80
Kuroshio current	Seki et al. 1972	3100
Subarctic Pacific	Seki et al. 1972	3000
NW Pacific central waters	Seki et al. 1972	9000
NW Pacific central waters	Seki et al. 1974	4000

Another type of data, measurements of the annual mean turnover time of total free amino acids in different marine environments, is also available (Table 10). As discussed earlier, total free amino acid concentration in each of these water types is maintained near to steady state by the microbial population and does not vary a lot. From the data of Table 2, the concentration can be grossly estimated as 0.5 μg-at N·liter^{-1} in coastal and neritic waters and 0.05-0.025 μg-at N·liter^{-1} in oligotrophic oceanic areas. Combining these estimations with those of Table 10, it is possible to calculate the rate of amino acid utilization by water types in the ocean (Table 11).

The data in Tables 9 and 11 can be used for further calculations by taking a mean value of 0.75 for the nitrogen mineralization ratio of amino acids (Hollibaugh 1978) (although this ratio has been shown in Fig. 10 to depend on the C/N ratio of the organic substrates taken up by bacteria). The calculations clearly show that bacterial mineralization of amino acids is responsible for most of the planktonic nitrogen regeneration in the photic zone in the neritic and coastal zones, but that in oligotrophic waters zooplankton excretion must play the dominant role. This is in agreement with the data of Table 7 and confirms the idea of Joiris (1977) and Joiris et al. (1982) that a fundamental difference exists between planktonic food

Table 11. Amino acid utilization by bacterial communities by water type of the world ocean.

	Turnover Time (hr)	Concentration (μg-at N·liter^{-1})	Utilization Rate[+] (g N·m^{-2}·yr^{-1})
Oligotrophic waters from anticyclonic eddies	4000 – 9000	0.05	0.06 – 0.15
Transitional waters	1000 – 3000	0.25	1 – 3
Equatorial divergence and subpolar zones	3000	0.25	1
Inshore waters	100 – 150	0.5	20 – 30
Neritic waters	25 – 100	0.5	30 – 122

[+] Calculated by considering a photic zone of 50 m for neritic and inshore waters and of 100 m for oceanic waters.

chains in oceanic systems dominated by grazing and coastal systems
dominated by the activity of microorganisms.

The amino acid utilization data in Table 11 also represent amino
acid production through macromolecular organic nitrogen hydrolysis.
In neritic waters, where the mean concentration of total proteins and
peptides probably ranges between 2.5 and 5 µg-at N·liter^{-1} (Table 2),
this implies a first order rate constant of degradation of about
0.2-0.02 day^{-1}. This is in reasonable agreement with the range found
experimentally for fresh, phytoplankton-derived, organic nitrogen
(Table 5). In oligotrophic areas, on the other hand, the concentration of dissolved, combined amino acids is about 0.3-0.5 µg-at
N·liter^{-1} (Table 2), implying a first order degradation constant
smaller than 0.0015 or at least two order of magnitude lower than
the values found for phytoplankton-derived organic nitrogen. This
suggests that an important part of combined amino acids dissolved in
these waters consists of refractory compounds.

REFERENCES

Akagi, Y., and N. Taga. 1980. Uptake of D-Glucose and L-Proline by
    oligotrophic and heterotrophic marine bacteria. Can. J.
    Microbiol. 26: 454-459.
Alexander, M. 1977. Introduction to Soil Microbiology. 2nd
    edition. Wiley, New York. 467 pp.
Andrews, P., and P. J. leB. Williams. 1971. Heterotrophic utiliza-
    tion of dissolved organic compounds in the sea. III. Measure-
    ment of the oxidation rates and concentrations of glucose and
    amino acids in sea water. J. Mar. Biol. Assoc. UK 51: 111-125.
Armstrong, F. A. J., P. R. Williams, and J. D. H. Strickland. 1966.
    Photooxidation of organic matter in seawater by UV radiation:
    analytical and other applications. Nature 211: 481-483.
Banoub, N. W., and P. J. leB. Williams. 1973. Seasonal changes in
    the organic forms of carbon, nitrogen and phosphorus in sea
    water at E$_1$ in the English Channel during 1968. J. Mar. Biol.
    Assoc. UK 53: 697-703.
Barvenik, F. W., and S. C. Malloy. 1979. Kinetic patterns of
    microbial amino acid uptake and mineralization in marine
    waters. Estuarine Coastal Mar. Sci. 8: 241-250.
Bianchi, A. J. M. 1980. Distribution quantitative et qualitative
    des populations bacteriennes a l'interface eau-sediment. In
    "Biogeochimie de la matiere organique a l'interface eau-sediment
    marin" Colloques Internationaux du CNRS, No. 293. Paris.
Billen, G. 1975. Nitrification in the Scheldt estuary (Belgium and
    the Netherlands). Estuarine Coastal Mar. Sci. 3: 79-89.
Billen, G. 1976. A method for evaluating nitrifying activity in
    sediments by dark ^{14}C-bicarbonate incorporation. Water Res.
    10: 51-57.
Billen, G. 1978. A budget of nitrogen recycling in North Sea

sediments off the Belgian coast. Estuarine Coastal Mar. Sci.
   11: 279-290.
Billen, G., C. Joiris, J. Wijnant, and G. Gillain. 1980. Concentration and microbiological utilization of small organic molecules
   in the Scheldt estuary, the Belgian coastal zone of the North
   Sea and the English Channel. Estuarine Coastal Mar. Sci. 11:
   279-294.
Brown, K. D. 1971. Maintenance and exchange of the aromatic amino
   acid pool in E. coli. J. Bacteriol. 106: 70-81.
Brown, C. M., D. S. Mac-Donald-Brown, and S. O. Stanley. 1972.
   Inorganic nitrogen metabolism in marine bacteria: nitrogen
   assimilation in some marine pseudomonas. J. Mar. Biol. Assoc.
   UK 52: 793-804.
Butler, E. I., S. Knox, and M. I. Liddicoat. 1979. The relationship
   between inorganic and organic nutrients in sea water. J. Mar.
   Biol. Assoc. UK 59: 239-250.
Carlucci, A. F., and J. D. H. Strickland. 1968. The isolation,
   purification and some kinetics studies of marine nitrifying
   bacteria. J. Exp. Mar. Biol. Ecol 2: 156-166.
Carpenter, E. J., C. C. Remsen, and S. W. Watson. 1972. Utiliza-
   tion of urea by some marine phytoplankters. Limnol. Oceanogr.
   17: 165-169.
Christensen, D., and T. H. Blackburn. 1980. Turnover of tracer
   (^{14}C, ^{3}H labelled) alanine in inshore marine sediments. Mar.
   Biol. 58: 97-103.
Christison, S., and S. M. Martin. 1971. Isolation and preliminary
   characterization of an extracellular protease of Cytophaga sp.
   Can. J. Microbiol. 17: 1207-1216.
Conway, H. L. 1977. Interactions of inorganic nitrogen in the
   uptake and assimilation of marine phytoplankton. Mar. Biol.
   39: 221-232.
Corner, E. D. S., and B. S. Newell. 1967. On the nutrition and
   metabolism of zooplankton. IV. The forms of nitrogen excreted
   by Calanus. J. Mar. Biol. Assoc. UK 47: 113-120.
Corner, E. D. S., C. B. Cowey, and S. M. Marshall. 1965. On the
   nutrition and metabolism of zooplankton. III. Nitrogen
   excretion by Calanus. J. Mar. Biol. Assoc. UK 47: 113-120.
Crawford, C. C., J. E. Hobbie, and K. L. Webb. 1974. The
   utilization of dissolved free amino acids by estuarine
   microorganisms. Ecology 55: 551-553.
Dawson, R., and K. Gocke. 1978. Heterotrophic activity in
   comparison to the free amino acid concentrations in Baltic Sea
   water samples. Oceanol. Acta 1: 45-54.
Degens, E. T. 1970. Molecular nature of nitrogenous compounds in
   sea water and recent marine sediments, pp. 77-106. In: D. W.
   Hood [ed.], Organic Matter in Natural Waters. University of
   Alaska, Publ. No. 1.
Derenbach, J. B., and P. J. leB. Williams. 1974. Autotrophic and
   bacterial production. Fractionation of plankton populations by
   differential filtration of samples from the English Channel.
   Mar. Biol. 25: 263-269.

Dubois, E., and M. Grenson. 1979. Methylamine/ammonia uptake
    systems in Saccharomyces cerevisiae: multiplicity and
    regulation. Mol. Gen. Genet. 175: 67–76.
Dugdale, R. C., and J. Goering. 1967. Uptake of new and regenerated
    forms of nitrogen in primary productivity. Limnol. Oceanogr.
    12: 196–206.
Eppley, R. W., E. H. Renger, E. L. Venrick, and M. M. Mullin. 1973.
    A study of plankton dynamics and nutrient cycling in the central
    gyre of the North Pacific Ocean. Limnol. Oceanogr. 18: 534–551.
Eppley, R. W., J. N. Rogers, and J. J. McCarthy. 1969. Half-satura-
    tion constant for uptake of nitrate and ammonium by marine
    phytoplankton. Limnol. Oceanogr. 14: 912–920.
Eppley, R. W., and L. Solorzano. 1969. Nitrate reductase in marine
    phytoplankton. Limnol. Oceanogr. 14: 194–205.
Eppley, R. W., and B. J. Peterson. 1979. Particulate organic matter
    flux and planktonic new production in the deep ocean. Nature
    282: 677–680.
Falkowski, P. G. 1975. Nitrate uptake in marine phytoplankton:
    comparison of half-saturation constants from seven species.
    Limnol. Oceanogr. 20: 412–417.
Falkowski, P. G., and R. B. Rivkin. 1977. The role of glutamine
    synthetase in the incorporation of ammonium in Skeletonema
    costatum. J. Phycol. 12: 448–450.
Fein, J. E., and R. A. McLeod. 1975. Characterization of neutral
    amino acid transport in a marine pseudomonas. J. Bacteriol.
    124: 1170–1190.
Fenchel, T. 1980. Suspension feeding in ciliated protozoa: feeding
    rates and their ecological significance. Microb. Ecol. 6:
    13–25.
Fraga, F. 1966. Distribution of particulate and dissolved nitrogen
    in the Western Indian Ocean. Deep-Sea Res. 13: 413–426.
Gits, J., and M. Grenson. 1967. Multiplicity of the amino acid
    permeases in Saccharomyces cerevisiae. III. Evidence for a
    specific methionine transporting system. Biochim. Biophys.
    Acta 135: 507–516.
Glenn, A. R. 1976. Production of extracellular proteins by
    bacteria. Ann. Rev. Microbiol. 30: 41–62.
Golterman, H. L. 1972. The role of phytoplankton in detritus
    formation. Mem. Ist. Ital. Idrobiol. 29(suppl): 89–103.
Grenson, M. 1966. Multiplicity of the amino acid permeases in
    Saccharomyces cerevisiae. II. Evidence for a specific lysine
    transporting system. Biochim. Biophys. Acta 127: 339–346.
Grenson, M., M. Mousset, J. M. Wiame and J. Bechet. 1966.
    Multiplicity of the amino acid permeases in Saccharomyces
    cerevisiae. I. Evidence for a specific arginine transporting
    system. Biochim. Biophys. Acta 127: 325–338.
Grenson, M., C. Hov, and M. Crabeel. 1970. Multiplicity of the
    amino acid permeases in Saccharomyces cerevisiae. IV.
    Evidences for a general amino acid permease. J. Bacteriol.
    103: 770–777.

Gundersen, K., and C. W. Mountain. 1973. Oxygen utilization and pH change in the ocean resulting from biological nitrate formation. Deep-Sea Res. 20: 1083-1091.

Guardiola, J., M. Defelice, T. K. Lopotowski, and M. Iacarino. 1974. Multiplicity of isoleucine, leucine and valine transport systems in E. coli: K_{12}. J. Bacteriol. 117: 382-392.

Hagel, P., and J. W. A. Van Rijn Van Alkemade. 1973. Eutrophication of the North Sea. International Council for the Exploration of the Sea. CM 1973/L: 22. Plankton Committee.

Hama, T., and N. Handa. 1980. Molecular weight distribution and characterization of dissolved organic matter from lake waters. Arch. Hydrobiol. 90: 106-120.

Hamilton, R. D., and J. E. Preslan. Observations on heterotrophic activity in the eastern tropical Pacific. Limnol. Oceanogr. 15: 395-401.

Harrison, W. G. 1978. Experimental measurements of nitrogen remineralization in coastal waters. Limnol. Oceanogr. 23: 684-694.

Harvey, H. W. 1945. The Chemistry and Biology of Sea Water. Cambridge Univ. Press, Cambridge.

Hellebust, J. A. 1965. Excretion of some organic compounds by marine phytoplankton. Limnol. Oceanogr. 10: 192-206.

Hellebust, J. A. 1970. The uptake and utilization of organic substances by marine phytoplanktoners, pp. 225-256. In: D. W. Hood [ed.], Organic Matter in Natural Waters. Institute of Marine Science Occasional Publication No. 1.

Hellebust, J. A. 1974. Extracellular products, pp. 838-863. In: W. D. P. Stewart [ed.], Algal Physiology and Biochemistry. Blackwell Scientific Publications, Oxford.

Henricksen, K. 1980. Measurement of in situ rates of nitrification in sediments. Microb. Ecol. 6: 329-337.

Hobbie, J. E., C. C. Crawford, and K. L. Webb. 1968. Amino acid flux in an estuary. Science (NY) 159: 1463-1464.

Hollibaugh, J. T. 1978. Nitrogen regeneration during the degradation of several amino acids by plankton communities collected near Halifax, Nova Scotia, Canada. Mar. Biol. 45: 191-201.

Hollibaugh, J. T. 1979. Metabolic adaptation in natural bacterial populations supplemented with selected amino acids. Estuarine Coastal Mar. Sci. 9: 215-230.

Hollibaugh, J. T., A. B. Carruthers, J. A. Fuhrman, and F. Azam. 1980. Cycling of organic nitrogen in marine plankton communities studied in enclosed water column. Mar. Biol. 59: 15-21.

Holm-Hansen, O., J. D. H. Strickland, and P. M. Williams. 1966. A detailed analysis of biologicall important substances in a profile off Southern California. Limnol. Oceanogr. 11: 548-569.

Iturriaga, R., and H. G. Hoppe. 1977. Observations of heterotrophic activities on photoassimilated organic matter. Mar. Biol. 40: 101-108.

Jannasch, H. W. 1968. Growth characteristics of heterotrophic bacteria in sea water. J. Bacteriol. 95: 722-723.

Jassby, A. D., and C. R. Goldman. 1974. Loss rates from a lake phytoplankton community. Limnol. Oceanogr. 19: 618-627.

Jawed, M. 1969. Body nitrogen and nitrogenous excretion in Neomysis rayii (Murdoch) and Euphansia pacifica (Hansen). Limnol. Oceanogr. 14: 748-754.

Jawed, M. 1973. Ammonia excretion by zooplankton and its signifi- cance to primary productivity during summer. Mar. Biol. 23: 115-120.

Johannes, R. E. 1968. Nutrient regeneration in lakes and oceans. Adv. Microbiol. Sea 1: 203-212.

Johannes, R. E., and K. L. Webb. 1965. Release of dissolved amino acids by marine zooplankton. Science 150: 76-77.

Joiris, C. 1977. On the role of heterotrophic bacteria in marine ecosystems: some problems. Helgol. Wiss. Meeresunters. 30: 611-621.

Joiris, C., G. Billen, C. Lancelot, M. H. Daro, J. P. Mommaerts, J. H. Hecq, A. Bertels, M. Bossicart, and J. Nijs. 1982. A budget of carbon cycling in the Belgian coastal zone: relative roles of zooplankton, bacterioplankton and benthos in the utilization of primary production. Neth. J. Sea Res. 16: 260-275.

Jones, K., and W. D. P. Stewart. 1969. Nitrogen turnover in marine and brackish habitats. II. The production of extracellular N by Colothrix scopulorum. J. Mar. Biol. Assoc. UK 49: 475.

Juttner, F., and T. Matuschek. 1978. The release of low molecular weight compounds by the phytoplankton in an eutrophic lake. Water Res. 12: 251-255.

Kasper, H. F., and K. Wuhrmann. 1978. Kinetic parameters and relative turnovers of some important catabolic reactions in digesting sludge. Appl. Environ. Microbiol. 36: 1-7.

Kay, W. W., and A. F. Gronlund. 1969. Proline transport by Pseudomonas aeruginosa. Biochim. Biophys. Acta 193: 444-445.

Kay, W. W., and A. F. Gronlund. 1969. Amino acid transport in Pseudomonas aeruginosa. J. Bacteriol. 97: 273-281.

Kay, W. W., and A. F. Gronlund. 1971. Transport of aromatic amino acids by Pseudomonas aeruginosa. J. Bacteriol. 105: 1039-1046.

Khaylov, K. M., and Z. Z. Finenko. 1970. Organic macromolecular compounds dissolved in sea-water and their inclusion into food-chains, pp. 6-18. In: J. H. Steele [ed.], Marine Food Chains. Oliver and Boyd, Edinburgh.

Kim, J., and C. E. Zobell. 1974. Occurence and activities of cell-free enzymes in oceanic environments, pp. 368-385. In: R. R. Colwell and R. Y. Morita [eds.], Effects of the Ocean Environment on Microbial Activity. University Park Press, Baltimore.

Knowles, G., A. L. Downing, and M. J. Barrett. 1965. Determination of kinetics constants for nitrifying bacteria in mixed culture with the aid of a computer. J. Gen. Microbiol. 38: 263-278.

Lancelot, C. 1979. Gross excretion rates of natural marine phyto- plankton and heterotrophic uptake of excreted products in the southern North Sea, as determined by short-term kinetics. Mar. Ecol. Prog. Ser. 1: 179-186.

Lancelot, C. 1982. Etude écophysiologique du phytoplancton de la zone côtière belge. Thesis, University of Brussels, Faculty of Science.

Larsson, U., and A. Hagström. 1979. Phytoplankton exudate release as an energy source for the growth of pelagic bacteria. Mar. Biol. 52: 199-206.

Lee, C., and J. L. Bada. 1975. Amino acids in Equatorial Pacific Ocean water. Earth Planet. Sci. Letts. 26: 61-68.

Lee, C. L., and J. L. Bada. 1977. Dissolved amino acids in the Equatorial Pacific, the Sargasso Sea and Biscayne Bay. Limnol. Oceanogr. 22: 502-510.

Lewin, J., and J. A. Hellebust,. 1975. Heterotrophic nutrition of the marine pennate diatom Navicula pavillardi. Can. J. Microbiol. 21: 1335-1342.

Liebezeit, G., M Bölter, I. F. Brown and R. Dawson. 1980. Dissolved free amino acids and carbohydrates at pycnocline boundaries in the Sargasso Sea and related microbial activity. Oceanol. Acta 3: 357-362.

Liu, M. S., and J. A. Hellebust. 1974. Uptake of amino acids by the marine centric diatom Cyclotella cryptica. Can. J. Microbiol. 20: 1109-1118.

Martin, Y. P. 1980. Succession écologique de communautés bactériennes au cours de l'évolution d'un écosystéme phytoplanktonique marin expérimental. Oceanol. Acta 3: 293-300.

Martin, Y. P., and M. Bianchi. 1980. Structure, diversity, and catabolic potentialities of aerobic heterotrophic bacterial populations associated with continuous cultures of natural marine phytoplankton. Microb. Ecol. 5: 265-279.

Mayzaud, P., and J. L. M. Martin. 1975. Some aspects of the biochemical and mineral composition of marine plankton. J. Exp. Mar. Biol. Ecol. 17: 297-310.

Meybeck, M. 1982. Nutrient (C,N,P) transport by world rivers. Am. J. Sci. 282: 401-450.

McCarthy, J. J. 1972. The uptake of urea by natural populations of marine phytoplankton. Limnol. Oceanogr. 17: 738-745.

McIssaac, J. J., and R. C. Dugdale. 1969. The kinetics of nitrate and ammonium uptake by natural populations of marine phytoplankton. Deep-Sea Res. 16: 45-57.

Miflin, B. J., and P. J. Lea. 1977. Amino acid metabolism. Ann. Rev. Plant Physiol. 28: 299-329.

Mitamura, O., and Y. Saijo. 1975. Decomposition of urea associated with photosynthesis of phytoplankton in coastal waters. Mar. Biol. 30: 67-72.

Morris, I. 1974. Nitrogen assimilation and protein synthesis, pp. 583-609. In: W. D. P. Stewart [ed.], Algal Physiology and Biochemistry. University of California Press.

Nalewajko, C., T. G. Dunstall, and H. Shear. 1976. Kinetics of extracellular release in axenic algae and in mixed algal-bacterial cultures: significance in estimation of total (gross) phytoplankton excretion rates. J. Phycol. 12: 1-5.

Nihoul, J. C. J., and C. Boelen [eds.]. 1976. Niveaux de pollution du réseau hydrographique et de la zone cotière belges. Recueil des données. Tome C: Yser et côte belge. Projet Mer, rapport final. Ministére de la politique Scientifique. Bruxelles.

North, B. B. 1975. Primary amines in California coastal waters: utilization by phytoplankton. Limnol. Oceanogr. 20: 20-27.

North, B. B., and G. S. Stephens. 1967. Uptake and assimilation of amino acids by Platymonas. Biol. Bull 133: 391-400.

Ogura, N. 1974. Molecular weight fractionation of dissolved organic matter in coastal seawater by ultrafiltration. Mar. Biol. 24: 305-312.

Otsuki, A., and T. Hanya. 1972. Production of dissolved organic matter from dead green algae cells. I. Aerobic microbial decomposition. Limnol. Oceanogr. 17: 248-264.

Paasche, E., and S. Kristiansen. 1982. Nitrogen nutrition of the phytoplankton in the Oslofjord. Estuarine Coastal Shelf Sci. 14: 237-249.

Painter, H. A. 1970. A review of literature on inorganic nitrogen metabolism in microorganisms. Water Res. 4: 393-450.

Pichot, G. 1980. Simulation du cycle de l'azote à travers l'écosystème pélagique de la Baie Sud de la Mer du Nord. Thesis, Université de Liège, Faculté des Sciences.

Piperno, J. R., and D. L. Oxender. 1968. Amino acid transport systems in E. coli K_{12}. J. Biol. Chem. 243: 5914-5920.

Pillai, T. N. V., and Gauguly, A. K. 1970. Nucleic acids in the dissolved constituents of sea water. Curr. Sci. 39: 501-504.

Pocklington, R. 1971. Free amino acids dissolved in North Atlantic Ocean waters. Nature 230: 374.

Pollock, M. R. 1962. Exoenzymes, pp. 121-178. In: I. C. Gunsalus and R. Y. Stanier [eds.], The Bacteria, Vol. IV. Academic Press.

Prochazkova, L. P., P. Blazka, and M. Kralova. 1970. Chemical changes involving nitrogen metabolism in water and particulate matter during primary production experiments. Limnol. Oceanogr. 15: 797-807.

Rahmanian, M., D. R. Claus, and D. L. Oxender. 1973. Multiplicity of leucine transport systems in E. coli K_{12}. J. Bacteriol. 116: 1258-1266.

Reichardt, W. J., J. Overbeck, and L. Steubing. 1967. Free dissolved enzymes in lake water. Nature 216: 1345-1347.

Reisner, G. S., R. K. Gering, and J. F. Thompson. 1960. The metabolism of nitrate and ammonia by Chlorella. Plant Physiol. 35: 48-52.

Remsen, C. C. 1971. The distribution of urea in coastal and oceanic waters. Limnol. Oceanogr. 16: 732-740.

Remsen, C. C., E. J. Carpenter, and B. W. Schroeder. 1972. Competition for urea among estuarine microorganisms. Ecology 53: 921-926.

Riley, J. P., and D. A. Seagar. 1970. The seasonal variations of the free and combined dissolved amino acids in the Irish Sea. J. Mar. Biol. Assoc. UK 50: 713-720.

Rittenberg, S. C. 1963. Marine bacteriology and the problem of mineralization, pp. 48-60. In: C. H. Oppenheimer [ed.], Symposium on Marine Microbiology, Thomas, Springfield.

Rogers, H. J. 1961. The dissimilation of high molecular weight organic substances, pp. 261-318. In: I. C. Gunsalus and R. Y. Stanier [eds.], The Bacteria, Vol. II. Academic Press, New York.

Roon, R. T., H. L. Even, P. Dunlop and F. L. Larimore. 1975. Methylamine and ammonia transport in Saccharomyces cerevisiae. J. Bacteriol. 122: 502-509.

Roth, M. 1965. Fluorimetric assay of some peptidases, pp. 10-18. In: R. Ruyssen and E. L. Vandenriesche [eds.], Enzymes in Clinical Chemistry. Elsevier, Amsterdam.

Russel, J. B., and R. L. Baldwin. 1979. Comparison of substrate affinities among several rumen bacteria: a possible determinant of rumen bacterial competition. Appl. Environ. Microbiol. 37: 531-543.

Schell, D. M. 1974. Uptake and regeneration of free amino acids in marine waters of Southeast Alaska. Limnol. Oceanogr. 19: 260-270.

Seki, H., T. Nakai and H. Otoba. 1972. Regional differences on turnover rate of dissolved materials on the Pacific Ocean at summer of 1971. Arch. Hydrobiol. 71: 79-89.

Seki, H., T. Nakai, and H. Otobe. 1974. Turnover rate of dissolved materials in the Philippine Sea at winter of 1973. Arch. Hydrobiol. 73: 238-244.

Seki, H., K. S. Shortreed, and J. G. Stockner. 1980a. Turnover rate of dissolved organic materials in glacially-oligotrophic and dystrophic lakes in British Columbia, Canada. Arch. Hydrobiol. 90: 210-216.

Seki, H., T. Terada, and S. Ichimura. 1980b. Steady state oscillation of uptake kinetics by microorganisms in mesotrophic and eutrophic water masses. Arch. Hydrobiol. 88: 219-231.

Seki, H., Y. Yamaguchi, and S. Ichimura. 1975. Turnover rate of dissolved organic materials in a coastal region of Japan at summer stagnation period of 1974. Arch. Hydrobiol. 75: 297-305.

Sharp, J. H. 1977. Excretion of organic matter by marine phytoplankton: do healthy cells do it? Limnol. Oceanogr. 22: 381-399.

Sieburth, J. McN., and A. Jensen. 1969. Studies on algal substances in the sea. II. The formation of gelbstoff by exudates of phaeophyta. J. Exp. Mar. Biol. Ecol. 3: 275-289.

Siegel, A., and E. T. Degens. 1966. Concentration of dissolved amino acids from saline waters by liquid-exchange chromatography. Science 151: 1098-1101.

Somville, M. 1978. A method for the measurement of nitrification rates in water. Water Res. 12: 843-848.

Somville, M. 1980. Etude écophysiologique de l'activité bactériénne dans l'estuaire de l'Escaut. These, Université Libre de

Bruxelles, Faculté des Sciences.

Somville, M., and G. Billen. 1983. A method for determining exo-
proteolytic activity in natural waters. Limnol. Oceanogr. 28:
190-193.

Stanley, P. M., and J. T. Stanley. 1977. Acetate uptake by aquatic
bacterial communities measured by autoradiography and
filterable radioactivity. Limnol. Oceanogr. 22: 26-37.

Stewart, W. D. P. 1963. Liberation of extracellular nitrogen by
two nitrogen fixing blue-green algae. Nature 200: 1020-1021.

Strayer, R. F., and J. M. Tiedje. 1978. Kinetic parameters of the
conversion of methane precursors to methane in a hypereutrophic
lake sediment. Appl. Environ. Microbiol. 36: 330-340.

Tempest, D. W., J. L. Meers, and C. M. Brown. 1970. Synthesis of
glutamate in Aerobacter aerogines by a hitherto unknown route.
Biochem. J. 117: 405-407.

Thomas, W. H., E. H. Reuger, and A. N. Dodson. 1971. Near surface
organic nitrogen in the eastern tropical Pacific Ocean. Deep-
Sea Res. 18: 65-67.

Tuschall, J. R., and P. L. Brezonik. 1980. Characterization of
organic nitrogen in natural waters: its molecular size, protein
content and interaction with heavy metals. Limnol. Oceanogr.
25: 495-504.

Vaccaro, R. F., and H. W. Jannasch. 1966. Variations in uptake
kinetics for glucose by natural populations in sea water.
Limnol. Oceanogr. 12: 540-542.

Van Bennekom, A. J., W. W. C. Gieskes, and S. B. Tijssen. 1975.
Eutrophication of Dutch coastal waters. Proc. R. Soc. Lond. B
189: 359-374.

Von Brand, T. H., H. W. Rakestraw, and C. E. Renn. 1937. The
experimental decomposition and regeneration of nitrogeneous
organic matter in sea water. Biol. Bull. 72: 165-175.

Walker, D. J., and P. R. Monk. 1971. Fate of carbon passing
through the glucose pool of rumen digests. Appl. Microbiol.
22: 741-747.

Webb, K. L., and R. E. Johannes. 1969. Do marine crustaceans
release dissolved amino acids? Comp. Biochem. Physiol. 29:
875-878.

Wheeler, J. R. 1976. Fractionation by molecular weight of organic
substances in Georgia coastal waters. Limnol. Oceanogr. 21:
846-852.

Wheeler, P., B. North, M. Littler, and G. Stephens. 1977. Uptake
of glycine by natural phytoplankton communities. Limnol.
Oceanogr. 22: 900-910.

Whitledge, T. E., and R. C. Dugdale. 1972. Creatine in sea water.
Limnol. Oceanogr. 17: 309-314.

Wiebe, W. J., and D. F. Smith. 1977. Direct measurement of
dissolved organic carbon release by phytoplankton and
incorporation by microheterotrophs. Mar. Biol. 42: 213-223.

Williams, P. J. leB. 1970. Heterotrophic utilization of dissolved
organic compounds in the sea. I. Size distribution and

relationship between respiration and incorporation of growth substances. J. Mar. Biol. Assoc. UK 50: 859-870.

Williams, P. J. leB. 1973. The validity of the application of simple kinetic analysis to heterogeneous microbial population. Limnol. Oceanogr. 18: 159-164.

Williams, P. J. leB. 1975. Biological and chemical aspects of dissolved organic material in sea water, pp. 301-363. In: J. P. Riley and G. Skirrow [eds.], Chemical Oceanography. Academic Press, New York.

Williams, P. J. leB., T. Berman, and O. Holm-Hansen. 1976. Amino acid uptake and respiration by marine heterotrophis. Mar. Biol. 35: 41-47.

Williams, P. J. leB., and R. W. Gray. 1970. Heterotrophic utilization of dissolved organic compounds in the sea. II. Observations on the responses of heterotrophic marine populations to abrupt increases in amino acid concentration. J. Mar. Biol. Assoc. UK 50: 871-881.

Williams, P. M. 1971. The distribution and cycling of organic matter in the ocean, pp. 145-163. In: S. D. Faust and J. V. Hunter [eds.], Organic Compounds in Aquatic Environments. Marcel Dekker, New York.

Wollast, R., and G. Billen. 1981. The fate of terrestrial organic carbon in the coastal area, pp. 331-359. In: Flux of Organic Carbon by Rivers to the Oceans, NRC. Carbon dioxide effects research and assessment program. Report of a Workshop, Woods Hole, Massachusetts, Sept. 21-25, 1980. U. S. Dept. of Energy. CONF-8009140.

Wright, R. T. 1974. Mineralization of organic solutes by heterotrophic bacteria, pp. 546-565. In: R. R. Colwell and R. Y. Morita [eds.], Effects of the Ocean Environment on Microbial Activities. University Park Press, Baltimore.

Wright, R. T., and J. E. Hobbie. 1966. Use of glucose and acetate by bacteria and algae in aquatic ecosystems. Ecology 47: 447-464.

Wuhrmann, K. 1968. Adaptationen bei Gesellschaften von Mikroorganismen in Wasser. Bibl. Microbiol. 4: 52-64.

A REVIEW OF MEASUREMENTS OF RESPIRATION RATES

OF MARINE PLANKTON POPULATIONS

P. J. leB. Williams

Department of Oceanography
Southampton University
Southampton SO9 5NH, England

INTRODUCTION

The biological cycle in the oceans may be viewed to be comprised of two fundamental processes: photosynthesis and respiration. The introduction by Steemann Nielsen of the ^{14}C-technique for measuring planktonic photosynthesis has enabled this particular process to be extensively and easily studied. As a consequence there is now a considerable body of data available of measurements of plankton photosynthesis. This contrasts with respiration for which there are still remarkably few direct measurements. Although the processes of respiration and photosynthesis may be regarded to be in balance and essentially equal on the oceanic scale, food chain processes cause them to be separated both in space and time. In contrast to photosynthesis, planktonic respiration may be expected to occur throughout the water column. Thus in any one situation, respiration cannot be taken to be simply equal to photosynthesis but needs to be measured independently. Furthermore respiration is not restricted to a single group of organisms but common to all. Thus the study of respiration is inherently more complex and extensive than photosynthesis from both the trophodynamic and geographical points of view. The distribution of respiration between the major planktonic groups (net zooplankton, microzooplankton, heterotrophic microorganisms, algae) will give insight into the trophic structure of the planktonic community. The vertical distribution of respiration in the water column can given accounts of the export to and fate of the products of phytoplankton photosynthesis in the deeper parts of the ocean.

The need to determine planktonic respiration has been long recognized and a variety of approaches have been evolved over the years. Essentially three general approaches may be recognized:

357

1)   determination of the spatial changes in concentration of a
      reactant (e.g., oxygen) or a product (e.g., carbon dioxide)
      of the respiratory process;
2)   measurement of a biochemical rate process associated with
      respiration such as the activity of the electron transport
      system (ETS);
3)   calculations from biomass, using established or presumed
      size-specific activity relationships.

Ammonia and phosphate, essential and limiting nutrients to the
algae, are produced as a consequence of respiration and in principle
can be used to infer respiration rates. In the present account they
will not be considered in any detail because determination of nitrogen and phosphate mineralization rates are specialized and major
topics in themselves. Harrison (1981) has recently produced a review
of nutrient regeneration.

MEASUREMENTS OF RESPIRATION RATE

The rate measurements will be considered under three headings:
1) oxygen consumption; 2) carbon dioxide production; and 3) ETS
measurements.

Calculations from biomass are intimately tied up with the
distribution of activity within the planktonic community and will be
considered along with this topic.

Measurement of Oxygen Consumption

Methodological Background. In common with other planktonic
processes that give rise to differences in concentration, the rate
of oxygen consumption may be followed either by in situ changes in
the environment (i.e., diel, seasonal or longer changes), or in
incubated samples. Incubating samples in glass bottles has the
advantage that a regime may be imposed upon them (e.g., they may be
incubated in the dark or light); outwardly the interpretation of the
results is therefore simpler. There has however been a longstanding
concern that maintaining a sample in a small bottle induces
artifacts. This may be due to the walls of the container; or the
small isolated volume may modify grazing by herbivores or exclude
them. Some aspects of the so-called "bottle effect" have been
discussed in an number of papers (see e.g., Sheldon and Sutcliffe
1978; Fuhrman and Azam 1980; Williams 1981a; Packard and Williams
1981). The alternative, i.e., following changes in the environment,
is not without its difficulties; most serious of which are the
inaccuracies in the data introduced by advective processes. In
addition there is exchange of carbon dioxide and oxygen with the
atmosphere. To some extent it has been possible to effectively
eliminate horizontal advective problems and to a lesser degree,

vertical ones, by the use of large enclosed ecosystems (see e.g., Steele et al. 1977). These, however, are generally expensive to deploy and presently have been mainly restricted to the study of inshore water columns. Both in the case of carbon dioxide and oxygen, the vertical variations in concentration can give information on the rates of respiration if the dynamics of the environment are known well enough. Whereas the potential of this approach is recognized, it raises mainly problems of a physical oceanographic nature and a discussion of these would be outside the competence of the present author but is dealt with by Jenkins in an accompanying article. Thus this chapter will mainly consider changes in incubated samples and observation of diel changes.

Away from oligotrophic areas, one may expect diel changes in oxygen concentration, due to plankton metabolism, of 50-250 μg $O_2 \cdot dm^{-3} \cdot day^{-1}$ (see e.g., Table 1). Assuming a standing oxygen concentration of 1000 μg $O_2 \cdot dm^{-3}$ this will amount to changes of 0.5-2.5% per day. Conventional oxygen electrodes presently have an accuracy of between 1-5% and thus can only be used in very favorable circumstances. The classic Winkler technique, with current refinements for end point detection, can give a precision of 0.1% to 0.03% (see Bryan et al. 1976; Williams and Jenkinson 1982).

Oxygen Consumption In Vitro. This approach essentially dates back to the work of Gaarder and Gran (1927), who used Winkler technique to determine changes in oxygen concentration in samples incubated in glass bottles. Dissolved oxygen flux measurements are somewhat unique, in that the basic methodology has not changed significantly since the early measurements. Thus, with certain provisos, all the historical data are potentially valid and compatible with contemporary observations. The oxygen technique was used by G. A. Riley in his pioneering work on plankton production. Although he was principally interested in photosynthesis, respiration data was obtained as a matter of course. His work was criticized by Steemann Nielsen (1952) for the long incubation times employed (up to five days). Steemann Nielsen's arguments were very persuasive, and as a consequence the early work of Riley appears to have been largely discounted. With a renewed interest in oxygen as a measurable parameter of plankton metabolism, the criticism of Riley's work needs reconsidering. Most of the objections center around the incubation of samples in bottles, dark bottles in the case of respiration. Bottle artifacts are complex and far from resolved and any conclusions regarding the accuracy, in the strict sense, of the results of dark bottle incubations presently are subjective. It is the author's prejudice and experience that changes in the rate of oxygen consumption, resulting from incubating samples for long periods (i.e., a day or a couple of days), are more likely to result in an underestimate rather than an overestimate of the instantaneous rate at the time of collection. Thus I would argue that until evidence arises to the contrary, Riley's early respiration data

Table 1. Measured rates of oxygen consumption (all rates as µg $O_2 \cdot dm^{-3} \cdot day^{-1}$).

Region	Depth Range Studied	Respiration Rate Mean	Respiration Rate Range	Number of Observations	Reference
Coastal, Shelf and Slope					
N.W. Atlantic	0-75	101	8-248	4	Pomeroy and Johannes 1968
	100-500	13	11-15		
N.W. Atlantic	1-25	53	48-74	6	Packard and Williams 1981
N.W. Scotland (Loch Ewe)	0-15	189	17-647	43	Williams, unpublished
Saanich Inlet	0-15	241	49-491	13	Williams, unpublished
Georges Bank	Surface	185	6-288	6	Riley 1941
Gulf of Maine and Georges Bank	1-10	132	3-190	10	Packard and Williams 1981
North Sea (S. Bight)	5-20*	c.200	115-270	4	Tijssen and Eijgenraam 1980
Oceanic Areas					
N.W. Atlantic	Surface	180	50-320	4	Riley 1939

Location	Depth (m)		Range	N	Reference
N.W. Atlantic	0–40	12	3–30	6	Hobbie et al. 1972
	100–500	0.8	0.1–2.4	6	
Tropical N. Atlantic	0–25*	320	–	4	Tijssen 1979
N.W. Caribbean	Mixed Layer*	213	139–344	4	Johnson et al. 1981
Antarctic	0–100	4.5	0–16	47	Pomeroy et al., pers. comm.
	200–800	0.42	0–2.2	9	
	2000–2500	0.26	0–2.1	5	
N.E. Pacific	0–75	44	38–50	2	Pomeroy and Johannes 1968
	100–500	8	5–11	3	
Gulf Stream	0–75	14	3–42	17	Pomeroy and Johannes 1968
	100–500	5.4	1.6–8	13	
	500–800	3	0.8–5	4	
Sargasso Sea	Surface	118	40–230	13	Riley 1939
Sargasso Sea	20–75	90	45–230	4	Williams and Jenkinson 1982

*Data derived from diel studies

should not be rejected out of hand but taken as a potential underestimate of plankton respiration. It is clear however that changes
to the microflora are occurring in samples incubated in glass bottles
(Eppley et al. 1973).

After a gap of almost three decades, interest was renewed in
oxygen flux measurements and ways of avoiding the long incubation
employed by the early workers were actively sought. Pomeroy and
Johannes (1966, 1968) concentrated the plankton to increase the
activity per unit volume and were able to determine respiration
rates over short periods using an oxygen electrode. The rates are
probably minimal ones, because it was subsequently shown (e.g.,
Holm-Hansen et al. 1970; de Souza Lima and Williams 1978) that
losses of activity occurred during the concentration step. An
alternative solution has been to increase the precision of the
oxygen determination by using instrumental methods to establish the
end point of the Winkler titration (see Talling 1973; Carritt and
Carpenter 1966; Bryan et al. 1976). Subsequently, automation of the
whole titration has been achieved by microprocessor control (e.g.,
Williams and Jenkinson 1982). These improvements in the Winkler
method have enabled measurements to be made in most surface waters,
with incubation times of a day to a couple of hours. A summary of
results of measurements of oxygen consumption of coastal and oceanic
waters is given in Table 1.

Measurements of In Situ Oxygen Change. Concerns, justified or
otherwise, over the accuracy of in vitro measurements of photosynthesis, led to studies of in situ oxygen changes. These type of
measurements have been used extensively and successfully in freshwater studies. Until recently in marine systems, in situ work was
mainly restricted to interpretation of deep oxygen profiles (e.g.,
Riley 1951; Munk 1966; Kuo and Veronis 1970; Wytkri 1962) or long
term, i.e., annual oxygen changes (e.g., Cooper 1933; Shulenberger
and Reid 1981). Only in the last few years have studies of diel
changes in oxygen concentration been attempted. The difficulties of
making these measurements should not be overlooked; the diel changes
may be very small, often much less than 1% and are easily blurred by
horizontal variability and advection.

The results of two studies have reported by Dutch Texel group
(Tijssen 1979; Tijssen and Eijgenraam 1980) and are illustrated in
Figure 1 and rates abstracted from them included in Table 1. Those
obtained for the tropical station would be regarded to be high, at
least as judged by the observed photosynthetic rates. Johnson et al.
(1981) made diel studies of up to 48 hours in duration at a drogue
station in the N.W. Caribbean and they also observed rates of oxygen
metabolism higher than one would conventionally expect for such an
area (see Figure 2 and Table 1). In Figure 3, data are presented
from studies in a 100 m^3 floating enclosure at Loch Ewe (details
of the environmental enclosure is given in Gamble et al. 1977).

The experiments were designed to test in vitro incubation procedures, by comparing the results obtained from them with observed in situ changes. In general, the agreement between the results of in situ and in vitro measurements of photosynthetically produced changes in oxygen concentration are very satisfactory. However there do appear to be some signs of a discrepancy between the measured in vitro and

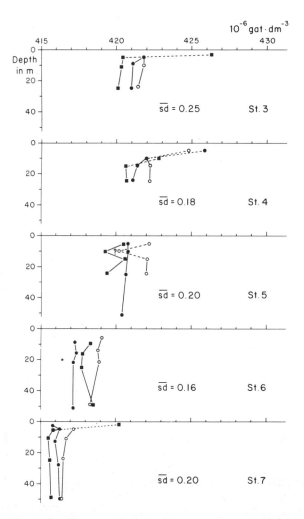

Figure 1a. Diurnal observations of oxygen concentrations in the tropical N. Atlantic Ocean. Oxygen concentrations versus depth in the mixed layer at 5 stations at 20° N, 18 November to 1 December 1978: first day morning cast, 0-4 hours after sunrise (●), first day night cast, 1-2.5 hours after sunset (○), and second day morning cast, 0-2 hours after sunrise (■). Each point represents the mean value obtained from 4 samples (from Tijssen 1979).

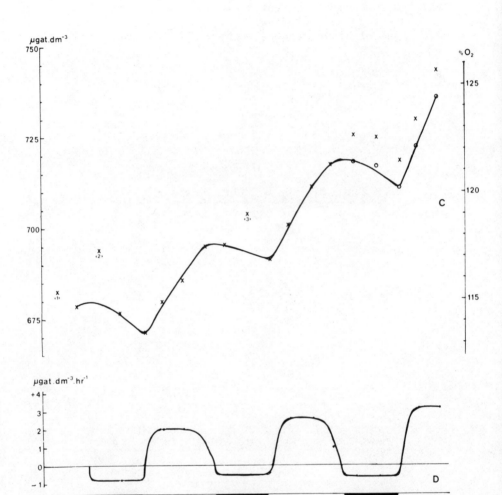

Figure 1b. Diurnal observations of concentrations (C) and rates of
change (D) of oxygen in April in the Southern Bight of the North
Sea. x - original values, o = shifted values (from Tijssen and
Eijgenraam 1980).

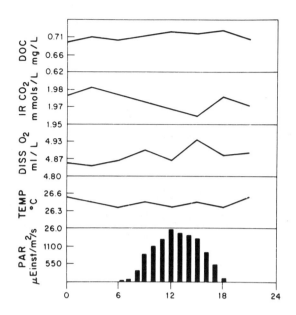

Figure 2.   Composite diel cycle for the N.W. Caribbean Sea.   IR CO$_2$: infrared determined ΣCO_2; PAR: photosynthetic active radiation (from Johnson et al. 1981).

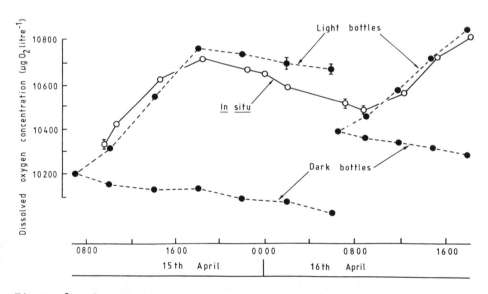

Figure 3.   Oxygen concentrations measured by in vitro incubations and in situ observation.   Data from DAFS, Loch Ewe research site, N.W. Scotland.   Experiment 15–16 April 1981, sample from 5 M.   (o) in situ measurements; (●) in vivo measurements (from Davies and Williams, in press).

in situ changes due to respiration. Where differences occur, greater
rates are observed in situ than in vitro, thus the traditional "wall
growth" effect cannot be offered as an explanation. However, it
would be premature to conclude from this limited set of observations
that containment effects were producing inaccuracies in in vitro
measurements of respiration. Parsons et al. (1981) published a
study of medium-term changes in oxygen concentration in an enclosure
driven into heterotrophic metabolism by "feeding" with an organic
carbon source (glucose). In that experiment (see Fig. 4) generally
good agreement was obtained between in situ oxygen changes and in
vitro rates measured with 24 hour incubations.

   Summary of Determined Rates of Respiratory Oxygen Consumption.
Table 1 represents a collation of plankton respiration rates, derived
from observation of oxygen concentration change over time-scales of a
fraction of a day to a few days. Most of the data derive from bottle
incubations; the small amount originating from diel observations has
been identified in Table 1. The data of Pomeroy and Johannes (1968)
and Hobbie et al. (1972) stand out as being low; it has been pointed
out earlier in the text that the technique of concentration of the
plankton by reverse filtration is known to give rise to underestimates of respiration. In oceanic water, higher rates were obtained
in the two sets of observations derived from in situ diel studies
than the in vitro measurements (see Table 1). However in view of

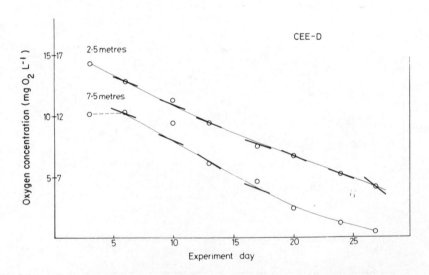

Figure 4. Oxygen decrease due to respiration: comparison of
observations of in situ dissolved oxygen concentration over 27 days
with decreases derived from 18 hour in vitro incubations (details in
Parsons et al. 1981). Continuous line follows in vivo observations
(open circles). Thick lines represent measured in vitro rates.

the general spread of the data and its paucity, it would seem hasty
to conclude any systematic difference between the in situ and in
vitro determined rates; the matter clearly needs more extensive
study.

Measurements of Carbon Dioxide Production

Methodological Considerations. The normal rates of plankton
metabolism in coastal and offshore waters give rise to daily carbon
changes of 1 μg $C \cdot dm^{-3}$ or less per day in oligotrophic systems to
perhaps as high as 250 μg $C \cdot dm^{-3}$ per day in euphotic environments.
The ambient carbon dioxide concentrations in the seas is in the
vicinity of 24,000 μg $C \cdot dm^{-3}$; thus at maximum ΣCO_2 changes in the
order of 1% may be expected. This is close to the precision with
which "direct" determinations of ΣCO_2 may be made by alkalinity
titration or by published CO_2-stripping techniques. Changes in ΣCO_2
will give rise to changes in pCO_2; the fractional change in pCO_2,
according to a relationship derived by Bolin and Eriksson (1959), is
12.5-fold greater than the fractional change in ΣCO_2, thus 10% or so
changes per day in pCO_2 may be anticipated. According to Wong et
al. (1975), pCO_2 changes may be measured with a precision of about
0.3%. Changes in ΣCO_2 also give rise to changes in pH. At the pH
of seawater (i.e., 8.2), it may be calculated that a ΣCO_2 change of
250 μg CO_2-$C \cdot dm^{-3}$ would give rise to a pH shift of about 0.05 of a
pH unit. With care, pH measurements can probably be made with a
precision of 0.001, thus this approach potentially has sufficient
precision to measure changes in regions of high productivity. The
article by Wong et al. (1975) contains a useful review of the
analytical accuracy of the measurable inorganic carbon parameters in
seawater.

The biological oceanographer has been inclined in the past to
shy away from precise chemical methods and instead to exploit the
sensitivity of radioisotope procedures. Two approaches have been
used to study plankton respiration based on radiocarbon methodology.
The first is an extension of the $^{14}CO_2$ productivity method; the
procedure relies upon the determination of the decrease of photoassimilated radiocarbon during a dark incubation which follows a
light period. This technique will measure processes mainly associated with algal respiration. The second radiochemical approach
concerns itself with microheterotrophic respiration and measures the
release of $^{14}CO_2$ from added ^{14}C-labelled organic compounds.

Measurements of ΣCO_2, pH and pCO_2 Changes. This general type
of approach has had more application in limnology: many lakes are
less well buffered with CO_2 than seawater, enabling the ΣCO_2 changes
to be more readily determined. The classical study in a marine
system was made by Cooper (1933). He determined the seasonal
changes in hydrogen ion concentration using pH indicator dyes.
The pH changes were used to calculate pCO_2 and ΣCO_2 variations and

to derive a figure for the productivity of the area. This data is
given in Table 2, along with productivity estimates derived from
changes in nitrate, phosphate and oxygen. The data are remarkable
considering the methods available at the time and speak highly for
the analytical skills of Cooper and his contemporaries.

Weichart (1980) recently published a study of diel pH changes
from which he calculated changes in ΣCO_2 (see Fig. 5). The diurnal
fluctuations due to the daytime net photosynthesis and nighttime
respiration are clearly seen. The mean net carbon production rate
over a 12-day period during the spring bloom was calculated to be
about 2 g $C \cdot m^{-2} \cdot day^{-1}$. Similar rates were calculated from changes
in inorganic phosphate concentrations. These would be regarded to
be high photosynthetic rates as they stand, but of course are only
the net photosynthetic rates. The gross photosynthetic rates and
the rates of respiration, estimated from the diurnal curves, were
often in excess of 10 g $C \cdot m^{-2} \cdot day^{-1}$ over the upper 50 meters. These
rates are not too far from the accepted maximum for aquatic environments (see e.g., Ryther 1959) and the rates in the near surface
waters appear to be equivalent to 500 μg $C \cdot dm^{-3} \cdot day^{-1}$, which would
be regarded to be high by most criteria.

Codispoti et al. (1982) observed a pCO_2 fall, due to a vernal
bloom in the Bering Sea, from 340 to 150 ppm over a seven week period
(April-late May). The ΣCO_2 fall over the first month, in the top
50 m, was in the region of 100-200 $\mu moles \cdot dm^{-3}$ which is equivalent
to about 40-80 μg $C \cdot dm^{-3} \cdot day^{-1}$, i.e., a moderate net photosynthetic
rate. The rates calculated from the pCO_2 changes are similar.

Table 2. Determined summer to winter concentration changes in
production related parameters (Cooper 1933).

	Winter-Summer Change ($\mu g \cdot dm^{-3}$)	Calculated Net Plankton (10^6 g wet weight $\cdot km^{-2}$)	Production (g $C \cdot m^{-2}$)
ΣCO_2	1400	1730	104
PO_4	5.8	1260	76
NO_3	88	1270	76
O_2	2250	1000	60

* Conversion from wet weight to carbon weight using Cooper's value
  of 6%.

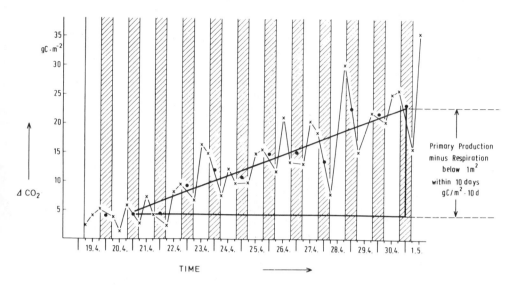

Figure 5.   Change of ΣCO_2 quantity (in g $C \cdot m^{-2}$) between the surface
and the depth below which no temporal pH change could be observed
from the beginning of the phytoplankton bloom. Photosynthetic and
respiratory changes were calculated from measurements of diurnal
changes in pH. The observations were made during the North Sea
FLEX experiment (from Weichart 1980).

These data give <u>net</u> production rates, the rates of respiration and
as a consequence <u>gross</u> photosynthesis are not known.

        Johnson et al. (1981) determined diurnal oxygen and ΣCO_2 changes
at a drogue station in the N.W. Caribbean. The observed ΣCO_2 changes
were calculated to be equivalent to respiration rates of 205 to 411
μg $C \cdot dm^{-3} \cdot day^{-1}$, the mean (n=5) value being 301 μg $C \cdot dm^{-3} \cdot day^{-1}$.
These rates are high and were recognized to be inconsistant with and
about four times greater than the carbon flux, implied from the
observed changes in oxygen concentration (oxygen data given in Table
1). Figure 2 reproduces their composite diurnal curves. In their
paper Johnson et al. (1981) present a useful summary of published ΔO_2
and ΔCO_2 studies and this is reproduced in part and with additions
in Table 3.

        <u>Radiochemical Procedures: Autotrophic Respiratory CO_2</u>
<u>Production</u>. Inevitably the question arises over the nature of the
population responsible for the observed respiration; the magnitude of
algal respiration (both light dependent and independent respiration)
has continued to be a matter of uncertainty. The issue arose during
the debate over the interpretation of the ^{14}C technique for measuring
primary production. During this period Steemann Nielsen introduced a

Table 3.   An annotated summary of some studies using O_2 and CO_2
           concentration changes to monitor ecosystem metabolism
           (abstracted from Johnson et al. 1981, with additions).

Location	Methods	Remarks	Reference
Salt and fresh waters	Direct or free water measurements, compared to bottle methods	In mixed communities, bottle techniques under-estimate metabolism	Odum & Hoskin 1958
Marine embayments, Texas Gulf Coast	In situ variation in O_2 and calculated ΣCO_2 changes	CO_2 uptake greatly exceeded O_2 production PQ's consistently <1.0	Park et al. 1958
Diverse waters	Empirical relationship between pH and ΣCO_2 changes	Useful at least for limited temperature and pH ranges	Beyers & Odum 1959
Strait of Georgia	Variation in O_2 and CO_2 followed over 9 days in a floating plastic sphere. ΣCO_2 determined directly by a modified Van Slyke method.	High rates of production, good agreement between CO_2 uptake and production; however, metabolism esti-mates from these data exceeded those derived from ^{14}C uptake measure-ments (PQ >1.0)	Antia et al. 1963
North Atlantic coastal waters	Seasonal changes in pCO_2 determined by IR analyser	Seasonal changes in pCO_2 converted to CO_2 uptake or production to estimate net productivity. CO_2 flux not always correlated with chlorophyll	Teal & Kanwisher 1966
Eniwetok Atoll	In situ variation in pH and an estimation of ΔpCO_2 from pH data	Significant changes in the surface ocean waters studied	Schmalz & Swanson 1969
Tropical Atlantic	Diurnal in situ variation in O_2	O_2 changes yielded a 4-fold increase over previous estimates of metabolism in open ocean waters	Ivanenkov et al. 1972
N.E. Pacific	GEOSECS data used to correlate O_2 and ΣCO_2 changes with depth	Good agreement between the two variables below 250 m, above ΣCO_2 deficit greatly exceeded O_2 excess	Ben-Yaakov 1972
Eniwetok Atoll	Variation in ΣCO_2 calculated from pH and alkalinity changes observed in waters passing over the reef flat	CO_2 system used to monitor a highly productive marine system	Smith 1973
Eniwetok Atoll	O_2 variation compared to the ΣCO_2 changes cited above (Smith 1973)	Good correlation between the two variables. Production estimated from O_2 changes slightly lower	Smith & Marsh 1973
Stuart Channel (Canadian Gulf Islands)	Followed monthly variation in O_2 and calculated ΣCO_2	Partitioned the observed changes into biological and atmospheric exchange components	Johnson et al. 1979
Atlantic Ocean and North Sea	Diurnal O_2 changes	Observed production of O_2 indicated significantly higher primary productiv-ity than previous ^{14}C studies of tropical waters	Tijssen 1979
North Sea	In situ changes in $\overline{\Sigma CO_2}$ derived from pH measurements	Good concurrence between observed PO_4 and calculated CO_2 changes. Net production rates high (c. 2 g $C \cdot m^{-2} \cdot day^{-1}$), gross production rates very high (c. 500 μg $C \cdot dm^{-3} \cdot day^{-1}$)	Weichart 1980
Coastal areas	In situ and in vitro O_2 changes and ^{14}C fixation measurements	Good agreement between in situ and in vitro O_2 changes. Good concurrence between in vitro ΔO_2 and ^{14}C fixation rates	Davies & Williams, in press

radiochemical approach to measure algal respiration (Steemann Nielsen 1955). The algae were labelled with ^{14}C during a light period and then respiration was determined as the decrease in ^{14}C in the cells during the subsequent dark period. Originally the method was used with cultures but subsequently it has been applied to natural plankton populations (Eppley and Sharp 1975; Smith 1977; Smith et al. 1977). This technique indicates that on occasions (see Table 4) a substantial proportion of algal primary production was found to be respired during the dark period.

It was recognized that there is a possibility that the measurements could include a component of bacterial as well as algal respiration. Organic material transferred from the algae to the bacteria, via the extracellular dissolved state, and then respired would also be determined. Various workers (Derenbach and Williams 1974; Smith et al. 1977; Wiebe and Smith 1977; Iturriaga and Hoppe 1977; Larsson and Hagström 1979; Lancelot 1979) have attempted to determine this process. One tentative conclusion (Williams 1981b) is that the rapid transfer of material from the algae to bacteria could amount to at least one-third of algal production.

Radiochemical Procedures: Heterotrophic Respiratory CO_2 Production. The question of the scale of heterotrophic processes in the plankton has been a matter of debate for some decades and still profoundly different views may be encountered in the literature (see Williams 1981a,b). The need to obtain some indication of the

Table 4. Measurements of algal respiration determined from $^{14}CO_2$ dark release: integrated values over euphotic zone.

Location	Respiration as % of 24 Hour $^{14}CO_2$ Fixation	Respiration Rate (mg $C \cdot m^{-2} \cdot day^{-1}$)	Number of Stations	Reference
Atlantic (off N.W. Africa)	13	262	38	Smith 1977
Pacific (off Peru)	74.6	1450	14	Smith 1977
Pacific (off Peru)	7.8	211	86	Smith, unpublished
Central N. Pacific	65	–	15	Eppley and Sharp 1975

magnitude of microbial processes led Parsons and Strickland (1961)
to introduce a radiochemical procedure employing added ^{14}C-labelled
organic compounds. By analogy with the [14]CO_2 technique they measured
uptake of the isotope. Subsequently some workers chose to determine,
alternatively or additionally, [14]CO_2 production from the added
organics (Kadota et al. 1966; Williams and Askew 1968; Hobbie and
Crawford 1969). Essentially answers to two quesions have been sought
with these techniques: could the rate of [14]CO_2 production from the
added organics be used to provide useful quantitative information on
bacterial respiration, secondly could simultaneous measurements of
[14]C uptake and [14]CO_2 production be used to indicate the conversion
efficiency of natural heterotrophic populations. The evidence from
the ^{14}C-methodology (see Williams 1970, 1973; Crawford et al. 1974)
implied, what at the time was rather surprising, high utilization
efficiencies (see Table 5). More recently alternative approaches
towards the determination of the growth efficiency of marine bacteria

Table 5. Extent of respiration of added ^{14}C-labelled substrates by
          natural marine populations.

Region	Substrate	% Respiration	Reference
English Channel, N.E. Atlantic, and Mediterranean	Glucose Amino acid mixture	24–49 13–34	Williams 1970
U.S. Estuary	Glucose Various amino acids	8–17 3–57	Crawford et al. 1974
North Sea	Glucose* Amino acid mixture*	29 25	Gocke 1976
Bahamas	Glucose Amino acid mixture	3.7–38 8–49	Williams and Yentsch 1976
California coastal water	Various amino acids	10–40	Williams et al. 1976
Baltic	Glucose Amino acid mixture	13–24 19–31	Dawson and Gocke 1978
North Sea	Glucose Various amino acids	50–80 50–84	Billen et al. 1980

* Average value for mixture

have challenged the conclusions drawn from radiochemical techniques
(see e.g., Newell et al. 1981).

In order to determine the mass of CO_2 produced, it is necessary
to measure the original concentration of the organic substrate under
study. This was, and still is to some extent, technically difficult
and has probably been a major factor limiting the usefulness and
application of the ^{14}C technique to study heterotrophy. A summary
of the results is given in Table 6. The data are mainly restricted
to amino acids; the number of data available are surprisingly few,
due in part to frequent practice of reporting the sum of uptake and
respiration rather than the individual measurement observations.

The full exploitation of the approach awaits a better
understanding of the type of compounds used by the heterotrophic
microflora and improvements in analytical technique to the point
that the analysis of seawater for individual organic comounds is a
semi-routine procedure.

Measurement of Electron Transport System Activity

This method (see Packard 1971; Packard et al. 1971) came out of
the general quest in the 1960-1970 era for methods to measure, or to
simply indicate the scale of, microbial activity in particular and
plankton respiration in general. The principle behind the technique
is reasonably straightforward; interpretation of the data, on the
other hand, may not be.

Methodological Background. The aim is to determine the
respiratory potential by assaying the activity of the terminal
electron transport system (ETS). To this end the plankton is
collected by filtration and a cell-free extract prepared by
homogenization. The ETS activity is then assayed by providing
electron donors (NADH, NADPH, succinate) and an artificial electron
acceptor (2-(p-iodophenyl)-3-(p-nitrophenyl)-5-phenyltetrazolium).
The extent of reduction of the tetrazolium compound is determined
colorimetrically. The principles and the details of the methodology
have been discussed by Christensen and Packard (1979). Problems are
encountered at two levels: first the electron transport system has
to be extracted from the cell completely with no loss of activity
and assayed under optimum conditions. This is a familiar problem in
enzymology for which potential solutions exist. The second
uncertainty concerns the interpretation of the determined enzyme
activity in terms of plankton respiration rates. In the intact
cell, respiration is controlled by the rate of phosphorylation; this
control is removed once the enzyme complex is extracted. Thus the
measured rates of the cell-free extracts are maximum potentials. It
turns out that calculating in situ respiration from this maximum
rate is the major obstacle in interpreting the results of the ETS
technique. The problem has essentially two solutions: 1) a

Table 6. Determined rates of respiration of organic substrates, as derived from ^{14}C-tracer studies in coastal and oceanic waters.

Region	Substrate	Determined Substrate Concentration ($\mu g\ dm^{-3}$)	Rate of Respiration		Reference
English Channel (summer)	Glucose	1-6	14-50[a]	0.2-1.1[b]	Andrews & Williams 1971
	Amino acids (total)	10-80	4-11[a]	1-3[b]	
Californian Coast	Amino acids (total)	5-9	1.5-7.9[a]	0.048-0.670[b]	Williams et al. 1976
	Amino acids (individual)	0.05-2.0	0.1-29[a]	0.005-0.315[b]	
Bahamas	Glucose	–	0.5-13[a]	–	Williams & Yentsch 1976
	Amino acids (total)	–	1.8-37[a]	–	

a % substrate day^{-1}
b μg C·dm^{-3}·day^{-1}

conversion factor may be presumed from the biochemical literature or derived from cultures of appropriate organisms (Packard et al. 1971; Packard 1979); 2) alternatively a conversion factor may be obtained from simultaneous measurement of ETS activity and respiration rates made on natural populations (see e.g., Packard and Williams 1981). With the first approach one encounters the criticism that there is no reason to expect that the physiological state of a culture will be the same as a natural population. The second approach is subject to the objection that there is no a priori reason to presume that a single conversion factor exists for all planktonic communities. Thus one may either derive a factor, with calculated limits, from a large set of intercomparisons from a number of environments, or alternatively establish one by making comparisons on site for the environment in question. One could argue that if it is necessary to calibrate the ETS technique on site against direct measurements of oxygen consumption, they why use the ETS technique in the first place? The arguments for using the ETS technique are that its relative simplicity enables it to be used to provide fine details of the distribution of metabolism, which is presented not feasible with direct determination of oxygen consumption. Secondly, the ETS method provides an estimate of the instantaneous rate with no in vitro effect. Thirdly, unlike the direct measurements of oxygen consumption, it has the potential sensitivity to be used in deep water. The method, however, cannot be calibrated for deep water samples directly by the oxygen technique and this can only be achieved by extrapolation from comparisons on surface water. Inevitably the greater the body of comparison data available, the more confidence one can have in the extrapolation.

Field Measurements of ETS Activity. Measurements of ETS activity have been made on a variety of organisms and locations. Table 7 includes a summary of data from marine systems, measurements on marine organisms and sediments have not been included for they would not seem to be relevant in the present context. Calculated respiration rates per unit volume vary from 8-169 $\mu g\ O_2 \cdot dm^{-3} \cdot day^{-1}$; rates per unit area, for which there is more data, vary from 0.05 to 13 g $O_2 \cdot m^{-2} \cdot day^{-1}$. The rates are generally in the same range as those determined from direct oxygen consumption measurements although the ETS data shows a greater variation.

DISTRIBUTION OF RESPIRATORY ACTIVITY WITHIN THE MARINE PLANKTON GROUPS

An appreciation of the distribution of respiratory activity within the marine food chain may be argued to be a major goal of biological oceanography. Without this knowledge, the distribution of remineralization, metabolic activity and organic flux within the planktonic community cannot be understood. Indeed, the virtual absence of this information until somewhat recently, could well mean

Table 7. A summary of measured electron transport system activity.

Region	Mean Rate/Volume in Euphotic Zone (as $\mu g\ O_2 \cdot dm^{-3} \cdot day^{-1}$)		N***	Rate/Area (as $g\ O_2 \cdot m^{-2} \cdot day^{-1}$)		Reference
	Observed ETS Activity*	Calculated Oxygen Consumption Rate**		Observed ETS Activity*	Calculated Oxygen Consumption Rate**	
Coastal, Shelf and Slope						
Saanich Inlet	595	–	–	2.7	0.9	Packard et al. 1973
Gulf of Maine	271	169	11	26.1	8.9	Packard & Williams 1981
N.W. Atlantic	–	59	12	17.5	5.5	Packard & Williams 1981
Upwelling Areas						
N.W. Africa (1971)	156	53	7	6.1	2.1	Packard et al. 1974
N.W. Africa (1974)	–	–	–	6.1	2.1	Packard 1979
Baja California (1972)	–	–	–	38	13	Packard et al. 1973
Baja California (1973)	–	–	–	5	1.7	Packard, unpublished
Peru Current (1969)	44	15	44	19	6.5	Packard et al. 1971
Peru Current (1977)	234	80	47	5.2	1.8	Setchell & Packard 1979
Oceanic and Other Areas						
Costa Rica Dome (1973)	–	–	–	2.3	0.8	Kuntz et al. 1975
Tropical N.E. Pacific (1972)	65	22	7	3.5	1.2	King et al. 1978
Mediterranean Sea	–	–	–	1.9	0.6	Slawyk et al. 1976
Californian Current (1972)	31	11	9	–	–	Packard et al. 1975
Tropical N.E. Pacific (1972)	36	12	5	–	–	Packard et al. 1975

* Where appropriate the correction given by Packard (1979) has been used to bring the data to a common basis.

** Values are converted to oxygen consumption rate using the coefficient (ETS/R = 2.92) determined by Packard and Williams (1981).

*** N is the number of observations.

that some of our present notions of the marine food chain metabolic structure are in error (see discussion by Williams 1981b). The natural temptation is to search for some model, or generalization, for the marine system as a whole, However, since it is debatable if we possess at the present sufficient data to produce a detailed picture for a single oceanic station, any attempt at the present to produce generalizations for planktonic systems as a whole must be seen to be at the best fool-hardy.

There appear to be two basic approaches towards the determination of the respiratory structure of planktonic populations: 1) to measure the distribution of activity against some parameter, that in a general or particular way, can be related to the trophic level (e.g., size of the organism); 2) to endeavour to calculate respiratory activity from measurements of the biomass distribution within the community. Both approaches are recognized to have severe limitations, no size fractionation technique is ideal: often it may not be possible to separate groups which are functionally totally dissimilar (e.g., heterotrophic and photosynthetic microorganisms). Size fraction procedures may separate prey from predator, source from sink and thus alter the rates of metabolism of the separated fractions. Calculations from biomass likewise run into severe problems, especially with microorganisms, where the determination of biomass can be difficult and ascribing metabolic rate is, in truth, largely a matter of guesswork. However, one of the fundamental matters of discussion is the distribution of respiration within the three major planktonic groups: the phytoplankton, microheterotrophs (i.e., bacteria and protozoa) and the metazoans. The data is probably good enough to give pointers to such broad questions.

Size Fractionation Studies

The published data on size fractionation of metabolic activity, although not extensive, probably serves as the best starting point. Size fractionations of natural planktonic populations have been undertaken using nets and filters. In the case of heterotrophic studies data are available in the form of ETS and oxygen consumption measurements, a summary of the data is attempted in Table 8. Early studies were mainly concerned with resolving between the activities of net zooplankton and that of the remainder of the population. Recently more ambitious studies have been made which have attempted to produce size fractions, or size horizons down to 1 μm, for the determination of respiratory activity (e.g., Williams 1981a; Packard and Williams 1981). The data are few but do reveal that often the bulk of the respiration of organisms caught in water samplers is associated not with the net zooplankton, but with organisms less than 100 μm, i.e., the algae, the micrometazoa and the microorganisms (bacteria and flagellates).

The high respiratory activity observed in the smaller size

Table 8. A summary of size fractionation studies of respiration.

Region	Method	Depth Range (meters)	Size Fraction (μm)	% Rate*	Reference
Coastal and Shelf Areas					
N.W. Scotland, Loch Ewe	Oxygen consumption	0 - 15	<150	84(4)	Williams 1981c
			< 50	90(10)	
			< 10	81(6)	
			< 5	76(9)	
Vancouver Island, Saanich Inlet	Oxygen consumption	0 - 15	<150	99(3)	Williams 1981c
			< 50	90(4)	
			< 10	81(4)	
			< 5	76(7)	
			< 1	60(6)	
Gulf of Maine Georges Bank	ETS	Surface	> 20	12(4)	Packard and Williams 1981
			20>S>8	23(4)	
			8>S>3	22(2)	
			3>S>1	26(4)	
			<1	45(4)	
Gulf of Maine Georges Bank	Oxygen consumption	Surface	<20	69(3)	Packard and Williams 1981
			<8	75(2)	
			<3	49(2)	
			<1	26(2)	

Oceanic Areas

W. Atlantic Hatteras–Sargasso Sea)	Oxygen consumption by concentrates	Surface-500	<366	97(6)	Pomeroy and Johannes 1966
E. Atlantic (off N.W. Africa, 1971)	ETS	Integral over euphotic zone	<215	81(5)	Packard et al. 1974
E. Atlantic (off N.W. Africa, 1974)	ETS	Integral over euphotic zone	<102	84(16)	Packard 1979
Tropical E. Pacific	ETS	Integral over euphotic zone	<212	47(14)	King et al. 1978

* Rate as % unfractionated sample (number of observations).

fractions raises the question of whether this activity is algal or
that of heterotrophic microorganisms. The question is a key one, for
if the respiration in the small size fractions were algal then the
algae would be seen to be substantially involved in the recycling of
nutrients and engaged in little or no net 1° production. This is
not a type of food chain structure that is given much consideration.
The loss of ^{14}C from the algae in dark has been used to produce
estimates of algal respiration (see Table 3). The data are, regret-
fully, few and variable, and they can only indicate the likely scale
of algal respiration. As discussed earlier the rates determined by
the technique could include processes other than algal respiration,
e.g., organic exudation by the algae, bacterial respiration of assimilated algal exudates. If, on the other hand, the respiration of
the small size fractions is mainly that of heterotrophic microorganisms, this then brings up the issue of the transfer of organic
material from the algae to the microheterotrophic community. This
has been discussed by Williams (1981b), who has endeavoured to argue
that if one is willing to accept high algal organic exudation rates,
then it is possible in principal to maintain a sufficient flow of
organic material from the algae to the microheterotrophs to sustain
the anticipated metabolism of the latter.

     Williams (1981a), working on populations from two coastal
environments, has claimed to have resolved the non-algal, non-net
zooplankton component of respiration using size fractionation
procedures. He concluded from his observations that the respiration
which he attributed to the non-photosynthetic microorganisms, was
frequently a major component of overall planktonic community respiration (see Fig. 6). These observations await confirmation by other
workers and until then it is not known how general such a phenomenon
is in the marine plankton food web.

Rates Derived from Biomass Calculations

     This approach has a long history although this does not neces-
sarily assure its worthiness. The calculation of the metabolic
structure of the planktonic community from biomass data is a tedious
and inevitably a very subjective exercise. The approach would seem
best restricted to complimenting or explaining observations, rather
than to confirm hypotheses or intuitions. Calculations from biomass
become progressively less certain as the size of the organism
becomes smaller. This is unfortunate because much of the debate and
difference of opinion between marine microbiologists and biological
oceanographers has been over the relative scale of the metabolism of
the smallest organisms: the bacteria and protozoa.

     Many of the early calculations were directed towards the
problem of inorganic nitrogen remineralization and the role of the
zooplankton in the process. Not uncommonly the activity attributed
to the microorganisms was that not accounted for by the calculated

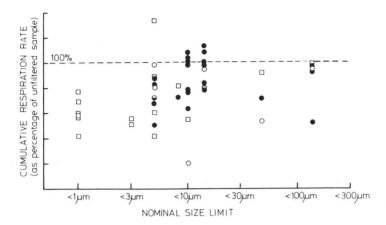

Figure 6. Distribution of respiratory activity with size. (□)
CEPEX, samples from bag; (○) Loch Ewe, samples from bag; (●) Loch
Ewe samples from outside bag. Data are expressed as cumulative
respiration up to various size limits, normalized against the rate
in the unfiltered sample. All the data points are for a single size
horizon and are not replicates.

zooplankton rate. One wonders how often the zooplankton estimate
was in fact an upper maximum rate rather than a median, i.e., the
bacterial component was seen to be small because of an overestimation
of zooplankton component. In very few cases is the data set detailed
enough to allow calculations to be made for the microbial community.
Table 9 is an attempt to assemble some of the published calculations
of the distribution of aspects of plankton catabolic activity
(nitrogen remineralization, respiration) within major planktonic
groups (see also Williams, in press). Inevitably there is quite a
range in the calculated distribution of activities. To some extent
this will be a consequence of the differing assumptions and calculation methods. However, one should reiterate the point made earlier
in the text that there is no a priori reason to presume a general
plankton food chain model. For example, there is a growing feeling
that there is a progressive change in the size structure of plankton
metabolism seawards (see e.g., Harrison 1980; Sieburth, this volume).
The calculations suggest that in general the proportion of plankton
metabolism that can be attributed to the bacteria, characteristically
falls in the region of one half to one third of overall metabolism.
The calculations furthermore imply that the algal component of
overall community respiration can be substantial and that of the
metazoan zooplankton on occasions comparatively small.

CONCLUSIONS

      Whereas in principle a number of approaches exist towards the

Table 9. Calculations of contribution of various planktonic types to community catabolism.

Aspect of Metabolism	Region	Summary of Approach and Observations	References
Respiration	Tropical upwelling	Computer simulation model, verified against biomass observations. Respiration apportioned: Carnivores 12.2% Omnivores 5.2% Herbivores 7.3% Micrometazoa 13% Phytoplankton 23% Protozoa 11% Bacteria 29%	Vinogradov et al. 1972
Respiration	Oligotrophic tropical ocean	As above: Carnivores 9% Herbivores 23% Phytoplankton 29% Protozoa 0.1% Bacteria 39%	Vinogradov et al. 1973
Inorganic nitrogen excretion	Offshore and oceanic areas	Data abstracted from various studies apportioned metabolism in the following manner: Zooplankton 7-60% (x = 32%) Microheterotrophs 13-93% (x = 54%) Bacteria 5-100% (x = 42%)	Harrison 1980
Respiration	Coastal (CEPEX contained ecosystem Saanich Inlet)	Detailed calculation of respiration rates from biomass records, estimated: Zooplankton 8% Phytoplankton 35-47% Protozoans 1-2% Bacteria 46-68%	Williams 1981c

problem of determining plankton respiration rates, the bulk of the
data to date originates from only two: the measurement of oxygen
consumption, and the electron transport system activity measurements.
Although some of the earlier ETS measurements gave rates lower than
the direct oxygen consumption measurements, they are usually of the
same scale and thus provide some mutual support, at least that the
rates they provide are of the correct magnitude. In coastal and
inshore regions, where the agreement between the ETS and oxygen
methods appears to be best, the measured and calculated rates of
respiration can be shown at times to match up, within the certainty
of the measurements, with the observed rates of photosynthesis
determined, either by the $^{14}CO_2$ or the oxygen technique. Thus in
these areas there is some general concordance between the measurements of plankton autotrophy and heterotrophy.

The situation is far less clear in oceanic water. First, no
direct comparisons exist between the two methods of determining
respiration in these areas, thus one can only draw provisional
conclusions from the studies by different workers made in areas of
presumed similar biological character. In areas known to be oligotrophic in character, the oxygen consumption studies of respiration
give rates in the range 10-100 μg $O_2 \cdot dm^{-3} \cdot day^{-1}$ (equivalent carbon
rates 3-30 μg $C \cdot dm^{-3} \cdot day^{-1}$) and the ETS technique gives corrected
rates in the same range (10-70 μg $O_2 \cdot dm^{-3} \cdot day^{-1}$). In similar
areas the rate of photosynthesis determined by the ^{14}C technique,
characteristically fall in the range 0.5-5 μg $C \cdot dm^{-3} \cdot day^{-1}$ (Steemann
Nielsen 1952; Menzel and Ryther 1960; Eppley et al. 1973; Sharp et
al. 1980; Eppley 1980). This disparity between the ^{14}C measurements
of photosynthetic production and the observed rates of respiration
may lend some credence to the views expressed by Sieburth (1977),
Gieskes et al. (1979), Eppley (1980) and others, that the ^{14}C technique is providing us with a low figure for phytoplankton production
(however, see Williams et al. 1983). As noted in the introduction,
the amount of data available on respiration is small, and there is
clearly a need to remedy this before attempting to use them to come
to any firm conclusion over the validity of the ^{14}C technique from
comparisons with respiration measurements.

ACKNOWLEDGMENTS

I would like to express my appreciation to Drs. L. R. Pomeroy,
W. J. Wiebe and J. P. Thomas for providing me with the unpublished
respiration data included in Table 1, to Ted Packard for scrutinizing
and correcting the ETS data and to Dr. W. O. Smith for unpublished
data included in Table 4.

REFERENCES

Andrews, P., and P. J. leB. Williams. 1971. Heterotrophic utiliza-
    tion of organic compounds in the sea. III. Measurement of the
    oxidation rates and concentrations of glucose and amino acids
    in sea water. J. Mar. Biol. Assoc. UK 51: 111-125.
Antia, N. J., C. D. McAllister, T. R. Parson, K. Stephens, and J. D.
    H. Strickland. 1963. Further measurements of primary produc-
    tion using a large-volume plastic sphere. Limnol. Oceanogr. 8:
    166-183.
Ben-Yaakov, S. 1972. On the CO_2-O_2 system in the northeastern
    Pacific. Mar. Chem. 1: 3-26.
Beyers, R. J., and H. T. Odum. 1959. The use of carbon dioxide to
    construct pH curves for the measurement of productivity.
    Limnol. Oceanogr. 4: 499-502.
Billen, G., C. Joiris, J. Winant, and G. Gillain. 1980. Concentra-
    tion and metabolism of small organic molecules in estuarine,
    coastal and open sea environments of the Southern North Sea.
    Estuarine Coastal Mar. Sci. 11: 279-294.
Bolin, B., and E. Eriksson. 1959. Changes in the carbon dioxide
    content of the atmosphere and sea due to fossil fuel combustion,
    pp. 130-146. In: B. Bolin [ed.], Rossby Memorial Volume.
    Rockefeller Press, New York.
Bryan, J. R., J. P. Riley, and P. J. leB. Williams. 1976. A Winkler
    procedure for making precise measurements of oxygen concentra-
    tion for productivity and related studies. J. Exp. Mar. Biol.
    Ecol. 21: 191-197.
Carritt, D. E., and J. H. Carpenter. 1966. Comparison and evalua-
    tion of currently employed modifications of the Winkler method
    for determining dissolved oxygen in seawater: a NASCO report.
    J. Mar. Res. 24: 286-318.
Christensen, J. P., and T. T. Packard. 1979. Respiratory electron
    transport activities in phytoplankton and bacteria: Comparison
    of methods. Limnol. Oceanogr. 24: 576-583.
Codispotti, L. A., G. E. Friederich, R. L. Iverson, and D. W. Wood.
    1982. Temporal changes in the inorganic carbon system of the
    southeast Bering Sea during spring 1980. Nature 296: 242-245.
Cooper, L. H. N. 1933. Chemical constituents of biological
    importance in the English Channel, November 1930 to January
    1932. Part II. Hydrogen ion concentration, excess base, carbon
    dioxide and oxygen. J. Mar. Biol. Assoc. UK 18: 729-751.
Crawford, C. C., J. E. Hobbie and K. L. Webb. 1974. The utilization
    of dissolved free amino acids by estuarine microorganisms.
    Ecology 55: 551-563.
Davies, J. M., and P. J. leB. Williams. 1984. Verification of [14]C
    and O_2 derived primary organic production measurements using an
    enclosed ecosystem. J. Plank. Res. In press.
Dawson, R., and K. Gocke. 1978. Heterotrophic activity in compari-
    son to the free amino acid concentration in Baltic sea water
    samples. Oceanol. Acta 1: 45-54.

Derenbach, J. B., and P. J. leB. Williams. 1974. Autotrophic and
    bacterial production: fractionation of plankton populations by
    differential filtration of samples from the English Channel.
    Mar. Biol. 25: 263-269.
Eppley, R. W. 1980. Estimating phytoplankton growth rates in the
    central oligotrophic oceans, pp. 231-242. In: P. Falkowski
    [ed.], Primary Productivity in the Sea. Plenum Press, New York.
Eppley, R. W., and J. H. Sharp. 1975. Photosynthetic measurements
    in the Central North Pacific. The dark loss of carbon in 24-hr
    incubations. Limnol. Oceanogr. 20: 981-987.
Eppley, R. W., E. H. Renger, E. L. Venrick, and M. M. Mullin. 1973.
    A study of plankton dynamics and nutrient cycling in the central
    gyre of the North Pacific Ocean. Limnol. Oceanogr. 18: 534-551.
Fuhrman, J. A., and F. Azam. 1980. Bacterial secondary production
    estimates for coastal waters of British Columbia, Antarctica,
    and California. Appl. Environ. Microbiol. 39: 1085-1095.
Gaarder, T., and H. H. Gran. 1927. Investigation of the production
    of plankton in the Oslo Fjord. Rapp. P.V. Cons. Int. Explor.
    Mer 42: 1-48.
Gamble, J. C., J. M. Davies, and J. H. Steele. 1977. Loch Ewe bag
    experiment 1974. Bull. Mar. Sci. 27: 146-175.
Gieskes, W. W. C., G. W. Kraay, and M. A. Baars. 1979. Current ^{14}C
    methods for measuring primary productivity: gross underestimates
    in oceanic waters. Neth. J. Sea Res. 13: 58-78.
Gocke, K. 1976. Respiration von gelosten organischen Verbindungen
    durch naturliche Mikroorganismen-Populationen. Ein Vergleich
    zwischen verschiedenen Biotopen. Mar. Biol. 35: 375-383.
Harrison, W. G. 1980. Nutrient regeneration and primary production
    in the sea, pp. 433-460. In: P. Falkowski [ed.], Primary
    Productivity in the Sea. Plenum Press, New York.
Hobbie, J. E., and C. C. Crawford. 1969. Respiration corrections
    for bacterial uptake of dissolved organic compounds in natural
    waters. Limnol. Oceanogr. 14: 528-532.
Hobbie, J. E., O. Holm-Hansen, T. T. Packard, L. R. Pomeroy, R. W.
    Sheldon, J. P. Thomas, and W. J. Wiebe. 1972. A study of the
    distribution and activity of microorganisms in ocean water.
    Limnol. Oceanogr. 17: 544-555.
Holm-Hansen, O., T. T. Packard, and L. R. Pomeroy. 1970. Efficiency
    of the reverse-flow filter technique for the concentration of
    particulate matter. Limnol. Oceanogr. 15: 832-835.
Itturiaga, R., and G.-G. Hoppe. 1977. Observations of heterotrophic
    activity in photoassimilated organic matter. Mar. Biol. 40:
    101-108.
Ivanenkov, V. N., V. V. Sapozhnikov, A. M. Chernyackova, and A. N.
    Jusarova. 1972. Rate of chemical processes in the photosyn-
    thetic layer of the tropical Atlantic. Oceanology 12: 207-214.
Johnson, K. M., C. M. Burney, and J. McN. Sieburth. 1981. Enigmatic
    marine ecosystem metabolism measured by direct diel CO_2 and O_2
    flux in conjunction with DOC release and uptake. Mar. Biol.
    65: 49-60.

Johnson, K. S., R. M. Pytkowicz, and C. S. Wong. 1979. Biological
    production and the exchange of oxygen and carbon dioxide across
    the sea surface in Stuart Channel, British Columbia. Limnol.
    Oceanogr. 24: 474-482.
Kadota, H., Y. Hata, and H. Miyoshii. 1966. A new method for esti-
    mating the mineralization activity of lake water and sediment.
    Mem. Res. Inst. Food Sci. Kyoto Univ. 27: 28-30.
King, F. D., A. H. Devol, and T. T. Packard. 1978. Plankton
    metabolic activity in the eastern tropical North Pacific.
    Deep-Sea Res. 25: 689-704.
Kuntz, D., T. T. Packard, A. Devol, and J. Anderson. 1975. Chem-
    ical, physical and biological observations in the vicinity of
    the Costa Rica Dome (January-February 1973). Tech. Ref. No.
    321, Dept. of Oceanography, Univ. of Washington. 187 pp.
Kuo, H. H., and G. Veronis. 1970. Distribution of tracers in the
    deep oceans of the world. Deep-Sea Res. 17: 29-46.
Lancelot, C. 1979. Gross excretion rates of natural marine phyto-
    plankton and heterotrophic uptake of excreted products in the
    Southern North Sea, as determined by short-term kinetics. Mar.
    Ecol. Prog. Ser. 1: 179-186.
Larsson, U., and Å. Hagström. 1979. Phytoplankton exudate release
    as an energy source for the growth of pelagic bacteria. Mar.
    Biol. 52: 199-206.
Menzel, D. W., and J. H. Ryther. 1960. The annual cycle of primary
    production in the Sargasso Sea off Bermuda. Deep-Sea Res. 6:
    351-367.
Munk, W. H. 1966. Abyssal recipes. Deep-Sea Res. 13: 707-730.
Newell, R. C., M. I. Lucas, and E. A. S. Linley. 1981. Rate of
    degradation and efficiency of conversion of phytoplankton debris
    by marine microorganisms. Mar. Ecol. Prog. Ser. 6: 123-136.
Odum, H. T., and C. M. Hoskin. 1958. Comparative studies on the
    metabolism of marine waters. Publ. Inst. Mar. Sci. Univ. Tex.
    5: 16-46.
Packard, T. T. 1971. The measurement of respiratory electron trans-
    port activity in marine phytoplankton. J. Mar. Res. 29: 235-244.
Packard, T. T. 1979. Respiration and respiratory electron transport
    activity in plankton from the Northwest African upwelling area.
    J. Mar. Res. 37: 711-742.
Packard, T. T., A. H. Devol, and F. D. King. 1975. The effect of
    temperature on the respiratory electron transport system in
    marine plankton. Deep-Sea Res. 22: 237-249.
Packard, T. T., D. Harman, and J. Boucher. 1974. Respiratory
    electron transport activity in plankton from upwelled waters.
    Tethys 6: 213-222.
Packard, T. T., M. L. Healy, and F. A. Richards. 1971. Vertical
    distribution of the activity of the respiratory electron trans-
    port system in marine plankton. Limnol. Oceanogr. 16: 60-70.
Packard, T. T., T. Moore, D. Harmon, A. Devol, and F. D. King. 1973.
    Respiratory electron transport activity in the euphotic zone
    plankton of the western Mediterranean Sea, North Atlantic Ocean,

and the North Pacific Ocean, pp. 201-207. In: J. J. MacIsaac
[ed.], Report of the working conference on a systems approach
to eutrophication problems in the eastern Mediterranean.
Special Report No. 53 of the Dept. of Oceanography, Univ. of
Washington, Seattle. 270 pp.

Packard, T. T., and P. J. leB. Williams. 1981. Respiration and
respiratory electron transport activity in sea surface seawater
from the northeast Atlantic. Oceanol. Acta 4: 351-358.

Park, K., D. W. Hood, and H. T. Odum. 1958. Diurnal pH variation
in Texas bays, and its application to primary production
estimation. Publ. Inst. Mar. Sci. Univ. Tex. 5: 47-64.

Parsons, T. R., L. J. Albright, F. Whitney, C. S. Wong, and P. J.
leB. Williams. 1981. The effect of glucose on the productivity of sea water: an experimental approach using controlled
aquatic ecosystems. Mar. Env. Res. 4: 229-242.

Parsons, T. R., and J. D. H. Strickland. 1961. On the production
of particulate organic carbon by heterotrophic processes in sea
water. Deep-Sea Res. 8: 211-222.

Pomeroy, L. R., and R. E. Johannes. 1966. Total plankton respiration. Deep-Sea Res. 13: 971-973.

Pomeroy, L. R., and R. E. Johannes. 1968. Occurrence and respiration of ultraplankton in the upper 500 metres of the ocean.
Deep-Sea Res. 15: 381-391.

Riley, G. A. 1939. Plankton studies III. The Western North
Atlantic, May-June 1939. J. Mar. Res. 2: 145-162.

Riley, G. A. 1941. Plankton studies IV Georges Bank. Bull. Bingham
Oceanogr. Collect. 7: 1-74.

Riley, G. A. 1951. Oxygen, phosphate, and nitrate in the Atlantic
Ocean. Bull. Bingham Oceanogr. Collect. 13: 1-126.

Ryther, J. H. 1959. Potential productivity of the sea. Science
130: 602-608.

Schmalz, R. F., and F. J. Swanson. 1969. Diurnal variations in the
carbonate saturation of seawater. J. Sediment. Petrol. 39:
255-267.

Setchell, F. W., and T. T. Packard. 1979. Phytoplankton respiration
in the Peru upwelling. J. Plankton Res. 1: 343-354.

Sharp, J. H., M. J. Perry, E. H. Renger, and R. W. Eppley. 1980.
Phytoplankton rate processes in the oligotrophic waters of the
central North Pacific Ocean. J. Plankton Res. 2: 335-353.

Sheldon, R. W., and W. H. Sutcliffe. 1978. Generation times of 3 h
for Sargasso Sea microplankton as determined by ATP analysis.
Limnol. Oceanogr. 23: 1051-1055.

Shulenberger, E., and J. L. Reid. 1981. The Pacific shallow oxygen
maximum deep chlorophyll maximum, and primary production,
reconsidered. Deep-Sea Res. 28: 901-919.

Sieburth, J. McN. 1977. International Helgoland Symposium:
Convenor's report on the informal session on biomass and
productivity of microorganisms in planktonic ecosystems.
Helgol. Wiss. Meeresunters. 30: 697-704.

Slawyk, G., H. J. Minas, and T. T. Packard. 1976. A further

investigation on the primary productivity in the divergent zone
near the French Mediterranean Coast. Int. Revue ges. Hydrobiol.
61: 373-381.

Smith, S. V. 1973. Carbon dioxide dynamics: a record of organic
carbon production, respiration, and calcification in the
Eniwetok reef flat community. Limnol. Oceanogr. 18: 106-120.

Smith, S. V., and J. A. Marsh. 1973. Organic carbon production on
the windward reef flat of Eniwetok Atoll. Limnol. Oceanogr.
18: 953-961.

Smith, W. O. 1977. The respiration of photosynthetic carbon in
eutrophic areas of the ocean. J. Mar. Res. 35: 557-565.

Smith, W. O., R. T. Barber, and S. A. Huntsman. 1977. Primary
production off the coast of Northwest Africa: excretion of
dissolved organic matter and its heterotrophic uptake.
Deep-Sea Res. 24: 35-47.

de Souza Lima, H., and P. J. leB. Williams. 1978. Oxygen consumption by the planktonic population of an estuary. Estuarine
Coastal Mar. Sci. 6: 515-521.

Steele, J. H., D. W. Farmer, and E. W. Henderson. 1977. Temperature structure in large marine enclosures. J. Fish. Res. Board
Canada 34: 1095-1104.

Steemann Nielsen, E. 1952. The use of radioactive carbon (^{14}C) for
measuring organic production in the sea. J. Cons. Explor. Mer
18: 117-140.

Steemann Nielsen, E. 1955. The interaction of photosynthesis and
respiration and its importance for the determination of ^{14}C-
discrimination in photosynthesis. Physiol. Plant. 8: 945-953.

Talling, J. F. 1973. The application of some electrochemical
methods to the measurement of photosynthesis and respiration in
fresh waters. Freshw. Biol. 3: 335-362.

Teal, J. M., and J. Kanwisher. 1966. The use of pCO_2 for the
calculation of biological production, with examples from waters
off Massachusetts. J. Mar. Res. 24: 4-14.

Tijssen, S. B. 1979. Diurnal oxygen rhythm and primary production
in the mixed layer of the Atlantic Ocean at 20°N. Neth. J. Sea
Res. 13: 79-84.

Tijssen, S. B., and B. Eijgenraam. 1980. Diurnal oxygen rhythm in
the Southern Bight of the North Sea: net and gross production
in April 1980 in a Phaeocystis bloom. ICES C.M. 1980/C:17.

Vinogradov, M. E., V. V. Menshutkin, and E. A. Shushkina. 1972. On
a mathematical simulation of a pelagic ecosystem in tropical
waters of the ocean. Mar. Biol. 16: 261-268.

Vinogradov, M. Y., V. F. Krapivin, V. V. Menshutkin, B. S. Fleyshman,
and E. A. Shushkina. 1973. Mathematical model of the functions
of the pelagic ecosystem in tropical regions (from 50th voyage
of the R/V Vityaz). Oceanology 13: 704-717.

Weichart, G. 1980. Chemical changes and primary production in the
Fladen Ground area (North Sea) during the first phase of the
spring phytoplankton bloom. "Meteor" Forsch.-Ergebnisse 22:
79-86.

Wiebe, W. J., and D. F. Smith. 1977. Direct measurement of
    dissolved organic carbon release by phytoplankton and
    incorporation by microheterotrophs. Mar. Biol. 42: 213-223.
Williams, P. J. leB. 1970. Heterotrophic utilization of dissolved
    organic compounds in the sea. I. Size distribution of popula-
    tion and relationship between respiration and incorporation of
    growth substances. J. Mar. Biol. Assoc. UK 50: 859-870.
Williams, P. J. leB. 1973. On the question of growth yields of
    natural heterotrophic populations, pp. 399-400. In: T. Rosswall
    [ed.], Modern Methods in the Study of Microbial Ecology. Bull.
    Ecol. Res. Comm. (Stockholm) 1973. Swedish Natural Science
    Research Council.
Williams, P. J. leB. 1981a. Microbial contribution to overall
    marine plankton metabolism: direct measurements of respiration.
    Oceanol. Acta 4: 359-364.
Williams, P. J. leB. 1981b. Incorporation of microheterotrophic
    processes into the classical paradigm of the planktonic food
    web. 15th European Symposium on Marine Biology, Kiel, F.G.R.
    Kiel. Meeresforsch. 5: 1-28.
Williams, P. J. leB. 1981c. Microbial contribution to overall
    plankton community respiration - studies in CEE's, pp. 305-321.
    In: G. D. Grice and H. R. Reeve [eds.], Marine Mesocosms:
    Biological and Chemical Research in Experimental Ecosystems.
    Springer-Verlag, Berlin.
Williams, P. J. leB. Bacterial production in the marine food chain:
    the emperor's new suit of clothes. In: M. J. Fasham [ed.],
    Flow of Energy and Material in Marine Ecosystems: Theory and
    Practice. Plenum, New York. In press.
Williams, P. J. leB., and C. Askew. 1968. A method of measuring the
    mineralization by microorganisms of organic compounds in sea
    water. Deep-Sea Res. 15: 365-375.
Williams, P. J. leB., T. Berman, and O. Holm-Hansen. 1976. Amino
    acid uptake and respiration by marine heterotrophs. Mar. Biol.
    35: 41-47.
Williams, P. J. leB., K. R. Heinemann, J. Marra, and D. A. Purdie.
    1983. Comparison of ^{14}C and O_2 measurements of phytoplankton
    production in oligotrophic waters. Nature (London) 305: 49-50.
Williams, P. J. leB., and N. W. Jenkinson. 1982. A transportable
    microprocessor-controlled precise Winkler titration suitable for
    field station and shipboard use. Limnol. Oceanogr. 27: 576-584.
Williams, P. J. leB., and C. S. Yentsch. 1976. An examination of
    photosynthetic production, excretion of photosynthetic products,
    and heterotrophic utilization of dissolved organic compounds
    with reference to results from a coastal subtropical sea. Mar.
    Biol. 35: 31-40.
Wong, C. S., R. D. Bellegay, and A. B. Cornford. 1975. Measurable
    inorganic carbon parameters in seawater, pp. 47-57. Spec. Tech.
    Publ. No. 573. Am. Soc. Testing and Materials, Philadelphia.
Wyrtki, K. 1962. The oxygen minimum in relation to oceanic circula-
    tion. Deep-Sea Res. 9: 11-23.

THE USE OF TRACERS AND WATER MASSES TO ESTIMATE RATES OF RESPIRATION

William J. Jenkins

Woods Hole Oceanographic Institution
Woods Hole, Massachusetts 02543

INTRODUCTION

One method of estimating oxygen utilization rates in the sea is to infer them from the spatial distributions of dissolved oxygen and other properties. That is, one obtains some measure of the "age" or velocity of a given parcel of water, and combining this with an observed deficit or gradient in dissolved oxygen concentration, one obtains an apparent oxygen utilization rate. The advantage of such an approach is that it represents a true in situ measurement; there is no experimental perturbation of the system. The disadvantage of such an approach is that it is often a rather model-dependent calculation and is therefore subject to potential ambiguities, inaccuracies and even fallacies due to the model used. However, in all fairness it should be noted that even quite unrealistic models may in fact give rather good estimates of oxygen utilization, largely due to the degree of similarity between the boundary conditions of oxygen and the tracer used to calibrate the model. Such "metaphoric" models have proven very powerful in this regard, and in large part form the basis of our quantitative understanding of the rates of chemical and biological processes in the oceans. After all, a model is really an abstraction of the essence of a system, and as such cannot contain all elements of reality. Philosophically, it may be argued that the more abstract the model, the more that is learned, whereas a too articulated model, although perhaps more successful in mimicing observations, may not have much predictive power.

The general approach of using property distributions to obtain oxygen utilization rates (O.U.R.) may be broken into two major categories: those which use radioactive tracers (as clocks) and those which use dynamic calculations. The latter use the field of mass in some way, assuming a balance of forces (generally geostrophic, potential vorticity conserving). The former may involve the use of steady-state tracer distributions (e.g., natural ^{14}C) and therefore may be Eulerian, or may use the time evolution of transient tracer distributions (e.g., bomb fallout), and therefore may be Lagrangian. As opposed to the dynamical methods, which utilize a balance of forces, the tracer methods generally rely on advection-diffusion balances, or in the case of the transient tracer analyses, the time varying advection-diffusion equations. In general form, they are

$$\frac{\delta C}{\delta t} = \vec{\nabla}(K\vec{\nabla}C) - \vec{V} \cdot \vec{\nabla}C + J(C,\vec{X}) \tag{1}$$

where C is the property of interest (in concentration units), K is the turbulent diffusivity tensor, $\vec{V}$ the fluid velocity, $J(C,\vec{X})$ the general source-sink term and $\vec{\nabla}$ is the vector gradient operator,

$$\vec{\nabla} = \hat{i}\,\frac{\delta}{\delta x} + \hat{j}\,\frac{\delta}{\delta y} + \hat{k}\,\frac{\delta}{\delta z}\ .$$

The left-hand side of equation (1) is the local (Eulerian) time rate of change of the property, and the right hand side represents the combined effects of turbulent diffusion, fluid flow (transport of material by advection) and in situ changes (consumption, production, decay). It should be noted that as with fluid velocity, the diffusivity tensor will exhibit spatial dependence, and indeed will rotate due to changes in the slopes of isopycnals. This latter effect can be rationalized in light of the fact that turbulent exchange is energetically more favorable along surfaces of constant density than across them. The former is in part attributable to the fact that processes such as velocity shear and frictional boundary conditions which contribute to the turbulent energy flux responsible for the exchange, vary spatially as well. For example, areas near the Gulf Stream, its extension and recirculation regions, are characterized by relatively high eddy kinetic energy levels and hence have a higher coefficient of lateral eddy exchange.

The source-sink term in general will also exhibit spatial dependence. Oxygen, for example, will have a strong source term in the photic zone, a relatively strong sink term immediately below it, and a decreasing sink term with depth below. Because of the dependence on available carbon, there will be a horizontal variation in $J(O_2)$ as well, with upwelling regions having higher consumption terms than relatively quiescent subtropical gyres. The concentration dependence of the J term is at most assumed to be restricted to first order kinetics (usually reserved for radiotracer decay)

$$J = -\lambda C$$

and it appears that zeroeth order kinetics are more appropriate.
Devol (1978) suggested a form for the oxygen consumption term:

$$J = \frac{-J_o C}{C + C_o}$$

where J_o is the asymptotic (an oxygen limited) consumption rate and
C_o is the half saturation constant. He obtained values of much less
than 0.1 ml liter^{-1} for C_o. This means that for most open ocean
environs, oxygen consumption is best represented by zeroeth order
kinetics.

Of course, it should be noted that some tracers, for example
^{14}C, require both zeroeth and first order terms, where we may have

$$J = J_o - \lambda C$$

where J_o is the _in situ_ production rate for ^{14}C by a combination of
$CaCO_3$ dissolution and oxidation of organic matter.

STEADY STATE TRACER ESTIMATES

Generally, the approach to obtaining $J(O_2)$ has been to use a
limited form of equation (1). This has been achieved by making
certain assumptions regarding spatial distributions, fluxes and the
physics leading to these distributions. A more popular version,
expressed by Wyrtki (1962), Munk (1966) and Craig (1969) is the one
dimensional advection diffusion model. There, lateral fluxes are
assumed to be small, the system is steady state ($\frac{\delta C}{\delta t} = 0$), and K to
be constant, viz.,

$$K \frac{\delta^2 C}{\delta Z^2} - w \frac{\delta C}{\delta t} - J(z) - \lambda C = 0 \tag{2}$$

The last term is zero for oxygen, and the last two terms are zero
for stable-conservative properties such as salinity or temperature.
For these last tracers, equation (2) reduces to a simple form whose
solution is the classical exponential profile,

$$C = C_1 + C_2 \, e^{-Z/\alpha}$$

where the scale height "α" is given by

$$\alpha = K/W.$$

This yields a ratio between the two properties but does not yield
absolute rates, and some other information must be sought if J, K,
and W are to be known. Modelling of the radiocarbon (^{14}C) distribu-

tions yields some estimate of K and W, and application of these values to the oxygen equation should yield $J(O_2)$.

Wyrtki (1962) did not have adequate ^{14}C data to take this step, but did, however, show that consistent fits to the oxygen minima could be obtained using

$$J = J_0 e^{-Z/Z'}$$

where $Z' \sim 300$ m. Munk (1966), on the other hand, did estimate values of W and K from ^{14}C profiles, but failed to include the zeroeth order part of J in the ^{14}C equation, that is, he did not account for in situ CO_2 production. Craig (1969), on the other hand, made a more complete study of the ^{14}C system, and correctly applied equation (2). In a later paper (Craig 1971) he obtained a value of $J/W = -0.7$ ml·liter^{-1}·km^{-1}, and $W = 5$ m yr^{-1}, which gives $J(O_2) = -0.004$ ml·liter^{-1}·yr. He also noted some evidence of a depth dependence (decrease with depth), but beyond that did not quantify the trend.

Kuo and Veronis (1970, 1973) used a time independent, vertically averaged version of equation (1) as a basis for a simplified-geometry numerical world ocean model. Although their approach was one of using spatially invariant K_H and $J(O_2)$, they obtained moderately consistent results for $J(O_2) \sim 0.002$ ml·liter^{-1}·yr. Fiadero and Craig (1981) using 3D models found similar values for the deep water.

GEOSTROPHIC ESTIMATES

Another means of "calibrating" the advective diffusive models is to assume a geostrophic balance, i.e., a balance between the horizontal pressure gradient force (as produced by a sloping density surface) and the Coriolis force (as produced by the fluid flow on a rotating surface). Thus one obtains a relation between the horizontal density gradient and the fluid velocity at right angles to that slope. However, it is important to note that in fact one must reference the sloping density against some truly horizontal surface. (The truly horizontal surface, if geostrophic balance prevails, must therefore be a depth of no motion.) That is, the geostrophic balance gives only the relative velocity (the "baroclinic component"), and thus expressed in differential form, the geostrophic equations are

$$\frac{\delta u}{\delta z} = \gamma \frac{\delta \rho}{\delta y} \ , \quad \frac{\delta v}{\delta z} = -\gamma \frac{\delta \rho}{\delta x} \tag{3}$$

where $\gamma = g/f\rho_0$. In order to obtain the absolute velocity, one must integrate equation (3), which introduces a constant of integration, called the "barotropic component". This constant of integration

represents an unknown offset in the velocity distribution. As will
be discussed below, this represents an important ambiguity in the
technique.

      In a classic paper, Riley (1951) used a three-dimensional
advection-diffusion model to estimate oxygen utilization and nutrient
regeneration rates in the Atlantic. As mentioned earlier for the
simple one-dimensional analysis, this gives only a ratio of the non-
conservative rates to the primary advective terms. To obtain the
actual rates, i.e., to calibrate the model, Riley first computed the
geostrophic flow field (a baroclinic or "relative" flow field) and
then estimated "depths-of-no-motion" to obtain the barotropic
component and convert the relative flow field into an absolute one.
Using the absolute flow fields in the AD model and the observed
salinity distributions, he then obtained the turbulent exchange
coefficients. This produced to a fully calibrated model, which was
in turn used to estimate the non-conservative terms, in particular
the oxygen utilization rate. These are given in Figure 1.

      While this approach represented a tremendous step forward, it
suffered from one underlying problem. The choice of the depth of no
motion was somewhat arbitrary, and sometimes very much based on
approximate and even crude balances. The nature of the geostrophic
calculation is, in fact, such that only relative velocities can be
calculated. In practice, however, the absolute velocity field
decreases almost exponentially with depth so that an error in the
choice of the depth of no motion leads to only a small error in the
barotropic component. This means that shallow velocity fields have
an inherently small relative uncertainty whereas the deeper flow
fields suffer much more in a relative sense. This basically means
that Riley's estimates are likely to be fairly accurate for shallow
surfaces, but may be substantially (relatively speaking) in error

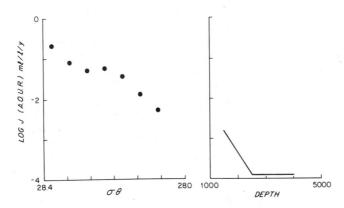

Figure 1. Riley's (1951) estimated oxygen consumption rates for
various surfaces.

deeper down.   There is, in fact, some evidence that he may have
underestimated the deep ventilation rates in his model and hence
underestimated abyssal O_2 consumption rates.   This is exemplified in
his seemingly low horizontal diffusivities at depth (e.g., 2 x 10^6
cm^2 s^{-1} for σ_θ = 27.5) compared to what is required for non-eddy-
resolving models (e.g., see Kuo and Veronis 1973) and his relatively
small Gulf Stream Recirculation (40 Sv compared to >100 Sv required
by Worthington 1976).   Nonetheless, for the shallow layers above the
oxygen minimum the relatively small barotropic uncertainty does not
seriously jeopardize the accuracy of the $J(O_2)$ estimates.

Riley also estimated what he believed to be an upper limit
estimate of 0.002 ml·$liter^{-1}$·yr^{-1} for $J(O_2)$ by relating the observed
mean AOU for North Atlantic Deep Water with estimated NADW formation
rates.   It is this value, in fact, which appears more consistent
with the steady-state tracer models of Kuo and Veronis (1970, 1973).

TRANSIENT TRACER ESTIMATES

More recently, we have been able to use observational data
from non-steady state tracers to look at transport processes and
ultimately for estimating $J(O_2)$.   As man's activities have generated
substances which are now entering the oceans by various pathways,
we are in effect "pulsing" the oceanic system and observing the
response.   The time-history, spatial pattern and other boundary
conditions of a particular transient tracer will determine what
aspects or processes of the ocean will be highlighted.   Since the
observed tracer distributions and their evolution represent the
interactions between the input-boundary functions and the advection
diffusion-transport operators, there is in general a certain degree
of ambiguity or "degeneracy".   That is, the "inverse calculation",
the backwards estimation of the transport processes from the tracer
pattern, is ill-behaved and may not always lead to unique or meaningful results.   Rather, it appears that a more realistic approach
is to use tracer diagnostic models (such as the time dependent
advection diffusion models) to test dynamic diagnostic models and
then ultimately, having solved the advection diffusion problem,
solve for $J(O_2)$.

Rooth and Ostlund (1972) used the observed penetration of
tritium into the North Atlantic subtropical main thermocline to
obtain estimates of the vertical turbulent diffusivity.   Tritium, the
heaviest isotope of hydrogen, is radioactive (half-life 12.45 yr) and
occurs in the ocean virtually solely as part of the water molecule.
As such it represents the ideal fluid tracer.   The natural inventory
of tritium (it is produced by cosmic ray spallation in the upper
atmosphere) was dwarfed by atmospheric nuclear weapons testing
production in the 1950's and 1960's.   This "bomb tritium" entered

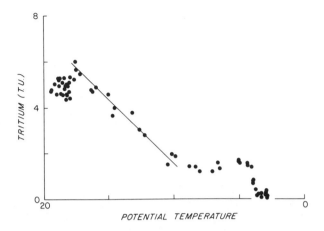

Figure 2. Tritium vs. potential temperature for a station near
Bermuda (occupied three times between March and July 1977). Note
the linear trend between about 10° and 18°C.

the oceans predominantly at high northern latitudes (the hemisphere
where most of the tests took place) and has since been carried into
the intermediate and deeper waters by a combination of mixing,
turbulent diffusion and fluid flow.

    The analysis of Rooth and Ostlund consisted of relating the
temperature and tritium distributions in such a way as to eliminate
the advective ($\vec{V}\nabla C$) term from the advection diffusion equation
(equation (1)). For tritium this left a balance between the decay
term, the diffusive term and the Eulerian time derivative. Noting
an apparently vanishingly small value for this last term over the
time of study (1968 to 1972), and assuming lateral effects to be
small, they obtained an upper limit estimate for K_v of ~ 0.2 cm^2 s^{-1}.
This study went only part of the way toward calibrating the terms in
the equation, because it did not solve for the advection terms and
in fact could not a priori discriminate between lateral and vertical
diffusion. Although a valuable contribution, Rooth and Ostlund's
analysis serves to underline the limitations of using just one
transient tracer.

    This can be further exemplified by noting that the tritium
distribution in the upper thermocline ($8° < \theta < 18°$) has evolved to a
point where it appears to be a linear function of potential
temperature. Figure 2, for example, shows the ^{3}H – θ relation for
the Bermuda Panulirus Station in the Sargasso Sea in 1977. Note
that the θ – ^{3}H relation between 10° and 18° is very nearly linear.
That is, for the upper thermocline,

$$C(^3H) \approx a + b\theta \tag{4}$$

where a and b are constants which within the gyre appear to be
spatially invariant. The significance of this co-relation can be
seen by the following. Writing equation (1) in a slightly more
compact form and eliminating the J term,

$$\frac{\delta c}{\delta t} = DC - \lambda C \tag{5}$$

and,

$$o = D\theta \tag{6}$$

where C is the tritium concentration, θ is the potential temperature,
λ is the radiodecay constant for tritium and D is the advection
diffusion operator

$$D = \nabla K\nabla(\ ) - \vec{V}\nabla(\ )$$

That is, D represents the combined effects of fluid advection and
turbulent diffusion, and is a linear differential operator. Note
that since potential temperature is a "stable", conservative, steady-
state tracer, it is affected only by E (equation (5)) and in effect,
the observed temperature distribution must result from a balance
between the individual terms within D.

Combining equations (4) and (6), and remembering that D is a
differential operator, we have

$$\begin{aligned}
DC &\approx D(a + b\theta) \\
&\approx bD\theta \\
&\approx 0
\end{aligned}$$

That is, equation (5) must reduce to

$$\frac{\delta C}{\delta t} \approx -\lambda C$$

or quite simply the radioactive decay equation. What this says is
that the distribution of tritium in the upper thermocline at this
point in time is governed more by radioactive decay than by advection
and diffusion. This bodes ill for using tritium alone in this
regime to calibrate our AD models for the purpose of determining
$J(O_2)$, and clearly points to the need for a multitracer approach.

Later, Jenkins (1977) used the ^{3}H - ^{3}He dating technique to
estimate $J(O_2)$ in the 18° water of the Sargasso Sea. The essence of
the technique is that as a parcel of water resides at the ocean
surface, tritium within it decays continuously but no excess ^{3}He
(the stable, inert daughter of tritium) occurs because it escapes to
the atmosphere. Once leaving the surface, however, the parcel

accumulates excess ^{3}He so that at some later time T, we have

$$T = \frac{1}{\lambda} \ln \left[1 + \frac{C(^3He)}{C(^3H)}\right] \quad .$$

The technique was limited to relatively shallow (young) water because although the approach is valid for short timescales (e.g., less than a decade; Jenkins 1980) it suffers from mixing non-linearities for longer timescales (Jenkins and Clarke 1976; Jenkins 1977, 1980). For the deeper, "older" waters, the simple closed system dating approach is inadequate and some multitracer approach needs to be implemented. The advantage of using more than one tracer is that, provided enough tracers are used, one can obtain an over-determined system of equations and test the physical model used. Further, ambiguities may be eliminated and sensitivity to different time- and space-scales may be enhanced. An example of this is the combination of tritium and its daughter ^{3}He. It has been shown (Jenkins 1980) that ^{3}He enhances sensitivity to timescales ranging from a few months to a decade, and due to its complementary (relative to tritium) boundary conditions, acts as a strong test of the physical model. Subsequently, it was demonstrated that a simple vertical advection-diffusion model is inadequate for the main thermocline (Jenkins 1980). It appeared that a simple isopycnal outcrop model, one consistent with the early concepts of Iselin (1936), and Montgomery (1938) and others, successfully described the ^{3}H and ^{3}He distributions and evolution (Jenkins 1980).

Accepting that the lateral mixing model portrayed the tracer distributions accurately, it was then possible to compute the $J(O_2)$ using observed AOU values and the ^{3}H – ^{3}He derived ventilation rates. These are shown in Figure 3, and for comparison, the results of Jenkins (1977) and Riley (1951) are included.

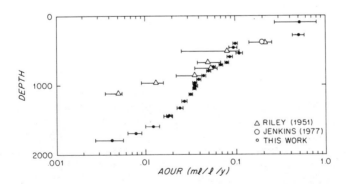

Figure 3. The ^{3}H – ^{3}He estimated oxygen utilization rates for the Sargasso Sea. Included for comparison are Riley's (1951) estimates normalized to the correct σ_θ – depth curve, and an earlier (Jenkins 1977) ^{3}H – ^{3}He estimate.

The results are in remarkably good agreement with Riley's in the top 1000 m, an agreement which is important in that the two approaches are completely independent. Below 1000 m, the results diverge decidedly. The difference can be defended on the basis of the primary problem of determining the barotropic component in the dynamical approach. An uncertainty of 0.1 cm s^{-1} is easily seen in Riley's deep velocities which results in a barotropic transport of the order of 10 Sv across 45°N and an uncertainty of much more than a factor of two in his $J(O_2)$. In fact it is not unexpected that his $J(O_2)$ estimates are low for the deeper waters both on the basis of his low K and V values (discussed earlier) and the fact that other tracer studies (e.g., Kuo and Veronis) lean toward higher values as well.

How generally useful the $J(O_2)$ values are, however, remains to be established. Clearly, variation with watermass (hydrographic) structure, surface production and perhaps other as yet unknown properties may play a role. The deep water estimates are likely the most universally applicable, whereas the shallower data may need careful consideration before "export". Further, the $J(O_2)$ estimates obtained by $^3H-^3He$ in the Sargasso Sea (Jenkins 1980) represent some kind of areal average in that the gyre recirculation timescale (a few years) is short relative to the oxygen ventilation-consumption timescale (a few decades). It is likely that they as such represent characteristic average values for a subtropical gyre, but confirmation of its universality is needed in future.

It is of interest to note that irrespective of the disagreement in the deeper waters, integration of the $J(O_2)$ curve yields an estimate of the net water column integrated oxygen consumption of about 9 moles $O_2 \cdot m^{-2} \cdot yr^{-1}$, corresponding to a net production of about 80 g $C \cdot m^{-2} \cdot yr^{-1}$ at the surface (Jenkins 1982a). This represents a firm lower limit to surface production since there exists a substantial degree of recycling of organic carbon in the upper layers. This suggests that the standard ^{14}C fixation techniques for determination of surface primary productivity may substantially underestimate the true productivity (Jenkins 1982a).

FUTURE DIRECTIONS

An additional development in the diagnostic dynamical approach used by Riley may soon follow from the Stommel and Schott (1978) Beta-Spiral technique. Whereas the classical geostrophic technique yields only a relative velocity, the Beta-Spiral calculation includes the conservation of density and potential vorticity to achieve closure and determine the absolute velocity. The theory relates the observed rotation of the geostrophic velocity vector with depth to the absolute velocity. Determination of the absolute velocity can lead to (by comparison with a conserved tracer field, e.g., salinity) the turbulent diffusion terms and ultimately to $J(O_2)$. Preliminary

analysis of the data for the shallow surfaces yields a $J(O_2)$ value
not unlike the Jenkins (1980) and Riley (1951) estimates, but some
further thought is required.

Large (North Atlantic basin) scale mapping of transient tracers
is currently under way, so that synoptic maps will eventually lead
to more realistic ventilation-transport models than the regional
model presented by Jenkins (1980) or the one- to two-dimensional
models of Wyrtki, Munk, Craig and Kuo and Veronis. In addition to
the enhanced coverage, the systematic measurement of tracers such as
^{85}Kr (half-life ~ 10 yr, produced by nuclear fuel reprocessing),
^{137}Cs (half-life ~ 30 yr, also produced by nuclear fuel reprocessing
and bomb fallout) and CCl_3F (a stable, inert refrigerant and spray
can propellant) will allow more sophisticated diagnostic models.

The development of highly detailed and realistic eddy resolving
Ocean General Circulation Models is paced more by technological
limitations of the computers than by theory, although the application of realistic physical and tracer boundary conditions is a major
challenge. The testing and calibration of these models using
transient tracers will be a powerful tool for refining our understanding of both ocean circulation and the consumption of oxygen in
the water column. The advent of CRAY-2 and perhaps application of
the CDC CYBER 205 will help.

Finally, it is becoming evident that the intermediate timescale
physical process responsible for modifying and maintaining the
observed tracer and oxygen distributions are in themselves time
varying. One need only recognize that the spectrum of climate
variations on these timescales is strong in long term fluctuations
to realize that the physical transport processes associated with
water mass formation and ventilation, i.e., heat flux and wind
stress, will also show this decade timescale variability. The 27
year hydrographic record at Bermuda shows a strong (factor of two)
modulation in the degree of water mass formation in the upper waters
(Jenkins 1982b). The nature of the tracer diagnostic models must be
such that this modulation is taken into account and the oxygen
database used to back calculate $J(O_2)$ must also account for this.
Clearly a mature theory must use the modern tracer measurements
against a backdrop of sound physical models and the much broader
base of hydrographic data.

CONCLUSION

It appears that the two approximately independent approaches of
physical diagnostic (e.g., Riley 1951) and tracer diagnostic (e.g.,
Craig 1971; Jenkins 1980) modelling give compatible results for the
upper part of the water column, whereas they disagree for the deeper
waters. The inconsistencies can be resolved on the basis of the non-

absolute nature of the geostrophic velocity computations and suggest that the tracer diagnostic approach is more likely correct. We have, at hand, estimates which are probably accurate to a factor of 1.5, and limited to certain geographic areas. In the future, as we gain more of an understanding of oceanic transport processes and perhaps their variability, we hope to considerably refine and expand on our quantitative understanding of the rates of oxygen utilization in the sea.

REFERENCES

Craig, H. 1969. Abyssal carbon and radiocarbon in the Pacific. J. Geophys. Res. 74: 5491–5507.

Craig, H. 1971. The deep metabolism: oxygen consumption in abyssal ocean water. J. Geophys. Res. 76: 5078-5-91.

Devol, A. H. 1978. Bacterial oxygen uptake kinetics as related to biological processes in oxygen deficient zones of the oceans. Deep–Sea Res. 25: 137–146.

Fiadero, M., and H. Craig. 1978. Three dimensional modeling of tracers in the deep Pacific Ocean: I. Salinity and oxygen. J. Mar. Res. 36: 323–355.

Iselin, C. O'D. 1936. A study of the circulation of the western North Atlantic. Pap. Phys. Oceanogr. 4: 101.

Jenkins, W. J. 1977. Tritium helium dating in the Sargasso Sea: a measurement of oxygen utilization rates. Science 196: 291–292.

Jenkins, W. J. 1980. Tritium and ^{3}He in the Sargasso Sea. J. Mar. Res. 38: 533–569.

Jenkins, W. J. 1982a. Oxygen utilization rates in the North Atlantic subtropical gyre and primary production in oligotrophic systems. Nature 300: 246–248.

Jenkins, W. J. 1982b. On the climate of a subtropical ocean gyre: decade timescale variations in water mass renewal in the Sargasso Sea. J. Mar. Res. 40 (Suppl.): 265–290.

Jenkins, W. J., and W. B. Clarke. 1976. The distribution of ^{3}He in the western Atlantic Ocean. Deep–Sea Res. 23: 481–494.

Kuo, H. H., and G. Veronis. 1970. Distribution of tracers in the deep oceans of the world. Deep–Sea Res. 17: 29–46.

Kuo, H. H., and G. Veronis. 1973. The use of oxygen as a test for an abyssal circulation model. Deep–Sea Res. 20: 871–888.

Montgomery, R. B. 1938. Circulation in the upper layers of the southern North Atlantic deduced with use of isentropic analysis. Pap. Phys. Oceanogr. Meteorol. 6: 55.

Munk, W. H. 1966. Abyssal recipes. Deep–Sea Res. 13: 707–730.

Riley, G. A. 1951. Oxygen, phosphate and nitrate in the Atlantic Ocean. Bull. Bingham Oceanogr. Coll. 13: 1–126.

Rooth, C. G., and H. G. Ostlund. 1972. Penetration of tritium into the North Atlantic thermocline. Deep–Sea Res. 19: 481–492.

Stommel, H., and F. Schott. 1977. The Beta-Spiral and the determination of the absolute velocity field from hydrographic

    station data. Deep-Sea Res. 24: 325.
Worthington, L. V.   1976.   On the North Atlantic circulation, pp. 75.
    The Johns Hopkins Oceanographic Studies No. 6, Baltimore.
Wyrtki, K.   1962.   The oxygen minima in relation to ocean circula-
    tion. Deep-Sea Res. 9: 11-28.

PROTOZOAN BACTERIVORY IN PELAGIC MARINE WATERS

John McN. Sieburth

Graduate School of Oceanography
University of Rhode Island, Bay Campus
Narragansett, Rhode Island 02882-1197

OLDER LITERATURE ON PROTOZOAN BACTERIVORY

The significance of bacteria and bacterial-grazing protozoa in
the decomposition of detritus and mineral cycling in the benthic
environment has been excellently reviewed by Fenchel and Harrison
(1976) and Fenchel and Jørgensen (1977) and demonstrated in a series
of experiments by Fenchel (1977). In reviewing the role of protozoa
in nutrient cycling and energy flow, Stout (1980) correctly states
that there is little information on the nutrients and the nutrient
cycling of planktonic protozoa in pelagic waters. Although the role
in the pelagic food chain of the protozooplankton and their grazing
of bacteria was recognized by Lohmann (1911), only recently is it
receiving renewed attention (Pomeroy 1974; Sieburth 1976; 1979;
Sieburth et al. 1978; Williams 1981). Such studies require a sound
taxonomic basis. Contemporary studies still rely heavily upon late
19th century (e.g., Kent 1880-81; Stokes 1888) and early 20th
century (e.g., Calkins 1901; Griessman 1914) monographs, although
this area of enquiry has recently been reopened (Fenchel 1982a;
Davis and Sieburth, unpublished). Before discussing the recent
developments, it is first necessary to draw upon the older literature
and build a foundation of what is known about the grazing of bacteria
by rhizopods, flagellates and ciliates in general, before looking at
the nature of in situ pelagic populations of bacterial prey and their
protozoan predators.

Rhizopod Bacterivory

The more conspicuous members of this group of oceanic protozoa
are the pelagic foraminifera, radiolaria and acantharia with tests
and/or spines of calcium carbonate, silicon dioxide, and strontium

sulfate, respectively. These relatively large forms feed primarily
on the larger phototrophs and even protozoa and metazoa (Anderson
and Bé 1976; Anderson et al. 1979; Swanberg 1979). Although the
benthic foraminifera are primarily algal feeders, bacteria are part
of their diet (Lee et al. 1966) and they appear to require bacteria
for sustained reproduction (Muller and Lee 1969). The smaller
single-chambered testate forms, the testaceans, that are common in
freshwaters and may be limited to the littoral zone in marine waters
(Sieburth 1979), are also regarded as algal feeders. This would
restrict the potential procaryote-consuming protozoa to the naked
forms, the amoebae and the heliozoans. The heliozoa are small forms
difficult to see among the detritus until one becomes familiar with
their "sun appearance" due to their fine filopodia radiating like
the sun's rays. They are commonly observed in seawater enrichments,
and in nearshore plankton samples. They are apparently rare in the
open ocean although they have been observed in oceanic waters (Davis
et al. 1978) and have been recently observed on marine snow (P. G.
Davis, personal communication).

This leaves the naked amoebae as the potentially most important
bacteria-eating rhizopods in pelagic ecosystems. Unless specially
stained, e.g., the difficult protargol procedure, marine amoebae blend
in with the detritus in preserved samples and are best detected by
enrichment culture. Lighthart (1969) plated seawater samples on agar
surfaces heavily seeded with a suspension of resting bacteria. This
procedure is limited by the volume of the innoculum and only detects
high populations of amoebae. Sawyer (1971, 1975) concentrated
particles from large volumes of nearshore seawater onto Millipore
filters that were then immersed in Petri dishes of seawater and
observed at intervals for the presence of amoebae. This procedure
indicates the diversity of the cultured amoebae but does not yield
population estimates. Davis et al. (1978) used reverse-flow
filtrators (Hinga et al. 1979) to gently concentrate particles from
large volumes of oceanic waters that were then serially diluted and
enriched with a rice grain or cube of agar to obtain Most Probable
Number (MPN) estimates of cultivable amoebae. These ranged from a
few per liter in subsurface waters to 10^3/ml of surface microlayer.

Living amoebae can be distinguished from debris with phase
contrast microscopy, but upon fixation their detection becomes
difficult. We need simple microscopic procedures to detect and
differentiate different groups of amoebae in preserved samples.
Procedures are also required for the estimation of grazing rates and
growth rates of amoebae in their natural microhabitats of surfaces
and seston. Until we have such procedures we will have no idea of
their abundance or distribution in the sea and can only presume that
they are essentially restricted to marine snow and other surfaces
undergoing intensive bacterial colonization. Cutler and Crump
(1920) used a procedure based on 2% hydrochloric acid (Cutler 1920)
to differentiate encysted from feeding protozoa, and unlike the

comparisons between flagellates and bacteria which showed no direct relationships, the populations of feeding amoebae were inversely related to bacterial populations. Danso and Alexander (1975) report that the amoebae (Hartmanella and Naegleria stopped feeding below concentrations of 10^6 or 10^7 cells/ml of Rhizobium meliloti and concluded that a reduction in the density of prey below such threshold concentrations in nature would regulate predation.

Flagellate Bacterivory

The existence of microflagellates that pass the usual plankton nets was recognized by Lohmann (1911) and Hentschel (1936), who used hand-powered centrifuges to concentrate them. Through the years small 2-5 µm colorless (non-pigmented) flagellates such as Bodo, Monas, Paraphysomonas, Rhynchomonas, Oicomonas, Actinomonas, and choanoflagellates as well as larger euglenids and dinoflagellates have been recognized as frequent components of bacterial enrichment cultures from coastal waters. These flagellates have been shown to graze upon bacteria (Hardin 1944; Haas and Webb 1979; Kopylov and Moiseev 1980). However, Pütter's (1909) hypothesis that many organisms may obtain their nutrition from the large pool of DOC, as well as the successes of culturists such as Droop (1970, 1973) in growing phagotrophs on bacteria-free liquid diets, have kept alive the belief that most flagellates earn their living this way (Beers et al. 1975). Even obvious bacteria-grazing ciliates like Uronema marinum can be grown axenically in a defined medium (Hanna and Lilly 1974). This does not mean that they can use appreciable amounts at the lower concentrations of DOC found in natural environments, especially in the presence of prey bacteria for which their feeding mechanisms have been developed. A reason why data on the grazing of bacteria by colorless flagellates has been slow in developing is that only recently have good methods using epifluorescence microscopy been developed for estimating in situ populations of the procaryote prey (Hobbie et al. 1977; Porter and Feig 1980) and their protist predators (Burney et al. 1981; 1982, Davis and Sieburth 1982; Haas 1982).

The bacterial-flagellate connection has been best demonstrated in benthic-detrital ecosystems. When Lighthart and Liston (1964) used sediment cores that they infused with organic substances as a model seabed, a bacterial flora devloped which in turn enriched a population of bacteria-eating protozoans. This observation was followed up with a survey of the bacteria and bacterial grazing protozoa and their relationship in Puget Sound and the waters of the adjacent Pacific Ocean (Lighthart 1969). In nearshore waters, bacteria forming colonies on agar ranged from 5/ml to 8×10^4/ml while the protozoa dominated by colorless flagellates that developed on the bacterial lawns on agar plates ranged from undetectable to approximately 10^3/ml; there was a regression of 580 culturable bacteria per culturable protozoan for both seawater and the sediments (Lighthart 1969). The succession of bacteria, flagellates, then ciliates in detrital enrichments in the laboratory has been well documented by Fenchel and Harrison (1976) although Johannes (1965) minimizes the role of flagellates as bacterial grazers and as a link in the microbial food chain. Little information is available on specific populations of heterotrophic flagellates in the plankton. An exception is the choanoflagellates which have been estimated at several hundred/ml in MPN cultures from the coastal waters of Norway (Throndsen 1969, 1970a,b) as well as from a variety of other locations (Leadbeater 1974).

Sorokin (1977) has correctly pointed out the high standing stock of oceanic bacteria and zooflagellates, however the use of darkfield microscopy to differentiate phytoflagellates from zooflagellates on the basis of detectable size, chloroplasts, and motion (Sorokin 1979) appears to severely overestimate the colorless flagellates. Sorokin's colleagues at Gelendzhik, Kopylov and Moiseev (1980), have used the nearshore waters of the Black Sea to show that the colorless flagellates passing 7-10 μm net and even 1.5 μm filters (30% passage) control the numbers of bacteria developing in seawater. A bacterial generation time of 0.51 to 0.63 days without zooflagellates increased to an apparent generation time of 0.83 to 1.64 days due to zooflagel- late predation. They observed flagellate doublings of 0.67 to 1.44 days and concluded that the zooflagellates consumed 50% of daily bacterial production.

Newell et al. (1981) have used ground-up cultures of a number of phytoplankton species to determine the rates and efficiencies of the microbial degradation of both soluble and particulate carbon. They noted that the dissolved organic component, some 34% of the total cell debris, had a 50% utilization time of 1.6 d while the remaining 66% of particulate carbon had a 50% utilization time of 11.6 d. They were surprised that after 3 days, the standing stock of bacteria only accounted for $1.86 \pm 0.76\%$ of the carbon supplied and represented a conversion efficiency of just 9.8%. When the bacteria were replaced by flagellates at day 6, their peak biomass represented only some $12.5 \pm 3.6\%$ of the bacterial biomass. Such calculations may be gross underestimates since, as Schleyer (1980) points out, biomass calculated from direct counts do not adequately reflect growth. An equal (Allen 1977) to twentyfold (Paerl 1978) biomass of bacterial production may be formed as extracellular polysaccharide (Geesey 1982) which if used by the flagellates would greatly increase their conversion efficiency.

Ciliate Bacterivory

There are only a few estimates of ciliate populations in pelagic waters. Beers et al. (1975) included a number of categories of cili- ates in their comprehensive survey of the microbial plankton of the North Pacific Gyre using the Utermöhl technique (Utermöhl 1958).

Mean concentrations for the non-loricate cilites were 65 to 92 cells/
100 ml sample in the upper 100 m or approximately 0.7 ciliate/ml.
Pace and Orcutt (1981) in discussing the enumeration of freshwater
ciliates state that formaldehyde poorly preserves ciliates while
Lugol's iodine stains debris as deeply as the ciliates. Conversely,
mercuric chloride is allegedly a good preservative while brom thymol
blue is supposed to stain ciliates more deeply than organic debris.
The effectiveness of this preservative and stain for ciliates in
marine samples remains to be confirmed. Sorokin (1977, 1980) states
that fresh seawater samples must be used within 1 h of sampling with
minimal handling to estimate ciliates as well as flagellates. He
found that in the Japan Sea, ciliates in the upper 100 m ranged
from 6 to 180 cells per 20 ml sample with a mean of 64 cells or 3.2
cells/ml. This is less than a five-fold difference in the ciliate
populations of the oligotrophic gyre waters studied by Beers et al.
(1975) and the coastal waters of the Japan Sea studied by Sorokin
(1977). These low ciliate populations of 0.7 to 3.2 cells/ml have
been obtained with 20-100 ml of seawater sample.

The mechanism of ciliate feeding will depend to a large degree
upon the microhabitat of the ciliates and their bacterial prey.
There are two apparent bacterial food sources: the large population
of small planktonic bacteria free in the water, or a smaller but
more dense population of larger epibacteria (Sieburth 1976, 1979)
developing in marine snow that is sporadically distributed in
pelagic waters and is akin to resuspended organic debris in nearshore waters. Recent studies on feeding mechanisms for ciliates
grazing on bacterial-sized particles presume that the bacteria are
in suspension (Fenchel 1980a,d). It is concluded that different
types of ciliates feed on different sized particles (Fenchel 1980a)
that are retained by a sieving mechanism (Fenchel 1980d). Based
upon the bacterial population required to sustain suspension feeding
ciliates, Fenchel (1980b,c) concludes that ciliates cannot exist at
all in oceanic waters. However, particle size selection and clearing
rates (Fenchel and Small 1980) may be of little significance if the
pelagic ciliates are mainly gleaners of bacterial lawns on organic
debris and on fouled surfaces. The growth rate of ciliates in the
marine microbenthos has been empirically related to size by Fenchel
(1968). This relationship has been tested for the bacteria-grazing
ciliates in a small freshwater pond by Taylor (1978b) who confirmed
that the dominant species are not opportunistic but have a strategy
for persistence based on growth rate and size unlike the other
trophic types of ciliates (Taylor 1979). The growth rates and
processes in pelagic communities of bacteria-grazing ciliates in
the sea are unknown.

Bacteria-grazing ciliates in fresh water may have a selectivity
for their bacterial prey. Different species of bacteria have been
found to promote growth, fail to support growth, or to be toxic
(Kidder and Stuart 1939; Burbanck 1942; Curds and Van Dyke 1966;

Barna and Weis 1973; Dive et al. 1974). In general gram negative
bacteria were good food, gram positive bacteria were poor food, while
pigmented species were toxic. These differences have been used to
explain the disappearance of gram negative enteric bacteria from
soil while gram positive bacteria such as Bacillus and Arthrobacter
persist (Burbanck 1942; Coler and Gunner 1969). Taylor and Berger
(1976) found that four cultures of ciliates were more selective in
their utilization of recent bacterial isolates from a pond than they
were for established laboratory cultures of bacteria.

Isolates of the marine scuticociliate Uronema have yielded
conflicting data on their preference for bacterial prey. Hamilton
and Preslan (1969) failed to observe a preference for different
bacterial strains nor a toxicity of pigmented bacteria reported for
freshwater ciliates. Berk et al. (1976) and Kumé (1979), however,
found a preference in their Uronema isolates for strains of Vibrio
compared to strains of Bacillus and Arthrobacter, respectively. A
further study of prey selection by bacteria-grazing marine ciliates
and flagellates is needed.

A number of studies have attempted to determine the threshold
concentrations of suspended prey bacteria required to maintain and
grow ciliates. For the freshwater ciliate Glaucoma fed a strain of
Pseudomonas, Harding (1937a) found that at concentrations up to 6 x
10^5 bacterial cells/ml, the rate of ciliate feeding was a function
of the concentration of food and that the rate of disappearance of
bacteria depended upon the concentrations of the ciliate. Between
bacterial concentrations of 6 to 70 x 10^5 cells/ml the ciliate
generation time was 2.3 h. When the ciliates were starved, food
vacuoles disappeared after 5 h, multiplication stopped after 6-12 h,
and although the cells became smaller and more slender, they could
withstand 30 days of starvation without death (Harding 1937b).
Hamilton and Preslan (1970) correctly point out that marine
phagotrophs dependent upon bacterial-sized particles must be either
capable of efficiently scavenging bacteria at concentrations
characteristic of the water or be dependent upon localized high
concentrations of bacteria as might be found on decaying organisms.
The pelagic Uronema of Hamilton and Preslan (1969) in flask culture
had a maximal growth rate (Um) of 0.147/h or a doubling time of 4.6 h
and a threshold of 3.2 x 10^6 bacteria/ml. When the continuous cultures of their ciliate were fed resting cells of Serratia marinorubra
(Hamilton and Preslan 1970), a similar Um was obtained but the
threshold was lower, ranging from 0.5 to 2.4 x 10^6 bacteria/ml with
an average of 1.5 x 10^6 bacteria/ml. Berk et al. (1976) used tube
cultures and determined that the estuarine ciliates (Uronema
nigricans and Potomacus pottsi required 10^6 to 10^7 bacteria/ml for
growth. Using their values of 10^3 to 10^4 bacterial cells/ml in the
water column of the Rhode River estuary and 10^4 to 10^6 bacterial
cells/g of wet sediment (which appear to be underestimates by at
least 2 orders of magnitude), they concluded that these ciliates

would not encounter levels of bacteria able to maintain growth.
Fenchel (1980b) studied 14 species of ciliates in terms of the
maximum rate of water cleared at low particle concentrations and the
maximum rate of ingestion at high particle concentrations. He found
that ciliates which feed on 1-5 μm particles compare favorably with
metazoan suspension feeders in their ability to concentrate dilute
suspensions of particles while species grazing on bacterial-sized
particles (0.2 to 1.0 μm) required bacterial concentrations that do
not occur in open ocean waters. Fenchel (1980c) was more specific,
stating that bacteria-grazing ciliates feeding on cells with a 0.1
μm^3 volume require a threshold population of 10^7 to 10^8 bacteria/ml
but that such bacteria in open lakes and oceans occur at a concentration of 10^5 to 10^6/ml, which is 2 orders of magnitude too low for
the ciliates to feed and survive.

One might imagine that in the absence of sufficient populations
of bacterial-sized prey, a bacterivorous ciliate might survive on the
microflagellates present at densities of hundreds to thousands per ml
of seawater. Ciliates, however, are very specific in their ability
to utilize different sized foods. A microflagellate 2 x 5 μm in size
has a biovolume over 2 orders of magnitude larger than a planktonic
bacterium. Species of bacterivorous ciliates that compete with the
bacterivorous flagellates for their bacterial-sized prey should not
be confused with the herbivorous and carnivorous species such as
oligotrichs preying upon the phototrophic and phagotrophic flagellates in the nanoplankton size fraction (Sieburth et al. 1978).

The apparent discrepancy between natural bacterial populations
and populations that maintain bacterivorous ciliates may not be this
large or may not even exist. Data given by Bick (1968) indicate
that the minimal concentrations of bacteria (direct counts) associated with freshwater ciliates usually ranged from 1.1 to 3.1 x 10^6
cells/ml, a usual value for natural waters. In fact the concept of
threshold values below which protozoa will not eat bacteria may not
be valid. Habte and Alexander (1978) studied the mechanisms whereby
bacteria preyed upon by protozoa can persist. They found that
non-dividing bacteria inhibited by antibiotics were grazed down to
< 10^3 cells/ml and even to zero by Tetrahymena pyriformis but in the
absence of antibiotics Klebsiella, Escherichia coli and Rhizobium
persisted around 10^6 cells/ml. Habte and Alexander concluded that
the persistence of bacteria with protozoa in solution and in soil is
governed by their capacity to reproduce and replace the cells consumed by predation. The sparse ciliate populations free in the
water may indicate that they are transients separated from the rich
bacterial habitats they require (Hamilton and Preslan 1970; Fenchel
1980b,c).

Prey Aggregation

The aggregation or flocculation of bacteria to concentrations that ciliates and flagellates can use may not be dependent just upon the sporadic occurrence of marine snow or freshly decaying seston that can support dense growth of bacteria. Even in rich bacterial habitats there appear to be mechanisms whereby protozoa actively flocculate or aggregate bacteria before or during ingestion. Reynoldson (1942) and Pillai and Subrahmanyan (1942) observed an inverse relationship between decreased bacterial activity and the abundance of the peritrichous ciliates Vorticella and Epistylis, respectively. The motile ciliate Paramecium caudatum was also shown to induce bacterial flocculation (Barker 1946). The secretion of "mucus" which can stick bacteria, colloids and debris together to form floccules was shown for the motile soil ciliate Balantiophorus minutus by Watson (1945) and for the sessile peritrich Carchesium by Sugden and Lloyd (1950). Curds (1963) used cultures of Paramecium caudatum and of peritrichs to elucidate the flocculation mechanisms. A polysaccharide, substance "P", whose monosaccharide constitutents are glucose and arabinose, was secreted by Paramecium caudatum and Vorticella microstoma into the medium. Suspended particles with a negative charge that absorbed substance "P" underwent a change of charge and aggregated to form floccules. An additional flocculating mechanism was a mucoprotein which coated egested particles that were produced by Paramecium caudatum, Vorticella microstoma, and Epistylis plicatilis.

     The concentration of bacteria is apparently beneficial to ciliates. Pond isolates of ciliates observed in shaken and static cultures by Taylor (1978a) were found to attain higher growth rates and lower half saturation prey densities on settled bacteria in static cultures than on suspended bacteria in shaken cultures. The microspatial heterogeneity in the distribution of ciliates in a small pond was shown by Taylor and Berger (1980), who found that the mean patch size was 1.5 to 2.0 cm while the distance between patches was 3 to 4 cm. One might regard the aggregation of bacteria by protozoa in active sludge plants and the microspatial heterogeneity of pond ciliates as being irrelevant to the marine situation. But Sorokin (1970) has reported an increasing aggregation of marine bacteria with depth as total numbers decreased in waters south of the Gilbert Islands and a similar increase in percent aggregated bacteria with decreasing total bacteria seaward from the Great Barrier Reef.

     The flocculation of bacteria is apparently not restricted to ciliates. The ubiquitous freshwater and soil flagellate Oicomonas termo, which can grow in seawater in the presence of bacteria (Hardin 1942), also causes a very marked flocculation of a number of bacterial species (Hardin 1943). Two or three days after the initiation of two membered cultures, floccules would occur with a maximal diameter of 3 mm. Each floccule consisted of a large central mass of bacteria with a few flagellates at the periphery. Johannes (1965) observed that in cultures of heterotrophic marine microflagellates, bacteria were significantly lower than in control

cultures and by the third day mucous particles containing many
bacteria developed while the flagellate population plateaued at 6 x
10^5 cells/ml for the following 6 days. Independently Aaronson (1973)
found that the phytoflagellate Oochromonas danica could force fluid
and particles over its side by flagellar current, where the bacteria
were aggregated, trapped on the surface and engulfed. Aggregation
and phagocytic activity required live, metabolically active bacterial
cells. Flocculation of algae and clay has also been demonstrated
and could be a site of bacterial colonization (Simon et al. 1982).

RECENT LITERATURE ON PROTOZOAN BACTERIVORY

The prey and predators in pelagic ecosystems are differentiated
by their trophic modes and their size (Wiegert and Owen 1971). In
order to have a uniform system of terminology and sizes which would
include the heterotrophic compartments, Sieburth et al. (1978) suggested an expanded system of size fractions in which the phototrophic
and non-phototrophic compartments could be distinguished by epifluorescence microscopy or flow cytofluorometry. In putting this proposal
into practice (Burney et al. 1981, 1982; Davis and Sieburth 1982 and
unpublished) it has been necessary to develop a jargon to describe
the phototrophs, saprotrophs and biotrophs (Sieburth, in press) that
occur in the smaller size fractions. These are shown in Table 1.
This scheme is based upon that of Sheldon and Parsons (1967) that
defined the commonly accepted term microplankton as the 20-200 µm
size fraction and nanoplankton as the 2-20 µm size fraction, but left
the bacterial sized particles < 2 µm in an open-ended group called
the ultra-nanoplankton. The next smallest grouping of 0.2 - 2.0 µm
has been named the picoplankton (Sieburth et al. 1977, 1978; Sieburth
1979), a term that is finding acceptance into the literature. The
use of epifluorescence microscopy permits the enumeration of total
cells and phototrophs, and the estimation of presumed heterotrophs
by difference (Davis and Sieburth 1982). This allows one to follow
changes in the population and calculated biomass of the main trophic
groupings shown in Table 1. For the discussion of bacterivory by
protozoa we will limit ourselves principally to the heterotrophic
components of the picoplankton and nanoplankton size fractions
which are dominated by bacteria and bacterivorous flagellates,
respectively.

Epifluorescence and Electron Microscopy

The dominant cells in the picoplankton are presumably organotrophic bacterial cells which get smaller as one progresses from
estuary to continental shelf to the open sea. This has been reported
previously by Hoppe (1976), Ferguson and Rublee (1976) and Watson et
al. (1977), among others. This is shown in transmission electron
micrographs of stained whole amounts in Figure 1 which use the
centrifugates of particles passing a 2 µm Nuclepore membrane. In

Table 1. Summary of the trophic and size categories of plankton suggested by Sieburth et al. (1978) and used by Burney et al. (1981, 1982), Davis et al. (unpublished) Davis and Sieburth (1982 and unpublished).*

Trophic Types	Plankton Diameter (μm)+						
	0.2	Picoplankton	2.0	Nanoplankton	20	Microplankton	200
Heterotrophs (fluorochromed)		H-pico		H-nano		H-micro	
		bacteria		smaller flagellates		larger flagellates	
				smaller rhizopods		larger rhizopods	
				smaller ciliates		larger ciliates	
Phototrophs (autofluorescent)		P-pico		P-nano		P-micro	
		chroococcoid cyanobacteria		smaller flagellates		larger flagellates	
		bacterial-sized eucaryotes		smaller diatom and dinoflagellates		larger diatoms and dinoflagellates	

* In practice, totals of each size category are obtained from fluorochromed preparations and the phototrophic counts obtained by autofluorescence are subtracted to yield the heterotrophic counts by difference (Davis and Sieburth 1982).

+ As determined by ocular micrometer.

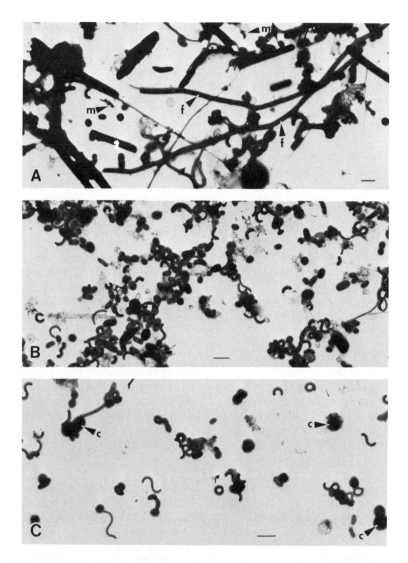

Figure 1. The diversity of morphology of natural populations of
bacteria in the picoplankton size fraction of marine waters as shown
by transmission electron microscopy of negatively stained preparations. A) Estuarine populations range from 30 μm bacterial filaments
(f) to minicells (m) approximately 0.1 μm in diameter. B) Shelf populations are dominated by coccobacillary forms with thin horseshoe-
shapes also present. C) Oceanic samples are primarily small sigmoid
or horseshoe-shaped cells and small cocci, and coccoid forms (c)
about 1.0 μm in diameter which are probably autotrophic forms.
Marker bars equal 1.0 μm.

the open sea the planktonic bacteria are mainly C-shaped forms,
which elongate to form S-shaped forms before division, as well as
small coccobacillary heterotrophs and larger coccoid forms that
include the chroococcoid cyanobacteria. The larger coccobacillary
heterotrophs are a minor component in the natural population, but
when confined in growth chambers in the absence of predators can
take over the culture. It is plausible that the natural populations
we examine represent what is out there and growing, however, it is
difficult to dismiss the nagging possibility that what we are
examining are the leftovers after the fat, ribosome-rich and
actively growing cells have been grazed by the protozoa. The
dominant oceanic forms are still present over the shelf and into the
estuary, but they become overwhelmed with larger forms. These
larger rods may be growing as free cells, but since the larger forms
appear inactive (Hoppe 1976) they may be inactive epibacteria
resuspended from sediment or released from fragile particles of
marine snow or other debris. The gram-negative and organotrophic
nature of these populations is shown in thin sections from both
estuarine and offshore populations in Figure 2. These are intact
and apparently functional cells but they are poor in ribosomes
characteristic of growing cultures (Maaløe and Kjeldgaard 1966).
The very small size of the cells of the dominant population in
oceanic waters speaks against their being motile by bacterial
flagella and they may be truly dependent upon Brownian movement and
diffusion for suspension and nutrition, respectively.

    In addition to the above organotrophs or saprotrophs, the larger
(1-2 μm) cells in the picoplankton size fraction are phototrophs,
both procaryotes and eucaryotes. The first of these forms to be
recognized as a potentially important food for the protozooplankton
were the chroococcoid cyanobacteria (Sieburth 1978; Johnson and
Sieburth 1979; Waterbury et al. 1979). Populations of these cells,
which range from 10^3 to 10^5/ml, can be readily observed in pelleted
picoplankton preparations (Fig. 2). They are ingested and digested
by protozoa such as the ciliate Uronema while being refractory to
digestion by copepods (Johnson et al. 1982). Our continuing studies
on the ultrastructure of the picoplankton also show that in both
nearshore and offshore populations, small phototrophic eucaryotes
< 2 μm are present at most stations, sometimes reaching populations
equal to those of the cyanobacteria (Fig. 2) (Johnson and Sieburth
1982). These consist of the flagellate Micromonas pusilla, a scale-
less prasinophyte common to nearshore waters, an aflagellated and
very small prasinophyte (0.75 μm) with scales that occurred at 15 of
21 sampling stations in the western North Atlantic, and non-flagellated chlorophytes like Chlorella and Nanochloris which have both
nearshore and offshore forms. An interesting observation on these
P-pico is that regardless of whether being procaryotic or eucaryotic,
they serve as food for the protozooplankton in the nanoplankton and
microplankton size fractions while being refractory to digestion by
copepods (Silver and Alldredge 1981; Johnson et al. 1982).

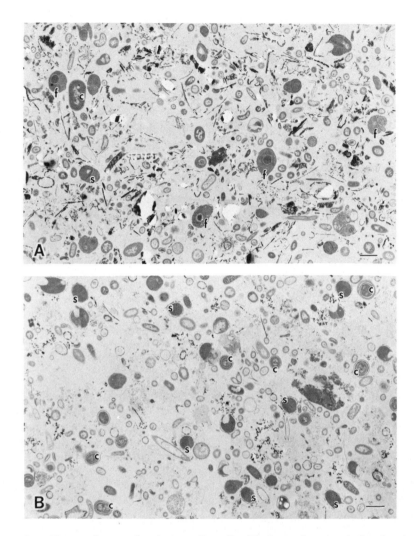

Figure 2. Natural populations of unicellular, bacterial-sized
eucaryotic phototrophs in the picoplankton from Narragansett Bay,
RI (A) and the Grand Banks (B), as shown by transmission electron
microscopy of thin sections. A) The nearshore estuarine sample has
high numbers of the microflagellate (f) Micromonas pusilla, with a
scaled, non-flagellated prasinophyte (s) and cyanobacteria (c) also
present. B) The offshore Grand Banks sample has higher populations
of cyanobacteria (c) and the scaled prasinophyte (s), with fewer M.
pusilla cells. Marker bars equal 1.0 μm.

    Although the non-phototrophic, bacterial-sized cells observed
in fluorochromed preparations with epifluorescence microscopy are
assumed to be organotrophs, and may often be dominated by them, the

bacterial population could also contain a significant fraction of
chemotrophs. Nitrifying bacteria, methane oxidizing bacteria and
cyanobacteria share similar cytomembranes and carboxysomes. The
Type II and Type III "shade type" cyanobacteria observed at a depth
of 100 m (Johnson and Sieburth 1979), the zone of a marked nitrite
maximum (Wada and Hattori 1971; Olson 1981a,b) as well as a methane
maximum (Scranton and Brewer 1977), have resisted all cultivation
attempts as phototrophs and may indeed be chemotrophs having
populations of 10^4 to 10^5 cells/ml. If this hypothesis can be
substantiated, then the chemotrophic picoplankton (C-pico) will have
to be accomodated not only in the trophic scheme in Table 1, but in
the data on bacterivory in Tables 2-7. Such appears to be the case.
While this manuscript was awaiting proofing (March 1983), obligate
methane-oxidizing bacteria which have an ultrastructure similar to
type III cells have been grown from the Sargasso Sea by the author.

The picoplankton prey occurring in oceanic waters, therefore,
probably consist of two types of cells, the smaller organotrophs
which occur towards the lower end of this size fraction and which
have populations between 10^5 and 10^6 cells/ml and a mean cell volume
near 0.04 μm^3, and the larger autotrophic cells (both phototrophic
and chemotrophic) with populations between 10^3 and 10^5 cells/ml and a
mean cell volume approaching or exceeding 0.2 μm^3. These two groups
differing in trophic mode and size may approach similar biomasses of
20 μg wet weight/liter.

Transmission electron microscopy of thin sections of pelleted
nanoplankton has not been as informative as that of the picoplankton.
Sections through the larger microorganisms do not always yield enough
information for trophic or taxonomic identification. Epifluorescence
microscopy shows that heterotrophic nanoplankton (H-nano) are usually
as numerous as the phototrophic nanoplankton (P-nano); both occur at
between 10^2 and 10^5 cells/ml (Davis et al., unpublished). Bacteria-
grazing protozoa can be cultured on bacteria mildly enriched in
seawater samples using rice or wheat grains to yield cultures of
bacteria-eating amoebae, flagellates and ciliates (Fig. 3). Most
Probable Number estimates using such cultural procedures, however,
only account for 0.1% of the in situ population in oceanic waters
and 0.1 - 26% in estuarine waters (Davis and Sieburth, unpublished).
One problem in characterizing the colorless (apochlorotic) flagellates by light microscopy is making sure they are not phytoflagellates containing chloroplasts. Such diagnosis is poor, especially
for the smaller cells. Sequential epifluorescence and electron
microscopy (Davis and Sieburth, unpublished) reliably indicates the
location of phototrophic and heterotrophic cells and provides
taxonomically useful information on cell morphology. Most of the
cells counted as H-nano by epifluorescence microscopy appear to be
flagellates (Fig. 4). This technique allows the detailed characterization of numerically dominant P-nano and H-nano species. Those
species that have evaded culture due to problems such as trace metal

toxicity (Brand et al. 1981; Knauer and Martin 1981) can now be
targeted and eventually brought into culture just as my laboratory
is doing for the picoplankton.

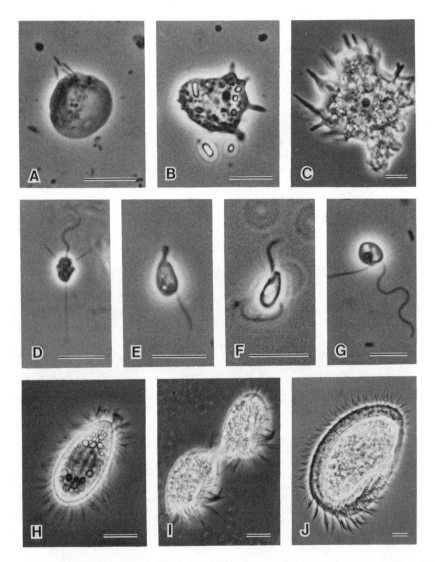

Figure 3. The diversity of bacterivorous protozoans obtained with
rice grain enrichments from nearshore and offshore waters, showing
amoebae (A-C), flagellates (D-G), and ciliates (H-J) in phase
contrast photomicrographs of living cells. These include species
of A) Platyamoeba, B) Paramoeba, C) Mayorella, D) Actinomonas, E)
Rynchomonas, F) Bodo, G) Paraphysomonas, H) Uronema, I) Euplotes,
and J) Peritromus. All marker bars equal 10 µm.

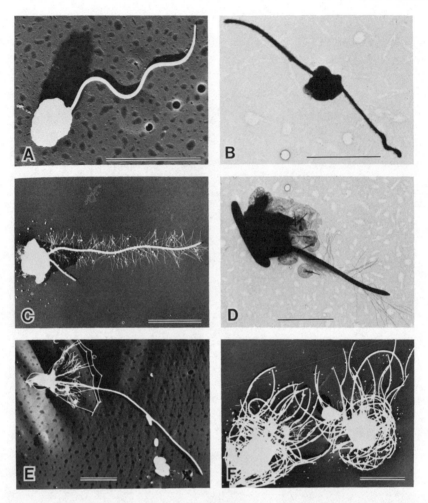

Figure 4. The diversity of bacterivorous flagellates observed by
transmission electron microscopy of negatively stained (B and D)
and shadowed cells from natural seawater (A,C,E,F) and enrichment
cultures (B,D). A) An unidentified monad; B) a species of Bodo:
C) a species of Paraphysomonas; D) Paraphysomonas imperforata; E)
Pleuorsiga minima Throndsen var. minuta Leadbeater; F) Acanthoepsis
unguiculata. Marker bar in D equals 1.0 µm; all other marker bars
equal 5.0 µm.

Distribution and Dynamics

     The relationship between bacterial prey and their flagellate
and ciliate predators is controlled in part by the nature and
temporal variation in the substrates used by the prey. The nature
of bacteria occurring as free cells in the water (planktonic

bacteria) and those associated with surfaces of aggregates such as marine snow (epibacteria) appear to be very different. The marked difference in size may represent just the difference between starved and growing populations (Maaløe and Kjeldgaard 1966) or it may represent differences between bacterial forms living on "dissolved organic matter" or on particulate organic matter (Sieburth 1976, 1979). Although planktonic bacteria vastly outnumber epibacteria, Ferguson and Rublee (1976) correctly point out that the epibacteria are much larger and have a greater biovolume, and that their importance is greater than their counts alone would indicate. Similarly, Linley and Field (1982) found that bacteria associated with particles in the coastal upwelling area of a kelp bed were only 5% of the population but up to 34% of the biomass. This is seen in the epifluorescent micrographs of DAPI stained planktonic bacteria free in the water and epibacteria on flakes of marine snow from the Sargasso Sea shown in Figure 5.

The contribution of 'free bacteria' relative to attached bacteria, appears to increase as one goes from freshwater to seawater and from the estuary to the open ocean. Bell and Albright (1981) studied the Fraser River Estuary in British Columbia and observed that although most of the bacterial biomass and activity (60%) was associated with the particles found in the river, there was decreasing attachment and activity down the estuary to the straits of Georgia (26 ‰ S) where attached bacteria which accounted for 15-39% of the total bacterial population only accounted for 4% of the heterotrophic activity. Sieburth and Davis (1982) reported that in the turbid estuary of Narragansett Bay the population of attached bacteria were 1.5 orders of magnitude lower than the free cells and decreased seaward until they were 2.5 orders of magnitude lower while the free cells decreased less than one order of magnitude. This is in agreement with Williams (1970) who showed that a major part of heterotrophic activity passed through Millipore filters while Azam and Hodson (1977) showed that more than 90% of heterotrophic activity in oceanic waters could pass 1 μm Nuclepore filters.

Although the dense microbial populations associated with flakes of marine snow or colloidal aggregates can be responsible for a disproportionate amount of activity (Kirchman and Mitchell 1982), the greater population and biomass of planktonic bacteria must be dependent upon and control in part the daily and seasonal fluctuations in "dissolved organic matter" and their labile fractions. The cyclical nature of usable "DOM" for bacteria appears to be closely related to the cyclical nature of primary production. Andrews and Williams (1971) measured the concentrations and oxidation rates of glucose and amino acids in the English Channel and concluded that bacterial activity and substrate concentration are interrelated, with the bacteria controlling substrate levels and only rarely permitting high levels to occur. Hobbie and Rublee (1977) found a strong correlation between primary production and bacterial activity while

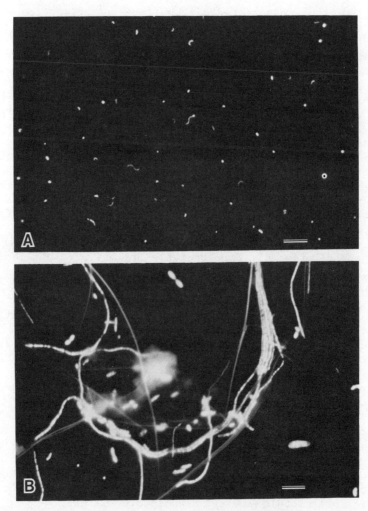

Figure 5. A comparison of the size and morphology of natural
populations of bacteria from the Sargasso Sea stained with DAPI and
observed with epifluorescence microscopy. A) The small planktonic
bacteria free in the water, with their characteristic C-, S- and
coccobacillary forms. B) The large epibacteria colonizing marine
snow consist of coccobacillary, rod and filamentous forms. Marker
bars equal 5.0 μm.

Sayler and Gilmour (1978) found that DOC enrichments caused a direct
and linear response by the bacteria. When the picoplankton size
fraction, presumably dominated by chemotrophic bacteria, was removed
from the predatory phagotrophs in the nanoplankton and incubated in
diffusion chambers in the dark while being bathed with the source

water there were discrete daily cycles of growth directly associated
with the photoperiod (Sieburth and Lavoie 1978; Sieburth 1979).
This also occurred in oceanic samples treated in a similar manner
(Sieburth et al. 1977). Comparisons of carbohydrate concentrations
at dusk and dawn at several stations in the North Atlantic indicated
a daytime accumulation of total carbohydrates and a night-time
utilization which coincided with a maxima in particulate ATP smaller
than 3 µm (Burney et al. 1979). While studying the distribution of
suspended bacteria in the neritic waters south of Long Island during
stratified conditions, Ferguson and Palumbo (1979) observed a
significant variation in the number of bacteria with the time of day
in which there was a marked increase during the night until daybreak
when numbers decreased into the early afternoon.

The ability to follow population trends of the P-nano and H-nano
and H-pico by epifluorescence microscopy (Davis 1982) has permitted
Burney et al. (1981) to compare the relationships of these populations to diel changes in dissolved carbohydrate concentrations in
the Sargasso Sea. The combined activity of the < 20 µm plankton
appeared to actively regulate dissolved carbohydrate concentrations.
Inverse trends dominated the relationships of TCHO and PCHO with
H-pico indicating that increasing bacterial numbers were associated
with a decrease in carbohydrates.

In a subsequent study Burney et al. (1982) observed changes in
picoplankton and nanoplankton with carbohydrate in the mixed layer of
the northwestern Caribbean Sea. Two different types of diel patterns
of dissolved carbohydrates were observed as shown in Figure 6. On
three days when there was no sustained CO_2 uptake, indicating no
sustained photosynthesis, TCHO and PCHO generally declined during
the afternoon and early evening while MCHO tended to increase due to
release from PCHO (see Figure 6B for example and note the absence of
an AM peak in phosphate). On two other days when sustained CO_2
uptake showed that photosynthesis occurred during the day, there
were large evening peaks in TCHO and PCHO probably resulting from
the release of recently produced PCHO by P-nano while MCHO remained
constant or declined, indicating minimal daytime utilization of PCHO
(see Figure 6A for example and note the presence of an AM peak in
phosphate). Fluctuations in H-pico populations were inversely
correlated with PCHO dynamics and directly related to MCHO variation,
possibly due to extracellular hydrolysis of PCHO to MCHO during
periods of rapid bacterial growth and net PCHO utilization. These
changes have also been observed to occur during the utilization of
the polysaccharide laminarin by natural populations of marine
bacteria in batch culture (Baxter 1982).

Apparent Relationships

Before dealing with the heterotrophic part of the microbial food
chain, it is necessary to put the relative proportion of phototrophic

to heterotrophic biomass into perspective. We are so conditioned
by the concepts developed by Slobodkin (1962) that a predator is
only 10% efficient and requires a biomass of prey 10 times its
biomass, that we ignore the fact that microorganisms in the size
range being discussed have a feeding efficiency (conversion factor)
closer to 40-60% (Payne and Wiebe 1978) or 60-70% (Calow 1977).

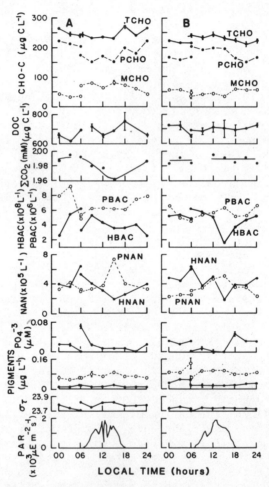

Figure 6. Comparison of the diel cycles of dissolved carbohydrates,
organic carbon and ΣCO_2 with those of the microbial plankton and
environmental parameters at one drift station (4) in the northwestern
Caribbean during a nonproductive Type I day (B) and a productive Type
II day (A). (Fig. 3 from Burney et al. 1982.) Note presence of PO_4
at start of Type II day and release of PO_4 associated with H-nano
increase on Type I day. Pigments are chlorophyll a (o--o--o) and
phaeopigment a (•—•—•).

Fenchel (1982b) found that yield for bacterivorous flagellates was
quite constant in nearshore isolates that had a net growth efficiency
of 60% on the basis of carbon. In comparing the trophic structure,
available resources and population density in terrestrial vs. aquatic
ecosystems, Wiegert and Owen (1971) state that "most stable terrestrial communities are dominated by autotrophs that are large, slow-
growing and have a low intrinsic rate of increase. First order
biophages (those eating plants directly) are usually more numerous,
smaller and/or faster growing than the plants...Contrast this situation with that of the typical open-water aquatic community where the
autotrophs (phytoplankton) have little structure, are more numerous,
smaller and have a higher biotic potential than their predators.
Complete elimination of these producers...by plankton feeders is
improbable....In aquatic ecosystems degree of resource exploitation
by first order biophages is high, ingestion sometimes approaching
100% of net primary production in the open ocean where the phytoplankton has a high biotic potential relative to its grazers."

The microflagellates have usually been considered phytoplankton
(Beers et al. 1975; Booth et al. 1982), with few exceptions (Wood
1955). Fenchel (1982d) has used epifluorescence microscopy to note
the quantitative changes in the bacterial prey and heterotrophic
flagellates at two day intervals over a month in the Limfjorden,
Denmark. The differential epifluorescence procedure of Davis
(1982a) has provided data on the populations of both phototrophic
and heterotrophic nanoplankters as well as the picoplankton (Burney
et al. 1981, 1982; Davis et al, unpublished). The latter includes
two yearly cycles in Narragansett Bay, RI, a number of diel cycles in
both nearshore and offshore waters as well as the vertical distribution of these populations at 32 stations. When quantifying the
heterotrophic flagellates and bacteria in the Limfjorden, Fenchel
(1982d) did not include the counts on the phototrophic flagellates
which are necessary to put the ratios of heterotrophs and phototrophs
into perspective. When data from EN-009 along the Atlantic seaboard
of the United States, which is representative, is used to calculate
the relative biomass of phototrophs to heterotrophs in both nearshore and offshore waters, we obtain the values shown in Table 2.
Despite an almost six-fold reduction in total biomass from the shelf
to the open sea, the proportion of picoplankton and nanoplankton in
the phototrophs and heterotrophs remains amazingly constant. The
phototrophs which account for a third of the total biomass appear to
support twice their biomass of "heterotrophs". Remember that a
substantial fraction of the H-pico could be C-pico.

A useful calculation in studying the prey and predators that
make up much of this heterotroph biomass may be the number of prey
per predator such as the 580 bacteria/predator value obtained by
Lighthart (1969) using culture techniques. The marked difference in
the biovolumes of the H-pico between nearshore and offshore waters
illustrated in Figure 1 is reflected in the number of H-pico per

Table 2. A comparison of the relative biomass of phototrophic and
heterotrophic microorganisms < 20 μm along the Atlantic
Seaboard.

	Biomass μg wet weight L^{-1}*					
	Phototroph			Heterotroph		
Province	-pico	-nano	subtotal	-pico	-nano	subtotal
Shelf	12 (5%)	74 (27%)	86 (32%)	115 (42%)	72 (26%)	187 (68%)
Open ocean	1 (5%)	14 (29%)	15 (31%)	20 (42%)	13 (27%)	48 (69%)

* This assumes a specific gravity of 1.0, that the biovolumes of
attached bacteria are 0.3 μm^3, H-pico are 0.09 μm^3 in nearshore
waters but 0.04 μm^3 in oceanic waters, P-pico procaryotes are
0.2 μm^3, P-pico eucaryotes are 0.4 μm^3, and that both P-nano and
H-nano have a mean biovolume of 24 μm^3. Data from 3 shelf and 3
ocean stations, EN-009.

H-nano shown in Table 3. These values obtained by direct cell counts
seem to be sharply divided into low values between 243 and 322 found
in estuarine, shelf and slope waters, with a mean of 275 and high
values of 1017 to 1691 with a mean of 1262 found in the Gulf Stream,
Sargasso, and Caribbean Seas. Note that the numbers of nanoplankton
which vary from environment to environment are quite evenly balanced
between P-nano and H-nano.

This is further illustrated in Table 4 which shows the ratios
of populations and biovolumes between the H-nano predators and the
H-pico prey in coastal and oceanic waters. The mean values of
H-pico prey per H-nano predator obtained from Table 2 and the biovolumes obtained from our samples as well as from Ferguson and
Rublee (1976) and Watson et al. (1977) are used in making these
calculations. If one assumes equal carbon contents per volume of
prey and predator, a system where prey growth is balanced by
predation, and a 50% conversion efficiency of consumed prey to
predator biomass, then in the oceanic environment every doubling of
H-pico would allow one doubling of H-nano. For the coastal waters
these same assumptions would lead to the conclusion that nearshore

Table 3. A summary of average populations of phototrophic and
          heterotrophic nanoplankton and heterotrophic picoplankton
          as well as the ratios of H-pico prey per potential H-nano
          predator for a number of marine environments (R/V Endeavor
          cruises 009, 2-22 June 1977, and 033, 22 Feb.- 18 Mar.
          1979). A marked increase in H-pico:H-nano ratio in
          offshore waters may reflect a major decrease in bacterial
          biovolume in offshore waters.

Environ-ment	Location	P-nano	H-nano	H-pico	H-pico: H-nano	Mean Ratio
		$\times 10^3 \text{ ml}^{-1}$		$\times 10^5 \text{ ml}^{-1}$		
	Estuarine	5.9	9.1	23.7	260	
Coastal	Shelf	3.1	3.0	9.6	322	275
	Slope (Cold Ring)	1.6	1.4	3.4	243	
	Gulf Stream	0.7	1.1	18.6	1691	
Oceanic	Sargasso Sea	0.8	0.6	6.1	1017	1262
	NW Caribbean	0.4	0.5	5.4	1080	

Table 4. A comparison between coastal and oceanic environments of
          the mean abundance ratios (from Table 3) and the calcula-
          ted biovolume ratios of the H-nano predators and the
          H-pico prey.

Environment	Abundance H-nano:H-pico	Biovolume (μm^3)* H-nano:H-pico
Coastal	1:275	24:33
Oceanic	1:1,262	24:50

* Biovolume calculations: H-nano mean cell vol. = 24 μm^3. Coastal
  H-pico = 15% attached cells of 0.3 μm^3 mean cell vol. and 85% free
  cells of 0.09 μm^3 mean cell vol.; Biovolume = (0.15)(0.3)(275) +
  (0.85)(0.09)(275) = 33 μm^3. Oceanic H-pico = 100% free cells of
  0.04 μm^3 mean cell vol.; Biovolume = (0,04)(1262) = 50 μm^3.

flagellates increase at a slower rate relative to bacterial turnover time (1.4 doublings of H-nano per doubling of H-pico). However, coastal H-nano growth rates could easily exceed those of oceanic populations. Coastal bacterial (prey) populations likely grow at considerably faster rates than oceanic populations, and coastal H-nano may supplement their prey with the large epibacteria associated with organic debris more abundant in coastal waters. Additionally, the conversion efficiency for oceanic populations may be lower, due perhaps to the smaller size and lower density of the prey requiring a greater energy expenditure by the predator.

The abundance and biovolume of different trophic groups in the picoplankton and their relative volumes available for the proto-zooplankton in three oceanic provinces are given in Table 5. For ease of comparison, H-pico has been normalized to 100. It is obvious that the major source of prey in all three environments is the H-pico which may include a significant C-pico component in offshore waters.

When one considers the possible prey for the flagellates and ciliates, there is more to consider than populations. One must also consider how much time it takes to obtain a biovolume necessary for one protozoan division assuming a 50% feeding efficiency. This is calculated in Table 6. Feeding rates of 1 cell/min for flagellates and 5 cells/min for ciliates seem reasonable from personal observation and from the literature. For the flagellates feeding solely on the minicells of oceanic H-pico, they could achieve their food requirements in some 20 hrs. But if larger actively growing bacteria approaching the size of epibacteria were available, e.g., C-pico, then populations one-tenth that of the minicells could provide the feeding requirements of the flagellates in less than three hours.

The six species of nearshore, bacterivorous flagellates charac-terized by Fenchel (1982a) with regard to morphology and feeding mechanisms have been used to study both growth (Fenchel 1982b) and starvation (Fenchel 1982c). He found that maintenance energy is very small (2-5% that of growing cells) and that yield was constant with a net efficiency of 60%. Specific growth rates ranged from 0.15 to 0.25/h with the smaller cells consuming 27 to 107 bacteria per hour. Thirty estuarine species and 12 oceanic species of bacterivorous flagellates were isolated and characterized taxonomically by Davis and Sieburth (unpublished). In addition to an ubiquitous oceanic flagellate previously undescribed which was put into a new genus, three new species were also described. Fourteen of these isolates (seven estuarine and seven oceanic) were used to estimate predation on actively growing bacteria (Davis and Sieburth, unpublished) rather than on the usual static bacterial cultures. This was achieved using the frequency of dividing/divided cells (FDDC) to estimate the rates of bacterial growth and predation. Similar clearing and feeding rates to those on static culture were obtained, indicating that both estuarine and oceanic isolates can effectively prey at concentrations as little as 10^5 cells/ml at average rates of 30 to 200 bacteria per flagellate per hour.

Table 5. The abundance and biovolumes of different trophic groups
in the picoplankton and their ratios available for the
predators of bacterial-sized prey in the protozooplankton
in three oceanic provinces. The heterotrophic picoplankton
(H-pico) have been normalized to 100, phototrophic picoplankton (P-pico) include procaryotic cyanobacteria and
eucaryotic microalgae.

Parameter	H-pico	P-pico Procaryote	P-pico Eucaryote
Mean cell volume (μm^3)	*	0.2	0.4
Estuary			
Abundance, cells ml^{-1}	3×10^6	5×10^3	1×10^4
Biovolume, μm^3 ml^{-1}	3×10^5	1×10^3	4×10^3
Relative volume	100	0.3	1.3
Shelf			
Abundance, cells ml^{-1}	1×10^6	5×10^4	5×10^3
Biovolume, μm^3 ml^{-1}	1×10^5	1×10^4	2×10^3
Relative volume	100	10	2.0
Open Sea			
Abundance, cells ml^{-1}	5×10^5	5×10^3	1×10^2
Biovolume, μm^3 ml^{-1}	2×10^4	1×10^3	4×10^1
Relative volume	100	5	0.2

*See Table 4 for biovolume assumptions.

Table 6.   The number of cells and time required to consume potential
           prey as a function of prey biovolume and 2x predator bio-
           volume (50% feeding efficiency) for flagellates (24 μm^3
           biovolume) and ciliates (3500 μm^3 biovolume) at feeding
           rates of 1 cell min^{-1} for flagellates and 5 cells min^{-1}
           for ciliates.

Predator	Prey	Number of cells/predator required for one division at specified feeding rate	Calculated time required for one division on diet
Flagellates	H-pico	1200	20 h
"	P-pico	240	4 h
"	Epibacteria*	160	2.7 h
Ciliates	H-pico	175,000	583 h
"	P-pico	35,000	116 h
"	Epibacteria*	23,333	77 h

* These are underestimates due to fragile nature of marine snow and
  inherent sampling problems.

     The case for ciliates is much different due to their vastly
greater biovolume which may be some 146-fold greater than that of
the phagotrophic flagellates. None of the bacterial sized prey are
present in sufficient bulk to permit a reasonable division rate.
Only the nanoplankters, which are not the normal prey for bacterivorous ciliates, are present in sufficient bulk. It should be
remembered, however, that these crude calculations assume equal
predation rates at all prey densities, which is unlikely. Higher
consumption rates must be achieved in the dense populations of prey
associated with marine snow in a manner similar to the higher rates
of growth and lower half saturation prey densities obtained on
settled bacteria by Taylor (1978a).

     The number of bacterivorous ciliates which the biovolume of
picoplankton in natural populations might support in coastal and
oceanic environments is given in Table 7. These "populations" of
ciliates would appear to be reasonable estimates compared to
populations enumerated by Beers et al. (1975) for the North Pacific
Gyre (0.7 ciliates/ml) and Sorokin (1977) for the Japan Sea (3.2
cells/ml).

If the pelagic ciliates, like those in nearshore waters, are associated with the surface of particles undergoing bacterial degradation such as marine snow (Alldredge 1979; Shanks and Trent 1979; Silver and Alldredge 1981; Silver and Bruland 1981) or detritus (Fenchel 1977, Fenchel and Harrison 1976, Fenchel and Jørgensen 1977), then only ciliates dissociated from such particles will be enumerated in such small portions of sample. Recent observations on marine snow in the Sargasso Sea (Caron et al. 1982) show that ciliates can occur up to populations of 10^4/ml of colonized seston and may be the reservoir for the low populations found free in the water. The species of ciliates associated with marine snow in the Sargasso Sea collected while diving by Paul Davis and David Carbon belong to genera associated with the seston in turbid coastal waters, such as Uronema, Euplotes and Peritromus, as well as a number of new species (Gene Small, personal communication).

Epilogue

We know very little about the nature and cycling of organic matter in the sea and how the latter is influenced by the tropho-dynamics of the nanoplankton and picoplankton. But it is a far cry from a biological desert (Ryther 1969) with ancient and refractory organic matter (Williams et al. 1969). Unravelling the temporal and spatial scenarios hour by hour during the diel cycle as the sun cues the different trophic groups to enter the stage, perform their necessary roles, then retire, is as exciting to imagine as any contrived stage play. I now want to venture into unknown areas, conjecture, and describe how I perceive the release, nature, and

Table 7. The number of planktonic ciliates equivalent to the bio-volumes of picoplankton available for coastal and oceanic environments. This assumes a biovolume of 3500 μm^3 per ciliate and a 50% food conversion.

	Compartment	Biovolume μm^3 ml^{-1}	Equivalent number of ciliates ml^{-1}
Shelf	P-pico	12,000	1.7
	H-pico	115,200	16.0
Open Ocean	P-pico	1,040	0.15
	H-pico	20,000	2.8

transformation of organic matter as it forms the microhabitats for
free and aggregated bacteria and how the bacterivorous protozoa
recover "lost soluble production" back to the food chain. The
purpose of this blatant speculation is to illustrate the kind of
information that I think we need in order to understand heterotrophic
microbiology in the open sea.

I envision a calm and flat sea that awakens at dawn to watch
the night shift of organotrophic bacteria and some of the protozoa
essentially retiring for the day as a result of photoinhibition or
inhibition produced during photosynthesis. Solar radiation that
penetrates some 80 m with levels of light useful for photosynthesis,
apparently also produces near UV radiation at inhibitory levels that
penetrates much of this depth (Smith and Baker 1979), impeding or
stopping the processes of organotrophic bacteria, chemotrophic
bacteria (Olson 1981b), and perhaps the protozoa. The apparent
inhibition of heterotrophic processes during the photoperiod may also
be due to the formation of toxic derivatives of phytosynthetically
produced oxygen such as superoxides and organic peroxides which are
quite labile and would disappear during the dark period. Such photic
and chemical inhibition would allow the photosynthesizers and their
released photosynthate to accumulate during the photoperiod, thereby
setting the stage for the night shift. A significant fraction of
the "dissolved" organic matter which accumulates during the photoperiod may exist in a colloidal state (Sharp 1973), perhaps as microfibrils (Leppard et al. 1977). As these are adsorbed onto the
planktonic bacteria and utilized, lowering concentrations of
substances such as carbohydrates down to threshold levels, the
bacteria and their associated extracellular polymeric slime would
provide nutrients for the protozoa.

The protozoa presumably release inorganic nitrogen and phosphorus at concentrations sufficient to permit photosynthesis to
start at daybreak (Fig. 6). The stage is then set for the dayshift
of actors. During the photoperiod the photosynthesizers take up the
carbon dioxide released during darkperiod respiration and reduce it
to form carbohydrates which are slowly released and accumulated
during the photoperiod which along with other released materials may
account for some 20 to 25% of primary production. During the
episodic activity of the protozooplankton reacting to accumulated
phototrophs, more polymeric organic matter, perhaps richer in
nitrogen, is released which may account for another 20 to 25% of
primary production, to yield a total of 40 to 50% production that
will be released as DOM.

The same solar radiation which provides energy for photosynthesis also warms the air and causes daytime sea breezes. These sea
breezes in turn set up Langmuir circulations and other mixing
processes that vertically advect particles including the polymeric
microfibrils in the colloidal organic carbon which appear to form

waterlogged gels or films that act like semi-submerged rafts near
the sea-air interface. Other particulate organic matter accumulating
at the sea-air interface would be low density material such as the
empty chitinous exoskeletons of the smaller metazooplankton, which
are frequently cast off during metamorphosis and would enrich chitin
digesting bacteria and phycomycetes. During the post-meridian period
as the sun gets lower on the horizon and darkness falls, the levels
of solar radiation drop below inhibitory levels and the accumulated
organic matter in the water column and in the surface films, which
may accumulate to grams of carbon per liter concentrations (Sieburth
et al. 1976), becomes vulnerable to the organotrophic bacteria. The
levels of "dissolved" organic carbon in the water column drops back
to minimal threshold levels by dawn. Distinctive microcolonies of
bacteria develop on the upper surface of the sea skin to support
thriving and diverse populations of flagellates and ciliates that
attach to the film or crawl through it grazing upon bacteria
(Sieburth, in press). The daytime sea breezes which help form the
new layers of surface films also provide the turbulence to sink
densely colonized patches of surface film which when submerged may
become flakes of marine snow. The surface microlayer can be thought
of as clouds of potential snowflakes, perhaps a primary source of
such material. This sloughed skin of the sea is then renewed by
Langmuir circulations and their upwelling currents which bring newly
released organic matter to the surface in a matter of hours. The
flakes of marine snow with their colonizers not only act as oases
for the larger protozooplankton and metazooplankton grazers but may
be instrumental in the release of phosphates (Barsdate et al. 1974)
essential for the next day's primary production and provide food for
grazing fish. The flakes of marine snow settling through the mixed
layer re-form at the denser water in the thermocline, the lower
boundary of the mixed layer. The latter must act as a site of
aggregation of secondary snow clouds which when dense enough or
through turbulence will release snowstorms of flakes to the aphotic
waters below the thermocline.

But it must be the planktonic populations that form the fabric
of life in the mixed layer. Between the sporadic occurrences of
snowflakes there are millions of unattached bacteria and thousands
of unattached flagellates per ml. The flakes of marine snow and
their microbiota are just embroidery on this basic fabric of life in
pelagic waters. Both the fabric and its ornamentation working in
excess of 50% feeding efficiency (Calow 1977; Payne and Wiebe 1978)
must be an efficient mechanism for returning "lost" soluble production back into the food chain.

Lindeman (1942) constructed a model which diagrammed energy
transfer from the producers to various orders of consumers with
waste materials from all trophic levels going to the decomposers.
An important improvement in this model was the recognition by Wiegert
and Owen (1971) that a distinction should be made between organisms

ingesting living cells (biophages) and those subsisting on dead organic matter (saprophages). Since both trophic types are decomposers this distinction permitted the development of a hypothesis to explain how the first order biophages are determined by the properties of the producers. A feature of this model is that each order of biophages depends upon the one above it. In the case of oceanic ecosystems the high biotic potential of the very small phytoplankters (31% of the biovolume) appears to support a two-fold heterotrophic biomass as biotrophs (biophages, 26% of the biovolume) and saprotrophs (saprophages, 42% of the biovolume) as summarized in Table 2. I believe that the saprotrophs, dominated by the organotrophic bacteria, are in a more or less equal partnership with the producers

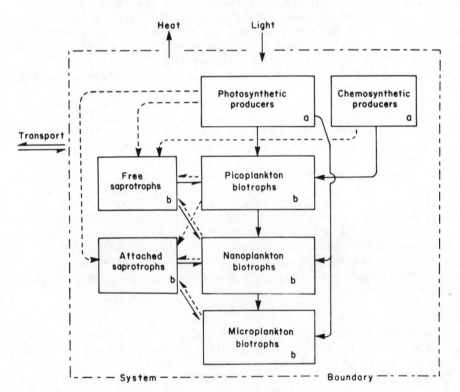

Figure 7. Trophic structure and transfer within the microbial plankton, adapted from the model of Wiegert and Owen (1971), which recognizes the differences between biotrophs which consume live cells and saprotrophs which consume dead organic matter. The primary producers (a), both photosynthetic and chemosynthetic, use the dissolved inorganic nutrients released from both the saprotrophs and the biotrophs (b), while the dissolved, colloidal, and particulate organic matter derived from the primary producers and biotrophs (---) supports the saprotrophs.

and biotrophs and that this should be reflected in a trophic transfer model. The studies and observations summarized in this review have been used to modify the basic trophic transfer model of Wiegert and Owen as shown in Figure 7. One major change is that chemosynthetic (chemotrophic) bacteria are given recognition as primary producers since a significant fraction of carbon cycling may be via chemotrophs such as methanogens and methane oxidizers. A second major change is that the various orders of biotrophs and saprotrophs are not serially dependent upon one another as in the Wiegert and Owen model, but rather the 2nd and 3rd order biotrophs may be dependent upon the 1st and 2nd order saprotrophs, respectively. The division of bacteria into the 1st order saprotrophs which live a "free" existence as planktonic bacteria utilizing "DOM" and the 2nd order saprotrophs attached to particles of marine snow or other seston is important as it points out that the biotrophs are not necessarily serially dependent upon each other but that trophic transfer can be sideways through the saprotrophs. Also note that the 3rd order biotrophs are dependent upon the 1st as well as the 2nd order biotrophs. This scheme emphasizes that all trophic levels and modes other than primary producers are "decomposers" which yield minerals necessary to sustain these photosynthetic and chemosynthetic microorganisms (as pointed out by Wiegert and Owen 1971, Fenchel and Harrison 1976, and Fenchel and Jørgensen 1977), that help sustain the top-heavy heterotrophic biomass in oceanic ecosystems dominated by bacteria and their protozoan predators.

ACKNOWLEDGMENTS

     The work reported on in this review has been made possible by the Biological Oceanography Program of the National Science Foundation through grants GA-41501X, DES-7401537, OCE-7681779, OCE-7814528, OCE-7826388, and OCE-8121881, and the combined efforts of my graduate students cited in the references and my research colleagues who helped proof the manuscript. Paul G. Davis is gratefully acknowledged for obtaining or supervising the collection of data on the picoplankton and nanoplankton, as is Paul W. Johnson for Figures 2-7 and for his usual great help in the preparation of the figures, and not least Jean Knapp who suffers through the handscript and retrieving of missed references.

REFERENCES

Aaronson, S. 1973. Particle aggregation and phagotrophy by
     Ochromonas. Arch. Mikrobiol. 92: 39-44.
Alldredge, A. L. 1979. The chemical composition of macroscopic
     aggregates in two neritic seas. Limnol. Oceanogr. 24: 855-866.
Allen, H. L. 1977. Experimental studies of dissolved organic

matter in a soft-water lake, pp. 477-527. In: J. Cairns [ed.],
Aquatic Microbial Communities. Garland Publ. Co., New York.

Anderson, O. R., and A. W. H. Bé. 1976. A cytochemical structure
study of phagotrophy in a planktonic foraminifer Hastigerina
pelagica (d'Orbigny). Biol. Bull. 151: 437-449.

Anderson, O. R., M. Spindler, A. W. H. Bé, and C. Hemleben. 1979.
Trophic activity of planktonic foraminifera. J. Mar. Biol.
Assoc. UK 59: 791-799.

Andrews, P., and P. J. leB. Williams. 1971. Heterotrophic utilization of dissolved organic compounds in the sea. III. Measurement of the oxidation rates and concentrations of glucose and
amino acids in sea water. J. Mar. Biol. Assoc. UK 51: 111-125.

Azam, F., and R. E. Hodson. 1977. Size distribution and activity
of marine microheterotrophs. Limnol. Oceanogr. 22: 492-501.

Barker, A. N. 1946. The ecology and function of protozoa in sewage
purification. Ann. Appl. Biol. 33: 314-325.

Barna, I., and D. S. Weis. 1973. The utilization of bacteria as
food for Paramecium bursaria. Trans. Am. Microsc. Soc. 92:
434-440.

Barsdate, R. J., R. T. Prentki, and T. Fenchel. 1974. Phosphorus
cycle of model ecosystems: significance for decomposer food
chains and effect on bacterial grazers. Oikos 25: 239-251.

Baxter, M. 1982. The response of marine bacteria to carbohydrate.
M.S. thesis, University of Rhode Island. 88 pp.

Beers, J. R., F. M. H. Reid, and G. L. Stewart. 1975. Microplankton
of the North Pacific Central Gyre. Population structure and
abundance, June 1973. Int. Rev. ges. Hydrobiol. 60: 607-638.

Bell, C. R., and L. J. Albright. 1981. Attached and free-floating
bacteria in the Fraser River Estuary, British Columbia,
Canada. Mar. Ecol. Prog. Ser. 6: 317-327.

Berk, S. G., R. R. Colwell, and E. B. Small. 1976. A study of
feeding responses to bacterial prey by estuarine ciliates.
Trans. Am. Microsc. Soc. 95: 514-520.

Bick, H. 1968. Autokologische und saprobiologische Untersuchungen
am Susswasserciliaten. Hydrobiologia 31: 17-36.

Booth, B. C., J. Lewin, and R. E. Norris. 1982. Nanoplankton
species predominant in the subarctic Pacific in May and June
1978. Deep-Sea Res. 29: 185-200.

Brand, L. E., R. R. L. Guillard, and L. S. Murphy. 1981. A method
for the rapid and precise determination of acclimated
phytoplankton reproduction rates. J. Plankton Res. 2: 193-201.

Burbanck, W. D. 1942. Physiology of the ciliate Colpidium colpoda.
I. The effect of various bacteria as food on the division rate
of Colpidium colpoda. Physiol. Zool. 15: 342-362.

Burney, C. M., K. M. Johnson, D. M. Lavoie, and J. McN. Sieburth.
1979. Dissolved carbohydrate and microbial ATP in the North
Atlantic: concentrations and interactions. Deep-Sea Res. 26:
1267-1290.

Burney, C. M., P. G. Davis, K. M. Johnson, and J. McN. Sieburth.
1981. Dependence of dissolved carbohydrate concentrations upon

small scale nanoplankton and bacterioplankton in the western Sargasso Sea. Mar. Biol. 65: 289-296.

Burney, C. M., P. G. Davis, K. M. Johnson, and J. McN. Sieburth. 1982. Diel relationships of microbial trophic groups and in situ dissolved carbohydrate dynamics in the Caribbean Sea. Mar. Biol. 67: 311-322.

Calkins, G. N. 1901. Marine protozoa from Woods Hole. Bull. US Fish. Comm. 21: 413-468.

Calow, P. 1977. Conversion efficiencies in heterotrophic organisms. Biol. Rev. 52: 385-409.

Caron, D. A., P. G. Davis, L. P. Madin, and J. McN. Sieburth. 1982. Heterotrophic bacteria and bacterivorous protozoa in oceanic macroaggregates. Science 218: 795-797.

Coler, R. A., and H. B. Gunner. 1969. Microbial populations as determinants in protozoan succession. Water Res. 3: 149-156.

Curds, C. R. 1963. The flocculation of suspended matter by Paramecium caudatum. J. Gen Microbiol. 33: 357-363.

Curds, C. R., and J. M. Van Dyke. 1966. The feeeding habits and growth rates of some freshwater ciliates found in activated sludge plants. J. Appl. Ecol. 3: 127-137.

Cutler, D. W. 1920. A method for estimating the number of active protozoa in soil. J. Agric. Sci. 10: 135-143.

Cutler, D. W., and L. M. Crump. 1920. Daily periodicty in the numbers of active soil flagellates: with a brief note on the relation of trophic amoebae and bacterial numbers. Ann. Appl. Biol. 7: 11-24.

Danso, S. K., and M. Alexander. 1975. Regulation of predation by prey density: the protozoan-Rhizobium relationship. Appl. Microbiol. 29: 515-521.

Davis, P. G. 1982. Bacterivorous flagellates in marine waters. Ph.D. dissertation, University of Rhode Island, Kingston.

Davis, P. G., D. A. Caron, and J. McN. Sieburth. 1978. Oceanic amoebae from the North Atlantic: culture, distribution, and taxonomy. Trans. Am. Microsc. Soc. 97: 73-88.

Davis, P. G., and J. McN. Sieburth. 1982. Differentiation of the phototrophic and heterotrophic nanoplankton populations in marine waters by epifluorescence microscopy. Ann. Inst. Oceanogr. 58(S): 249-260.

Dive, D., C. Dupont, and H. Leclerc. 1974. Nutrition holozoique de Colpodium campylum aux depens de bacteriés pigmentées ou synthétisant des toxines. Protistologica 10: 517-525.

Droop, M. R. 1970. Nutritional investigation of phagotrophic protozoa under axenic conditions. Helgol. Wiss. Meeresunters 20: 272-277.

Droop, M. R. 1973. Nutrient limitation in osmotrophic protista. Am. Zool. 13: 209-214.

Fenchel, T. 1968. The ecology of marine microbenthos. III. The reproductive potential of ciliates. Ophelia 5: 123-136.

Fenchel, T. 1977. The significance of bactivorous protozoa in the microbial community of detrital particles, pp. 529-544. In: J.

Cairns [ed.], Aquatic Microbial Communities. Garland Publ.,
    New York.
Fenchel, T. 1980a. Suspension feeding in ciliated protozoa:
    functional response and particle size selection. Microb.
    Ecol. 6: 1-11.
Fenchel, T. 1980b. Suspension feeding in ciliated protozoa:
    feeding rates and the ecological significance. Microb. Ecol.
    6: 13-25.
Fenchel, T. 1980c. Relation between particle size selection and
    clearance in suspension-feeding ciliates. Limnol. Oceanogr.
    25: 733-738.
Fenchel, T. 1980d. Suspension feeding in ciliated protozoa:
    structure and function of feeding organelles. Arch.
    Protistenk. 123: 239-260.
Fenchel, T. 1982a. Ecology of heterotrophic microflagellates. I.
    Some important forms and their functional morphology. Mar.
    Ecol. Prog. Ser. 8: 211-223.
Fenchel, T. 1982b. Ecology of heterotrophic microflagellates. II.
    Bioenergetics and growth. Mar. Ecol. Prog. Ser. 8: 225-231.
Fenchel, T. 1982c. Ecology of heterotrophic microflagellates.
    III. Adaptations to heterogeneous environments. Mar. Ecol.
    Prog. Ser. 9: 25-33.
Fenchel, T. 1982d. Ecology of heterotrophic microflagellates. IV.
    Quantitative occurrence and importance as bacterial consumers.
    Mar. Ecol. Prog. Ser. 9: 35-42.
Fenchel, T., and P. Harrison. 1976. The significance of bacterial
    grazing and mineral cycling for the decomposition of particulate
    detritus, pp. 285-299. In: J. M. Anderson and A. MacFadyen
    [eds.], The Role of Terrestrial and Aquatic Organisms in
    Decomposition Processes. Blackwell Scientific Publ., Oxford.
Fenchel, T., and B. B. Jørgensen. 1977. Detritus food chains of
    aquatic ecosystems: the role of bacteria. Adv. Microb. Ecol.
    1: 1-58.
Fenchel, T., and E. B. Small. 1980. Structure and function of the
    oral cavity and its organelles in the hymenostome ciliate
    Glaucoma. Trans. Am. Microsc. Soc. 99: 52-60.
Ferguson, R. L., and P. Rublee. 1976. Contribution of bacteria to
    standing crop of coastal plankton. Limnol. Oceanogr. 21:
    141-145.
Ferguson, R. L., and A. V. Palumbo. 1979. Distribution of suspended
    bacteria in neritic waters south of Long Island during strati-
    fied conditions. Limnol. Oceanogr. 24: 697-705.
Geesey, G. G. 1982. Microbial exopolymers: ecological and economic
    considerations. Am. Soc. Microbiol. News 48: 9-14.
Griesmann, K. 1914. Uber marine flagellaten. Arch. Protistenk.
    32: 1-78.
Haas, L. W. 1982. Improved epifluorescence microscopy for observing
    planktonic microorganisms. Ann. Inst. Oceanogr. 58(S): 261-266.
Haas, L. W., and K. L. Webb. 1979. Nutritional mode of several
    non-pigmented microflagellates from the York River estuary,

Virginia. J. Exp. Mar. Biol. Ecol. 39: 125-134.

Habte, M., and M. Alexander. 1978. Mechanisms of persistence of low numbers of bacteria preyed upon by protozoa. Soil Biol. Biochem. 10: 1-6.

Hamilton, R. D., and J. E. Preslan. 1969. Cultural characteristics of a pelagic marine hymenostome ciliate, Uronema sp. J. Exp. Mar. Biol. Ecol. 4: 90-99.

Hamilton, R. D., and J. E. Preslan. 1970. Observations on heterotrophic activity in the eastern tropical Pacific. Limnol. Oceanogr. 15: 395-401.

Hanna, B. A., and D. M. Lilly. 1974. Growth of Uronema marinum in chemically defined medium. Mar. Biol. 26: 153-160.

Hardin, G. 1942. An investigation of the physiological requirements of a pure culture of the heterotrophic flagellate Oikomonas termo, Kent. Physiol. Zool. 15: 466-475.

Hardin, G. 1943. Flocculation of bacteria by protozoa. Nature 151: 642.

Hardin, G. 1944. Physiological observations and their ecological significance: A study of the protozoan Oikomonas termo Kent. Ecology 25: 192-201.

Harding, J. P. 1937a. Quantitative studies on the ciliate Glaucoma. I. The regulation of the size and the fission rate by the bacterial food supply. J. Exp. Biol. 14: 422-430.

Harding, J. P. 1937b. Quantitative studies on the ciliate Glaucoma. II. The effects of starvation. J. Exp. Biol. 14: 431-439.

Hentschel, E. 1936. Allgemeine Biologie des Sudatlantisschen Ozeans. Deutsche Atlantische Expedition METEOR, Band XI. Walther de Gruyter, Berlin. 344 pp.

Hinga, K. R., P. G. Davis, and J. McN. Sieburth. 1979. Enclosed chambers for the convenient reverse flow concentration and selective filtration of particles. Limnol. Oceanogr. 24: 536-540.

Hobbie, J. E., R. J. Daley, and S. Jasper. 1977. Use of Nuclepore filters for counting bacteria by fluorescence microscopy. Appl. Environ. Microbiol. 33: 1225-1228.

Hobbie, J. E., and P. Rublee. 1977. Radioisotope studies of heterotrophic bacteria in aquatic ecocystems, pp. 441-476. In: J. Cairns [ed.], Aquatic Microbial Communities. Garland Publ. Co., New York.

Hoppe, H.-G. 1976. Determination and properties of actively metabolizing heterotrophic bacteria in the sea, investigated by means of microautoradiography. Mar. Biol. 36: 291-302.

Johannes, R. E. 1965. Influence of marine protozoa on nutrient regeneration. Limnol. Oceanogr. 10: 434-442.

Johnson, P. W., and J. McN. Sieburth. 1979. Chroococcoid cyanobacteria in the sea: A ubiquitous and diverse phototrophic biomass. Limnol. Oceanogr. 24: 928-935.

Johnson, P. W., and J. McN. Sieburth. 1982. In situ morphology and occurrence of eucaryotic phototrophs of bacterial size in the picoplankton of estuarine and oceanic waters. J. Phycol. 18: 318-327.

Johnson, P. W., H.-S. Xu, and J. McN. Sieburth. 1982. The utiliza-
    tion of chroococcoid cyanobacteria by marine protozooplankters
    but not by calanoid copepods. Ann. Inst. Oceanogr. 58(S):
    297-308.
Kent, W. W. 1880-1881. A Manual of the Infusoria. Vol. I, III.
    David Bogue, London.
Kidder, G. W., and C. A. Stuart. 1939. Growth studies in ciliates.
    I. The role of bacteria in the growth and reproduction of
    Colpoda. Physiol. Zool. 12: 329-340.
Kirchman, D., and R. Mitchell. 1982. Contribution of particle-
    bound bacteria to total microheterotrophic activity in five
    ponds and two marshes. Appl. Environ. Microbiol. 43: 200-209.
Knauer, G. A., and J. H. Martin. 1981. Primary production and
    carbon-nitrogen fluxes in the upper 1,500 m of the northeast
    Pacific. Limnol. Oceanogr. 26: 181-186.
Kopylov, A. I., and E. S. Moiseev. 1980. Effect of colorless
    flagellates on the determination of bacterial production in
    seawater. Dokl. Akad. Nauk SSSR Biol. Sci. 252: 272-274.
Kumé, T. 1979. Feeding patterns of marine ciliates fed on the
    heterotrophic bacteria. La Mer 17: 109-116.
Leadbeater, B. S. C. 1974. Ultrastructural observations on nano-
    plankton collected from the coast of Yugoslavia and the Bay of
    Algiers. J. Mar. Biol. Assoc. UK 54: 179-196.
Lee, J. J., M. E. McEnery, S. Pierce, H. D. Freudenthal, and W. A.
    Muller. 1966. Tracer experiments in feeding littoral foramini-
    fera. J. Protozool 13: 659-670.
Leppard, G. G., A. Massalski, and D. R. S. Lean. 1977. Electron-
    opaque microscopic fibrils in lakes: their demonstration, their
    biological derivation and their potential significance in the
    redistribution of cations. Protoplasma 92: 289-309.
Lighthart, B. 1969. Planktonic and benthic bacteriovorous protozoa
    at eleven stations in Puget Sound and adjacent Pacific Ocean.
    J. Fish. Res. Board Canada 26: 299-304.
Lighthart, B., and J. Liston. 1964. Design, operation, and prelim-
    inary test results of a model sea-bed microbial ecosystem.
    Bacteriol. Proc. G133, p. 38.
Lindeman, R. L. 1942. The trophic-dynamic aspect of ecology.
    Ecology 23: 399-417.
Linley, E. A. S., and J. G. Field. 1982. The nature and ecological
    significance of bacterial aggregation in a nearshore upwelling
    ecosystem. Estuarine Coastal Shelf Sci. 14: 1-11.
Lohmann, H. 1911. Uber das Nannoplankton und die Zentrifugierung
    kleinster Wasserproben zur Gewinnung desselben in lebendem
    Zustande. Int. Rev. ges. Hydrobiol. Hydrogr. 4: 1-38.
Maalpe, O., and N. O. Kjeldgaard. 1966. Control of Macromolecular
    Synthesis. W. A. Benjamin, Inc., New York and Amsterdam. 284
    pp.
Muller, W. A., and J. J. Lee. 1969. Apparent indispensability of
    bacteria in foraminiferan nutrition. J. Protozool. 16: 471-478.

Newell, R. C., M. I. Lucas, and E. A. S. Linley. 1981. Rate of degradation and efficiency of conversion of phytoplankton debris by marine microorganisms. Mar. Ecol. Prog. Ser. 6: 123-136.

Olson, R. J. 1981a. ^{15}N tracer studies of the primary nitrite maximum. J. Mar. Res. 39: 203-226.

Olson, R. J. 1981b. Differential photoinhibition of marine nitrifying bacteria: a possible mechanism for the formation of the primary nitrite maximum. J. Mar. Res. 39: 227-238.

Pace, M. L., and J. D. Orcutt Jr. 1981. The relative importance of protozoans, rotifers and crustaceans in a freshwater zooplankton community. Limnol. Oceanogr. 26: 822-830.

Paerl, H. W. 1978. Microbial organic carbon recovery in aquatic ecosystems. Limnol. Oceanogr. 23: 927-935.

Payne, W. J., and W. J. Weibe. 1978. Growth yield and efficiency in chemosynthetic microorganisms. Ann. Rev. Microbiol. 32: 155-183.

Pillai, S. C., and V. Subrahmanyan. 1942. Role of protozoa in the activated sludge process. Nature 150: 525.

Pomeroy, L. R. 1974. The ocean's food web, a changing paradigm. BioScience 24: 499-504.

Porter, K. G., and Y. S. Feig. 1980. The use of DAPI for identifying and counting aquatic microflora. Limnol. Oceanogr. 25: 943-948.

Pütter, A. 1909. Die Ernahrung der Wassertiere und der Stoffhaushalt der Gewasser. Gustav Fischer, Jena. 168 pp.

Reynoldson, T. B. 1942. Vorticella as an indicator organism for activated sludge. Nature 149: 608-609.

Ryther, H. H. 1969. Photosynthesis and fish production in the sea. Science 166: 72-76.

Sawyer, T. K. 1971. Isolation and identification of free-living marine amoebae from upper Chesapeake Bay, Maryland. Trans. Am. Microsc. Soc. 90: 43-51.

Sawyer, T. K. 1975. Marine amoebae from surface waters of Chincoteague Bay, Virginia: two new genera and nine new species within the families Mayorellidae, Flabellulidae, and Stereomyxidae. Trans. Am. Microsc. Soc. 94: 71-92.

Sayler, G. S., and C. M. Gilmour. 1978. Heterotrophic utilization of organic carbon in aquatic environments. J. Environ. Qual. 7: 385-391.

Schleyer, H. H. 1980. A preliminary evaluation of heterotrophic utilisation of a labelled algal extract in a subtidal reef environment. Mar. Ecol. Prog. Ser. 3: 223-229.

Scranton, M. I., and P. G. Brewer. 1977. Occurrence of methane in the near-surface waters of the western subtropical North-Atlantic. Deep-Sea Res. 24: 127-138.

Shanks, A. L., and J. D. Trent. 1979. Marine snow: microscale nutrient patches. Limnol. Oceanogr. 24: 850-854.

Sharp, J. H. 1973. Size classes of organic carbon in seawater. Limnol. Oceanogr. 18: 441-447.

Sheldon, R. W., and T. R. Parsons. 1967. A continuous size

spectrum for particulate matter in the sea. J. Fish. Res.
    Board Canada 24: 909-915.
Sieburth, J. McN. 1976. Bacterial substrates and productivity in
    marine ecosystems. Ann. Rev. Ecol. Syst. 7: 259-285.
Sieburth, J. McN. 1978. Bacterioplankton: nature, biomass, activity
    and relationships to the protist plankton. J. Phycol.
    14(Suppl.): 31.
Sieburth, J. McN. 1979. Sea Microbes. Oxford University Press,
    New York. 491 pp.
Sieburth, J. McN. In press. Microbiological and organic-chemical
    processes in the surface and mixed layers. In: P. S. Liss and
    W. G. N. Slinn [eds.], Air-Sea Exchange of Gases and Particles.
    NATO Advanced Study Inst. Ser. D. Reidel Publ. Co., Dordrecht,
    Holland.
Sieburth, J. McN., and P. G. Davis. 1982. The role of heterotrophic
    nanoplankton in the grazing and nurturing of planktonic bacteria
    in the Sargasso and Caribbean Seas. Ann. Inst. Oceanogr.
    58(S): 285-296.
Sierburth, J. McN., K. M. Johnson, C. M. Burney and D. M. Lavoie.
    1977. Estimation of in situ rates of heterotrophy using diurnal
    changes in dissolved organic matter and growth rates of pico-
    plankton in diffusion culture. Helgol. Wiss. Meeresunters 30:
    697-704.
Sieburth, J. McN., and D. M. Lavoie. 1978. A non-standard approach
    to heterotrophy: ATP estimation of natural populations of
    selectively filtered bacterioplankton and their growth rates on
    in situ water in diffusion-culture, pp. 77-94. In: First
    American-Soviet Symposium on Biological Effects of Pollution on
    Marine Organisms, US EPA-600/9-78-007.
Sieburth, J. McN., V. Smetacek and J. Lenz. 1978. Pelagic ecosystem
    structure: heterotrophic compartments of the plankton and their
    relationship to plankton size fractions. Limnol. Oceanogr. 23:
    1256-1263.
Sieburth, J. McN., P.-J. Willis, K. M. Johnson, C. M. Burney, D. M.
    Lavoie, K. R. Hinga, D. A. Caron, F. W. French, III, P. W.
    Johnson, and P. G. Davis. 1976. Dissolved organic matter and
    heterotrophic microneuston in the surface microlayers of the
    North Atlantic. Science 194: 1415-1418.
Silver, M. W., and A. L. Alldredge. 1981. Bathypelagic marine snow:
    deep-sea algal and detrital community. J. Mar. Res. 29:
    501-530.
Silver, M. W., and K. W. Bruland. 1981. Differential feeding and
    fecal pellet composition of salps and pteropods, and the pos-
    sible origin of the deep-water flora and olive green "cells."
    Mar. Biol. 62: 263-273.
Simon, S. A., T. J. McIntosh, and R. Latorre. 1982. Mutual floc-
    culation of algae and clay: evidence and implications. Science
    216: 63-66.
Slobodkin, L. B. 1962. Growth and Regulation of Animal Populations.
    Holt, Rinehart and Winston, New York. 184 pp.

Smith, R. C., and K. S. Baker. 1979. Penetration of UV-B and bio-
  logically effective dose-rates in natural waters. Photochem.
  Photobiol. 29: 311-323.
Sorokin, Y. I. 1970. Aggregation of marine bacterioplankton. Dokl.
  Akad. Nauk SSSR Biol. Sci. 190: 337-339.
Sorokin, Y. I. 1977. The heterotrophic phase of plankton succes-
  sion in the Japan Sea. Mar. Biol. 41: 107-117.
Sorokin, Y. I. 1979. Zooflagellates as a component of the Community
  of eutrophic and oligotrophic waters in the Pacific Ocean.
  Oceanology 19: 316-319.
Sorokin, Y. I. 1980. A chamber for the quantitative recording of
  protozoa and nanoplankton organisms under field conditions.
  Hydrobiol. J. 16: 74-75.
Stokes, A. C. 1888. A preliminary contribution toward a history of
  the fresh-water infusoria of the United States. J. Trenton
  Nat. Hist. Soc. 1(3): 71-345.
Stout, J. D. 1980. The role of protozoa in nutrient cycling and
  energy flow. Adv. Microb. Ecol. 4: 1-50.
Sugden, B., and L. Lloyd. 1950. The clearing of turbid waters by
  means of the ciliate Carchesium: a demonstration. J. Proc.
  Inst. Sew. Purif. 1: 16-26.
Swanberg, N. R. 1979. The ecology of colonial radiolarians: their
  colony morphology, trophic interactions and associations,
  behavior, distribution, and the photosynthesis of their
  symbionts. Ph.D. thesis, WHOI/MIT Joint Program, Woods Hole,
  Massachusetts.
Taylor, W. D. 1978a. Growth responses of ciliate protozoa to the
  abundance of their bacterial prey. Microb. Ecol. 4: 207-214.
Taylor, W. D. 1978b. Maximum growth rate, size and commoness in a
  community of bactivorous ciliates. Oecologia 36: 263-272.
Taylor, W. D. 1979. Sampling data on the bactivorous ciliates of a
  small pond compared to neutral models of community structure.
  Ecology 60: 876-883.
Taylor, W. D., and J. Berger. 1976. Growth responses of cohabiting
  ciliate protozoa to various prey bacteria. Can. J. Zool. 54:
  1111-1114.
Taylor, W. D., and J. Berger. 1980. Microspatial heterogeneity in
  the distribution of ciliates in a small pond. Microb. Ecol. 6:
  27-34.
Throndsen, J. 1969. Flagellates of Norwegian coastal waters. Nytt.
  Mag. Bot. 16: 161-216.
Throndsen, J. 1970a. Salpingoeca spinifera sp. nov., a new plank-
  tonic species of the Craspedophyceae recorded in the Arctic.
  Br. Phycol. J. 5: 87-89.
Throndsen, J. 1970b. Marine planktonic Acanthoecaceans (Craspedophyceae). Nytt. Mag. Bot. 17: 103-111.
Utermöhl, H. 1958. Zur Vervollkommung der Quantitativen Phytoplank-
  ton-Methodik. Mitt. Int. Ver. Limnol. 9: 1-38.
Wada, E., and A. Hattori. 1971. Nitrite metabolism in the euphotic
  layer of the central North Pacific Ocean. Limnol. Oceanogr.
  16: 766-772.

Waterbury, J. B., S. W. Watson, R. R. L. Guillard, and L. E. Brand.
    1979. Widespread occurrence of a unicellular, marine,
    planktonic cyanobacterium. Nature 277: 293-294.
Watson, J. M. 1945. Mechanisms of bacterial flocculation caused by
    protozoa. Nature 155: 217.
Watson, S. W., T. J. Novitsky, H. L. Quinby, and F. W. Valois. 1977.
    Determination of bacterial number and biomass in marine environ-
    ments. Appl. Environ. Microbiol. 33: 940-954.
Wiegert, R. G., and D. F. Owen. 1971. Trophic structure, available
    resources and population density in terrestrial vs. aquatic
    ecosystems. J. Theor. Biol. 30: 69-81.
Williams, P. J. leB. 1970. Heterotrophic utilization of dissolved
    organic compounds in the sea. I. Size distribution of popula-
    tion and relationship between respiration and incorporation of
    growth substrates. J. Mar. Biol. Assoc. UK 50: 859-870.
Williams, P. J. leB. 1981. Incorporation of microheterotrophic
    processes into the classical paradigm of the planktonic food
    web. Kiel. Meeresforsch. Sonderheft 5: 1-29.
Williams, P. M., H. Oeschger, and P. Kinney. 1969. Natural radio-
    carbon activity of the dissolved organic carbon in the north-
    east Pacific Ocean. Nature 224: 256-258.
Wood, E. J. F. 1955. Fluorescent microscopy in marine microbiology.
    J. Cons. Int. Explor. Mer 21: 6-7.

EFFECT OF GRAZING: METAZOAN SUSPENSION FEEDERS

C. Barker Jørgensen

Zoophysiological Laboratory A
University of Copenhagen
Universitetsparken 13, DK-21 Copenhagen Ø, Denmark

INTRODUCTION

Suspension feeding is primarily an adaptation to utilize the primary production in aquatic habitats, which is predominantly in the form of phytoplankton, particularly in the sea. The microscopic phytoplankton algae are present in dilute suspensions in the water masses. Utilization of the phytoplankton therefore involves processing large volumes of the surrounding water which, besides the algal cells, also contains other types of suspended matter, including bacteria and other heterotrophs, as well as dissolved organic molecules. Metazoan suspension feeders may therefore exert an effect by directly grazing upon the bacterioplankton or by acting as competitors for the dissolved organic molecules that constitute bacterial food.

The impact of a population of suspension feeders on its surroundings depends upon the rates at which the members of the population clear the water of the various types of suspended particles and dissolved organic molecules. Significant effects of metazoan suspension feeding on the bacterioplankton can only be expected in habitats that support high densities of certain types of suspension feeders. This paper attempts to evaluate the effect of suspension feeding in such selected habitats.

METHODS OF MEASURING RATES AND EFFICIENCIES OF WATER PROCESSING

The rates at which suspension feeders clear the surrounding water of suspended particles and dissolved organic molecules are determined by the rate at which water is transported to the feeding

445

structures, and by the efficiencies with which the structures retain particles, as expressed in a retention spectrum of particle sizes, and absorb organic molecules.

Rate of water transport can be determined directly in animals of suitable size and anatomy. Thermistor flow meters have been developed to measure rates at which water leaves the exhalant apertures of suspension feeders, such as spongia, bivalves, and ascidians (McCammon 1965; Reiswig 1971).

In other suspension feeders the rates of water processing must be determined indirectly from the rates at which animals clear the surrounding water of particles. When suspended particles cannot pass the particle-retaining structures, particle clearance volume equals rate of water transport through the structures.

Efficiency of particle retention can be determined by measuring clearance as a function of particle size. The volume of water cleared of particles, as a percentage of the volume filtered, expresses the efficiency of retention.

In attempts to assess the importance as a food source of small suspended organisms, such as bacteria, or the impact of suspension feeding on such organisms, it is important to know how efficiently the particles are removed from the water passing the feeding organs. In such studies clearance determinations are useful.

Uptake of dissolved organic molecules can also be measured from the rates at which suspension feeders clear the surrounding water of the molecules. Clearance of organic molecules varies with concentration, and the uptake of organic molecules can be described in terms of Michaelis-Menten kinetics.

Most measurements of particle clearances have been made photometrically with meters of low sensitivity that required unnaturally high concentrations of algae and other particles.

Clearances at low normal concentrations of particles have also been made photometrically by using long light paths (Jørgensen and Goldberg 1953; Riisgård and Møhlenberg 1979) and by using radioactively labelled phytoplankton cells (Jørgensen and Goldberg 1953; Chipman and Hopkins 1954). It was, however, the introduction of the electronic particle counter that constituted the breakthrough in measuring rates at which suspension feeders clear the water of particles of various sizes and at natural concentrations. Unfortunately, the lower end of the size range that can be counted with the conventional models does not extend much below a micron. The technique thus covers only the upper range of the bacterioplankton which is the concern of the present volume. Determinations of the rates at which suspension feeders clear the water of bacterioplankton

therefore involve the tedious procedure of direct counting of
bacteria or the use of labelled bacteria, and as a consequence few
measurements are available. Most of what can be concluded about
effects of metazoan suspension feeding on the bacterioplankton is
therefore tentative and preliminary.

CHARACTERISTICS OF SUSPENSION FEEDING

Suspension feeding is typically automatic and non-selective,
but suspension feeders are generally sensitive to environmental
changes. They respond to adverse conditions by reducing or
discontinuing normal processing of water. The disregard of this
fact explains the wealth of records of low and varying rates of
water processing and low efficiencies of particle retention.

The relationship between concentration of suspended particulate
matter and the rates at which suspension feeders clear the suspension
is of special interest. Most studies on particle clearance in
suspension feeders have been made with unnaturally high concentrations of suspended algae or other particles, and it is regularly
found that clearance decreases with increasing particle concentration. With the introduction of the electronic particle counter it
became possible routinely to measure the rates at which suspension
feeders clear the surrounding water of particles at low, natural
concentrations. Such measurements on representatives of various
suspension feeders, including bivalves, tunicates, and polychaetes,
indicate that undisturbed suspension feeders do process the
surrounding water at relatively constant rates that express innate,
genetically determined capacities of the water processing structures.
Presumably, the capacity for water processing has evolved as an
adaptation to prevailing concentrations of suspended food in the
habitat.

MECHANISMS OF FEEDING, EFFICIENCY ON BACTERIA

Suspension feeders have adopted one of three main types of
structures, setae, cilia and flagella, and secretions forming
filters, e.g. mucus filters, in order to extract particles suspended
in the surrounding water.

Setae

Setae are characteristic of suspension feeding crustaceans,
such as barnacles, copepods, euphauseans, and branchiopods, such as
Artemia, and cladocerans.

The feeding structures in suspension feeding copepods and
euphaseans seem adapted to deal especially with large phytoplankton,

and they only inefficiently retain the small phytoplankton cells.

Branchiopods can retain particles down to bacterial size, but little is known about retention efficiencies of particles within the range covered by the bacterioplankton.

Flagella

Spongia transport the surrounding water through their body by means of the flagellae of the choanocytes united in chambers. The choanocytes carry so-called collars, which are rings of villi around the centrally placed flagella. The villi are 0.13 μm thick and the distance between neighboring villi is 0.12 μm (Fjerdingstad 1961). The active flagellum will produce a water current away from the collar and thus establish a pressure gradient across the collar. It is generally assumed that the collars constitute the main filter of the water current passing the sponge, and that the water is drawn through the spaces between the villi by the pressure drop produced by the flagella.

The collar may be treated as a filter consisting of parallel cylinders, and its hydromechanical resistence may be estimated from an equation expressing the relation between velocity of the current and the pressure drop across the filter:

$$\Delta p = \frac{8 \pi \mu U}{b (1 - 2\ln\tau + 1/6\tau^2 - 1/144\tau^4)}$$

where $\tau = \pi d/b$. U is the undisturbed velocity, d is the diameter of the cylinder, i.e. 0.13 μm for the villi, b is the distance between centers of neighboring cylinders, 0.25 μm, and μ is the dynamic viscosity (Jørgensen 1981a).

The pressure drop is directly proportional to the velocity U. This velocity can be calculated from the rate of flow through the sponge and the area of the surfaces of the collars. Reiswig (1971, 1974) measured flow rates in marine sponges in situ, and he counted the choanocyte chambers and measured the dimensions of the collars (1975). From these data it can be calculated that U is about 3 μm·s^{-1}, at which velocity of the water the pressure drop across the collar amounts to only about 0.03 mm H$_2$O. This low pressure drop is compatible with the concept that practically all water that passes through the choanocyte chambers is filtered by the choanocyte collars, and thus cleared of particles as small as 0.1 μm, including the bacterioplankton.

The sponges thus constitute an example of a type of suspension feeders where structural features seem to permit conclusions concerning the particle retaining properties of the feeding structure. In most other suspension feeders it has not been possible to predict

efficiencies of particle retention exclusively from structural
studies.

Cilia

It is widely believed that ciliary feeding is intrinsically
associated with mucus that acts to trap particles suspended in the
feeding current produced by the cilia, but it now seems that several
ciliary feeders, including bryozoans, phoronoids, brachiopods,
sabellid polychaetes, planktotrophic larvae of bottom invertebrates,
and bivalves, feed solely by means of cilia. In recent years several
hypotheses have been forwarded to account for the retention of
particles in ciliary suspension feeders. The hypotheses include
ciliary sieving mechanisms (Gilmour 1978; Fenchel 1980), ciliary
reversal in response to encounters with particles (Strathmann et al.
1972), and transfer of particles from feeding currents to surface
currents by fluid mechanical forces Jørgensen 1981a, 1982b). A fluid
mechanical model was suggested to explain particle retention in the
gills of the mussel Mytilus edulis and other suspension feeding
bivalves; it may also apply to other types of ciliary feeders.

Function of the bivalve gill. Most studies on mechanisms of
suspension feeding in bivalves have been made on species posessing
eulaterofrontal cilia or cirri, that is, species in which three types
of ciliary tracts can be distinguished on the gill filaments, the
frontal ciliary tract, the rows of laterofrontal cirri, and the
lateral ciliary tracts (Fig. 1). The laterofrontal cirri are widely
assumed to strain suspended particles from the through current
produced by the lateral ciliary bands, and to transfer the particles
to the band of cilia on the frontal surface of the filaments. The
straining function of the laterofrontal cirri was inferred from
their filter-like structure, which arise from the tips of the
individual cilia producing side branches about one µm apart along
the cirri (Moore 1971; Owen 1974; Owen and McCrae 1976). The
structure of the cirri thus approximates that of a filter consisting
of parallel cylinders, and the equations relating pressure drop and
velocity of water passing the filter can be applied. It turned out
that in Mytilus edulis the pressure drop needed across the cirri if
water passes between the branching cilia amounted to about 1.3 mm
H_2O, or several times higher than pressure drops measured across the
whole gill. It was therefore concluded (Jørgensen 1981a) that in
Mytilus edulis, and probably also in other suspension feeding
bivalves possessing eulaterofrontal cirri, the unrestricted flow of
water through the gill takes place at such low pressures that water
cannot pass between the branching cilia. Consequently the eulaterofrontal cirri move water rather than filter it.

Suspension feeding bivalves generally produce two main systems
of water currents, the through currents that pass between the
filaments and the surface currents that initially pass along the

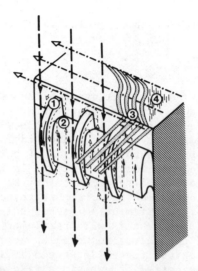

Figure 1. <u>Mytilus edulis</u>. Diagram of gill filament showing ciliary
tracts and water currents. 1, Phase of effective stroke of metachronal wave of band of lateral cilia; 2, phase of recovery stroke of the
wave; 3, tract of laterofrontal cirri alternatingly shown during
resting phase crossing the interfilamentary space, and at the end of
the effective stroke normal to the frontal surface of the filament;
4, band of frontal cilia. Heavy arrows: ▬▬▶ , indicate interfilamentary through current; –·–·–·⇢, indicate surface current along
frontal surface of the filaments. Light arrows: ───⇢, indicate
directions of movements of the surface enveloping the metachronally
beating band of lateral cilia; ─────⇢, indicate oscillating water
currents produced by the metachronal wave.

frontal surface of the filaments and ultimately leads to the mouth
and esophagus. Particle retention implies that particles are transferred from the through currents into the surface current where the
two systems meet at the entrance to the interfilamentary spaces.
This region of contact between the two current systems constitutes a
complex three-dimensional pattern of laminar flow characterized by
steep velocity gradients. In this region particles suspended in the
through current become exposed to transverse forces that cause the
particles to migrate normally to the direction of flow. It is
suggested that it is these viscous forces that are responsible for
the transfer of particles into the surface currents, and thus for
particle retention in suspension feeding bivalves (Jørgensen 1981a,
1982b).

    <u>Efficiency of particle retention</u>. Determinations of particle
retention spectra in a number of bivalve species, representing
different gill types, show that the efficiency of retention starts

to decline at particle diameters of about 3 μm, or more in bivalves
with gills that lack laterofrontal cirri. The efficiency declines
gradually with decreasing particle size, but the relations between
particle size and retention efficiency is not well established at
sizes $\leq$ 1 μm (Møhlenberg and Riisgård 1978). Similar relations
between decreasing particle size and retention have been found in
other ciliary feeders, such as the veliger larva of <u>Mytilus</u> <u>edulis</u>
(Riisgård et al. 1980) and the polychaete, <u>Sabella</u> <u>pavonina</u>
(unpublished observations.)

The efficiency with which ciliary feeders retain bacterioplankton may vary between species. Thus, Wright et al. (1982) found that
under conditions in the laboratory where the bivalve <u>Geukensia</u>
<u>demissa</u> cleared bacterioplankton at a rate amounting to 42% the
clearance of phytoplankton, <u>Mytilus</u> <u>edulis</u> did not measurably clear
the water of bacteria. However, the mussels were processing water
at low rates which may affect particle retention. Webber estimated
the retention efficiency of bacteria by <u>M</u>. <u>edulis</u> by comparing
bacterial counts in the inhalant and exhalant water. Retention of
1.0 x 1.2 μm <u>Escherichia</u> <u>coli</u> amounted to 16 $\pm$ 6%, whereas bacterioplankton was not retained (David L. Webber, personal communication).

Flagella and Ciliary Filter Efficiency

It is noteworthy that the particle retaining mechanism in
sponges apparently is based on filtration of the water through about
0.1 μm filters whereas the water flowing through the bivalve gill
seems to bypass the much coarser filters of the laterofrontal cirri.
Presumably the highly efficient filtering in sponges depends upon
low velocity of water flow at the level of the filters, some $\mu m \cdot s^{-1}$.
This low rate again depends upon the large filtering area, correspond-
ing, <u>e</u>.<u>g</u>., to about 500 cm^2 per cm^3 of body in <u>Haliclona</u> <u>permollis</u>
(Reiswig 1975). This may be compared with the mussel, <u>Mytilus</u>
<u>edulis</u>, in which water passes the gill filaments at velocities of
some $mm \cdot s^{-1}$, whereas the gill area amounts to about 5 cm^2 in a mussel
of one g soft body mass (Jørgensen 1981a).

Mucus Filters

Suspension feeding by means of filtration of the surrounding
water through mucus filters is typical of practically all tunicates,
and it has been also adopted by many gastropods and by some polychaetes. Within the tunicates the mechanism presumably originated
as a muco-ciliary feeding mechanism. Ciliated ostia in the pharynx
produce the water current which is filtered by a mucus net which is
continuously secreted by the endostyle and drawn across the inner
wall of the pharynx by ciliary activity. This principle applies to
the sessile ascidians and also to some groups of the pelagic
thaliaceans. In the salps the mucus net still functions as the
filter that traps suspended particles in the water, but the water

current is produced by means of muscular contractions of the body
wall, valves rectifying the flow of water through the mucus net.

The appendicularians represent an interesting variant of the
tunicate feeding pattern. In this group special structures serve to
concentrate the suspended particles in the surrounding water so that
the amount of water actually filtered by the pharyngeal mucus net is
correspondingly reduced. In Oikopleura this concentration takes
place within a structure often termed the feeding filter. The size
spectrum of particles that remain in suspension to enter the pharynx
is determined by a net-like structure that separates the water
currents entering and leaving the feeding filter. Electronmicroscopical preparations indicate that the net consists of rectangular
meshes of dimensions 0.1 x 0.8 μm. The feeding filters thus seems
to prevent even the smallest bacterioplankton from escaping the
suspension.

It is not known accurately how much the feeding filter
concentrates the passing suspension. However, from the rate at which
Oikopleura processes the surrounding water and from the dimensions of
the spiracles which produce the pharyngeal water current it can be
estimated that the pharyngeal current is in the order of one percent
of the volume of the water processed. Particles that pass the
incurrent filters of the house and are retained within the feeding
filter may thus enter the pharynx at a concentration about 100 times
that at which they are present in the water outside the house.

The efficiency with which the tunicate mucus nets retain
particles has been determined both from the particle retention
spectra and from measurements of mesh size in electronmicroscopical
studies of the nets. In ascidians, retention spectra show that
particles of sizes down to 2-3 μm are completely retained, but only
about 70% of 1 μm particles (Randløv and Riisgård 1979). The
electronmicroscopical studies of ascidian mucus filters revealed net
structures of rectangular meshes with dimensions of 0.4 x 0.7 μm in
Ciona intestinalis and 0.3 x 0.6 μm in Phallusia mammillata, the
filaments of the meshwork being about 0.02 μm thick (Flood and
Fiala-Médioni 1979).

According to the electronmicroscopical studies the ascidian
mucus net should therefore efficiently retain bacterioplankton.
The discrepancy between functional and structural estimates of the
porosity of the nets may be due to shrinkage of the nets during
preparation for electronmicroscopy.

Early observations of salps in nature and in the laboratory
showed that they retained particles smaller than 1 μm, but the
efficiency of retention was not known (Madin 1974; Harbison and
Gilmer 1976). More recently retention spectra have been determined
in Cyclosalpa spp. (Harbison and McAlister 1979). There was little

Table 1.   Salps: Retention efficiency of mucus net.[a]

Species	Body length mm	Size in µm of particles retained		
		0%	50%	100%
Cyclosalpa floridana				
Aggregate	7-15	1.0	1.5	2.2
Solitary	10-30	0.7	1.5	2.2
C. affinis				
Aggregate	20-60	1.0	2.5	4.0
Solitary	80-150	2.3	3.3	4.5
C. polae				
Solitary	20-60	0.7	2.0	3.5

[a]From Harbison and McAlister 1979.

difference in retention efficiency between the aggregate and
solitary forms, but efficiency varied with size of the species.
Thus the smallest particles to be completely retained by the small
C. floridana were about 2 µm, and 4-5 µm in the large C. affinis.
The threshold size for retention was about 1 µm in C. floridana and
C. polae, and about 2 µm in the large solitary C. affinis (Table 1).

UPTAKE OF DISSOLVED ORGANIC MOLECULES

Marine invertebrates possess mechanisms for transporting small
organic molecules, such as glucose and amino acids, from the outside
medium across the apical membrane of the epidermal cells (Stephens
1982).   Rates at which suspension feeders may clear the surrounding
water of amino acids have been measured in some bivalves.   At
natural, µmolar, concentrations of amino acids, 20% or more may be
removed from the water passing the mantle cavity under conditions of
unrestricted flow (Mytilus californianus, Wright and Stephens 1978;
Mytilus edulis, Jørgensen 1983).   Suspension feeding bivalves may
thus clear the ambient water of small organic molecules much more
efficiently than of bacteria.   This high clearance of amino acids in
M. edulis is characteristic of mussels adapted to amino acid free
seawater, and lower clearances may apply in mussels that are adapted
to the prevailing concentrations in the ambient water (Jørgensen
1983).

IMPACT OF GRAZING, EFFECTS ON MICROBIAL POPULATIONS

Effects of suspension feeding on bacterioplankton depend upon the density of the grazing populations. Populations of benthic suspension feeders can attain much greater densities than can populations of pelagic suspension feeders. This is because pelagic suspension feeders are practically stationary relative to the surrounding water. Maximum densities of the populations are therefore determined by the rate at which food is being produced in the surrounding water. The densities that populations of benthic suspension feeders can reach will depend upon the rates at which water loaded with food particles passes the populations. It can therefore be expected that benthic populations of suspension feeders from coastal and estuarine zones generally represent greater potential grazing effects on the bacterioplankton than do pelagic suspension feeders.

PELAGIC COMMUNITIES

The water-processing activity of the most important pelagic suspension feeders, copepods and euphasians, is presumably of negligible effect on the bacterioplankton because the feeding structures do not retain such small particles. The most likely members of pelagic communities to exert significant grazing pressures on the bacterioplankton should be sought among the suspension feeders that utilize mucus filters or ciliary mechanisms, such as appendicularians, salps, and pelagic larvae of benthic invertebrates.

Appendicularians regularly predominate in the zooplankton, may multiply to peak population densities, and have been held to effectively retain particles down to sizes below those of the bacterioplankton. This latter belief is based on the misconception that it is the feeding filter that determines the efficiency of particle retention (Flood 1978; Harbison and McAlister 1979; King et al. 1980; Alldredge 1981). As already mentioned, the efficiency with which appendicularians retain small particles is determined by the properties of the pharyngeal mucus filter.

Studies on the rates at which Oicopleura dioica clears the surrounding water of particles have provided values for clearance of phytoplankton algae (flagellates and small diatoms) that are 2-3 times higher than clearance values for bacterioplankton (Fig. 2). This agrees with the finding that the efficiency of retention declines at particle sizes < 1 μm of bacterial size (G.-A. Paffenhöfer, personal communication), but disagrees with King and Hollibaugh (cited in King et al. 1980) who found that appendicularians cleared the water as efficiently of bacterioplankton as of flagellates. Alldredge (1981) measured in situ the rate at which O. dioica cleared suspensions of

plastic beads (Fig. 2). The clearances measured may underestimate
the volumes of water processed because the size-range of the beads
from 2 to 15 μm, with a mean size of 12 μm, seems to extend beyond
the range for optimal retention (G.-A. Paffenhöfer, personal
communication).

    Alldredge also measured in situ rates at which another
appendicularian, Stegasoma magnum, cleared the suspensions of
plastic beads. The clearances obtained on three different days
varied widely. Calculated for a 10 μg C animal the values ranged
from 2 to 1200 ml·d^{-1} with an estimated mean value of about 370
ml·d^{-1}.

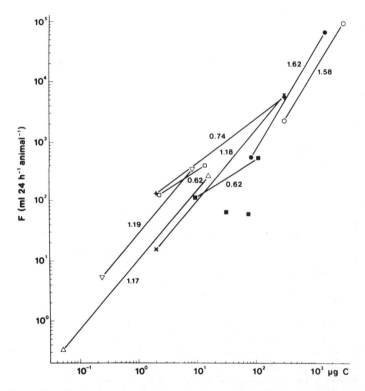

Figure 2. Zooplankton tunicates. Relationship between body mass
(μg C) and rate of water-processing (ml·animal^{-1}·d^{-1}) in thaliaceans
and appendicularians. ● , Pegea confederata, aggregates; ○, P. confederata, solitary; x, Thalia democratica, aggregates; +, Dolioletta
gegenbauri, gonozoids; △, Oicopleura dioica, clearance of bacterioplankton; ▽, clearance of algae; □, clearance of plastic beads; ■,
Stegosoma magnum. Figures indicate exponent b in the equation: Body
mass (μg C) = a·F^{b}. Sources: Harbison and Gilmer 1976; Paffenhöfer
1976; Deibel 1979; King et al. 1980; Alldredge 1981.

Table 2. Zooplankton water processing capacities: Percentage of
ambient water mass cleared per day.

Species	Biomass µg dry org. body mass l^{-1}	Percentage cleared	Locality
Appendicularians:			
Oicopleura dioica	10-12	9-10	Saanich Inlet, B.C.[a]
Oicopleura dioica	ca. 13	38	
Stegosoma magna	1-3	5-13	Gulf of California[b]
Thaliaceans:			
Salpa aspera	ca. 20	13	Sargasso Sea[c]
Bivalve larvae:			
Mytilus edulis	54-71	40-50	Isefjord, Denmark[d]

[a] King et al. 1980
[b] Alldredge 1981
[c] Wiebe et al. 1979
[d] Jørgensen 1981c

The generally higher clearances obtained in S. magnum than in
O. dioica, processing suspensions of 2-15 µm plastic beads, may be
correlated with the larger mesh size of the incurrent filter of S.
magnum (25 µm) than of O. dioica (15 µm).

Population Clearances

King et al. (1980) calculated the volumes of water that
populations of O. dioica cleared of bacterioplankton in the
Controlled Ecosystem Populations Experiment (CEPEX), conducted
in Saanich Inlet, British Columbia. They found that at maximum
population densities the appendicularians cleared 9-10% of the
surrounding water of bacterioplankton daily (Table 2).

Alldredge (1981) made comparable studies in the Gulf of
California. She found that at the highest population densities
O. dioica processed about 40% of the water masses in 24 h of
particles that were cleared as efficiently as 2-15 µm plastic
beads. Estimates for S. magnum amounted to 13% (Table 2).

Thaliaceans (salps and doliolids) constitute important and
often dominant parts of oceanic zooplankton. Sometimes salps

produce swarms that are believed to be capable of grazing down
available phytoplankton (Fraser 1962; Heron 1972; Madin 1974). The
mucus net by which salps feed can retain particles down to bacterial
sizes, however exact data on the potential impact of salps upon the
other components of the plankton are lacking. An attempt is made in
the present account to assess the impact of a bloom of salps in the
Northern Sargasso Sea, August 1974 (Wiebe et al. 1979). At one
station the number of salps, mainly <u>Salpa aspera</u>, amounted to 65 per
m^3 in the surface water during night.   This is considered as an
extremely large zooplankton biomass of about 9 μg C$\cdot$l^{-1}. This
biomass is based upon an estimate of of 140 μg C as the mean carbon
content of the salps. An individual of <u>Salpa maxima</u> of an equal
mass processes 1960 ml$\cdot$d^{-1} (Harbison and Gilmer 1976). Accordingly,
a population of 65 salps m^{-3} would process 13% of the surrounding
water per day. It is thus suggested that even under bloom conditions
the water-processing activity of salps has only minor effects on the
phytoplankton, and presumably negligible effects on the bacterioplankton, which is less efficiently retained.

Zooplankton Larvae (Bivalves)

Many benthic invertebrates reproduce by means of suspension-
feeding pelagic larvae. Within the temperate zones, breeding is
usually seasonally determined. If large populations are brought to
spawn simultaneously this may result in high densities of larvae in
the zooplankton. Temperature seems to be the most important
environmental factor that determines the annual reproductive cycle,
for instance in the common mussel, <u>Mytilus edulis</u>. A severe winter
will tend to synchronize spawning the subsequent spring. In the
Isefjord, Denmark, an area with rich mussel beds, such synchronized
spawning introduced about 3000 larvae per liter of water, evenly
distributed throughout the water column of 5 m. The cohort of
mussel larvae remained one month in the zooplankton, until they
metamorphosed. During this period of time the biomass of the larvae
amounted to 54-71 μg dry organic matter per liter, and the larvae
cleared 40-50% of the surrounding water daily of small food
particles, probably mostly flagellates (Jørgensen 1981c). The
volume of water cleared of bacteria is smaller. How much smaller
depends upon the efficiency with which bivalve larvae remove bacteria
from the water they process, which remains to be determined.

BENTHIC COMMUNITIES

Benthic communities of high density require habitats characterized by high rates of water exchange. Examples are sponges on coral
reefs and mussel beds within the intertidal or sublittoral zones.

<u>Sponges</u> are dominant suspension feeders on coral reefs. Reiswig
(1973, 1974) determined water-processing capacities of sponges <u>in</u>

situ, as well as population densities, on coral reefs in Discovery
Bay, Jamaica.

Sponge biomass reached maximum values on the fore-reef-slope
platform outside the bay area, at about 30-50 m depth. Here, total
sponge biomass was estimated to be about 3 liters·m^{-2}. At both
sides of this 20 m broad zone the density rapidly declined.

Sponges typically filtered a volume of water equal to their
own volume in 10 s (Reiswig 1974). A sponge biomass of 3 liters·m^{-2}
accordingly filters 26 m^3·m^{-2}·d^{-1}, and this is presumably also the
volume cleared of bacterioplankton (see above).

In stagnant water, sponge populations of densities about
3 liters·m^{-2} would thus clear a volume of water corresponding to
the 1 m bottom layer in 0.92 h, and would deplete the water of
bacteria and other particles in about 4 h. At a 24 h generation
time of bacterioplankton, the balance between production and removal
by the sponges would be reached at a flow velocity of only about
0.01 m·min^{-1}, or several orders of magnitude lower than the currents
passing the sponge zone. Water processing of the sponge zone is
therefore of negligible importance in the control of
bacterioplankton growth.

Bivalves are dominant suspension feeders in many coastal
habitats, and they may reach high population densities (Table 3).
Beds of mussels (Mytilus edulis) represent by far the highest
population densities reached by any suspension feeding organisms.
Densities of about 1 kg dry body mass m^{-2} are typical of mussel beds
from various types of habitats, both from the tidal zone (Dare 1976;
Wright et al. 1982) and from sublittoral habitats (Dare 1976;
Jørgensen 1980; Anders Randløv, personal communication).

The rates at which mussel beds process the surrounding water
can be estimated from the size frequency distribution of mussels on
the beds and the relation between size and clearance. Table 4 lists
some estimates made for maximum densities of mussel beds from various
intertidal and sublittoral localities. Rates of water-processing
amounting to about 10 m^3·m^{-2}·h^{-1} seem to be typical.

Mussel beds dominated by 5-7 cm long mussels may keep a layer
of water about 0.5 m deep circulated by means of the water currents
they produce. Such a layer is processed in 3.6 min. In the absence
of water currents the mussels may thus recirculate the layer of
water immediately above the bed about 400 times daily and deplete
the water of both particulate and dissolved matters, including
oxygen. In stagnant water mussel beds die. Survival depends upon
water currents, and the densities to which the mussel beds can build
up presumably depends upon the rates of water exchange over the beds.

Table 3.   Bivalves: Maximum population densities.

Species	Habitat	Dry wt of soft tissues $kg \cdot m^{-2}$
Mytilus edulis[a]	Intertidal	0.1-1.4
	Sublittoral	1.4
Geukensia (Modiolus) demissa[b]	Salt marsh	0.03
Cardium edule[a]	Intertidal	0.40
Ostrea edulis[a]	Intertidal	0.20
Mya arenaria[a]	Sublittoral	0.06

[a] Dare 1976
[b] Kuenzler 1961.

Table 4.   Maximum rates of water processing of beds of Mytilus edulis.

Locality	Habitat	Clearance[a] $m^3 \cdot m^{-2} \cdot h^{-1}$	Source
England	Intertidal	7	Dare 1976
England	Sublittoral	12	Dare 1976
Denmark	Sublittoral	7	Jørgensen 1980
		8	Anders Randløv, personal communication

[a] Calculated from the equation   Clearance $(l \cdot h^{-1})$ = 7.45 W (dry flesh weight)$^{0.66}$ (Møhlenberg and Riisgård 1979).

It may be of interest to estimate the relationship between the velocity at which water flows across mussel beds and the effects the water-processing activity of the mussels exerts on the various constitutents of the water.

As an example we may choose a 50 m wide mussel bed capable of mixing the water to a height of 0.5 m. If the mussels process the water at a rate of 8 $m^3 \cdot m^{-2} \cdot h^{-1}$ a 1 m broad transsectional area of the bed will process 7 $m^3 \cdot min^{-1}$, or 28% of the accessible volume of 25 m^3 over the bed. From the rate of water-processing and efficiencies of retention, the relations between pass time (t) and decrease in concentration of the various suspended or dissolved components of the passing water can be calculated from the equation:

$$t = \ln (C_0/C_t) \cdot M/m$$

where C_0 and C_t are upstream and downstream concentrations, M is the volume of water accessible to the mussels, and m the volume cleared per min.

The effects have been calculated as percentages of constituents removed from the passing water; the retention efficiencies are 100% for phytoplankton, 20% for dissolved amino acids, and 1% for bacterioplankton. It may be seen from Figure 3 that this model mussel bed will practically completely deplete the water of phytoplankton at flow velocities of a few meters per min, velocities at which significant amounts of dissolved amino acids, but only negligible amounts of the bacterioplankton are removed. The bacterioplankton becomes heavily grazed only at the very low velocities of < 0.1 $m \cdot min^{-1}$. At typical velocities of water passing an intertidal mussel bed, around 10-20 $m \cdot min^{-1}$, the model predicts that up to half the phytoplankton is removed during the passage against some 20% of the dissolved amino acids, whereas the effect on the bacterioplankton is negligible. The predictions of the model are in reasonable agreement with Wright et al. (1982), whose data have been used to calculate the percentage of phytoplankton removed from the water passing a 46.5 m wide intertidal mussel bed. The percentages ranged from 77-24 at velocities ranging from 7-20 $m \cdot min^{-1}$ (Fig. 3). There was no measurable change in number of bacteria in the water passing the mussel bed.

It is thus indicated that within a restricted range of flow velocities over mussel beds the mussels may practically deplete the water of substrate for growth of the bacterioplankton without interfering directly with the bacteria. Presumably, such situations only apply for restricted periods to mussel beds exposed to the tidal cycles, or to sublittoral mussel beds that depend on water currents generated by the winds.

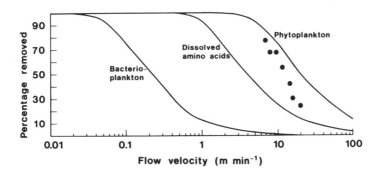

Figure 3. Mytilus edulis. Estimated relationship between flow
velocity of water across a mussel bed and removal of various
constituents from the water. The measured percentage removal of
phytoplankton, ●, is calculated from Wright et al. (1982).

Geukensia (Modiolus) demissa is so far the only example of a
bivalve that may affect the bacterioplankton. In a New England salt
marsh the G. demissa population filtered a volume of water in excess
of the tidal volume (Jordan and Valiela 1982). If G. demissa clear
the water of bacterioplankton with about 50% efficiency, as found by
Wright et al. (1982), the population would clear about half a tidal
volume of its bacteria. It is thus suggested that in salt marshes
dense populations of G. demissa may exert significant grazing
pressure on the bacterioplankton.

CONCLUSIONS

Water-processing mechanisms of suspension feeding metazoans
seem primarily adapted to exploit phytoplankton production. In most
groups the feeding structures efficiently retain suspended particles
down to a few microns which is the low end of the range of phytoplankton cell sizes. Below these particle sizes the efficiency of
retention mostly gradually declines, although the retention spectra
of the various types of suspension feeders are usually poorly
determined.

The sponges may be the only metazoan suspension feeders in which
the lower size limit for complete retention extends to below the
smallest members of the bacterioplankton. But significant retention
of bacterioplankton has also been found in an appendicularian,
Oicopleura dioica, and a bivalve, Geukensia demissa. The low
porosity of the sponge filters is correlated with low velocities of
the water passing the filter. The sponge body is predominantly an
aggregate of choanoflagellates and thus represents a protozoan type
of feeding mechanism scaled to metazoan proportions.

In other types of metazoan suspension feeders, the water passes the feeding structures at relatively high velocities. It is noteworthy that in O. dioica, which seems to efficiently retain bacterioplankton, a special device reduces the velocity at which the water passes the filters by a factor of about 100. In G. demissa the mechanism which extends the retention spectrum down into the size range of the bacterioplankton is not finally understood.

The generally low retention efficiency adds an extra link to the food webs that encompass bacterioplankton and metazoan suspension feeders. This link is the protozoan suspension feeders, of which the microflagellates are probably the most important (Fenchel 1982).

REFERENCES

Alldredge, A. L. 1981. The impact of appendicularian grazing on natural food concentrations in situ. Limnol. Oceanogr. 26: 247-257.
Chipman, W. A., and J. G. Hopkins. 1954. Water filtration by the bay scallop, Pecten irradians, as observed with the use of radioactive plankton. Biol. Bull. 107: 80-91.
Dare, P. J. 1976. Settlement, growth and production of the mussel, Mytilus edulis L., in Morecambe Bay, England. Fish. Invest. Ser. II 28: 1-25.
Deibel, D. R. 1979. Laboratory and field studies on the feeding, growth and swarm dynamics of neritic tunicates from the Georgia Bight. Ph.D. dissertation, Univ. of Georgia, Athens, Georgia.
Fenchel, T. 1980. Relation between particle size selection and clearance in suspension feeding ciliates. Limnol. Oceanogr. 25: 733-738.
Fenchel, T. 1982. Ecology of heterotrophic microflagellates. IV. Quantitative occurrence and importance as bacterial consumers. Mar. Ecol. Prog. Ser. 9: 35-42.
Fjerdingstad, E. J. 1961. The ultrastructure of choanocyte collars in Spongilla lacustris (L.). Z. Zellforsch. 53: 645-657.
Flood, P. R. 1978. Filter characteristics of appendicularian food catching nets. Experientia 34: 173-175.
Flood, P. R., and A. Fiala-Médioni. 1979. Filter characteristics of ascidian food trapping mucous films. Acta Zool. (Stockh.) 60: 271-272.
Fraser, J. H. 1962. The role of ctenophores and salps in zooplankton production and standing crop. Rapp. Cons. Int. Explor. Mer. 153: 121-123.
Gilmour, T. H. J. 1978. Ciliation and function of the food-collecting and waste-rejecting organs of lophophorates. Can. J. Zool. 56: 2142-2155.
Harbison, G. R., and R. W. Gilmer. 1976. The feeding rates of the pelagic tunicate Pegea confederata and two other salps. Limnol. Oceanogr. 21: 517-528.

Harbison, G. R., and V. L. McAlister. 1979. The filter-feeding
    rates and particle retention efficiencies of three species of
    Cyclosalpa (Tunicata, Thaliacea). Limnol. Oceanogr. 24:
    875–892.
Heron, A. C. 1972. Population ecology of a colonizing species: the
    pelagic tunicate Thalia democratica. I and II. Oecologia 10:
    269–312.
Jordan, T. E., and I. Valiela. 1982. A nitrogen budget of the
    ribbed mussel, Geukensia demissa, and its significance in
    nitrogen flow in a New England salt marsh. Limnol. Oceanogr.
    27: 75–90.
Jørgensen, B. B. 1980. Seasonal oxygen depletion in the bottom
    waters of a Danish fjord and its effect on the benthic
    community. Oikos 34: 68–76.
Jørgensen, C. B. 1981a. A hydromechanical principle for particle
    retention in Mytilus edulis and other ciliary suspension
    feeders. Mar. Biol. 61: 277–282.
Jørgensen, C. B. 1981b. Feeding and cleaning mechanisms in a
    suspension feeding bivalve, Mytilus edulis. Mar. Biol. 65:
    159–163.
Jørgensen, C. B. 1981c. Mortality, growth, and grazing impact of a
    cohort of bivalve larvae, Mytilus edulis L. Ophelia 20:
    185–192.
Jørgensen, C. B. 1982. Fluid mechanics of the mussel gill: The
    lateral cilia. Mar. Biol. 70: 275–281.
Jørgensen, C. B. 1983. Patterns of uptake of dissolved amino acids
    in mussels (Mytilus edulis). Mar. Biol. 73: 177–182.
Jørgensen, C. B., and E. D. Goldberg. 1953. Particle filtration in
    some ascidians and lamellibranchs. Biol. Bull. 105: 477–489.
King, K. R., J. T. Hollibaugh, and F. Azam. 1980. Predator-prey
    interactions between the larvacean Oikopleura dioica and
    bacterioplankton in enclosed water columns. Mar. Biol. 56:
    49–57.
Keunzler, E. J. 1961. Phosphorus budget of a mussel population.
    Limnol. Oceanogr. 6: 400–415.
Madin, L. P. 1974. Field observations on the feeding behavior of
    salps (Tunicata: Thaliacea). Mar. Biol. 25: 143–147.
McCammon, H. M. 1965. Filtering currents in brachiopods measured
    with a thermistor flowmeter. Ocean Sci. Ocean Engin. 2:
    772–779.
Møhlenberg, F., and H. U. Riisgård 1978. Efficiency of particle
    retention in 13 species of suspension feeding bivalves.
    Ophelia 17: 239–246.
Møhlenberg, F., and H. U. Riisgård 1979. Filtration rate, using a
    new indirect technique, in thirteen species of suspension-feed-
    ing bivalves. Mar. Biol. 54: 143–148.
Moore, H. H. 1971. The structure of the latero-frontal cirri on
    the gills of certain lamellibranch molluscs and their role in
    suspension feeding. Mar. Biol. 11: 23–27.
Owen, G. 1974. Studies on the gill of Mytilus edulis: the eulatero-

frontal cirri. Proc. R. Soc. Lond. Ser. B. 187: 83-91.

Owen, G., and J. M. McCrae. 1976. Further studies on the latero-
    frontal tracts of bivalves. Proc. R. Soc. Lond. Ser. B. 197:
    527-544.

Paffenhöfer, G.-A. 1976. On the biology of appendicularia of the
    southeastern North Sea, pp 437-455. In: G. Persoone and E.
    Jaspers [eds.], 10th European Symposium on Marine Biology.
    Vol. 2. Universa Press, Belgium.

Randløv, A., and H. U. Riisgård. 1979. Efficiency of particle
    retention and filtration rate in four species of ascidians.
    Mar. Ecol. Prog. Ser. 1: 55-59.

Reiswig, H. M. 1971. In situ pumping activities of tropical
    demospongiae. Mar. Biol 9: 38-50.

Reiswig, H. M. 1973. The aquiferous systems of three Jamaican
    demospongiae. Bull. Mar. Sci. 23: 191-226.

Reiswig, H. M. 1974. Water transport, respiration and energetics
    of three tropical marine sponges. J. Exp. Mar. Biol. Ecol. 14:
    231-249.

Reiswig, H. M. 1975. The aquiferous systems of three marine
    demospongiae. J. Morphol. 145: 493-502.

Riisgård, H. U., and F. Møhlenberg. 1979. An improved automatic
    recording apparatus for determining the filtration rate of
    Mytilus edulis as a function of size and algal concentration.
    Mar. Biol. 52: 61-67.

Riisgård, H. U., A. Randløv and P. S. Kristensen. 1980. Rates of
    water processing, oxygen consumption and efficiency of particle
    retention in veligers and young post-metamorphic Mytilus
    edulis. Ophelia 19: 37-47.

Stephens, G. C. 1982. The role of uptake of organic solutes in
    nutrition of marine organisms. Am. Zool. 22: 611-733.

Strathmann, R. R., T. L. Jahn, and J. R. C. Fonseca. 1972.
    Suspension feeding by marine invertebrate larvae: Clearance of
    particles by ciliated bands of a rotifer, pluteus, and trocho-
    phore. Biol. Bull. 142: 505-519.

Wiebe, P. H., L. P. Madin, L. R. Haury, G. R. Harbison, and L. M.
    Philbin. 1979. Diel vertical migration by Salpa aspera and
    its potential for large-scale particulate organic matter
    transport to the deep-sea. Mar. Biol. 53: 249-255.

Wright, R. T., R. B. Coffin, C. P. Ersing, and D. Pearson. 1982.
    Field and laboratory measurements of bivalve filtration of
    natural marine bacterioplankton. Limnol. Oceanogr. 27: 91-98.

Wright, S. H., and G. C. Stephens. 1978. Removal of amino acid
    during a single passage of water across the gill of marine
    mussels. J. Exp. Zool. 205: 337-352.

THE USE OF LARGE ENCLOSURES IN MARINE MICROBIAL RESEARCH

J. M. Davies

DAFS Marine Laboratory
Aberdeen, Scotland, U. K.

INTRODUCTION

The merits and drawbacks of the use of large enclosures in marine research have been extensively and recently (Grice and Reeve 1982) reviewed. In all but a few experiments the ecological role of microorganisms has been ignored and the central effort has concentrated on the food web from phytoplankton to primary carnivores.

However, the fairly recent development of techniques capable of measuring microbial activity in the sea and the suggestion (Joiris 1977; Sieburth et al. 1977) that bacterial production may be much greater than was first imagined (Steele 1974) and may in fact exceed by ten times the primary production as determined by the ^{14}C technique (Joiris 1977; Johnson et al. 1981) has opened up interest in the microbial food web.

If the evidence for these high rates of microbial activity is accepted, then it becomes difficult to explain how the total primary production measured by the ^{14}C technique can account for both the heterotrophic (microbial) and the zooplankton food requirements. Investigations into the trophodynamics of microbial food webs have centered on measurements of activity and growth over diel cycles (Burney et al. 1982) in a defined body of water.

However, one of the overriding problems associated with marine ecology is that advection processes in the open sea make time series of measurements on the same population difficult if not impossible. Attempts have been made to do this using drogue stations (e.g., Ryther et al. 1971); although this approach may reduce the problem of horizontal advection, the problem of vertical mixing remains.

An alternative is the use of a large enclosed body of water
(Grice and Reeve 1982) in which the horizontal advection processes
are essentially eliminated and vertical ones are markedly reduced
due to dampening of turbulent energy transfer by the wall of the bag
(Steele et al. 1977). The appropriateness of large enclosures for
the study of the lower end of the food web is illustrated in Figure 1
which attempts to relate the various types of marine investigations
to the appropriate temporal and spatial scales of the plants and
animals being studied. It shows that enclosures are able to
accommodate the temporal scales needed to study microorganisms,
phytoplankton, and perhaps even zooplankton but are obviously unable
to cover the spatial scales of phytoplankton and zooplankton although
they may cover temporal and spatial scales of bacteria. They are
certainly suitable for studying under near natural conditions the
trophic relationships between the soluble organic carbon pool, the
bacteria, the ciliates and microflagellates, the phytoplankton and
the zooplankton over periods of up to 100 days.

Most of the investigations into the role of microorganisms
carried out in enclosed ecosystems have been concerned with the
effects (Vaccaro et al. 1977) or fate (Lee et al. 1982; Topping et
al. 1982) of particular pollutants; these have been recently reviewed
(Gillespie and Vaccaro 1981). The experiments designed to study the
trophic role of microorganisms may be conveniently considered under
three broad headings.

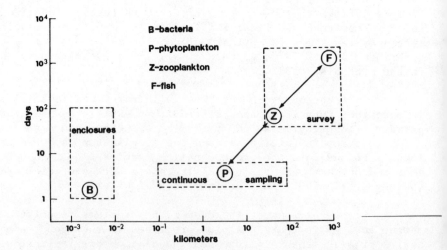

Figure 1. Relation between various types of marine investigation
and the spatial and temporal scales of the plants and animals being
studied. (From Menzel and Steele 1978.)

DISCUSSION

Production and Utilization of Soluble Organic Material

The excretion of soluble organic material by phytoplankton under natural conditions is now widely accepted as a real phenomenon (Mague et al. 1980) and not a methodological artifact (Sharp 1977). The real questions now become:

(a) how much soluble organic material is produced;
(b) how rapidly and efficiently is this material consumed;
(c) how much of it finds its way up the food chain.

In an attempt to answer the first question, Davies et al. (1980)

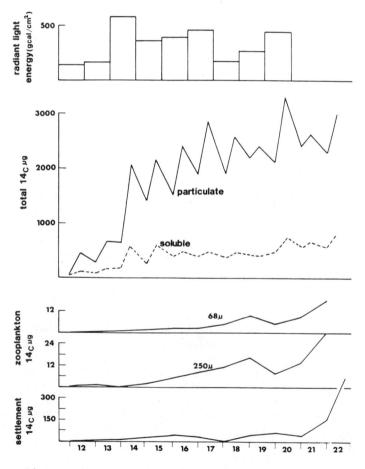

Figure 2. ^{14}C concentration in soluble organic material, phytoplankton, zooplankton and settled material along with daily irradiance for the 10-day experiment.

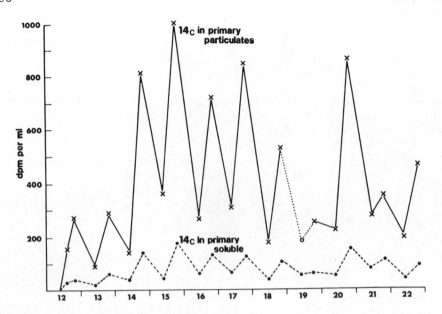

Figure 3. ^{14}C at $2^1/_2$ m depth in soluble organic pool and in particulate phytoplankton.

dosed a 100 m^3 bag with 1 Curie of ^{14}C bicarbonate. The ^{14}C label was then followed over a period of ten days.

The ^{14}C concentration in the phytoplankton, zooplankton, and the settled material, integrated over the whole bag volume, are shown in Figure 2 along with the daily irradiance. The most interesting feature, which is shown to greatest effect at 2.5 m (Fig. 3), is the production and utilization of particulate and soluble ^{14}C organic carbon in the light-dark cycle (samples were collected and filtered at approximately 9:30 a.m. and 7:00 p.m.). It was apparent that during the day the rate of production of soluble organic material exceeded the rate of utilization by bacteria but that during the dark the bacteria consumed all or nearly all of the available soluble ^{14}C organic material.

To estimate the true soluble production over 24 hours, a simple solution would be to consider soluble production to be occurring only during active photosynthesis during daylight and bacterial consumption to be constant over 24 hours (Fig. 4). This leads to a simple correction giving a revised true soluble production rate. An example of this for the ten day experiment described is shown in Table 1. This sort of correction may be taken a step further, for example Figure 5, in which a night time excretion rate is assumed as a percentage (R%) of the daylight excretion rate. If for example one assumes that this is 30%, then a further "real" excretion value

In a simple system allowing for <u>continuous heterotrophic uptake</u> and assuming <u>no</u>
<u>soluble excretion during the night</u>

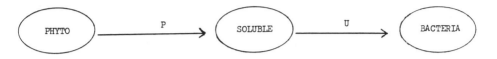

$$K = \text{Total daytime production} = P + (U \times \frac{C}{D})$$

P = Net ^{14}C production (dpm) of soluble organic material during the daylight period
    (C hours)

U = Net ^{14}C loss (dpm) of soluble organic material during the dark period (D hours)

Uptake of soluble organic material by phytoplankton and excretion by bacteria are
assumed to be negligible.

Figure 4. Estimation of time (24 h) soluble excretion rate from a
simple model of continual bacterial uptake.

Table 1. Soluble excretion rates as measured (% Sol) over the 10
             days of the experiment and with correction for (a) dark
             uptake by bacteria (Corr % Sol) (Fig. 4), and (b) dark
             uptake by bacteria and dark excretion by phytoplankton
             (Sol 24 h) (Fig. 5).

Day	Nutrient Addition↓ 1	2	3	Nutrient Addition↓ 4	5	6	7	Nutrient Addition↓ 8	9	10	11
% Sol	21	24	28	61	6	12	12	3	12	36	33
Corr % Sol	31	36	41	86	12	21	28	7	22	61	54
% Sol 24 h	59	41	106	104	21	50	36	14	38	108	111

% Sol $\bar{x}$ = 20.2%

Corr % Sol $\bar{x}$ = 36.2%

% Sol 24 h $\bar{x}$ = 63%

Correction to total soluble excretion assuming <u>night-time excretion</u> rate equal to R% of the day-time excretion rate.

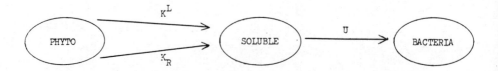

From figure 4   Total soluble excretion $= K_1^L = P + (U \times \dfrac{C}{D})$

But, Uptake$_R$ (U_1) = nett uptake (U) + night-time excretion $(K_R = (\dfrac{R}{100} \times K_1^L))$

ie $U_1 = U + (\dfrac{R}{100} \times K_1^L \times \dfrac{D}{C})$

Now Total excretion, $K_2^L = P + (U_1 \times \dfrac{C}{D})$

Therefore $K_R = (\dfrac{R}{100} \times K_2^L)$

and $U_2 = U + (\dfrac{R}{100} \times K_2^L \times \dfrac{D}{C})$

Hence $K_3 = P + (U_2 \times \dfrac{C}{D})$

and so on until $K_n^L = P + (U_{n-1} \times \dfrac{C}{D})$ when $U_n - U_{n-1} = 0.01$ or less

Then, the true daily excretion $K^{24} = K_n^L + (\dfrac{R}{100} \times K_n \times \dfrac{D}{C})$

Figure 5. Estimation of time (24 h) soluble excretion rate assuming a dark period excretion rate of 30% for light period rate.

is produced which is three times the measured rate (Table 1).

   Thus these sort of corrections which in a simple manner allow for the simultaneous utilization of soluble production by bacteria lead to a revised rate of soluble production that is almost double the measured rate of exudation.

   The chemical nature of the soluble material produced has been investigated in 3 m^3 enclosures by Brockmann et al. (1979). They compared the production of free carbohydrates and amino acids with the detailed phytoplankton succession over a 25-day period (Fig. 6). Glucose and lysine occurred in the highest concentration, and the peak of carbohydrate production appeared to be associated with the diatom bloom while the peak free amino acid production closely followed the dinoflagellate bloom.

The form of the soluble organic material may have a most important effect on the rate at which it is utilized. Parsons et al. (1981), for example, showed that the rate at which glucose was utilized in an enclosed water column was dependent upon the availability of inorganic nitrogen. For substrates containing nitrogen, such as amino acids, this presumably would not be a restriction on their rate of utilization. Thus future work on the production of soluble organic material from phytoplankton excretion or from zooplankton "sloppy feeding" should pay some attention to the nature of the organic material being produced.

Nutrient Cycling

Most interest has centered on the cycling of inorganic nitrogen since biological productivity in the sea often seems to be regulated

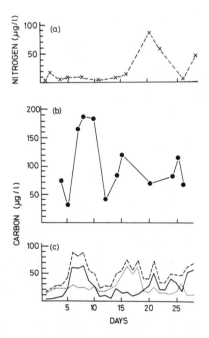

Figure 6.   Concentration of dissolved amino acids (a) and carbohydrates (b) compared to calculated biomass of phytoplankton (c): continuous line, diatoms; dotted line, dinoflagellates; dashed line, total phytoplankton.   (From Brockman et al. 1979.)

by the supply of inorganic nitrogen. According to Dugdale and Goering (1967), primary production in the sea may be considered as either new production (supported by nitrate supplied to the euphotic layer by advection from deep water) or regenerated production supported by ammonia produced by bacterial and higher forms within the euphotic zone. This source of nitrogen may account for 90% of the inorganic nitrogen used by phytoplankton in subtropical seas and about 60% in coastal and near shore waters.

The classical view of nutrient cycling in the sea is that of zooplankton feeding and excretion of ammonia (Harris 1959; Butler et al. 1970; Steele 1974). However, attempts to relate zooplankton grazing and excretion to the phytoplankton nutrient requirements have often shown a discrepancy (see Williams, this volume), and thus the role of microorganisms in nutrient cycling became of interest. A number of enclosure experiments have been designed to investigate this problem, usually (Harrison and Davies 1977; Hattori et al. 1980) using an isotopic labelling technique.

Several approaches to the problem have been used. The most common is that first used by Harrison and Davies (1977) in which inorganic nitrogen labelled with ^{15}N was introduced into the whole bag and its flux through the food web measured. The regeneration of ammonia was measured by isotope dilution and a budget for the nitrogen in the enclosure was calculated. Similar experiments have been carried out subsequently (Hattori et al. 1980; Koike et al. 1982), and the conclusions are very similar to those from the first experiment, which were that the excretion by copepods accounted for only 47% and 11% of the flux into the ammonia pool during two periods of investigation while ammonia input by mineralization of seston accounted for 45% and 81% respectively during the same periods.

Using a different approach to study the same problem Hollibaugh et al. (1980) measured the turnover rates of primary amines in large plastic enclosures in Saanich Inlet. From these measurements they determined the flux of dissolved free amino acids which was shown to be related to the amount of primary production. Comparison of the amino acid flux rates with changes in ammonia concentration in deep water suggested that dissolved amino acid degradation could account for 60% of the flux into the ammonia pool and thus indicated that a large fraction of the community carbon and nitrogen flux passes through the bacterioplankton.

Energy Transfer Through Bacteria

Enclosed ecosystems seem to offer great promise for investigations into the energy transfer in the lower food web. This is because (Fig. 1) the experimental time scales required to study the interaction of soluble organic material – bacteria – microzooplankton are short enough, about ten days, so that many of the potential problems associated with containment experiments (e.g., wall growth, pond effect) will not occur. Furthermore, the volume of the containers is such that there is every likelihood that trophic interactions studied will be reasonably representative of those in the open sea.

Two major experments have been carried out in enclosed ecosystems to look at the trophodynamics of the lower end of the food web. The first of these was a study of the predator-prey inter- actions between the larvacean <u>Oikopleura oioca</u> and bacterioplankton (King et al. 1980; King 1982). Laboratory-derived measurements of

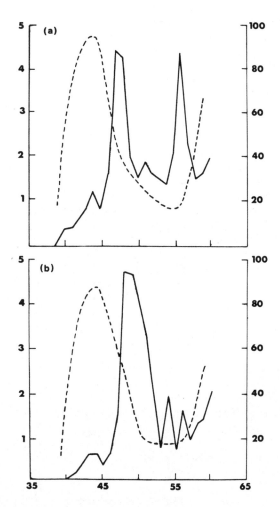

Figure 7. Biomass of larvaceans (continuous line, left ordinate) and bacterioplankton (dashed line, right ordinate). (From King et al. 1980.)

the larvacean feeding rate were used to estimate the impact of the
larvaceans on the bacterial population within the enclosure.   They
found that at times the bacteria formed up to 50% of the larvacean
daily food ration.   However, despite the apparent classic Lotka-
Volterra prey-predator relationship (Fig. 7), they concluded that
larvacean grazing could not have limited bacterial biomass and that
bacterial production must have been controlled by factors such as
the level of dissolved organic matter.   These experiments indicated
the possibility of an alternative pathway (Fig. 8) effectively by-
passing both phagotrophic flagellates and probably the herbivorous
copepods by substituting larvaceans between bacterioplankton and
primary carnivores.

The second experiment designed specifically to look at the
lower part of the food web involved the addition of small quantities
of glucose (1-5 mg/l) to enclosed water columns (Parsons et al. 1981)
and a comparison of the autotrophic, heterotrophic, secondary, and
tertiary prodution in these enriched water columns with those left
to develop naturally.   There were three major conclusions from this
experiment.   Photosynthesis (mg $C/m^2/day$) and the standing stock of
phytoplankton, as measured by chlorophyll a, showed a marked
depression in the enclosures receiving additions of glucose.   This
effect occurred after the first addition of glucose and continued
for the 20 days while the glucose additions were made.   The ability
of the heterotrophic organisms to suppress autotrophic production
appeared to be manifested by the availability of inorganic nitrogen
and was believed to have been caused by heterotrophic organisms
out-competing the autotrophs for the limited supply of nitrate in
the presence of glucose.

The transfer efficiency of glucose to heterotrophic particulate
organic carbon, measured as production of heterotrophic POC divided
by the sum of this figure plus the respired carbon, was 35% at a
level of 1 mg/l glucose and 45% at additions of 5 mg/l glucose.

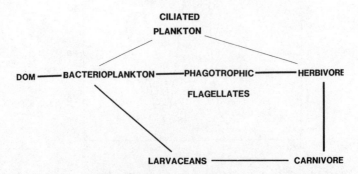

Figure 8.   Simplified heterotrophic food web.   Thickness of line
denotes importance of pathways.   (From King et al. 1980.)

These figures were lower than those derived from short term ^{14}C
glucose incubations, which gave apparent assimilation efficiencies
of about 80%.   However the transfer efficiencies measured (35-45%)
are similar to values summarized by Calow (1977), who reports a
range of 30-65% efficiency for bacteria on various substrates.   The
differences between the short term assimilation efficiencies and
longer term tranfer efficiencies might be accounted for by natural
bacterial mortality, grazing mortality, and changes in bacterial
species with time.

The transfer of heterotrophic production to secondary and
tertiary trophic levels was measured by a size fractionation of
the ^{14}C glucose assimliated in the short term (6 hrs) incubations.
The size fractions responsible for the greatest assimilition of
^{14}C glucose (Table 2, taken from Parsons et al. 1981) were found to
be the two smallest size fractions (< 3 μ).   However, there was a
significant accumulation in the largest size fraction (> 100 μ) which
either reflected a lack. of appreciable biomass of intermediate size
or a direct carbon transfer from the < 3 μ to the > 100 μ organisms.
The size spectrum of particulate material in the enclosed ecosystems
showed a progressive accumulation of a large amount of particulate
material in the < 3 μ size category which was shown to be mostly
bacteria.   The dominance of bacteria, reaching 80% of the total
bacterial plus phytoplankton biomass in the enclosure with the
highest level of glucose addition, was more consistent with the
interpretation of an uptake of glucose by heterotrophic bacteria and

Table 2.   Size fractionation of ^{14}C glucose uptake added to
enclosures to which glucose had been added at 5 mg/1
(in CPM).   (From Parsons et al. 1981.)

Size of Fraction (μm)	Date		
	6/26	6/27	6/28
0.2 - 3	710	421	805
0.4 - 3	307	244	484
3 - 12	14	3.4	12
12 - 35	9	11	8
35 - 100	36	36	52
100	25	20	38

a transfer of this material directly to a size fraction of 100 μ or greater, the latter being representative of the zooplankton.

The effect of glucose enrichment on secondary and tertiary (carnivorous jellies) trophic levels appeared to be an enhanced yield of zooplankton (benthic larvae and copepods) and of carnivorous jellies in the two enclosures dosed with glucose and supporting large heterotrophic populations when compared to the predominantly autotrophic ecosystem. This increased tertiary production was attributed to a bloom of benthic larvae in the two glucose-dosed enclosures, speculatively supported by the enhanced heterotrophic production, which did not occur in the autotrophic enclosure.

The importance of heterotrophic processes in marine food webs in enclosed ecosystems has also been amply demonstrated by a series of size fractionation experiments (Williams 1982). Calculations from enclosure experiments suggested that roughly 45% of overall respiration was bacterial, with the remainder mainly algal. However the presence of the walls of the enclosure and the removal of most of the horizontal and vertical advection and its effect on phytoplankton physiology present real possibilities for differences between the enclosed and natural bodies of water, and these must be investigated. Thus, perhaps the more difficult question to answer is whether the microorganisms contribute to the same extent in the open sea. However, evidence is beginning to accumulate (Sieburth et al. 1977; Burney et al. 1982) from open sea studies which also points to a very large microbial contribution to overall marine food chain energetics.

CONCLUSIONS

There has been a recent surge of interest in the quantitative role that microorganisms play in marine food webs. This has been centered in three different areas of work, enclosures (Parsons et al. 1981), frontal systems (Pingree et al. 1976), and drogue stations in open sea (Sieburth et al. 1977), but all have had similar objectives and used similar techniques. In each case it has been recognized that the sampling frequency must be matched to the growth rates of the organisms being investigated so that many of the studies have involved diel sampling based upon a two to six hour sampling period.

The central problem, the discrepancy between primary production measured by ^{14}C and heterotrophic production measured by oxygen consumption (Joiris 1977) or carbohydrate flux (Sieburth et al. 1977; Burney et al. 1982), has still to be resolved. However, recent research in enclosures and in the open sea has opened up the possibility of alternative pathways functioning alongside the classical primary production – herbivore pathway. If it is accepted that true soluble production is underestimated by conventional ^{14}C techniques and that the particulate ^{14}C measured by conventional techniques (0.45 µ membrane) may be a mixture of the particulate ^{14}C incorporated autotrophically by phytoplankton and heterotrophically by bacteria consuming soluble ^{14}C carbon excreted, then we could move some way towards resolving the problem. Obviously the scales of the autotrophic – heterotrophic pathways would be dependent on many factors such as zooplankton grazing pressure, nutrient availability, degree of water column vertical mixing, etc., and one would imagine that in most natural circumstances autotrophic processes would overshadow heterotrophic processes.

REFERENCES

Brockmann, U. H., K. Eberlein, H. Junge, D. E. Maier-Reimer, and D. Siebers. 1979. The development of a natural plankton population in an outdoor tank with nutrient-poor sea water. II. Changes in dissolved carbohydrates and amino acids. Mar. Ecol. Prog. Ser. 1: 283-291.

Burney, C. M., P. G. Davis, K. M. Johnson, and J. McN. Sieburth. 1982. Diel relationships of microbial trophic groups and in situ dissolved carbohydrate dynamics in the Caribbean Sea. Mar. Biol. 67: 311-322.

Butler, E. I., E. D. S. Corner, and S. M. Marshall. 1970. On the nutrition and metabolism of zooplankton. VII. Seasonal survey of nitrogen and phosphorus excretion of Calanus in the Clyde Sea area. J. Mar. Biol. Assoc. UK 50: 525-560.

Calow, P. 1977. Conversion efficiencies of heterotrophic organisms. Biol. Rev. 52: 385-409.

Davies, J. M., I. E. Baird, L. C. Massie, and A. P. Ward. 1980. Chemical and productivity measurements in an enclosed water column. CM 1980/E:49 Marine Environmental Quality Committee. Ref. Biological Oceanography Committee. DAFS Marine Laboratory, Aberdeen, Scotland, U.K.

Dugdale, R. C., and J. J. Goering. 1967. Uptake of new and regenerated forms of nitrogen in primary productivity. Limnol. Oceanogr. 12: 196-206.

Gillespie, P. A., and R. F. Vaccaro. 1981. Heterotrophic microbial activity in experimentally perturbed marine ecosystems, pp. 209-226. In: R. A. Geyer [ed.], Marine Environmental Pollution, Vol. II: Dumping and Mining. Elsevier, Holland.

Grice, G. D., and M. R. Reeve. 1982. Marine Mesocosms – Biological and Chemical Research in Experimental Ecosystems. Springer-Verlag, New York.

Harris, E. 1959. The nitrogen cycle in Long Island Sound. Bull. Bing. Ocean. Coll. 17: 31-65.

Harrison, W. G., and J. M. Davies. 1977. Nitrogen cycling in a marine planktonic food chain: Nitrogen fluxes through the principal components and the effects of adding copper. Mar. Biol. 43: 299-306.

Hattori, A., I. Koike, M. Ohtou, J. J. Goering, and D. Boissear.
    1980. Uptake and regeneration of nitrogen in controlled
    aquatic ecosystems and the effects of copper on these
    processes. Bull. Mar. Sci. 30: 431-443.
Hollibaugh, J. T., A. B. Carruthers, J. A. Fuhrman and F. Azam.
    1980. Cycling of organic nitrogen in marine plankton communi-
    ties studied in enclosed water columns. Mar. Biol. 59: 15-21.
Johnson, K. M., C. M. Burney, and J. McN. Sieburth. 1981.
    Enigmatic marine ecosystem metabolism measured by direct diel
    CO_2 and O_2 flux in conjunction with DOC release and uptake.
    Mar. Biol. 65: 49-60.
Joiris, C. 1977. On the role of heterotrophic bacteria in marine
    ecosystems. Some problems. Helgol. Wiss. Meeresunters. 30:
    611-621.
King, K. R. 1982. The population biology of the larvacean
    Oikopleura dioica in enclosed water columns, pp. 341-352 . In:
    G. D. Grice and M. R. Reeve [eds.], Marine Mesocosms -
    Biological and Chemical Research in Experimental Ecosystems.
    Springer-Verlag, New York.
King, K. R., J. T. Hollibaugh, and F. Azam. 1980. Predator-prey
    interactions between the larvacean Oikopleura dioica and
    bacterioplankton in enclosed water columns. Mar. Biol. 56:
    49-57.
Koike, I., A. Hatton, M. Takahashi, J. J. Goering, K. Iseki, and P.
    K. Bienfang. 1982. The use of enclosed experimental ecosystems
    to study nitrogen dynamics in coastal waters. And phytoplankton
    species' responses to nutrient changes in experimental enclos-
    ures and coastal waters, pp. 291-304. In: G. D. Grice and M.
    R. Reeve [eds.], Marine Mesocosms - Biological and Chemical
    Research in Experimental Ecosystems. Springer-Verlag, New York.
Lee, P. F., K. Hinga, and G. Almquist. 1982. Fate of radiolabelled
    polycyclic aromatic hydrocarbons and pentachlorophenol in
    enclosed marine ecosystems, pp. 123-136. In: G. D. Grice and
    M. R. Reeve [eds.], Marine Mesocosms - Biological and Chemical
    Research in Experimental Ecosystems. Springer-Verlag, New York.
Mague, T. H., E. Firberg, D. J. Hughes, and I. Morris. 1980. Extra-
    cellular release of carbon by marine phytoplankton: a physiolog-
    ical approach. Limnol. Oceanogr. 25: 262-279.
Menzel, D. W., and J. H. Steele. 1978. The applications of plastic
    enclosures to the study of pelagic marine biota. Rapp. P.-v.
    Reun. Cons. int. Explor. Mer 173: 7-12.
Parsons, T. R., L. J. Albright, F. Whitney, C. S. Wong, and P. J.
    leB. Williams. 1980-81. The effect of glucose on the
    productivity of seawater: An experimental approach using
    controlled aquatic ecosystems. Mar. Environ. Res. 4: 229-242.
Pingree, R. D., P. M. Holligan, G. T. Mardell, and R. N. Head.
    1976. The influence of physical stability on spring, summer
    and autumn phytoplankton blooms in the Celtic Sea. J. Mar.
    Biol. Assoc. UK 56: 845-873.
Ryther, J. H., D. W. Menzel, E. M. Hulbert, C. J. Lorensen, and N.

Corwin. 1971. The production and utilisation of organic matter in the Peru coastal current. Inv. Pesq. 35: 43-59.

Sharp, J. M. 1977. Excretion of organic matter by marine phytoplankton. Do healthy cells do it? Limnol. Oceanogr. 22: 381-399.

Sieburth, J. McN., K. M. Johnson, C. M. Burney and D. M. Lavoie. 1977. Estimation of the in situ rates of heterotrophy using diurnal changes in dissolved organic matter and growth rates of picoplankton in diffusion culture. Helgol. Wiss. Meeresunters. 30: 565-574.

Steele, J. H. 1974. The Structure of Marine Ecosystems. Harvard Univ. Press, Cambridge, MA.

Steele, J. H., D. M. Farmer, and E. W. Henderson. 1977. Circulation and temperature structure in large marine enclosures. J. Fish. Res. Board Can. 34: 1095-1104.

Topping, G., I. M. Davies, and J. M. Pirie. 1982. Processes affecting the movement and speciation of mercury in the marine environment, pp. 167-180. In: G. D. Grice and M. R. Reeve [eds.], Marine Mesocosms - Biological and Chemical Research in Experimental Ecosystems. Springer-Verlag, New York.

Vaccaro, R. F., F. Azam, and R. E. Hodson. 1977. Response of natural marine bacterial populations to copper: Controlled ecosystem pollution experiment. Bull. Mar. Sci. 27: 17-22.

Williams, P. J. leB. 1982. Microbial contribution to overall plankton community respiration - Studies in enclosures, pp. 305-322. In: G. D. Grice and M. R. Reeve [eds.], Marine Mesocosms - Biological and Chemical Research in Experimental Ecosystems. Springer-Verlag, New York.

Williams, P. J. leB. 1983. A review of measurements of respiration rates of marine plankton populations. In: J. E. Hobbie and P. J. leB. Williams [eds.], Heterotrophic Activity in the Sea. Plenum Press, New York.

LIPID INDICATORS OF MICROBIAL ACTIVITY IN MARINE SEDIMENTS

S. C. Brassell and G. Eglinton

Organic Geochemistry Unit
University of Bristol, School of Chemistry
Cantock's Close, Bristol BS8 1TS, England

INTRODUCTION

Detailed analyses at the molecular level of the organic matter
present in natural systems, such as oceanic waters and sediments,
have been possible only in recent years with the advent of powerful
chromatographic and spectrometric techniques. In this article, we
shall address the subject of this conference from our particular
viewpoint – that of the organic geochemist. In the widest sense,
organic geochemistry is concerned with the fate of carbon compounds
in space and through time. It offers unique insights into microbial
processes and activity from studies of the nature and fate of the
organic matter contributed to, and resident in, certain marine
sediments. Although much information on proteins, carbohydrates and
other highly significant classes of biomolecules has been collected,
by far the most coherent picture is given by the lipids – the storage
fats and membrane components with hydrophobic structures generally
resistant to degradation and with carbon skeletons well suited to
preservation as 'chemical fossils' (Eglinton and Calvin 1967).

In this article we will review the information presently
available concerning the nature and distribution of the lipid
compounds in the marine system, especially in the bottom sediments
and the underlying sedimentary columns. We shall seek to relate
these lipids to microbial activity – degradation and input – for
evidence of such activity.

The transformation pathways and products of the biochemical
metabolism of lipids in aquatic systems are generally poorly
understood. Following the death of the organism that synthesized
them, biolipids may be exposed to many potential agencies for their

481

alteration.   Such lipid reworking can occur at all stages of their
sedimentation within the water column.   Since the vast majority of
organic matter (typically > 95%) is recycled within the euphotic zone
(Deuser 1971) there is considerable remetabolism of organic matter in
the upper portion of the water column.   In general, it is the lipids
which survive complete degradation during this stage and reach the
sediment, albeit in altered form.   Recent investigations of sediment
trap particulates (Wakeham et al. 1980) explored the changes at
different stages of sedimentation and distinguished between pelagic
and benthic alteration processes.   Cultures of marine organisms and
incubation experiments designed to investigate the fate of specific
lipids in the environment provide points of reference in attempts to
evaluate the role and contributions of various organisms in the
generation of sedimentary organic matter.

The natural marine system is highly complex and its organic
matter is affected by numerous processes, such as photo-oxidation,
evaporation, solution and particle association, in addition to the
aspects of microbial degradation discussed herein.   Furthermore, the
organic matter itself can be absorbed in, adsorbed on or trapped in
particulate debris and also become bound into the polycondensate
molecules that constitute the bulk of the sedimentary organic matter.
Such insoluble organic material forms at the very earliest stages of
diagenesis as protokerogen and undergoes subsequent structural·modi-
fication and further condensation to give kerogen itself, which, in
turn, changes and evolves in response to further diagenesis.   The
influence and possible role of microbial processes in early-stage
kerogen formation are uncertain, but they may be highly significant.
In this paper we focus on the lipid evidence for the activity and
imprint of microbes on sedimentary organic matter.   Our approach is
that of the natural product chemist, aiming to identify discrete
molecules and assess their significance within a framework of
knowledge concerning the lipids of organisms, model systems and
present day environments (Fig. 1).

DISCUSSION

The Record of Microbial Metabolism of Lipids: An Example

The processes determining the fate of phytoplankton lipids may
be considered as a typical example of those that influence the
survival and degree of alteration of lipids in the oceanic system.
First, phytoplankton may be ingested by zooplankton and undergo
modification by microbes of the gut prior to excretion as a fecal
pellet.   Capsular fecal pellets are often rich in organic matter and
therefore can act as a good food source and host for microbes.
Within them, both aerobic and anaerobic bacterial activity may thrive
in discrete microenvironments, restricted by water pressures and such
variables.   On reaching the sediment, the lipids of fecal pellets,

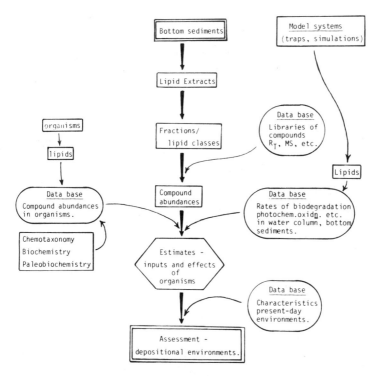

Figure 1. Scheme for the assessment of depositional environments of aquatic sediments by molecular organic geochemistry.

like those of all the other sedimented material, may undergo further microbial alteration by the benthic bacterial community. However, despite such opportunities for microbial modification, several classes of phytoplankton lipids, such as sterols (Volkman et al. 1980b), appear to survive in unaltered form. This survival is perhaps aided by the rapidity of fecal pellet transport. Other lipids not known to occur in phytoplankton also appear in fecal pellets and may be derived from zooplankton or produced by microbial action within the zooplankton gut or fecal pellet itself.

Lipid Resistance to Microbial Attack

The resistance of lipids to microbial attack is highly variable as are the rates of lipid alteration processes. The preferential degradation of shorter chain n-alkanes (e.g., $n-C_{17}$) relative to higher homologues is well documented (Johnson and Calder 1973; Brassell et al. 1978 and references therein) and is confirmed by the record of oil biodegradation both in crude oil reservoirs and in the environment following crude oil spills. Similarly, alkenoic acids are more susceptible to degradation in sediments than alkanoic acids

(Gaskell et al. 1976 and references therein). In a marine system
(e.g., Walvis Bay) it has been suggested that algal carboxylic acids
are replaced by bacterial carboxylic acids at an early stage of
burial (Boon et al. 1975), illustrating the general lability of this
class of lipid. Among sterols it appears that in the water column
Δ^{22} and $\Delta^{24(28)}$ components are less stable than their saturated
counterparts (Gagosian and Heinzer 1979), a postulate supported by
the recognition of an oxidation product of a $\Delta^{24(28)}$ stenol,
cholest-5-en-24-al-3β-ol in seawater. This compound, however, does
occur in various marine organisms (Carlson et al. 1978) and may
therefore represent a direct biological input from such a source.
The $\Delta^{5,7}$ stenols are probably more susceptible to microbial attack
than Δ^5 stenols, as suggested by their limited occurrence in seawater
and sediments (a single identification of each; Gagosian and Heinzer
1979; Brassell 1980), although cholesta-5,7-dien-3β-ol is widely
distributed in marine animals.

Sedimentary Lipids: Origins

As discussed earlier the lipids of marine sediments are not
necessarily those originally biosynthesized but may have undergone
various structural modification during their pre- and post-deposi-
tional history. Their basic carbon skeletons, however, often remain
unaffected and certain of their features, such as their carbon
number distributions, homology ranges and stereochemistry, are often
conserved or only gradually changed by diagenetic processes. The
recognition of intact and altered lipids is an important facet in
understanding the processes of microbial activity in aquatic systems.

Unaltered lipids. These components are intact biolipids. They
include many compounds that are relatively resistant to microbial
degradation, such as the long chain n-alkanes of higher plant waxes,
and some labile lipids that are preserved by a combination of environmental factors, such as high productivity and an oxygen minimum
zone, which have limited microbial degradation of the organic matter.
In addition, specific lipids found in marine bottom sediments are
biosynthesized by microbes inhabiting the sediment, and therefore
represent those lipids generated in situ. Acyclic isoprenoid alkanes
derived from methanogenic bacteria (Brassell et al. 1981) are an
example of such autochthonous sediment inputs. Indeed, it remains
open whether such compounds found in the lipid extracts of sediments
are present within, and extracted from, the sediment matrix and its
constituent dead organisms, or from bacteria still living within the
sediment. In some instance, the age (Cretaceous, ca. 125 million
years) and sub-bottom depth (>1200 m) of sediments containing such
compounds presumably precludes the possibility that methanogenic
bacteria are still living within them, but for many younger,
shallower sediments, these two origins seem equally probable.

Altered lipids. These components are those lipids that retain

sufficient structural information that their biological origin can
still be recognized despite minor or even substantial modification.
During transfer from their parent organism to a sediment, lipids can
undergo a diverse range of chemical transformations (hydroxylation,
dehydrogenation, reduction, dehydration, aromatisation, decarboxylation, oxidation and ring opening) and yet retain their biologically
inherited distinctiveness. For example, the dehydration of a sterol
does not affect the methyl substitution or stereochemistry of its
side chain, which are more significant indicators of its origin.

An important facet of lipid alteration not considered in detail
here is their possible incorporation or binding with kerogen,
processes that may also be reversible. Evidence for the presence of
biolipid structures in kerogen is provided by a number of analytical
techniques, notably pyrolysis-GC and pyrolysis-GC-MS. Such studies
and, in addition, the chemical degradation of kerogens, can also
provide information concerning the nature of the bonds linking the
lipids to the insoluble kerogen matrix.

Lipids in Marine Sediments: Source Indicators

The lipids found in bottom marine sediments may be ultimately
derived from marine algal, bacterial or terrestrial sources. To a
considerable degree, the structural specificity inherent in biochemical lipid synthesis enables these different origins of sedimentary
lipids to be distinguished. In addition, specific lipids may provide
evidence for microbial reworking of, for example, a terrestrial input
of organic matter.

Differences between the lipid composition of various organisms
stem, in part, from the biochemical requirements of the individual
habitats and lifestyles of different biota. For example, the composition of the lipids that act as rigidifiers of membranes responds to
changes in temperature. Also, terrestrial higher plants need protective waxes to shield against desiccation.

The lipids that we at present believe are indicative of particular marine or terrestrial contributions of organic matter are listed
in Table 1; those attributed to microbial origins or reworking are
discussed below.

Lipids of Microbial Origin

Many lipids found in bottom marine sediments are thought to be
of direct microbial origin, in that they are only known to be synthesized by bacteria. Simple branched alkanoic and alkenoic acids were
among the first compounds recognized as potential biological markers
for bacterial inputs to sediments, and many classes of compound have
subsequently been added to this list (Table 2). The origin of the
ubiquitous extended hopanoids found in sediments and crude oils was

Table 1. Proposed lipid indicators of direct marine and terrestrial inputs to oceanic sediments.[a]

(a) MARINE[b]

Straight-chain
  Alkanes (C_{15}, C_{17} and C_{19})
  Alkenes ($C_{37:3}$ and $C_{38:3}$)
  Alkadienones (C_{37}-C_{39})
  Alkatrienones (C_{37}-C_{39})
  Alkanols (C_{14}-C_{22})
  Alkenols (C_{14}-C_{24})
  Alkenoic acids ($C_{20:5}$ and $C_{22:6}$)
  Methyl ester (C_{36})
  Ethyl ester (C_{36})

Acyclic isoprenoid
  Phytol

Steroids
  Sterols (C_{21}-C_{26}, C_{30}, C_{31})
  23,24-Dimethylsterols (C_{29}, C_{30})
  $\Delta24(28)$-Sterols (C_{28}-C_{30})
  22,23-Methylenesterols (C_{29}-C_{31})
  4-Methylsterols (e.g. dinosterol)

Carotenoids
  Diatoxanthin
  Fucoxanthin

(b) TERRESTRIAL[c]

Straight-chain (C_{23}-C_{33} range)
  Alkanes[d]
  Alkanones[d]
  Alkanols[d]
  Alkenoic acids[e]
  Polyhydroxy carboxylic acids
    e.g. 10, 16-dihydroxyhexadecanoic acid

Iso- and Anteiso-Branched Components (C_{27}-C_{33})
  Alkanes

Diterpenoids
  Various alkanes, e.g. fichtelite
  Aromatic hydrocarbons, e.g. retene, simonellite
  Acids, e.g. dehydroabietic acid

Steroids
  24R-Ethyl stenols and stanols

Triterpenoids
  Many alkanones and alkenones
    e.g. friedelone, oleanone
  Many alkanols, e.g. oleanol

[a] These proposed lipid indicators are compiled from various authors (Eglinton and Hamilton 1967; Volkman et al. 1980; Brassell et al. 1978, 1980, and references therein; Simoneit 1977; Devon and Scott 1972; Tibbetts 1980; Cardoso et al. 1977).
[b] Principally lipids of algal origin; excludes inputs from marine bacteria.
[c] Principally lipid components of higher plants.
[d] With odd/even preference; [e] With even/odd preference

Table 2.   Proposed lipid indicators of direct bacterial inputs to
           oceanic sediments.[a]

Straight-chain
    Alkanes ($C_{14}-C_{28}$)[b]
    Alkenoic acids ($\Delta^{11}-C_{18:1}$)

Iso- and Anteiso-Branched components ($C_{10}-C_{22}$ range)
    Alkanes (also other monomethyl branched alkanes)
    Alkanols
    Alkanoic acids
    Alkenoic acids

Acyclic Isoprenoids
    Alkanes e.g. 2,6,10,15,19-Pentamethyleicosane
                  Squalane
                  Lycopane[c]
                  Other Irregular isoprenoids, including head-to-head
                  linked alkanes and diglyceryl ethers

Triterpenoids
    Extended hopanoids ($> C_{30}$)
    Hopenes[d]
    Fernenes[e]

Cyclopropanoid Compounds
    Alkanes
    Alkanoic acids

---

a These indicators, with the exception of the instances cited below,
  are found in sediments, but have only been recognized as constituents of bacteria (compiled from Youngblood et al. 1971; Cranwell
  1973; Boon 1978; Holzer et al. 1979; Ourisson et al. 1979; Brassell
  et al. 1980, 1981).
b With no odd/even preference.
c Lycopane has recently been suggested as a sediment input from
  methanogens although it has yet to be recognized in these
  organisms (Brassell et al. 1981).
d Hopenes are not uniquely synthesized by bacteria; they are also
  constituents of other classes of organism (e.g., Gelpi et al.
  1970) and are affected by diagenetic isomerization (Ensminger
  1977). Many hopenes appear, however, to derive from bacterial
  inputs to sediments (Brassell et al. 1981).
e The supposition that fernenes are terrestrial markers (Wardroper
  1979; Brassell et al. 1980) has been superceded by their recognition in a bacterium (Howard 1980), suggesting that fernenes
  reflect bacterial inputs to sediments (Brassell et al. 1981).

unclear until the recognition of their presumed precursors, various polyhydroxybacteriohpanes, in prokaryotes (Ourisson et al. 1979). It remains uncertain exactly how the extended hopanoid acids, alcohols and ketones found in shallow sediments arise from this limited range of precursors, although progress is being made (see below). Until recently, fernenes had only been recognized in ferns and their presence in marine sediments was therefore provisionally taken as an indicator of terrestrial input. It now appears, however, that they can be synthesized by bacteria (Howard 1980), and may therefore represent microbial rather than terrestrial contributions to sediments (Brassell et al. 1981).

The various acyclic isoprenoid alkanes cited in Table 2 appear to be uniquely synthesized by archaebacteria and therefore probably reflect an imprint of methanogenic bacteria to the sediments (Brassell et al. 1981).

Indeed, over recent years the prominence of lipids of microbial origin in sediments has become increasingly apparent. For example, virtually all of the major branched and cyclic hydrocarbons in immature sediments from the Japan Trench can originate from bacteria (Fig. 2; Brassell et al. 1981).

Lipids Attesting Microbial Modification

There are two principal bases for assigning a given lipid transformation as microbial: either it has been demonstrated that the modification can be accomplished by microbes under laboratory conditions or the structural alteration in question appears to be selective in some way that cannot be explained in physico-chemical terms. Hence, some lipid alteration processes that occur in the environment are deemed to be microbial in that physico-chemical transformations would not be expected to produce the discrepancies between the distributions of products and precursors observed. A prime example is the dehydration of 5α-stanols to Δ^2-sterenes (Dastillung and Albrecht 1977; Gagosian and Farrington 1978). In marine sediments, Δ^2-sterenes derived from 4-desmethylsterols with a wide variety of side chains occur, whereas 4-methylster-2-enes have yet to be conclusively recognized despite the abundance and frequent dominance of dinosterol (4α, 23,24-trimethyl-5α-cholest-22-en-3β-ol) among sedimentary sterols (Fig. 3). It seems plausible to rationalize this discrepancy between 4-desmethyl- and 4α-methylsterols in terms of microbial enzymatic systems for 3β-ol dehydration that cannot accomodate 4α-methyl substitution.

Table 3 cites further examples of transformation mechanisms that are believed to be microbially mediated.

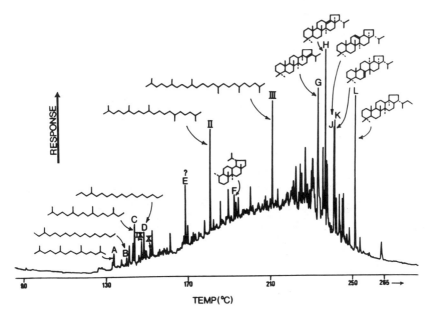

Figure 2. Gas chromatogram (GC) of the branched/cyclic hydrocarbons of a Pleistocene diatomaceous clay/silty claystone (DSDP Leg 57, sample 57-440B-8-4) from the Japan Trench. Peaks: A, pristane; B, 3-methyl-heptadecane; C, phytane; IX, phytenes; D, 4-methyl-octa-decane; X, 2-methyloctadecane; E, $C_{25}H_{44}$ bicyclic alkadiene of unknown structure; II, 2,6,10,15,19-pentamethyleicosane; F, tetracyclic alkane related to lupane; III, squalane; G, hop-17 (21)-ene; H, neohop-13(18)-ene; J, neohop-12-ene; K, fern-7-ene; L, 17β(H),21β(H)-homohopane. GC conditions: 20 m OV-1 wall-coated glass capillary, programmed from 50 to 265°C at 4°C per min after spitless injection at ambient temperature. A prominent feature of the GC trace is the apparent 'hump' which in crude oils and polluted sediments would be attributed to a background of thermally matured hydrocarbons. In this sample, however, the concentrations of other individual hydrocarbons (for example, steranes and hopanes) commonly associated with the 'hump' is small; hence, it is uncertain whether it represents highly biodegraded thermally matured or natural inputs or, alternatively, shipboard contamination with drilling lubricant. (Reproduced from Brassell et al. 1981).

Microbial Lipids - The Geological Record

Two important structural classes of microbial lipids found in sediments are acyclic isoprenoids of archaebacterial origin and hopanoids derived from procaryotes. The former class represents a relatively recent discovery whereas the latter group has a more extensive history in both bio- and geochemical terms. The ubiquity of hopanoids in sediments and crude oils (Van Dorsselaer et al. 1974)

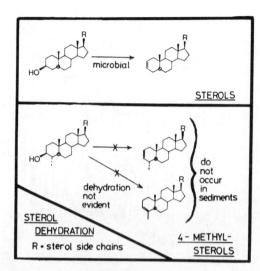

Figure 3. Postulated microbially mediated dehydration of sterols in water columns and sediments (Derived in part from Dastillung and Albrecht 1977; Gagosian and Farrington 1978).

suggested widespread natural product precursors, contrary to the limited biological occurrence of such structures prior to the recognition of polyhydroxybacteriohopanes (Ourisson et al. 1979; Fig. 4). Hopanoids have now been shown to be widely distributed among bacteria, probably functioning as membrane rigidifiers for these procaryotes in much the same way as sterols act in eucaryotes (Ourisson et al. 1979). The recent discovery of series of C_{36} to C_{40} 22(R)- and 22(S)-17α(H), 21β(H) − hopanes in a bitumen distillate (Rullkotter and Philp 1981) suggests that other, as yet unrecognized, precursors of sedimentary hopanoids may also be found in organisms. Hence, geolipids provide evidence of biolipid precursors unknown in organisms.

The schemes shown in Figures 4a and b represent attempts to describe the geological fate of tetrahydrobacteriohopane(I) as understood at present. The postulated early-stage diagenetic pathways involving defunctionalization are based solely on the occurrence of hopanoids in immature sediments, since no radiolabelled incubation studies have been carried out (cf. sterols, Gaskell and Eglinton 1975; Edmunds et al. 1980), whereas the later-stage diagenetic isomerization of hopanes is better understood (Van Dorsselaer et al. 1974; Ensminger 1977; Mackenzie et al. 1980). In brief, oxidation of tetrahydroxybacteriohopane(I) would give rise to 17β(H), 21β(H) − bishomohopan-32-ol(II) and 17β(H), 21β(H)-bishomohopanoic acid(III), which are the dominant hopanoid alcohol and acid, respectively, in immature sediments. Decarboxylation of the C_{32} acid would be expected to generate 17β(H), 21β(H) − homohop-29(31)-ene, (Dastillung et al. 1977) which may, in turn, be a precursor of 17β(H), 21β(H) −

Table 3.   Assessment of microbial activity in sediments: Examples
            of lipid products formed by microbial attack on other
            lipid precursors.

Precursor	Product	Comment
Sterol	Sterene	Microbial dehydration (see text)
Stenol	Steroid acids	Products are intermediates in the bacterial degradation of stenols to 17-oxo-steroids (Charney and Herzog 1967)
3-Oxytriterpenoids	A-ring degraded components	Microbial or photochemical degradation (Corbet et al. 1980)?

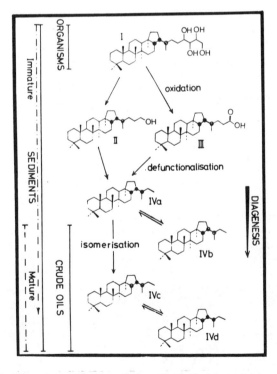

Figure 4a.   Sedimentary fate of bacterial hopanoids.   Postulated
diagenetic scheme.

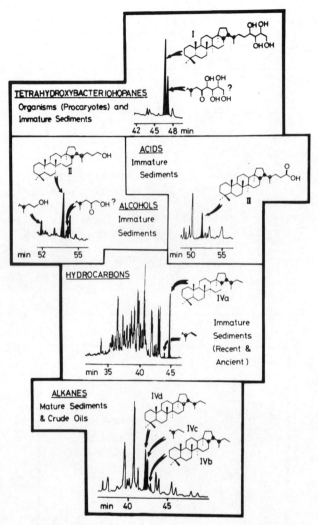

Figure 4b.   Sedimentary fate of bacterial hopanoids.   Typical
examples of sedimentary occurrences of specific extended hopanoids,
shown as partial capillary gas chromatograms and as a m/z 191 mass
chromatogram (bottom diagram).

homohopane (IVa), the major triterpane of immature sediments (e.g.,
Brassell 1980; Brassell et al. 1980 and 1981).   The 17β(H), 21β(H)
configuration in hopanoids, however, is not thermally stable; it
appears, intially, to undergo diagenetic isomerization to 17β(H),
21α(H), (IVb) and at higher maturity levels to 17α(H), 21β(H), (IVc
and d) a transformation which is later followed by the conversion of
17β(H), 21α(H) - hopanes to their 17α(H), 21β(H) isomers via 17β(H),
21β(H) components.   These geochemical processes reflect the relative

thermodynamic stability of the three series ($\alpha\beta > \beta\alpha > \beta\beta$; Van Dorsselaer et al. 1974; Ensminger 1977; Mackenzie et al. 1980).   The overall scheme represents one prominent diagenetic pathway for hopanoids; there are others that lead, for example, to the formation of aromatic components.   Despite the diagenetic changes discussed above, much of the hopanoid carbon skeleton remains unaltered and can therefore be recognized as an input from microbes to oils and mature sediments.

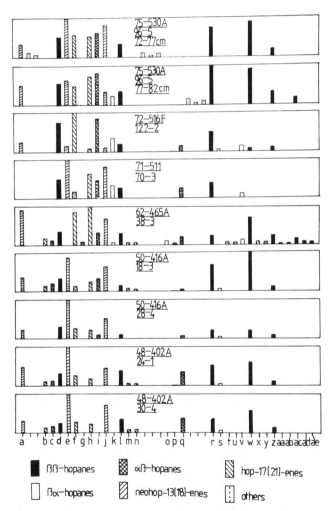

Figure 5.   Histograms of relative abundances (Based on m/z 191 mass fragmentograms) of hopanes and hopenes in a variety of Cretaceous DSDP sediments from the South Atlantic (75-530A, 72-516F and 71-511), Pacific (62-465A) and North Atlantic (50-416A and 48-402A) oceans (Brassell et al., in press).

As an example of the preservation of microbial lipid signatures, Figure 5 shows histograms of the relative abundances of hopanoids in a variety of Cretaceous sediments, notably black shales (Brassell et al., in press). Two points are particularly relevant. First, these lipids preserve a record of the bacterial inputs to sediments laid down up to 125 million years ago and second, there are marked similarities between the distribution of geographically diverse samples. Whether this latter observation is largely a function of (i) the similarity of microbial populations or (ii) diagenetic processes that obscure the original signatures and environmental differences has yet to be resolved.

In summary, the interpretation of hopanoids as molecular markers of microbial contributions to sediments has stemmed initially from consideration of their geological occurrence. It is to be hoped that a better appreciation and understanding of microbial lipid compositions will lead to the search for key compounds in geological materials. Indeed, it is this approach that has been adopted in investigations of archaebacterial acyclic isoprenoids in sediments (Chappe et al. 1980, 1982; Fig. 6). The unique feature of the glyceryl ethers shown in Figure 6 is their head-to-head linkage of isoprene units, a biosynthetic peculiarity of archaebacteria, in which lipids function as structural components of biomembranes. Such linkages indicate the origin of these compounds, which have been extracted from many immature sediments, even some of Cretaceous age (Chappe et al. 1982), a fact that illustrates the remarkable geological survival of these compounds, like the hopanoids. Archaebacterial lipids are also important in petroleum geochemistry because their isoprenoid moeities, with the head-to-head linkages that indicate their bacterial origin, occur widely in crude oils (Moldowan and Seifert 1979; Albaiges 1980). It appears that the archaebacterial glyceryl ethers are bound into kerogen during diagenesis and released as alkanes during its subsequent catagenetic breakdown (Fig. 6).

The widespread occurrence of both hopanes and isoprenoid alkanes derived from archaebacterial lipids in petroleums testifies to the importance of the role of microbial organic matter as an initial ingredient of sediments that later gave rise to crude oils.

A further consideration in the context of the geological record of microbial lipids is the conservation of the lipid constituents of a given biological family over geological time. It is uncertain whether the lipids of an individual species develop or remain constant over such a time scale (i.e., millions of years). In addition, the extent of inheritance of lipid compositions through species evolution is unknown, as are the compositions of extinct organisms. It can only be assumed that the lipids of extinct species were similar to those of their extant counterparts, and that morphological similarities of fossils are matched at the molecular level. For example, long chain alkatrienones and alkadienones have only been

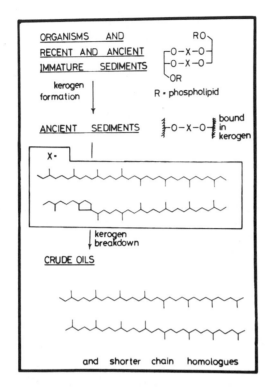

Figure 6. Schematic postulated sedimentary fate of Archaebacterial glyceryl ethers. (Compiled from Chappe et al. 1980, 1982; Moldowan and Seifert 1979; Albaiges 1980). Whether present as free phospholipids in sediment extracts or bound as ethers in kerogen, the acyclic isoprenoid moeity if isolated as an alkane, following laboratory cleavage (BBr$_3$) of the other linkages. In oils, these acyclic isoprenoids occur as alkanes, having been liberated from kerogen by maturation processes (catagenesis).

identified in the Quaternary coccolithophore Emiliania huxleyi; does their occurrence in Plicene sediments reflect inputs from the extinct coccolithophores E. ovata or Pseudoemiliania lacunosa? It is doubtful whether such questions can be satisfactorily answered, hence the assumption that the lipid compositions of organisms are conserved is used in the uniformitarian sense that the present is the key to the past. In the long term, however, the lipid compositions of ancient sediments may yet be found to provide the best record of molecular evolution.

Sediment Trap Measurements

The possibility of distinguishing lipid reactions taking place in the water column from those occurring in the sediment has been

realized by the advent of sediment trap collections. At present, few
detailed organic geochemical studies of trap particulates have been
undertaken, but recent preliminary results (Wakeham et al. 1980)
illustrate the ways in which such investigations can resolve or
clarify certain aspects of water column processes, which seawater
cannot or have not. For example, they found marked changes in the
lipid composition of the particulate material in the traps at
different depths in the water column, whereas there is only a slight
decrease in the total carbon flux through the water column at these
depths. Hence, it is the detailed lipid composition that provides
clues of microbial activity in the oceanic water column, not the
gross parameters of total carbon, etc.

The recognition of dinosterol (4α,23,24-trimethyl-5α-cholest-
22-en-3β-ol) and other 4-methylsterols among sediment trap particulates (Wakeham et al. 1980) illustrates that these components can be
derived from pelagic sources and therefore supports their origin from
dinoflagellates (Boon et al. 1979) rather than from benthic bacteria
(Dastillung et al. 1980), although they need not necessarily originate from the same source in all instances. The identification of
Δ^2-sterenes and $\Delta^{3,5}$- steradienes in trap particulates (Wakeham et
al. 1980) suggests that they form relatively fast and supports the
idea of the microbial dehydration of sterols.

Recognition of the source organisms of lipids found in the
natural environment is severely complicated by the inherent difficulties of reproducing natural systems in the laboratory. The
dependence of lipid composition of an organism on its food and
various growth conditions such as temperature and light results in
inevitable differences between the lipids found in cultured species
and in their naturally living counterparts. However, studies of the
effects of changes in laboratory growth conditions can help to
rationalize and to predict these differences.

Food can be expected to influence the biochemical action of
organisms and thence their lipid composition. It seems probable
that a microorganism cultured in a sugar-rich medium will prefer the
more "nutritious" food to lipids that are more difficult to degrade
and therefore will biochemically become "tuned" to the "richer" foodstuffs. In the natural environment, however, such sugars may not be
as abundant, forcing the microbe to live off lipids. Indeed, the
organic matter available to successive hierarchies of bacteria
colonizing discrete horizons of the sediment sequence will gradually
become less amenable to degradation.

A significant problem in the analysis of the hydrocarbon
composition of an organism is that of avoiding contamination from
such ubiquitous components of the environment. It is made more
difficult by the fact that hydrocarbons are often only minor
constituents of organisms, so that even low extraneous amounts of

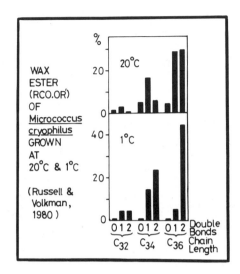

Figure 7.   Comparison of the wax ester profiles of <u>Micrococcus</u>
<u>cryophilus</u> grown at 20°C and 1°C.   The composition of wax esters of
<u>M. cryophilus</u> is measured by capillary GLC.   The values of Δ^9 and Δ^{11}
isomers have been combined for each particular chain length.   To make
comparisons easier, results are shown for the major wax ester chain
lengths; these accounted for about 95% of the total wax esters.
(Redrawn from Russell and Volkman 1980.)

such components can easily upset the natural compositional balance.

Growth conditions are also an important influence on the lipid
composition of organisms.   Depending on their growth temperature, the
degree of unsaturation of bacterial wax esters (Russell and Volkman
1980; Fig. 7) and of alkenoic acids (Kawamura and Ishiwatari 1981)
vary.   Evidence from various sedimentary localities suggests that
the unsaturation of long-chain unsaturated ketones may also depend
on water temperatures.   Hence, the alkatrienones derived from coccolithophores are more prominent in the sediments laid down beneath the
colder waters off Japan than in those beneath the tropical water off
Middle America (Fig. 8).   Biological lipid composition also change
with the growth stage of the organism.   Hence, the motile, sessile
and coccolith-bearing stages of <u>Emiliania huxleyi</u> possess slight
differences in their relative proportions of individual long chain
unsaturated ketones (Volkman et al. 1980a).   There are also marked
differences between the sterol composition of a planktonic brown alga
in its exponential and stationary growth phases, with Δ^5-sterols
dominant in the former and 5α-stanols dominant in the latter (Ballantine et al. 1979).   In general, lipid analyses show that 5α-stanols
are more abundant in sediments than in organisms, although the fact
that most organisms are harvested for analysis during their
exponential growth stage could presumably account, at least in part,

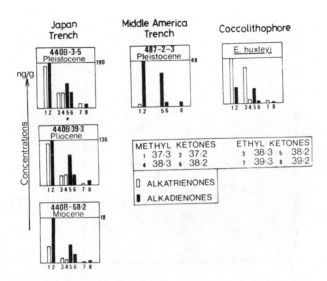

Figure 8. Concentrations of C_{37}, C_{38} and C_{39} long chain unsaturated ketones in recent and ancient marine sediments and in the living coccolithophore Emiliania huxleyi. Samples are from the Japan Trench (Hole 440B) and the Middle America Trench (Hole 487).

for such discrepancies in view of the evidence above. Furthermore, it seems probable that organisms will make their most significant contribution to underlying sediments when they are sessile and when in their stationary growth stage.

Linked to the changes in lipid composition with growth stages, are seasonal fluctuations in the biotic populations. Clearly, in the temperate zones, the foodstuffs available to an organism from the natural system may be regulated by annual cycles. Hence, in sedimentary environments where fine laminations or varves occur, the supply of organic matter to the benthic microbial population is undoubtedly changing cyclically and it may be that microbial activity is greater during the more productive summer months when the rain of organic debris reaching the sediment is much greater than during the winter period.

Lipids of Organisms and Sediments: Reasons for Differences

There are several factors that contribute to the significant discrepancies between the known lipid distributions of organisms and those found in sediments. First, since the sediment receives inputs from a multitude of sources, its complexity alone makes the unraveling of individual contributing organisms difficult. Second, the inherent problems associated with the culturing of organisms means that there inevitably are differences between the lipid compositions of cultured and natural biota which may lead to misinterpretations of inputs. Third, there are the wealth of lipid degradation reactions that can occur between the original synthesis and the subsequent deposition of an organism. In instances where there appears to be good correlation between the known lipids of a specific organism and a sediment, this may signify a lack of significant microbial alteration. However, such an observation may stem from the biochemically refractory nature of the compound, rather than an absence of microbial agents. For example, the distributions of long straight chain unsaturated ketones found in <u>Emiliania</u> <u>huxleyi</u> and numerous marine sediments may show good matches in several instances because the compounds are difficult to degrade.

CONCLUSIONS

Molecular organic geochemistry offers the unique opportunity to investigate characteristics of the microbial signature of past environments of deposition. There is a need, however, for suitable lipid parameters to be identified from the study of present-day systems. In particular, the role of different active components, such as the pelagic and benthic microflora, in terms of their alteration of, and contributions to, sedimentary organic matter needs to be understood.

Two potentially valuable approaches to the understanding of the role and significance of microorganisms in sediments have yet to be explored. First, a survey of lipid and lipid precursors in bacteria is needed to provide a basis for the investigation of the occurrence of relevant compounds in sediments. Second, the study of specific microbial processes should permit the prediction of their effect on lipids and the identity of any alteration products generated. Sulfate reducing and methanogenic bacteria would be prime candidates for such studies, but other, as yet unknown, organisms may be highly significant. There is also the need to search for specific microorganisms in water columns and sediments in order to determine their populations and their contributions to, and impact on, sediment lipids.

At present no environment has been fully studied or documented in terms of the lipid composition of its organisms, fecal pellets and other debris, water column particulates and bottom sediments. Although such research is in progress, there are major problems in the time scale of sampling, since seasonal, annual and longer-term cyclic and fluctuating effects perturb and change the oceanic system. Nevertheless, there is a major need to evaluate the fate of lipids in such a system so that the importance and role of individual processes can be assessed.

Major Questions

1. What is the depth extent of lipid alteration by bacteria in

a sediment? What are the principal factors that govern it?

2. How significant are single cultures of organism to the geochemist attempting to understand sedimentary distributions?

3. How important are environmental constraints, such as the sediment matrix, food sources and temperatures on benthic microbes?

4. What experiments can be designed to best test whether a specific lipid transformation is microbial or not?

ACKNOWLEDGMENTS

We thank the Natural Environment Research Council for financial support (GR3/2951 and GR3/3758) and are grateful to Keri Edmunds for access to unpublished data and to colleagues for helpful discussions. We appreciate help with GC-MS analyses from Ann Gowar.

REFERENCES

Albaiges, J. 1980. Identification and geochemical significance of long chain acyclic isoprenoid hydrocarbons in crude oils, p. 19. In: A. G. Douglas and J. R. Maxwell [eds.], Advances in Organic Geochemistry 1979. Pergamon Press, Oxford.

Ballantine, J. A., A. Lavis, and R. J. Morris. 1979. Sterols of the phytoplankton - effects of illumination and growth stage. Phytochem. 18: 1459.

Boon, J. J. 1978. Molecular biogeochemistry of lipids in four natural environments. Ph.D. thesis, University of Delft.

Boon, J. J., J. W. De Leeuw, and P. A. Schenck. 1975. Organic geochemistry of Walvis Bay diatomaceous ooze. I. Occurrence and significance of the fatty acids. Geochim Cosmochim. Acta 39: 1559.

Boon, J. J., W. I. C. Rijpstra, F. DeLange, J. W. De Leeuw, M. Yoshioka, and Y. Shimizu. 1979. The Black Sea sterol - A molecular fossil for dinoflagellate blooms. Nature 277: 159.

Brassell, S. C. 1980. The lipids of deep-sea sediments: their origin and fate in the Japan Trench. Ph.D. thesis, University of Bristol.

Brassell, S. C., G. Eglinton, J. R. Maxwell, and R. P. Philp. 1978. Natural background of alkanes in the aquatic environment, p. 69. In: O. Hutzinger et al. [eds.], Aquatic Pollutants, Transformation and Biological Effects. Pergamon Press, Oxford.

Brassell, S. C., P. A. Comet, G. Eglinton, P. J. Isaacson, J. McEvoy, J. R. Maxwell, I. D. Thomson, P. J. C. Tibbetts, and J. K. Volkman. The origin and fate of lipids in the Japan Trench, p. 375. In: A. G. Douglas and J. R. Maxwell [eds.], Advances in Organic Geochemistry 1979. Pergamon Press, Oxford.

Brassell, S. C., A. M. K. Wardroper, I. D. Thomson, J. R. Maxwell, and G. Eglinton. 1981. Specific acyclic isoprenoids as biological markers of methanogenic bacteria in marine sediments. Nature 290: 693.

Brassell, S. C., V. J. Howell, A. P. Gowar, and G. Eglinton. In
    press. Lipid geochemistry of Cretaceous sediments recovered by
    the Deep Sea Drilling Project. In: M. Bjorøy et al. [eds.],
    Advances in Organic Geochemistry 1981. Heydon and Son, London.
Cardoso, J. N., G. Eglinton, and P. J. Holloway. 1977. The use of
    cutin acids in the recognition of higher plant contribution to
    recent sediments, p. 273. In: R. Campos and J. Goni [eds.],
    Advances in Organic Geochemistry 1975. ENADIMSA, Madrid.
Carlson, R. M. K., S. Popov, I. Massey, C. Delseth, E. Ayanoglu, T,
    H. Varkony, and C. Djerassi. 1978. Minor and trace sterols in
    marine invertebrates. VI. Occurrence and possible origins of
    sterols possessing unusually short hydrocarbon side chains.
    Biorg. Chem. 7: 453.
Chappe, B., P. Albrecht, and W. Michaelis. 1982. Polar lipids of
    Archaebacteria in sediments and petroleums. Science 217: 65.
Chappe, B., W. Michaelis, and P. Albrecht. 1980. Molecular fossils
    of Archaebacteria as selective degradation products of kerogen,
    p. 265. In: A. G. Douglas and J. R. Maxwell [eds.], Advances
    in Organic Geochemistry 1979. Pergamon Press, Oxford.
Charney, W., and H. L Herzog. 1967. Microbial Transformations of
    Steroids. Academic Press, New York.
Corbet, B., P. Albrecht, and G. Ourisson. 1980. Photochemical or
    photourimetic fossil triterpenoids in sediments and petroleum.
    J. Amer. Chem. Soc. 102: 1171.
Cranwell, P. A. 1973. Branched-chain and cyclopropanoid acids in a
    recent sediment. Chem. Geol. 11: 307.
Dastillung, M., and P. Albrecht. 1977. Δ^2-sterenes as diagenetic
    intermediates in sediments. Nature 268: 678.
Dastillung, M., P. Albrecht, and G. Ourisson. 1980. Aliphatic and
    polycyclic ketones in sediments. C_{27}-C_{35} ketones and aldehydes
    of the hopanes series. J. Chem. Res.(S) 166: (M) 2325.
Dastillung, M., P. Albrecht, and M. J. Tissier. Hydrocarbures
    saturés est insaturés des sédiments, p. 209. In: Géochemie
    Organique des Sédiments Marin Profond. Orgon I: Mer de Norvège.
    CNRS, Paris.
Deuser, W. G. 1971. Organic carbon budget of the Black Sea. Deep-
    Sea Res. 18: 995.
Devon, T. K., and A. I. Scott. 1972. Handbook of Naturally-
    Occurring Compounds-II. Terpenes. Academic Press, New York.
Edmunds, K. L. H., S.C. Brassell, and G. Eglinton. 1980. The short-
    term diagenetic fate of 5α-cholestan-3β-ol in situ radiolabelled
    incubations in algal mats, p. 427. In: A. G. Douglas and J. R.
    Maxwell [eds.], Advances in Organic Geochemistry 1979. Pergamon
    Press, Oxford.
Eglinton, G., and M. Calvin. 1967. Chemical fossils. Sci. Amer.
    216: 32.
Eglinton, G., and R. J. Hamilton. 1967. Leaf epicuticular waxes.
    Science 156: 1322.
Ensminger, A. 1977. Evolution de composés polycycliques sédimen-

taires. Ph.D. thesis, University of Strasbourg.

Gagosian, R. B., and J. W. Farrington. 1978. Sterenes in surface
    sediments from the southwest African shelf and slope. Geochim.
    Cosmochim. Acta 42: 1091.

Gagosian, R. B., and F. Heinzer 1979. Stenols and stanols in the
    oxic and anoxic waters of the Black Sea. Geochim. Cosmochim.
    Acta 43: 471.

Gaskell, S. J., and G. Eglinton. 1975. Rapid hydrogenation of
    sterols in a contemporary lacustrine sediment. Nature 254: 209.

Gaskell, S. J., M. M. Rhead, P. W. Brooks, P. W., and G. Eglinton.
    1976. Diagenesis of oleic acid in an estuarine sediment.
    Chem. Geol. 17: 319.

Gelpi, E., H. Schneider, J. Mann, and J. Oro. 1980. Olefins of high
    molecular weight in two microscopic algae. Phytochem. 8: 603.

Holzer, G., J. Oro, and T. G. Tornabene. 1979. Gas chromatographic/
    mass spectrometric analysis of neutral lipids from methanogenic
    and thermoacidophilic bacteria. J. Chromatogr. 186: 795.

Howard, D. L. 1980. Polycyclic triterpenes of the anaerobic photo-
    synthetic bacterium Rhodomicrobium vanielli. Ph.D. thesis,
    University of California, Los Angeles.

Johnson, R. W., and J. A. Calder. 1973. Early diagenesis of fatty
    acids and hydrocarbons in a salt marsh environment. Geochim.
    Cosmochim. Acta 37: 1943.

Kawamura, K., and R. Ishiwatari. 1981. Polyunsaturated fatty
    acids in a lacustrine sediment as a possible indicator of
    paleoclimate. Geochim. Cosmochim. Acta 45: 149.

Mackenzie, A. S., R. L. Patience, J. R. Maxwell, M. Vandenbroucke,
    and B. Durand. 1980. Molecular parameters of maturation in
    the Toarrian Shales, Paris Basin, France. I. Changes in the
    configurations of acyclic isoprenoid alkanes, steranes and
    triterpanes. Geochim. Cosmochim. Acta 44: 1709.

Moldowan, J. M., and W. K. Seifert. 1979. Head-to-head linked
    isoprenoid hydrocarbons in petroleum. Science 204: 169.

Ourisson, G., P. Albrecht, and M. Rohmer. 1979. The hopanoids.
    Palaeochemistry and biochemistry of a group of natural products.
    Pure Appl. Chem. 51: 709.

Rullkotter, J., and R. P. Philp. 1981. Extended hopanes up to C_{40}
    in Thornoton bitumen. Nature, 292: 616.

Russell, N. J., and J. K. Volkman. 1980. The effect of growth
    temperature on wax ester composition in the psychrophilic
    bacterium Micrococcus cryophilus ATCC15174. J. Gen. Microbiol.
    118: 131.

Simoneit, B. R. T. 1977. Diterpenoid compounds and other lipids in
    deep-sea sediments and their geochemical significance. Geochim.
    Cosmochim. Acta 41: 463.

Tibbetts, P. J. C. 1980. The analysis of carotenoids and their use
    as environmental and palaeoenvironmental indicators. Ph.D.
    thesis, University of Bristol.

Van Dorsselaer, A., A. Ensminger, C. Spyckerelle, M. Dastillung, O.
    Sieskind, P. Arpino, P. Albrecht, G. Ourisson, P. W. Brooks,

S. J. Gaskell, B. J. Kimble, R. P. Philp, J. R. Maxwell, and
    G. Eglinton. 1974. Degraded and extended hopane derivatives
    (C_{27} to C_{35}) as ubiquitous geochemical markers. Tetrahedron
    Letts. 1349.

Volkman, J. K., G. Eglinton, E. D. S. Corner, and J. R. Sargent.
    1980a. Novel unsaturated straight chain C_{37}-C_{39} methyl and ethyl
    ketones in marine sediments and a coccolithophore Emiliania
    huxleyi, p. 219. In: A. G. Douglas and J. R. Maxwell [eds.],
    Advances in Organic Geochemistry 1979. Pergamon Press, Oxford.

Volkman, J. K., E. D. S. Corner, and G. Eglinton. 1980b. Transfor-
    mations of biolipids in the marine food web and in underlying
    bottom sediments, p. 185. In: R. Daumas [ed.], Biogéochemie de
    la Matiere Organique a l'Interface Eau-Sédiment Marin. CNRS,
    Paris.

Wakeham, S. G., J. W. Farrington, R. B. Gagosian, C. Lee, H. DeBaar,
    G. E. Nigrelli, B. W. Tripp, S. O. Smith, and N. H. Frew. 1980.
    Organic matter fluxes from sediment traps in the equatorial
    Atlantic Ocean. Nature 286: 798.

Wardroper, A. M. K. 1979. Aspects of the geochemistry of polycyclic
    isoprenoids. Ph.D. thesis, University of Bristol.

Youngblood, W. W., M. Blumer, R. R. L. Guillard, and F. Fiore. 1971.
    Saturated and unsaturated hydrocarbons in marine benthic algae.
    Mar. Biol. 8: 190.

ASPECTS OF MEASURING BACTERIAL ACTIVITIES IN THE DEEP OCEAN

Holger W. Jannasch

Woods Hole Oceanographic Institution
Woods Hole, Massachusetts   02543

Studies in deep sea biology commonly deal with depths well below the light-affected surface waters and the nutrient and temperature discontinuities of the thermocline. If the 1000 meter depth horizon is arbitrarily taken as the start of the deep sea, it still constitutes 75% of the ocean by volume and 88% by area. Of these figures, 0.1% and 0.2% respectively correspond to the deep sea trenches from 6000 meters to the greatest depth of approximately 11,000 meters. Although dependent on primary input of energy, carbon, and nutrients from surface water and land run-off, the deep sea contributes considerably by its sheer volume to the global biochemical cycles of elements. Much of this activity must be attributed to microorganisms, but their low and discontinuous rates of metabolism make direct measurements more difficult in the pelagic and deep ocean than in other parts of the biosphere.

The major environmental factors controlling microbial activities in the deep sea are low nutrient levels, temperature and pressure. The term oligotrophic was originally applied to designate low-nutrient waters but it is used more recently also to describe organisms with certain physiological characteristics that can be interpreted as adaptations to low nutrient habitats (Jannasch 1978; Poindexter 1981). The common occurrence of both low-temperature-adapted "psychrophilic" (Morita 1976; Baross and Morita 1978) and pressure-adapted "barotolerant" and "barophilic" bacteria (ZoBell 1968; Morita 1972; Wirsen and Jannasch 1975; Yayanos et al. 1979) is characteristic for the deep sea.

The bacterial activity in the deep sea is largely chemo-heterotrophic since, in the absence of light, the flow of nutrients and biochemical energy depends on the sedimentation of particulate

505

organic matter from the phototrophically productive surface waters and, to a lesser degree, on the transport of dissolved materials by currents and diffusion. The flow of particulate organic carbon to the deep sea has been estimated in a number of recent studies to be in the range of 20 to 800 $mg/m^2/year$ (Honjo 1978; Hinga et al. 1979; Brewer et al. 1980). Chemo-autotrophic bacteria, using CO_2 as the primary source of carbon and reduced inorganic compounds (such as H_2, H_2S, NH_3, NO_2^-, Fe^{2+} and Mn^{2+}) as electron sources, have also been found in the deep sea (Tuttle and Jannasch 1976; Ruby et al. 1981) but, with exception of the hydrothermal vent populations (see below), they probably play a minor role in the total transformation of organic carbon.

Reviews on deep sea microbiology were written by ZoBell (1968, 1970), and Morita (1976). Two books on barobiology edited by Sleigh and MacDonald (1972) and by Brauer (1972) contain valuable information on microorganisms. Treatises on hydrostatic pressure as an ecological, physiological and molecular parameter in microbial metabolism have been written by Marquis and Matsumura (1978) and Landau and Pope (1980). In view of these extensive literature surveys on deep sea microbiology, the references given in the present paper are kept to a minimum.

In accordance with the special purpose of this workshop, the following discussion will focus on a few selected aspects pertinent to the study of microbial activities in the deep sea environment. In particular these aspects are: (A) the significance of measuring potential activities, (B) the effects of decompression prior to activity measurements, and (C) chemoautotrophic bacteria in a newly discovered eutrophic deep sea habitat. The discussion will be limited to studies wtih natural populations. It does not include physiological and molecular effects of deep sea conditions which can only be studied with pure culture isolates.

"POTENTIAL" MICROBIAL ACTIVITY

Biomass Determination

Because of the oligotrophic characteristics of offshore and deep waters, much emphasis has been placed on certain techniques that offer high sensitivities. Epifluorescence microscopy (Zimmerman and Meyer-Reil 1974; Daley and Hobbie 1975) represents a decisive improvement over the use of acridine orange in transmission microscopy as originally introduced by Strugger (1949). The low statistical significance of surveying filters at high magnification with scanning electron microscopy may be overcome by computer programming (Krambeck et al. 1981). Sensitive photomultipliers are used in measuring the firefly bioluminescence reaction in ATP and other nucleotide determinations (reviewed by Karl 1980). Cellularly bound lipopolysaccharides are determined with high precision by using extracts of Limulus amebocytes (Watson et al. 1977). All these approaches to determine microbial biomass have their individual shortcomings as, for instance, the difficulty in estimating cell mass from microscopic counts, the problem of distinguishing between procaryote and eucaryote nucleotides in the firefly assay, or the question of including undecomposed lipopolysaccharides of dead cells and excluding gram-positive organisms by the Limulus assay. What these techniques have in common, however, is the problem of relating the measurements of various cell components (X) to the total mass or the amount of organic carbon, i.e., the unknown or variable X/C ratio.

A recurrent question is the actual significance of the various types of data on bacterial biomass or standing crop (a) as a substantial food source for filtering organisms or (b) as a valid indicator of microbial biosynthetic activity. The former depends to a large degree on the proportion between individually suspended cells and those attached to detritus particles or fecal pellets. This proportion will most likely be very different in various parts and depths of the ocean. In the absence of dissolved and metabolizable organic carbon, growth will be limited to those cells that are attached to undecomposed particulate matter. Yet, direct observations do not support this notion consistently.

Relatively large numbers of viable cells have been reported, starting with ZoBell's early studies (ZoBell 1968), from the uppermost deep sea sediments where the cells may accumulate by sedimentation. Even if these numbers are expressed as organic carbon, the biomass is extremely low if considered as a food source. Furthermore, patchiness can be expected to increase with depth where the more or less uniformly distributed small and slowly sinking particles are replaced by fewer fast-sinking larger chunks of organic debris (see below).

Since macroorganisms do not exist long in the absence of a substantial rate of metabolism, their mere existence and abundance can be taken as the basic indicator of metabolic activity. The occurrence of microorganisms, however, due to their size and dispersability in combination with their ability to survive long periods of unfavorable growth conditions, is not necessarily accompanied by any measurable metabolic activity. Relatively high microbial counts can commonly be obtained from oligotrophic environments where growth is unlikely. If streaked on nutrient agar, at least part of the natural bacterial population grows quickly without noticeable lag. The results of biomass determinations, therefore, are often difficult to interpret, especially where a constant input of cells occurs such as by particle sedimentation in the deep sea. The measurement of in situ activity has to be envisioned as an independent problem.

Activity Determination

Measuring metabolic activities of natural populations of chemo-heterotrophic microorganisms is done by various techniques using radiotracer transformations. In contrast to CO_2 representing the common carbon source for all photo- and chemo-autotrophic organisms, no single substrate can be used for measuring the chemo-heterotrophic activity of a natural microbial population. A multitude of assimilable and dissimilable organic compounds may be available to and used in succession or simultaneously by such populations in seawater. The choice of a suitable labeled substrate and its concentration for measuring conversion rates in a water sample imply a rather artificial situation, and the results have, therefore, been interpreted as potential rates. The added substrates imply a stimulation of ATP production and growth which is apt to obscure natural activity rates. However, data of such relative measurements can be obtained at varying temperature and pressure and will yield valuable information on growth of natural microbial populations under deep sea conditions.

Attempts to measure the unchanged activity of natural microbial populations, i.e., avoiding the stiumlation of ATP production, is based on the use of radiolabeled precursers in nucleotide synthesis such as adenine and thymidine (Karl 1979, 1981; Moriarty 1983). Another means of trying to overcome the growth-stimulating addition of tracer substrates, at least partially, has been the aim of keeping the concentration as low as possible, preferably near the natural level of a substrate if analyzable in seawater. This concept can be challenged on two points: (1) the nature of steady state substrate levels and (2) the existence of threshold concentrations.

Steady State Substrate Levels

A natural system is, kinetically speaking, an open system. Although no chemostat conditions exist (complete mixing, constant dilution rate, etc.), the concentration of the growth controlling substrate must be interpreted as representing the left-over concentration which is not indicative for the amount of substrate used for growth. In an abstract form indicating a steady state situation:

$$P = (S_i - S)\, k$$

where P is the product of the substrate conversion, e.g., biomass or standing crop, S_i the incoming substrate concentration, S the left-over concentration, and k a conversion or yield coefficient.

For our purpose, the incoming substrate concentration S_i is the most important figure in this equation and can theoretically be estimated from measurements of P and S, and an assumed yield coefficient - if P was indeed representing a population of equally

active cells. This, however, is hardly the case in natural populations, especially not in oligotrophic habitats such as the deep
sea. Any concentration of an added tracer substrate, therefore,
must be perceived as S_i which, as long as it is growth limiting, is
proportional to P but has no effect on S (Herbert et al. 1956).
Hence, keeping the concentration of tracer substrate as close as
possible to the level of its natural occurrence, for example
1-10 µg/l glutamate (Lee and Bada 1975) or 13-38 µg/l free amino
acids (Dawson and Gocke 1978), has little meaning from the kinetic
point of view. When Dawson and Gocke added a mixture of ^{14}C-labeled
amino acids in 0.2 µg/l quantities to the above mentioned background
concentration, they obtained uptake rates of 1.7-76.4 µg/l/day.
These unexpectedly high rates, if not questionable for technical
reasons (discussion by Dawson and Gocke 1978), must be interpreted
to the effect that the larger part of the in situ amino acid
concentration represented S_i rather than S. In other words, no
steady state existed.

Threshold Levels of Substrate

     The second argument, that casts doubt on effectively eliminating
the potential nature of radiotracer studies by using in situ concentrations concerns the existence of threshold levels. In many cases
the measured left-over concentrations of a metabolizable substrate,
in natural water as well as in the chemostat, are much higher than
theoretically possible. This indicates that they are no longer
growth limiting. For example, in a chemostat experiment (Jannasch
1967, 1979), S_i, the incoming growth limiting substrate concentration, was lowered stepwise. As a result, P, the population density
followed proportionally, while S, the steady state substrate
concentration, remained unchanged as expected. When a certain low
level of S_i was reached, however, its consumption by the cells was
affected. At that point, S increased to a "threshold" level
indicating that another factor in the system had become limiting.

     Specifically, the steady state population of a marine Pseudo-
monas was lowered stepwise by lowering the lactate concentration in
the seawater medium (Jannasch 1979). Unexpectedly the population
was washed out in the presence of a still substantial lactate
concentration. It was found that the clearly microaerophilic
organism, in the presence of an unchanged rate of aeration, was able
to maintain its growth rate only at high population densities by
reducing the redox potential to an optimum level for growth. At low
population densities, the organism was no longer able to do so and
the increasing redox potential became inhibitory and succeeded
lactate as the growth limiting factor. As a proof, the addition of
ascorbic acid lowered the threshold concentration of lactate
considerably. According to such observations, it is to be expected
that environmental factors, not only such as dissolved oxygen, but
probably also temperature, hydrostatic pressure, etc., will affect

threshold concentrations of metabolizable substrates below which
they can no longer be utilized economically.

Patchiness

All the concepts and studies discussed so far are based on the
general assumption that growth proceeds at a variable but more or
less continuous rate in the presence of an equally continuous energy
input in the form of a dissolved substrate. The situation may be
quite different in the deep sea as a highly oligotrophic environment
that is far removed from the photosynthetically productive surface
waters. The transport of the organic carbon and energy sources
proceeds primarily by sedimentation of solid materials as carried,
i.e., "marine snow", mostly described as a complex mixture of
organic inorganic detrital conglomerates. The size distribution of
these particles will shift with depth toward larger, faster sinking
particles. Decomposition of the sinking particles will enhance this
effect. In other words, small particles will never reach the deep
sea while very large pieces, such as fish carcasses, will arrive
virtually undecomposed. Examples for the latter have been observed
from the research submersible ALVIN (Jannasch 1978).

In principle, the heterogeneity or patchiness of food supply
increases with depth as the overall concentration decreases. This is
reflected in the distribution and feeding strategy of the macro- and
microfauna (Haedrich and Rowe 1977). From a microbiological point
of view, this increasing heterogeneity with depth supports the above
argument favoring the determination of potential microbial activities, because it is likely that these activities occur predominantly
on particles, i.e., at locally high substrate concentrations. As
these concentrations are reduced to near-threshold levels, the rates
of activity become exceedingly slow and approach zero.

In conclusion, being candid about the problems of measuring
microbial activities in oligotrophic environments makes potential
measurements, i.e., the use of relatively high concentrations of
tracer substrates, acceptable. It appears to mimic the actual input
of natural carbon and energy sources, S_i in the kinetic sense, more
accurately than the use of near in situ concentrations based on the
assumption of a continuous supply at a constant level. Furthermore,
the approach is useful for studying the relative effects of
temperature, pressure, etc., on the activities of natural microbial
populations. Using the data of organic carbon flux to the deep sea
(Honjo 1978; Hinga et al. 1979; Brewer et al. 1980), results of
potential microbial activity measurements might lead to reasonable
estimates of total or global turnover rates of organic carbon.

One afterthought concerns the preferable use of mixed labeled
substrates such as casamino acids (Wirsen and Jannasch 1975;
Jannasch and Wirsen 1982a). The concentrations of the individual

compounds are still lower than those that might induce catabolic
repression. In other words, the simulataneous utilization of the
individual substrates is likely.

Intestinal Tracts as Microbial Niches

In principle, the transformation and decomposition of organic
matter (after autolysis) in the various parts of the biosphere is
carried out by both free-living microorganisms and ingesting animals,
the latter often assisted by microorganisms living in or passing
through the intestinal tracts. The oligotropy and particle distribution characteristic for the deep sea appear to favor the role of food
gathering animals. A large portion of the above mentioned microbial
activity at relatively high substrate concentrations may indeed occur
in intestinal habitats of scavengers as particular ecological niches.

So far recent isolations of barophilic bacteria were successful
mainly when material from the intestinal tracts of deep sea amphipods
(Deming et al. 1981) or from whole amphipods (Dietz and Yayanos
1978; Yayanos et al. 1979, 1981) was used as an innoculum. This
limited occurrence, if not proven otherwise by different isolation
techniques, might be explained by the fact that pressure adaptation
can take place only in actively growing populations.

EFFECTS OF DECOMPRESSION

While it has never been a technical problem to retrieve deep sea
samples at unchanged in situ temperature, pressure retaining samplers
have not been used until recently (Jannasch et al. 1973, 1976; Tabor
and Colwell 1976; Jannasch and Wirsen 1977; Tabor et al. 1981).
Prior to this work, extensive studies have been done on the effect of
pressure on deep sea bacteria (ZoBell 1968; Morita 1972; Marquis and
Matsumura 1978; Landau and Pope 1980) isolated after decompression
of the original sample. Since many of such isolates were insensitive
to repeated compression and decompression, the general feeling
prevailed that, unlike the rise of temperature beyond a critical
degree, decompression may not be lethal to deep sea bacteria and
deactivation by decompression is always reversible. Proof of this
fact has to await work with pure cultures obtained in the absence of
decompression.

The majority of the organisms isolated after a brief decompres-
sion-recompression cycle exhibited barotolerance over a wide range
of pressures, but they grew best at 1 atm. Some isolates, however,
grew only, or grew better, at pressures higher than 1 atm (Fig. 1),
a characteristic termed barophilism by ZoBell and Johnson (1949).
The presently known barophilic isolates were obtained from relatively
nutrient-rich deep sea habitat, namely the intestinal tracts of
amphipods (see above; Dietz and Yayanos 1978; Yayanos et al. 1979,

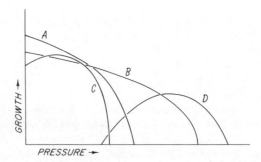

Figure 1. Scheme of microbial growth responses to pressure based on
data available at this time: A and B showing barotolerance varying
by degree and range; C and D showing barophilic characteristics as
defined by increased rates of growth at pressures higher than 1 atm.
The obligate barophilic response (D) is defined by the inability to
grow at 1 atm. The quantitative relationships between the responses
A to D and the shape of the curves are still most arbitrary in this
scheme since, in the few studies done so far, the types of growth
media and measurements used differed greatly (from Jannasch and
Wirsen 1982b).

1981; Deming et al. 1981). Growth of these organisms is strongly
affected by decompression, and in some cases perhaps irreversibly.
ZoBell (1968) states: "Considerable difficulty has been experienced
in trying to maintain barophilic bacteria in the laboratory. Most
of the enrichment cultures from the deep sea lose their viability
after two or three transplants to new media [infering decompression-
recompression cycles], although some barophiles have survived in
sediment samples for several months when stored at in situ pressures
and temperatures. Even at refrigeration temperatures, the deep sea
barophiles die off much more rapidly at 1 atm than when compressed
to in situ pressures."

     In natural populations of deep sea bacteria the barophilic
behavior is not readily apparent. Whenever bottles with various
kinds of dissolved or solid organic substrates were inoculated and
incubated on the deep sea floor (at depths between 1800 and 3900 m
by the research submersible ALVIN, Jannasch and Wirsen 1973), growth
was invariably slower than in the control bottles brought back to
the surface for incubation at 1 atm and at an in situ temperature of
approximately 3°C. These in situ incubation experiments are limited
to single endpoint measurements and, therefore, were supplemented by
time-course measurements in the absence of decompression using a
pressure-retaining sampler in combination with a prepressurized
culture chamber (Jannasch et al. 1973, 1976; Jannasch and Wirsen
1977).

     The use of defined substrates in these studies revealed that

the effect of pressure on the incorporation and respiration of radiolabeled organic carbon depends to a large degree on the biochemical
nature of the particular carbon and energy source (Jannasch and
Wirsen 1982a). While, for instance, the conversion of glutamate
invariably showed a considerable inhibition by pressure, the
conversion of acetate showed the least, in some cases no effect
whatsoever (Fig. 2). The incorporation and respiration of labeled
carbon from glucose and casamino acids was reduced to a lesser
degree than that of glutamate. Besides the effects of pressure on
the rates of conversion, the total utilization of the substrates was
also greatly reduced. On decompression, conversion rates as well as
total utilization increased without an appreciable lag. Similar
results were reported by Tabor et al. (1981) who developed a
different decompression-free sampling and culture system.

The following reasons may be found responsible for the effect of
pressure on natural populations of deep sea bacteria: (1) different
responses of the metabolism of the substrates on the biochemical and
enzymatic level, (2) shifts in the predominance of different physiological types within the mixed population and their individual
response to pressure, and (3) a combination of the two. It is to be
expected that the bacterial population of the deep sea is composed
of both "indigenous" species, more or less adapted to deep sea
conditions, and surface derived organisms constantly carried in by
sedimenting organic and inorganic materials. The extent of the
growth advantage conferred by pressure and temperature adaptation of
the baro- and psychrophilic organisms over the baro- and psychro-
tolerant species is difficult to estimate at present date. One way
to interpret the observations made so far (Fig. 2) is to assume the
predominance of non-adapted species in numbers obscuring the slight
growth advantage of the adapted species. Even if barophilic bacteria
were present in relatively high numbers, their activity at in situ
pressure will still be quite reduced as compared to that of the
non-barophilic portion of the natural population in a decompressed
sample.

CHEMOAUTOTROPHIC EUTROPHICATION AT DEEP SEA HYDROTHERMAL VENTS

So far this discussion was based on the claim that the deep sea
is principally an oligotrophic habitat. The microbial activity is
primarily controlled by the food supply from photosynthesis in
surface waters and, to a lesser degree, by the temperatures and
pressures characteristic for the deep sea. This concept was
confirmed by nutrient enrichment studies in situ and by laboratory
experiments with the specific sampling and culturing equipment
mentioned above.

Recently an exception to this rule was discovered. Copious
populations of invertebrates were found clustered around hydrothermal

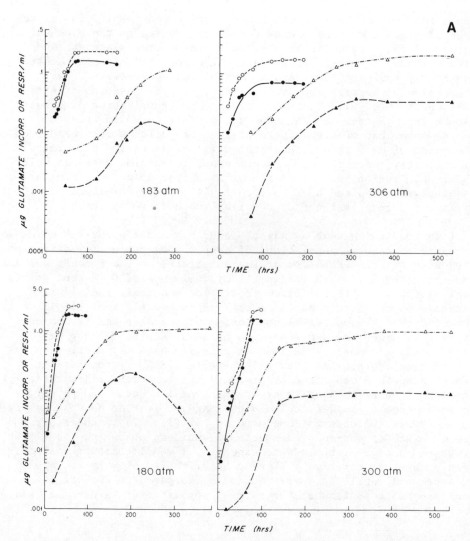

Figure 2A.   Respiration (o and △) and incorporation (● and ▲) of radiolabeled carbon from Na-glutamate at two levels of concentration (Top approx. 0.5 μg/ml; Bottom approx. 5.0 μg/ml) in seawater sampled from two depths and incubated at <u>in situ</u> pressure (triangles, undecompressed samples) and at 1 <u>atm</u> (circles, decompressed samples) (from Jannasch and Wirsen 1982b).

vents at depths of 2500 to 2600 m (Ballard 1977; Lonsdale 1977; Corliss et al. 1979) representing small but highly eutrophic ecosystems based on bacterial chemosynthesis (Jannasch and Wirsen 1979; Karl et al. 1980).

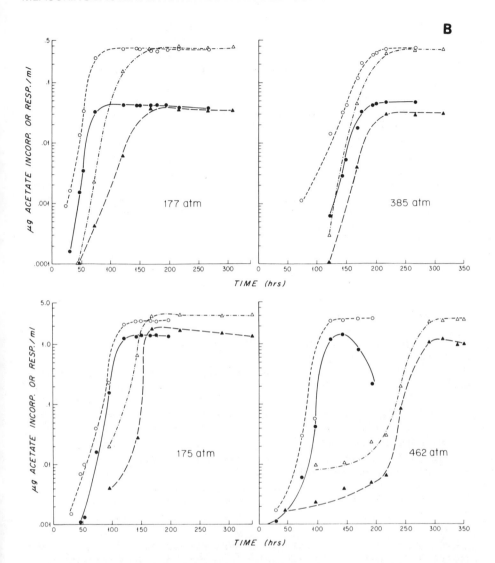

Figure 2B.   Respiration (○ and △) and incorporation (● and ▲) of
radiolabeled carbon from Na-acetate at two levels of concentration
(Top approx. 0.5 µg/ml; Bottom approx. 5.0 µg/ml) in seawater sampled
from two depths and incubated at in situ pressure (triangles,
undecompressed samples) and at 1 atm (circles, decompressed samples)
(from Jannasch and Wirsen 1982b).

     Of all inorganic materials that can be chemosynthetically
oxidized, reduced sulfur compounds appear to be quantitatively
predominant in the emitted vent waters.   The reduction of sulfate
occurs during the passage of seawater through hot parts of the

earth's crust (Edmond et al. 1979; Mottl et al. 1979). Thus, the
resulting hydrogen sulfide is produced by geothermal energy. As a
corollary, the chemosynthetic conversion of inorganic to organic
carbon based on hydrogen sulfide oxidation as the source of energy
can be termed a primary production. This definition of primary
production is based on the origin of energy and disregards the fact
that free oxygen as the electron sink required in chemosynthesis is
a product of photosynthesis. It is likely that other chemosynthetic
organisms, such as hydrogen-, ammonium-, nitrite-, iron-, and
manganese-oxidizing bacteria contribute to this production.

The bacterial oxidation of methane is most likely another source
of particulate organic carbon. It is indicated by intracellular
membrane systems in procaryotic cells found by transmission electron
microscopy in microbial mats near the vents as well as in symbiotic
tissues (see below) of certain invertebrates (Jannasch and Wirsen
1981). This fine structure of procaryotes is characteristic of
methylotrophic and ammonium-oxidizing bacteria. Substantial amounts
of methane have been found in various vent waters (Welhan and Craig
1979), but the question on its origin by geothermic reduction or by
methanogenic bacteria has not yet been answered.

At the present stage of our knowledge, the chemosynthetic
production appears to occur at three major locations within the vent
system, in surface mats outside of the vents, and by symbiosis with
animals.

(1) As filter samples and ATP measurements indicate (Karl et
al. 1980), the turbidity of water emitted from some vents is largely
due to a suspension of microbial cells rather than particulate
elemental sulfur or other amorphous materials. Filter-feeding
mussels are most commonly distributed near the opening of vents that
emit turbid water. Microbial growth is assumed to take place in the
subsurface vent system where oxygenated ambient seawater meets with
the reduced hydrothermic fluid at favorable temperatures.

(2) Surfaces in the immediate surrounding of the vents, exposed
intermittently to reduced vent water and oxygenated seawater, are
covered with multi-layered microbial mats (Jannasch and Wirsen 1981).
Scanning electron microscopy shows a variety of known and unknown
filamentous forms including Beggiatoa and unpigmented analogues to
Calothrix. Transmission electron micrographs reveal mostly procaryotic cells with walls typical for gram-negative bacteria. The
composition of manganese/iron deposits in these mats indicate their
origin from vent water, but an active chemosynthetic precipitation
could not yet be demonstrated.

(3) The newly described giant clam (Clayptogena magnifica; Boss
and Turner 1980) and vestimentiferan tube worms (Riftia pachyptila;
Jones 1981) occur mostly in non-turbid vent waters. They appear to

represent an earlier and a later evolutionary stage of symbiosis
between chemosynthetic bacteria and invertebrates. Procaryotic
symbionts exist in the gills as well as within the gill tissue of
the clams and appear to form the central tissue, the "trophosome",
of the tube worms (Cavanaugh et al. 1981). Felbeck (1981) presented
enzymatic evidence for the chemosynthetic activity. These symbiotic
systems, which are now also reported for shallow water marine
invertebrates, may well represent the most efficient form of chemosynthetic production at the deep sea vents.

According to the earlier statement that specific deep sea
adaptations will occur preferentially in nutrient-enriched habitats,
one might expect to isolate barophilic bacteria from deep sea vents.
This may still be true not withstanding the fact that the first
isolate of an obligately chemosynthetic, sulfur-oxidizing bacterium,
Thiomicrospira sp., was found to be barotolerant rather than barophilic. Only those organisms growing within the subsurface vent
system will be exposed to high temperatures. Isolated from vents
with emission temperatures of 8° to 12°C, the above organisms were
not thermophilic, but exhibited optimal growth at 28°C. Thermophilic
methane-producing bacteria were successfully isolated from "hot
smoker" vent systems (Leigh and Jones 1983; J. Baross, unpublished,
personal communication).

The only source of fixed nitrogen in these highly productive
habitats appears to be the normal nitrate concentration of deep sea
water. It was to be expected and has indeed been found (unpublished)
that nitrogen fixation occurs in the chemosynthetic bacteria. The
peculiar abundance of colorless, cyanobacteria-like trichomes
(Jannasch and Wirsen 1981) can also be taken as an indication of
possible nitrogen fixation.

Any estimate of the extent of these highly productive, eutrophic
deep sea habitats and their actual input of organic carbon into the
deep and pelagic ocean may be premature at this time. In order to
explain the manganese/magnesium anomaly in the ion composition of
seawater it has been theorized (Edmond et al. 1979) that the total
amount of seawater may percolate through the earth's crust once every
7 million years. The temptation to follow up this conjecture by
estimating the amount of geothermal energy spent for chemosynthetic
CO_2-reduction is thwarted by a number of unknown variables such as
the proportion between chemical and biological sulfur oxidation, the
rate of vent formation and die-off, the presence or absence of subsurface seawater mixing, etc. At this time, we are still in the
exploratory phase of research on this unique deep sea ecosystem. A
summarizing paper on the present knowledge of microbial processes
(Jannasch 1983) is included in a comprehensive treatise on the
biology, geochemistry and geology of deep sea hydrothermal vent
systems.

SUMMARY

Measuring microbial activities in the deep sea poses a number of problems that are characteristically different from those encountered in surface waters. With increasing depth and oligotrophy, the heterogeneity of microbial cells with respect to their metabolic activity will also increase, while the number of cells does not necessarily reflect this change. The food supply consists of fast sinking particles which reach the deeper waters largely undecomposed. The decrease of total food supply with depth is combined with locally and relatively high concentrations of substrates, resulting in an increasing patchiness. This heterogeneity of microbial niches includes the intestinal tract of food gathering invertebrates and fishes. Adaptations to deep sea temperatures and pressures, as in psychrophilic and barophilic bacteria, are found mainly where active growth takes place. This was demonstrated by successful isolations of pressure-adapted bacteria from invertebrate guts. Rich invertebrate populations discovered in the immediate vicinity of active hydrothermal vents represent naturally eutrophic deep sea habitats based on chemosynthetic primary production largely by sulfur oxidizing bacteria. The bulk of this production appear to occur in symbiotic associations between certain invertebrates and chemosynthetic microorganisms. Forthcoming bacterial isolations from these eutrophic and partly temperate high pressure habitats can be expected to explain typical deep sea characteristics of microbial metabolism.

ACKNOWLEDGMENTS

Some of the research reported and time for the preparation of this paper was supported by the National Science Foundation Grants OCE79-19178 and OCE81-24253 and by the Office of Naval Research Grant 71.80. Contribution No. 5029 of the Woods Hole Oceanographic Institution.

REFERENCES

Ballard, R. D. 1977. Notes on a major oceanographic find. Oceanus 20: 35-44.

Baross, J. A., and R. Y. Morita. 1978. Microbial life at low temperatures: ecological aspects, pp. 9-71. In: J. D. Kushner [ed.], Microbial Life in Extreme Environments. Academic Press, New York.

Boss, K. J., and R. D. Turner. 1980. The giant white clam from the Galapagos Rift Calyptogena magnifica species novum. Malacologia 20: 161-194.

Brauer, F. W. [ed.]. 1972. Barbiology and the Experimental Biology of the Deep Sea. North Carolina Sea Grant Program., Univ. North Carolina, Chapel Hill. 428 pp.

Brewer, P. G., Y. Nozaki, D. W. Spencer, and A. P. Fleer. 1980. Sediment trap experiments in the deep North Atlantic: isotopic and elemental fluxes. J. Mar. Res. 38: 703-728.

Cavanaugh, C. M., S. L. Gardiner, M. L. Jones, H. W. Jannasch, and J. B. Waterbury. 1981. Procaryotic cells in the hydrothermal vent tube worm Riftia pachyptila Jones: possible chemoautotrophic symbionts. Science 213: 340-341.

Corliss, J. B., J. Dymond, L. I. Gordon, J. M. Edmond, R. P. von Herzen, R. D. Ballard, K. Green, D. Williams, A. Bainbridge, K. Crane, and T. H. van Andel. 1979. Submarine thermal springs on the Galapagos Rift. Science 203: 1073-1083.

Daley, R. J., and J. E. Hobbie. 1975. Direct counts of aquatic bacteria by a modified epifluorescence technique. Limnol. Oceanogr. 20: 875-882.

Dawson, R., and K. Gocke. 1978. Heterotrophic activity in comparison to free amino acid concentrations in Baltic seawater samples. Oceanol. Acta 1: 45-54.

Deming, J. W., P. S. Tabor, and R. R. Colwell. 1981. Barophilic growth of bacteria from intestinal tracts of deep-sea invertebrates. Microb. Ecol. 7: 85-94.

Dietz, A. S., and A. A. Yayanos. 1978. Silica gel media for isolating and studying bacteria under hydrostatic pressure. Appl. Environ. Microbiol. 36: 966-968.

Edmond, J. M., C. Measures, R. E. McDuff, L. H. Chan, R. Collier, B. Grant, L. I. Gordon, and J. B. Corliss. 1979. Ridge crest hydrothermal activity and the balances of the major and minor elements in the ocean: The Galapagos data. Earth Planet. Sci. Lett. 46: 1-18.

Felbeck, H. 1981. Chemoautotrohic potentials of the hydrothermal vent tube worm, Riftia pachyptila (Vestimentifera). Science 213: 336-338.

Haedrich, R. L., and R. T. Rowe. 1977. Megafaunal biomass in the deep sea. Nature 29: 141-142.

Herbert, D., R. Elsworth, and R. C. Telling. 1956. The continuous culture of bacteria: a theoretical and experimental study. J. Gen. Microbiol. 14: 601-622.

Hinga, K. R., J. McN. Sieburth, and G. R. Heath. 1979. The supply and use of organic material at the deep sea floor. J. Mar. Res. 37: 557-579.

Honjo, S. 1978. Sedimentation of materials in the Sargasso Sea at a 5367 m deep station. J. Mar. Res. 36: 469-492.

Jannasch, H. W. 1967. Growth of marine bacteria at limiting concentrations of organic carbon in seawater. Limnol. Oceanogr. 12: 264-271.

Jannasch, H. W. 1978. Experiments in deep-sea microbiology. Oceanus 21: 50-57.

Jannasch, H. W. 1979. Microbial ecology of aquatic low nutrient habitats, pp. 243-260. In: M. Shilo [ed.], Strategies of Microbial Life in Extreme Environments. Life Sci. Res. Rept. 13, Verlag Chemie, Weinheim/New York.

Jannasch, H. W. 1983. Microbial processes at deep sea hydrothermal
    vents. In: P. A. Rona et al., [eds.], Hydrothermal Processes at
    Sea Floor Spreading Centers. Plenum Press, New York. In press.
Jannasch, H. W., and C. O. Wirsen. 1973. Deep-sea microorganisms:
    in situ response to nutrient enrichment. Science 180: 641-643.
Jannasch, H. W., and C. O. Wirsen. 1977. Retrieval of concentrated
    and undecompressed microbial populations from the deep sea.
    Appl. Environ. Microbiol. 33: 642-646.
Jannasch, H. W., and C. O. Wirsen. 1979. Chemosynthetic primary
    production at East Pacific sea floor spreading centers.
    Bioscience 29: 592-598.
Jannasch, H. W., and C. O. Wirsen. 1981. Morphological survey of
    microbial mats near deep-sea thermal vents. Appl. Environ.
    Microbiol. 41: 528-538.
Jannasch, H. W., and C. O. Wirsen. 1982a. Microbial activity in
    undecompressed and decompressed deep sea water samples. Appl.
    Environ. Microbiol. In press.
Jannasch, H. W., and C. O. Wirsen. 1982b. Microbiology of the deep
    sea. In: G. T. Rowe [ed.], The Sea, Vol. 8, Deep Sea Biology.
    John Wiley, New York. In press.
Jannasch, H. W., C. O. Wirsen, and C. D. Taylor. 1976. Undecom-
    pressed microbial populations from the deep sea. Appl. Environ.
    Microbiol. 32: 360-367.
Jannasch, H. W., C. O. Wirsen, and C. L. Winget. 1973. A bacterio-
    logical, pressure-retaining, deep-sea sampler and culture
    vessel. Deep-Sea Res. 20: 661-664.
Jones, M. L. 1981. Riftia pachyptila, n. gen., n. sp., the vesti-
    mentiferan worm from the Galapagos Rift geothermal vents
    (Pogonophora). Proc. Biol. Soc. Washington. 93: 1295-1313.
Karl, D. M. 1979. Measurement of microbial activity and growth in
    the ocean by rates of stable ribonucleic acid synthesis. Appl.
    Environ. Microbiol. 39: 850-860.
Karl, D. M. 1980. Cellular nucleotide measurements and applica-
    tions in microbial ecology. Microbiol. Rev. 44: 739-796.
Karl, D. M. 1981. Simultaneous measurements of rates of RNA and
    DNA syntheses for estimating growth and cell division of aquatic
    microbial populations. Appl. Environ. Microbiol. 42: 802-810.
Karl, D. M., C. O. Wirsen, and H. W. Jannasch. 1980. Deep-sea
    primary production at the Galapagos hydrothermal vents.
    Science 207: 1345-1347.
Krambeck, C., H. J. Krambeck, and J. Overbeck. 1981. Microcomputer-
    assisted biomass determination of plankton bacteria on scanning
    electron micrographs. Appl. Environ. Microbiol. 42: 142-149.
Landau, J. V., and D. H. Pope. 1980. Recent advances in the area
    of barotolerant protein synthesis in bacteria and implications
    concerning barotolerant and barophilic growth. Adv. Aquat.
    Microbiol. 2: 49-76.
Lee, C., and L. Bada. 1975. Amino acids in equatorial ocean water.
    Earth Planet. Sci. Lett. 26: 61-68.
Leigh, J. A., and W. J. Jones. 1983. A new extremely thermophilic

methanogen from a submarine hydrothermal vent. Am. Soc.
Microbiol., Proc. Ann. Meeting. In press.

Lonsdale, P. 1977. Clustering of suspension-feeding macrobenthos
near abyssal hydrothermal vents at oceanic spreading centers.
Deep-Sea Res. 24: 857-863.

Marquis, R. E., and P. Matsumura. 1978. Microbial life under
pressure. In: D. J. Kushner [ed.], Microbial Life in Extreme
Environments. Academic Press, New York.

Moriarty, D. J. W. 1983. Measurements of bacterial growth rates in
some marine systems using the incorporation of tritiated thymidine into DNA. In: J. E. Hobbie and P. J. leB. Williams [eds.],
Heterotrophic Activity in the Sea. Plenum Press, New York.

Morita, R. Y. 1972. Pressure. 1. Bacteria, fungi and blue-green
algae, pp. 1361-1388. In: O. Kinne [ed.], Marine Ecology, Vol.
1. John Wiley Interscience Publications.

Morita, R. Y. 1976. Survival of bacteria in cold and moderate
hydrostatic pressure environments with special reference to
psychrophilic and barophilic bacteria, pp. 279-298. In: T.
Gray and J. R. Postgate [eds.], Survival of Vegetative Microbes.
26th Symp. Soc. Gen. Microbiol., Cambridge Univ. Press.

Mottl, M. J., H. D. Holland and R. F. Corr. 1979. Chemical exchange
during hydrothermal alteration of basalt by seawater - II.
Experimental results for Fe, Mn, and sulfur species. Geochim.
Cosmochim. Acta 43: 869-884.

Poindexter, J. S. 1981. Oligotrophy: feast and famine existence.
Adv. Microb. Ecol. 5: 67-93.

Ruby, E. G., C. O. Wirsen and H. W. Jannasch. 1981. Chemolithotrophic sulfur-oxidizing bacteria from the Galapagos Rift
hydrothermal vents. Appl. Environ. Microbiol. 42: 317-324.

Sleigh, M. A., and A. G. MacDonald [eds.]. 1972. The Effects of
Pressure on Organisms. Symp. Soc. Exper. Biol. XXVI. Academic
Press, New York. 516 pp.

Strugger, S. 1949. Fluoreszenzmikroskopie und Mikrobiologie. M.
and H. Schaper, Hannover.

Tabor, P. S., and R. R. Colwell. 1976. Initial investigations with
a deep ocean in situ sampler. Proc. MTS/IEEE OCEANS '76,
Washington, D.C., 13D-1-13D-4.

Tabor, P. S., J. W. Deming, K. Ohwada, H. Davis, M. Waxman, and R.
R. Colwell. 1981. A pressure-retaining deep ocean sampler and
transfer system for measurements of microbial activity in the
deep sea. Microb. Ecol. 7: 51-65.

Tuttle, J. H., and H. W. Jannasch. 1976. Microbial utilization of
thiosulfate in the deep sea. Limnol. Oceanogr. 21: 697-701.

Watson, S. W., T. J. Novitsky, H. L. Quinby, and F. W. Valois.
1977. Determination of bacterial number and biomass in the
marine environment. Appl. Environ. Microbiol. 33: 940-946.

Welhan, J. A., and H. Craig. 1979. Methane and hydrogen in East
Pacific Rise hydrothermal fluid. Geophys. Res. Lett. 6: 829.

Wirsen, C. O., and H. W. Jannasch. 1975. Activity of marine
psychrophilic bacteria at elevated hydrostatic pressures and

low temperatures. Mar. Biol. 31: 201-208.

Yayanos, A. A., A. S. Dietz, and R. Van Boxtel. 1979. Isolation of a deep-sea barophilic bacterium and some of its growth characteristics. Science 205: 808-810.

Yayanos, A. A., A. S. Dietz and R. Van Boxtel. 1981. Obligately barophilic bacterium from the Mariana Trench. Proc. Nat. Acad. Sci. 78. In press.

Zimmerman, R., and L.-A. Meyer-Reil. 1974. A new method for fluorescence staining of bacterial populations on membrane filters. Kiel. Meeresforsch. 30: 24-27.

ZoBell, C. E. 1968. Bacterial life in the deep sea. Bull. Misaki Mar. Biol. Inst., Kyoto Univ. 12: 77-96.

ZoBell, C. E. 1970. Pressure effects on morphology and life processes of bacteria, pp. 85-130. In: H. M. Zimmerman [ed.], High Pressure Effects on Cellular Processes. Academic Press, New York.

ZoBell, C. E., and F. H. Johnson. 1949. The influence of hydrostatic pressure on the growth and viability of terrestrial and marine bacteria. J. Bacteriol. 57: 179-189.

BACTERIAL BIOMASS AND HETEROTROPHIC ACTIVITY IN SEDIMENTS

AND OVERLYING WATERS

Lutz-Arend Meyer-Reil

Institut für Meereskunde an der Universität Kiel
Abteilung Marine Mikrobiologie
Kiel, Federal Republic of Germany

INTRODUCTION

Most studies of aquatic microbial ecology have been concerned with the water column. In comparison, little is known about the microbiology of sediments, although sediments are unquestionably an important part of coastal ecosystems. From the high bacterial biomass, which is about equal to the faunal standing crop (Dale 1974), it has been concluded that the bacteria have an important role in the nutrient cycles and the food web. Much of the research has been concentrated on nutrient cycles (nitrogen, sulfur) and methanogenesis, which will not be discussed in this connection. However, information on the microbiology of sediments (bacterial biomass, activity) is still limited. Most of the earlier publications deal with the enumeration and isolation of bacteria using the agar plate technique (e.g., Westheide 1968; Stevenson et al. 1974; Boeye et al. 1975; Litchfield et al. 1976; Rheinheimer 1977). However, bacteria growing on agar plates account for only a small fraction of the total number of bacteria present in sediments. A more direct insight into bacterial colonization and biomass in sediments was obtained by scanning electron microscopy (Weise and Rheinheimer 1978) or epifluorescence microscopy (e.g., Dale 1974; Griffiths et al. 1978; Meyer-Reil et al. 1978; Jones 1980; Meyer-Reil and Faubel 1980). From these studies, the diversity of bacteria and their comparatively high biomass became obvious. In evaluating the role of bacteria as mineralizers and biomass producers, the determination of bacterial activity becomes important. However, only in a limited number of studies has the bacterial uptake of dissolved organic substrates been measured (Wood 1970; Harrison et al. 1971; Hall et al. 1972; Christian and Wiebe 1978; Griffiths et al. 1978; Hanson and Gardner 1978; Meyer-Reil et al. 1978, 1980; Litchfield et

al. 1979; Novitsky and Kepkay 1981). Reports on the degradation of
particulate organic matter by bacteria in sediments is even sparser
(Ayyakkannu and Chandramohan 1971; Maeda and Taga 1973; Oshrain and
Wiebe 1979; Sayler et al. 1979; King and Klug 1980, Meyer-Reil 1981).

In this paper, data reported in the literature concerning
bacterial biomass, uptake of dissolved organic substrates and
decomposition of particulate organic matter are summarized. For
comparisons, microbiological data from the water directly overlying
the sediments (if available) are included. Based on unpublished
results, factors related to bacterial biomass and activity and data
on the contribution of bacteria to the carbon cycle in sediments and
overlying waters are discussed.

MICROBIAL COLONIZATION OF SEDIMENTS

According to investigations of Meadows and Anderson (1966) and
Weise and Rheinheimer (1978), microorganisms colonize depressions
and crevices of particles where they are protected against
mechanical damages. This applies especially to bacteria inhabiting
beach sediments exposed to wave action and tidal activity. In these
sediments, more than 95% of the bacteria are attached to particles
(Meyer-Reil et al. 1978). However, in deeper sediments, the number
of bacteria "free-floating" in the interstitial water increases
considerably (up to 50% of the total number; Weise and Rheinheimer
1979). The abundance of bacteria is greatly increased in the
vicinity of deposits of detritus; small colonies prevail. The
individual cells seem to be closely related to each other according
to their morphology. The flora is of a high diversity: rods, cocci
of different sizes and shape. The mode of the attachment to
particles varies strongly. Most of the cells have direct contact
with the particle surface, other cells have indirect contact via
detrital material. Slime connections between the individual cells
are frequent, demonstrating the high bacterial extracellular
production in sediments.

BACTERIAL NUMBER AND BIOMASS

Variations in bacterial numbers in different types of sediments
are in the range of 1×10^8 to 507×10^8 cells g^{-1} dry weight sediment, a range already demonstrated by the investigations of Dale
(1974). It is interesting to note that quite different types of
sediments such as sandy beaches of the Kiel Bight, coral beaches of
the Philippines and muddy sediments from Antarctica (water depth
1951 m) contain comparable numbers of bacteria (cf. Table 1). From
plate count studies, the general decrease of bacteria with increasing sediment depth could be demonstrated (Rheinheimer 1980).
However, even in the 1,100 cm horizon of a sediment profile from

Antarctica, 11 x 10^8 cells g^{-1} sediment could be detected, only
in significantly less than in the surface horizon (Meyer-Reil,
unpublished data). The question of the physiological state of the
bacteria in this 10,000 year old sediment is still unresolved.
Epifluorescence microscopy reveals bacteria of an average size of
0.2 to 0.4 µm which could be starvation forms resting for time
ranges that we could not even think about in ecological studies.
In sediment profiles from Antarctica, inhomogeneities in bacterial
numbers could be detected which are paralleled by inhomogeneities in
concentrations of organic matter (total organic matter, proteins,
carbohydrates). These are obviously a reflection of pronounced
differences in sedimentation from the water column and/or differences
in decomposition rates of the organic material in the sediment.

From the data in Table 1, one can infer that bacterial numbers
are positively correlated with organic matter content of the sediment
and negatively correlated with grain size, which is obviously caused
by the greater surface available in fine than coarse sediments.

There are very few data on microbial biomass in sediments
published in the literature. Biomass calculations carried out by
epifluorescence microscopy require homogenization of the sediment
samples either by Ultra-Turrax (Gunkel 1964) or sonication (Weise
and Rheinheimer 1978) and a size determination of the individual
cells (Meyer-Reil 1977). Other dispersion techniques yield much
lower cell numbers and biomass values (Meyer-Reil et al. 1978). The
microscopic estimation of cell dimensions is very time-consuming and
is to a certain degree subjective. However, epifluorescence
microscopy and scanning electron microscopy (Krambeck et al. 1981)
are the only direct approaches to obtain biomass estimations which
provide fairly reliable information of the contribution of bacterial
carbon to the food web. Figure 1 gives an example of the variety of
bacteria present in sediments. Bacterial carbon is in the range of
5-452 µg g^{-1} dry weight sediment (cf. Table 2). Low values were
found in beach sediments, high values in shallow water sediments
(Kiel Bight; water depth 10 m), tropical coral beach sediments
(Mactan; Philippines) and subtidal sea grass sediments (Moreton Bay;
Queensland). Compared to the total organic matter content of the
sediment, bacterial biomass is low (0.2-2%). This is consistent
with the observations of Dale (1974) and Meyer-Reil et al. (1980) who
found that bacterial carbon accounted for 1.2 and 1.5%, respectively,
of the organic carbon of sediments, although much higher values
(10-14%) were reported by Moriarty (1980). The values obtained by
Wood (1970), that bacterial carbon represents 0.04-0.27% of the
organic carbon, are probably too low, since staining of samples with
erythrosine generally yield lower estimates of cell numbers and
consequently lower biomass values than staining with acridine orange
and analysis by epifluorescence microscopy.

An interesting parameter to characterize a bacterial population

Table 1. Comparison of data of bacterial numbers in relationship to organic matter and grain size in sediments. Average of the values reported in literature are summarized.

Location	Date	Temperature (°C)	Sediment Type	Depth (cm)	Bacterial number (10^8 g^{-1})	Organic matter (mg g^{-1})	Grain size (μm)	Remarks
Petpeswick Inlet, Nova Scotia[a]	6/3/72	–	Intertidal flats	0	99.70	3.8 [1]	19.1	7 stations, different horizons,
	8/30/72		Intertidal flats	5	1.17	0.1 [1]	132.5	maximum/minimum values
Beaufort Sea, Alaska[b]	Summer 75/76	1.2/–0.1	Arctic sediment	Surface	6.6/106	–	–	33/11 stations
	Winter 76	–1.9	Arctic sediment	Surface	10	–	–	13 stations
Kampinge, Baltic Sea[c]	6/1-5/77	15	Sandy beach	0	14.2	2.7	277	4 stations
Kiel Bight, Baltic Sea[d]	11/21/74	–	Sandy (14 m)	0	14.02	5.4	410	1 station
Halifax Harbor, Nova Scotia[e]	–	2	Muddy, aerobic/ anaerobic (24 m)	0-100	5-30	2-3 [1]	–	1 station, different horizons, different sampling dates
Windermere, South Basin[f]	4/79	–	Littoral sediment	2.9	290	–	–	Organic matter and particle
			Profundal sediment	2.1	340	–	–	size given for different sediment fractions
Blelham Tarn	4/79	–	Littoral sediment	3.6	280	–	–	
			Profundal sediment	1.2	119	–	–	
Kiel Fjord/ Kiel Bight, Baltic Sea[g,h]	7/4-13/77	21.0	Sandy beach	0	10.3	5.9	293	12 stations, 3 sampling dates
	3/6-11/78	3.0	Sandy beach	0	4.5	5.1	355	
	11/6-11/78	9.0	Sandy beach	0	15.3	11.7	282	
Mactan, Philippines[i]	4/2/80	27	Coral beach	0	14.7	–	–	3 stations
Kiel Bight, Baltic Sea[h]	5/30/80	8	Sandy (10 m)	0-2	8.8	12.7	–	
				2-4	6.7	8.1	–	
				4-6	0.7	7.2	–	
	5/28/80		Muddy, anaerobic (28 m)	0-2	84.7	164.4	–	
				2-4	69.0	157.7	–	
				4-6	86.8	160.3	–	
Antarctica[h]	1/26/81	2	Muddy (1951 m)	0	14.2	109.4	–	1 station, different horizons
				100	28.6	105.8	–	
				400	6.5	39.8	–	
				800	14.6	98.6	–	
				1100	11.1	82.5	–	

[1]Organic carbon (% by weight)
[a]Dale 1974; [b]Griffiths et al. 1978; [c]Meyer-Reil et al. 1978; [d]Weise and Rheinheimer 1978; [e]Kepkay et al. 1979; [f]Jones 1980; [g]Meyer-Reil et al. 1980; [h]Meyer-Reil unpublished; [i]Meyer-Reil 1981

is the biomass to number ratio, referring to the mean biomas per cell, which was found to be between 0.9×10^{-14} and 3.4×10^{-14} g C $cell^{-1}$ depending upon the sediments investigated (cf. Table 2). Again, high mean biomass values per cell were found in shallow water sediments (Kiel Bight; water depth 10 m) and tropical coral beach sediments (Mactan; Philippines); low values were found in beach sediments (Kiel Fjord, Kiel Bight). For samples taken in summer, the biomass of an average bacterial cell was found to be twice as high in the sediment (2.1×10^{-14} g C) as in the water above (1.1×10^{-14} g C) and three times as high as in the surface water of the Kiel Bight (0.7×10^{-14} g C) (Meyer-Reil 1977; Meyer-Reil et al. 1978).

Although bacterial numbers, biomass, mean biomass per cell in the interstitial water, and the water overlying the sediments were found to be in the same order of magnitude, pore water generally yields higher values. This is related to the role of interstitial water as a mediator between sediments and overlying waters that govern the exchange of nutrients between both environments (cf. below). However, compared to the sediment, bacterial numbers and biomass in the pore water and the water overlying the sediments were, respectively, two and three orders of magnitude lower (Meyer-Reil, unpublished data).

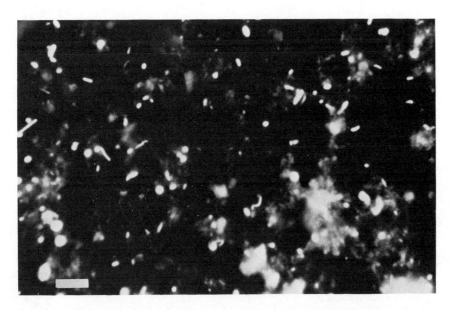

Figure 1. Epifluorescence photograph of bacteria from a shallow water sediment (water depth 10 m) of the Kiel Bight (Baltic Sea; FRG). The sediment sample was homogenized by sonication, diluted, filtered onto Nuclepore polycarbonate filter (pore size 0.2 μm) and stained with acridine orange. Bar represents 3 μm.

Table 2. Comparison of data of bacterial numbers, biomass and biomass/number ratio in relationship to the organic matter content in sediments. Average and range (in parenthesis) of the values reported in literature are summarized.

Location	Date	Temperature (°C)	Sediment type	Bacterial number (N) (10^8 g^{-1})	Bacterial biomass (B) (10^{-3} mg C g^{-1})	B/N ratio (10^{-11} mg C cell^{-1})	Organic matter (OM) (mg g^{-1})	Bacterial biomass (% OM)
Kampinge, Baltic Sea	6/1-5/77	15	Sandy beach[c]	14.2(7.5-27.2)	26.3(17.6-45.5)	1.9(1.7-2.4)	2.7(2.3-3.2)	1.9
Kiel Fjord/Kiel Bight, Baltic Sea	7/4-13/77	21.0	Sandy beach[d]	10.3(5.7-20.2)	20.4(11.0-36.7)	2.1(1.6-2.8)	5.9(3.3-9.6)	0.69
	3/6-11/78	3.0	Sandy beach[b]	4.5(1.4-11.8)	5.6(1.6-14.1)	1.2(1.0-1.4)	5.1(2.0-7.5)	0.22
	11/6-11/78	9.0	Sandy beach	15.3(4.5-32.1)	12.9(4.9-33.3)	0.9(0.6-1.1)	11.7(3.2-26.1)	0.22
Kiel Bight, Baltic Sea	3/3/ - 5/8/80	0.8-4.9	Sandy (10m)[a]	2.0(1.4-3.5)	5.8(3.9-10.3)	3.0(2.7-3.2)	8.1(7.1-10.3)	0.14
Mactan	4/2/80	27	Coral beach[e]	14.7(11.4-17.4)	50.0(39.0-59.0)	3.4	-	-
Moreton Bay, Queensland	-	-	Sub-tidal sea-grass sediment[f] (0,1,20cm)	7(6-8)[1]	240(110-370)	-	2.3(2.1-2.5)[2]	10.3[3]
	-	-	mid-intertidal sea-grass sediment (0,1,20cm)	10(6-15)[1]	203(100-320)	-	1.6(1.1-2.0)[2]	12.3[3]
	-	-	Intertidal sand flat (0,2,20cm)	9(6-10)[1]	50(10-80)	-	0.35 (0.29-0.43)[2]	13.7[3]
Elbe Estuary	2/2/80	3	Mud[g]	31.1	52.4	1.7	-	-
Kiel Bight	8/8/80	12	Mud[g]	507	452	0.9	-	-

[1] cells x 10^8 μg^{-1} muramic acid; [2] mg organic C g^{-1} sediment; [3] % organic carbon
[a] Meyer Reil 1981; [b] Meyer-Reil, unpublished; [c] Meyer-Reil et al. 1978; [d] Meyer-Reil et al. 1980; [e] Meyer-Reil et al.

UPTAKE OF DISSOLVED ORGANIC CARBON

Bacteria take up dissolved organic substrates, a process which is poorly understood at present. Substrate capturing, nutrient adsorption, and transport seem to be the most important processes involved. In nutrient-depleted waters, the capture of substrate by bacteria seems to proceed via high-affinity active transport, whereas in nutrient-rich waters, capture of substrates may involve chemotaxis and an active transport with reduced affinity to the substrates. Uptake systems seem to be specific for the substrates involved (Geesey and Morita 1979).

Different approaches have been used to follow the uptake of dissolved organic substances by bacteria in sediments. Most of the studies were carried out with slurried (suspended in sea water) sediments. Very few investigations relate to "undisturbed" sediments. Hall et al. (1972) incubated undisturbed sediment cores in environmental chambers connected to a purging train to collect the $^{14}CO_2$ evolved. Incorporation of ^{14}C was measured as $^{14}CO_2$ liberated after combustion of the sediment. In a similar way, Meyer-Reil (1978) incubated undisturbed sediment cores in environmental chambers consisting of a normal filtration apparatus. For incubations, the interstitial water originally present in the sediment cores was replaced by interstitial water spiked with ^{14}C-labeled substrates. Incubation times ranged between 1 and 9 minutes. Both respiration and incorporation could be determined, the latter after combustion of the sediment cores in a Tri-Carb sample oxidizer (Packard). In recent studies, the assay device has been simplified by using plastic syringes with a teflon piston equipped with a glass sintered disc thus allowing the incubation in the field under in situ conditions. The system works perfectly in sandy sediments; in muddy sediments, however, there are problems with the penetration of spiked pore water into the compact fine-grained sediments.

For sediments, it is important to incubate the samples under in situ redox potential conditions. This applies especially to anaerobic sediments (Christian and Wiebe 1978; Hanson and Gardner 1978; Novitsky and Kepkay 1981). In some investigations (cf. Table 3), sediments (anaerobic?) suspended in sea water were incubated under aerobic conditions. Information of the in situ redox potential are mostly lacking. However, the change from anaerobic (in situ) to aerobic (laboratory) conditions may drastically alter the uptake pattern of dissolved organic substrates by bacteria as it was pointed out already by Novitsky and Kepkay (1981) working with sediments from Halifax Harbor.

Hall et al. (1972) and Meyer-Reil (1978) showed that the gross uptake of ^{14}C-glucose by bacteria in undisturbed sediment cores was approximately one order of magnitude lower than uptake in disturbed cores. Diffusion processes alone, as mentioned by Hall et al.

(1972), can not be responsible for the lower bacterial uptake in undisturbed sediment cores, since the assay device used by Meyer-Reil (1978) allows an even distribution of label within the sediment core. More probably, the disruption of the sediments fine structure (composed of various microenvironments) may be responsible for the higher bacterial uptake activity. Cells and groups of cells are separated from each other, thus, for example, lowering the influence of competition and inhibition.

Ansbaek and Blackburn (1980) injected ^{14}C-labeled fatty acids into segments of undisturbed sediment cores and followed the disappearance of the label. This is probably the most reliable method for studying net substrate turnover in sediments.

In most of the studies, the Michaelis-Menten model has been used for the calculation of heterotrophic activity parameters. The limitations have been discussed in detail elsewhere (e.g., Wright 1973; Krambeck 1979). In very few investigations, the tracer technique was applied in combination with the determination of the natural substrate concentration (Hanson and Gardner 1978; Meyer-Reil 1978; Meyer-Reil et al. 1978; Meyer-Reil et al. 1980). From these studies, information of actual uptake rates (flux) of dissolved organic substrates can be expected, although the question arises, what amount of the substrate determined as "free-dissolved" is actually available for bacteria. In our own experiments, we incubated natural sediments over 14 days with frequent measurements of the concentration of glucose. Following a decrease during the first week of incubation, the glucose concentration remained fairly unchanged for the rest of the incubation period at a level of approximately one third of the initial concentration. This may be an indication that only two thirds of the natural "free-dissolved" glucose is actually available for the bacteria. The problem was discussed in detail by Gocke et al. (1981), who studied the availability of dissolved glucose to heterotrophic bacteria. In addition to data on concentration, much more information is required of the structure and status of the "dissolved" organic fraction of organic matter in waters and sediments. This again stresses the necessity of a close cooperation between microbiologists and chemists in future research.

The choice of the labeled substrate used in the investigations should reflect its ecological importance in the natural environment. In aerobic sediments, carbohydrates and amino acids (in this order?) may be the substrates of choice. In anaerobic sediments, however, short-chain fatty acids obviously play a more important role for characterizing the predominant metabolic pathways in bacteria. It is an unresolved discussion among microbiologists whether the importance of a substrate can be derived from its natural concentration. One may argue that high natural concentrations of substrates may reflect their low ecological relevance: the substrates are not used adequately. On the other hand, low concentrations of individual

natural substrates may indicate a rapid turnover, thus characterizing
their ecological importance. Glucose is a good example. In marine
inshore sediments of the Baltic Sea, it is present in fairly high
concentrations (Meyer-Reil et al. 1980; cf. Table 3), whereas in
lake sediments, glucose concentrations are extremely low (Overbeck,
personal communication).

Most frequently, ^{14}C-glucose (partially and uniformly labeled)
was used for the determination of bacterial uptake rates and turnover time in the sediment. In a few investigations, acetate and
glutamic acid were applied. The uptake of glycine, alanine, aspartic
acid, urea and lactic acid was investigated only occasionally (for
literature cf. Table 3).

Almost all of the investigations of the uptake of dissolved
organic substrates by bacteria in waters and sediments ignored the
possibility of excretion of metabolic products by bacteria.
Recently, Novitsky and Kepkay (1981) have shown that, at least under
anaerobic conditions, metabolic products released into the medium
constitute a large portion of the bacterial substrate uptake. The
presence of these low molecular weight (obviously excreted)
materials was most pronounced in and near the aerobic/anaerobic
transition zone where a peak in the uptake of labeled dissolved
organic substrates (lactate, glutamate, glucose) was observed.

Table 3 summarizes some of the data obtained for turnover time
and respiration (percent of the total uptake) of dissolved organic
carbon by bacteria in natural aerobic and anaerobic sediments.
Besides environmental factors, information on the kind of incubation
(compared to in situ conditions) as well as natural substrate
concentrations (if available) were included. Unfortunately, the
evaluation of some data is limited by the lack of important information (temperature, sediment type, incubation conditions, substrate
concentration). Furthermore, Table 3 was supplemented by data on
the uptake of dissolved organic substances by bacteria in the water
overlying the sediments (Harrison et al. 1971; Meyer-Reil et al.
1978; Meyer-Reil, unpublished data).

From Table 3 it becomes obvious that turnover times range
between a few minutes and some hundred hours, a range which is
not surprising taking into account the pronounced differences in
environmental parameters such as location, sampling date, temperature, sediment type (salt marshes, lake and marine sediments;
aerobic and anaerobic), and natural substrate concentration. As
already mentioned, methodological differences in the kind of
incubation, the choice of labeled substrate, and the approach used
for the calculation of the data will certainly restrict any detailed
comparisons. However, some more general conclusions from the data
available can be derived.

Table 3.   Comparison of data of the uptake of dissolved organic
           substrates by bacteria in sediments and overlying waters.
           Average and range (in brackets) of the values reported in
           literature are summarized.

Location	Date	Temp. (°C)	Environment	Aeration Status Original	Incubation	Substrate Type	Conc.
Pamlico River, North Carolina	Sept 68 May 69	4.1-29	Estuary sediment (2.5-7.5 cm core)	aerobic anaerobic aerobic anaerobic	aerobic aerobic aerobic aerobic	Gluc Gluc Ace Ace	– – – –
Upper Klamath Lake, Oregon	–	–	Lake sediments (2.7 m depth) Overlying water	– aerobic	aerobic	Gluc/ Ace Gluc/ Ace	– –
Marion Lake Br. Columbia	June 71– May 72	0-20	Shallow lake sediments (0-3 cm core)	anaerobic aerobic (aerobic: 0-1 cm)	aerobic aerobic aerobic	Glu Gly Ace	50 50 –
Duplin Estuary, Sapelo Island, Georgia	April– Sept. 75	18-28	Salt marsh (1-5 cm)	anaerobic		Glu	–
Beaufort Sea, Alaska	Summer 75/76 Winter 76	– -0.1-1.2 – -1.9	Sediment Water Sediment Water	– – – –	aerobic aerobic aerobic aerobic	Glut Glut Glut Glut	– – – –
Georgia	Oct. 76	–	Salt marsh (10 cm)	anaerobic anaerobic		Ala Asp	150 (2.4-550)[2] 24 (0.81-66)[2]
Kampinge Baltic Sea	June 77	15	Sandy beach (0-2 cm) Overlying water	aerobic aerobic		Glu Glu	62 (34-122) 41 (16-76)
New York Bight	August 76	–	Sediment (0-5 cm)	– –	aerobic aerobic	Asp Urea	– –
Limfjord	Spring 78 Autumn 77	2.0 7.0	Inshore sediment (10 m depth)	anaerobic anaerobic		Ace Ace	0.1-0.6[3]
Kiel Fjord/ Kiel Bight, Baltic Sea	July 77 March 78 Nov. 78	21.0 19.5 3.0 2.5 9.0 9.9	Sandy beach Overlying water Sandy beach Overlying water Sandy beach Overlying water	aerobic aerobic aerobic aerobic aerobic aerobic		Glu Glu Glu Glu Glu Glu	106 (22-318) 32 (9.4-80) 149 (32-329) 117 (26-428) 181 (4.3-356) 59 (4-282)
Halifax Harbor, Nova Scotia	–	2	Sediment (0-100 cm) (24 m depth)	aerobic (0-4 cm) anaerobic (40-100 cm) aerobic anaerobic aerobic anaerobic		Glu Glu Glut Glut Lact Lact	– – – – – –

[1] Glucose mineralization
[2] pmol cm^{-3}
[3] μmol l^{-1}
[4] Net turnover in mmol m^{-2} d^{-1} (derived from disappearance of label)

Turnover time (h)	Respiration (% total uptake)	Literature	Remarks
0.22	8.6	Wood 1970	5 stations,
0.03 (0.02–0.05)	2.3 (1.8–3.3)		9 sampling dates
1.4	–		
0.11 (0.04–0.18)	–		
2.3–0.75[1]	–	Harrison et al. 1971	
220–250[1]	–		
0.23 (0.06–0.36)	22 (18–31)	Hall et al. 1972	1 station
			12 sampling dates
0.67 (0.18–1.1)	63 (45–87)		
0.13 (0.42–0.3)	13 (9.1–16)		
1.3 (0.16–3.2)	–	Christian and Wiebe 1978	2 stations
			6 sampling dates
–	37–23	Griffiths et al. 1978	32/11 stations
–	59–46		50/16 stations
–	39		14 stations
–	85		23 stations
43 (9.3–71)	22 (13–30)	Hanson and Gardner 1978	4 stations
96 (19–253)	43 (31–58)		
0.4	6.5 (4–6)	Meyer-Reil et al. 1978	1 station
20 (13–30.4)	33 (27–39)		4 sampling dates
8.4 (2.4–19)[1]	–	Litchfield et al. 1979	3 stations,
112 (50–205)[1]	–		replicate samples
7.2[4]	–	Ansbaek and Blackburn 1980	–
20[4]	–		
0.5 (0.3–0.7)	7.7 (4.4–5.1)	Meyer-Reil et al. 1980	12 stations,
4.3 (2.8–9.1)	37 (30–42)		3 sampling dates
5.1 (1.7–14)	2.0 (0.4–4.2)	Meyer-Reil, unpublished	
366 (55.2–500)	16 (4–27)		
2.5 (0.2–5.3)	6.8 (4.8–8.9)		
68 (13–160)	37 (28–49)		
–	30–65	Novitsky and Kepkay 1981	Sediment cores
–	2–80		taken at different
–	35–95		seasons of the year
–	35–80		
–	35–100		
–	35–100		

In anaerobic sediments, acetate and glucose showed a rapid turnover. Whereas Wood (1970) revealed a faster turnover time for glucose, Hall et al. (1972) found acetate to be more rapidly turned over. However, both studies were carried out with anaerobic sediments which were incubated aerobically. The rapid turnover of glucose was confirmed by Christian and Wiebe (1978) who incubated their samples under in situ anaerobic conditions. The turnover of glycine (Hall et al. 1972), alanine, aspartic acid (Hanson and Gardner 1978), and urea (Litchfield et al. 1979) was much slower (up to 205 h). For aerobic beach sediments, the turnover of glucose was found to be between 0.2 and 13.5 h, depending upon the season. Natural glucose concentrations in these wave-washed beaches fluctuate strongly between 4.3 and 356 μg l^{-1} (Meyer-Reil et al. 1978, 1980). The average concentration of glucose is still two to three times higher than the values reported by Hall et al. (1972); however, number and range of observations in the latter study were not presented.

Respiration of organic substrates varied strongly (between 0.4 and 100%). However, the values measured concentrate on the lower end of the scale (around 20%; cf. Table 3). According to Wood (1970), a higher percentage of glucose is respired in aerobic sediments than in anaerobic sediments (average of 8.6 and 2.3%, respectively). In contrast, working with anaerobic sediments, Hall et al. (1972) found 22% of the total uptake of glucose to be respired. The difference in the data of Wood (1970) is attributed to the author's use of differently labeled glucose. Applying uniformly labeled glucose, the average respiration of 6%, summarized from investigations of Meyer-Reil et al. (1978, 1980) is in good agreement with the data reported by Wood (1970) who employed glucose-6-^{14}C. Besides glucose, only alanine seems to be respired to a lower percentage (average 22%; Hanson and Gardner 1978). Other substrates, such as glycine, glutamic acid, aspartic acid and lactic acid are respired to a much higher degree.

During their investigation in the Beaufort Sea, Alaska, Griffiths et al. (1978) found a higher percentage of glutamic acid respired in winter than in summer. The authors attribute the differences to deficiencies of nutrients required for biosynthesis in winter, which may result in a larger percentage of glutamic acid used for the energy requirements of the cells. However, in our own investigations, a drastic decrease in glucose respiration could be demonstrated for both water and sediment samples taken in winter (cf. Table 3), which can be simply explained by the lower energy requirements of bacteria at lower temperatures.

In the water overlying the sediment, turnover times of dissolved organic substrates are one to two orders of magnitude higher (Harrison et al. 1971; Meyer-Reil et al. 1978; Meyer-Reil, unpublished data). Differences seem to be most pronounced for

samples taken in winter. Between 16 and 37% of the total uptake of
glucose in the water overlying the sediment was respired (cf. Table
3). These values are similar to those generally reported for
bacterial respiration in the water column (cf. e.g., Gocke 1976).

Based on the data available, respiration of labeled glucose by
bacteria was considerably lower in sediments than in the water above
(cf. Table 3). If we assume that on an average 10% of the total
uptake of glucose is respired (water: 30%), the incorporation would
reflect a 90% efficiency which seems to be unreasonably high. One
may argue that sediments are characterized by a broader spectrum and
higher concentrations of organic substrates than overlying waters.
Other substrates besides glucose may be respired preferentially.
Bacteria could release fermentation products instead of or as well
as CO_2. The latter hypothesis is pointed out by Wood (1970) who
found a higher respiration of glucose in aerobic than in anaerobic
sediments. However, more recent data (not available at the time of
completion of this manuscript) imply that the low respiration rates
found in sediments could also be the result of isotope dilution
(King and Klug 1982). Furthermore, the authors pointed out that
glucose mass flow could not be calculated from end product formation
since the specific activity of added ^{14}C-glucose was significantly
diluted by pools of intracellular glucose and glucose metabolites
(King and Klug 1982).

DECOMPOSITION OF PARTICULATE ORGANIC MATTER

In sediments, the overwhelming portion of organic carbon is
present as particulate carbon. In sandy beaches of the Kiel Fjord
and Kiel Bight, particulate organic carbon amounts to 1.4 mg g^{-1} dry
weight sediment; dissolved organic carbon is roughly two orders of
magnitude lower (13.5 x 10^{-3} mg ml^{-1} pore water; Meyer-Reil et al.
1978, cf. Fig. 2). The strong local and seasonal variations of
particulate organic matter indicate a considerable decomposition,
obviously mostly carried out by extracellular enzymes. These are
produced in living cells, transported through the membranes, and
activated outside the cells. Dead and decaying organisms (bacteria,
algae, meiofauna, macrofauna, etc.) contribute to the release of
extracellular enzymes as well. Unless the enzymes are released in
close contact to the substrate, their fate is uncertain. Processes
like adsorption, decomposition and denaturation greatly influence
the efficiency of the enzymes released. Furthermore, the product of
the enzymatic reaction must be available for the bacterial cells.
Besides the above mentioned factors, competition for the product may
further reduce the availability of the product for the bacteria
releasing the enzymes. From the energetic point of view, the
continuous excretion of enzymes by bacteria must be doubted (Burns
1980). Our own investigations have failed to detect free-dissolved
enzymes (α-amylase, β-D-glucosidase, proteolytic enzymes) in the

interstitial water. Most of the enzymes seem to be adsorbed to
particles. Studies carried out by Burns (1980) have demonstrated
that extracellular enzymes remain active in soil by the formation
of humic acid-enzyme complexes. These enzymes may react with a
substrate to form a product which then can be taken up by the
bacterial cells. The product itself may induce the production of
bacterial enzymes which are secreted into the medium. In this model
(Burns 1980), the extracellular enzymes adsorbed to humic acid
complexes may act as starter enzymes inducing the production of
bacterial enzymes in the presence of sufficient concentrations of
suitable substrates.

        In the literature, there is only limited information on the
decomposition rates of particulate organic material by bacteria in
waters and sediments and the dependency of the decomposition rates
upon environmental parameters such as natural substrate concentration
and bacterial biomass. However, such data are urgently needed, since
by bacterial extracellular enzymatic decomposition, products of the
primary production become available for higher trophic levels.

        A variety of extracellular enzymes have been detected as
secretion products of culturable marine periphytic bacteria (Corpe
and Winters 1972; Corpe 1974) as well as cell-free enzymes in waters
and sediments (Kim and ZoBell 1974): proteinase, agarase, alginase,
esterase, β-glucosidase, phosphatase, succinate dehydrogenase,
amylase. The authors have already suggested the importance of these
enzymes in the decomposition of particulate organic matter.

        For the determination of enzymatic activity rates certain
requirements have to be fulfilled: linear response of activity over
time, absence of induction and substrate exhaustion phenomena, and
minimal changes of substrate concentration over the incubation
period. Homogenization of the sediment leads to a sharper response
of enzymatic activity due to a more even distribution and unmasking
of enzymes in the homogenized samples, although the activity pattern
of homogenized and non-homogenized samples is similar. Furthermore,
possible interactions of living bacteria with the enzymatic reaction
can be reduced in homogenized samples.

        Extracellular enzymatic activity seems to be greatest in
littoral sediments (sulfhydrolase; King and Klug 1980). This
becomes especially obvious when comparing specific activity rates of
α-amylase (based on the protein concentration of the sediment) in
littoral and sublittoral samples from the Kiel Fjord and the Kiel
Bight. Clay sediments seem to have higher activities than sandy
ones (phosphatase: Ayyakkannu and Chandramohan 1971) although for
deoxyribonuclease no evident difference in activity was found in
sandy-mud and mud-type sediments (Maeda and Taga 1973). Sulfhydrolase as determined by King and Klug (1980) showed pronounced variations with sediment depth which relate to ATP distribution. This

could be confirmed for α-amylase in sediment cores of the Kiel Bight, although pronounced profiles may be disturbed by biological processes such as bioturbation. One would expect to obtain some kind of relationship between enzymatic activity and substrate concentration. For marine sediments, Ayyakkannu and Chandramohan (1971) showed a positive relationship between total phosphate and phosphatase. In marine sediment cores, deoxyribonuclease varied with deoxyribonucleic acid, whereas in surface sediments, no apparent correlation existed (Maeda and Taga 1973). During the breakdown of the algal spring bloom in the Kiel Bight, concentrations and decomposition rates of carbohydrates (activity of α-amylase, β-D-glucosidase) and proteins (proteolytic enzymes) were closely related. The enzymatic degradation causes a drastic shift in the protein to carbohydrate ratio, obviously due to a more rapid initial decomposition of carbohydrates. To my knowledge, this was the first time that a close connection between sedimentation of organic material from the water column and enzymatic decomposition in the sediment could be followed (Meyer-Reil 1981).

The derivation of the extracellular enzymes responsible for the degradation of the particulate organic matter is an unresolved question. A correlation between enzymatic activity and ATP as determined by Oshrain and Wiebe (1979) for arylsulfatase and by Sayler et al. (1979) for alkaline phosphatase seems to indicate that the enzymes are associated with or derived from living (bacterial?) cells. This is supported by Ayyakkannu and Chandromohan (1971) who found a positive relationship between phosphatase activity and phosphate solubilizing bacteria in marine sediments. However, Maeda and Taga (1973) failed to demonstrate significant correlations between deoxyribonuclease and deoxyribonucleic acid hydrolyzing bacteria. During the above mentioned study in sediments of the Kiel Bight, most of the enzymes responsible for the initial decomposition of particulate organic material obviously originated from the lysis of plankton cells. As shown by the increases in ATP, many cells reached the sediment as intact cells. Further investigations carried out at different stations in the Kiel Fjord and Kiel Bight indicate that the activity rates of α-amylase differ considerably in their pattern of relationships to concentration of carbohydrates and bacterial biomass. Laboratory experiments have shown that the activity of α-amylase is rapidly inducible (time scale a few hours) and is dependent upon the concentration of dissolved organic substrates available (Daatselaar and Harder 1974; Little et al. 1979; Meyer-Reil, unpublished data). However, much more data are needed to draw general conclusions concerning the regulation of enzymatic activities in natural sediments. Of particular interest is the decomposition of particulate organic matter under anaerobic conditions. For arylsulfatase, the anaerobic production is likely (Oshrain and Wiebe 1979). The activity of sulfhydrolase was shown to be sensitive to the presence of oxygen. This enzyme may play an important role in the sulfur cycle by the maintenance of an active

population of sulfate reducers when inputs of sulfate into the
sediment are low (King and Klug 1980).

FACTORS RELATED TO BACTERIAL BIOMASS AND ACTIVITY

Until recently, the existence of short-term fluctuations (time
scale a few hours) has been ignored by microbiologists, mainly
because of time restrictions in processing frequent samples.
However, frequent sampling of one and the same body of water has
demonstrated that bacterial biomass and uptake activity of dissolved
organic substrates undergo strong short-term fluctuations or even
rhythms, which are paralleled by corresponding rhythms of organic
chemical parameters (Meyer-Reil et al. 1979). This could be
confirmed for shallow water sediments of the Kiel Bight. Bacterial
biomass, uptake rates of glucose, and decomposition rates of
carbohydrates (α-amylase activity) followed a one maximum curve with
increasing values during the morning, a maximum between noon and
early afternoon and decreasing values towards late afternoon and
night. Bacterial parameters were paralleled by fluctuations in ATP
and meiofauna activity (Faubel and Meyer-Reil 1981). Fluctuations
in total organic matter content, proteins, and carbohydrates were
less pronounced. This corresponds to investigations of Meyer-Reil
et al. (1979), who showed that against an almost constant background
of total organic matter, fluctuations mainly comprise individual
components of organic matter (carbohydrates, amino acids). The
control mechanisms of the diurnal rhythms are still unknown, but a
close relation to primary production and the release of organic
substances has to be assumed.

In a number of publications, the close interrelationships among
sediment grain size, organic matter content, and saprophytic bacteria
have been demonstrated (ZoBell 1938; Wood 1970; Hall et al. 1972;
Hargrave 1972; Dale 1974; Litchfield et al. 1976; Meyer-Reil et al.
1978, 1980; Reichgott and Stevenson 1978; Weise and Rheinheimer
1978). However, bacterial biomass and activity were a part of only a
few studies. The aim of an intensive study carried out in the Kiel
Fjord and Kiel Bight was to demonstrate "key" parameters influencing
bacterial biomass and activity (Meyer-Reil et al. 1980). Variations
of microbiological parameters (total number and biomass, saprophytic
bacteria, uptake rates and turnover time of glucose) were compared to
a total of 22 chemical and physico-chemical parameters in the sediment and overlying water. As it could be demonstrated, the correlation pattern of microbiological parameters in the sediment and overlying water showed principal differences, part of which could be
traced back to the individual features of both systems investigated.
Whereas in the sediment, bacterial biomass and activity were closely
related to salinity, microbiological parameters in the overlying
water were greatly influenced by temperature. Among the dissolved
inorganic nutrients, ammonia and nitrate paralleled the variation

pattern of microbiological parameters in the sediment; in the water
above, however, nitrite appeared to have the most influence on
bacterial biomass and activity. In both systems, variations in
concentrations of chlorophyll a, glucose, fructose, ribose and total
monosaccharides greatly influenced microbiological parameters;
however, the importance of total dissolved organic matter as a
regulating factor is restricted to the water overlying the sediment.
Bacterial biomass and activity were closely related to the total
organic matter content of the sediment; variations of individual
fractions of organic matter (particulate carbon and nitrogen) did
not parallel the variation of microbiological parameters. In the
water above the sediment, variations of particulate carbon and
nitrogen were closely tied to variations in bacterial biomass and
activity. To summarize this study, salinity, ammonia, nitrate,
dissolved monosaccharides, organic matter content, and chlorophyll a
may be regarded as "key" parameters influencing bacterial biomass
and activity in the sediment. In the water overlying the sediment,
however, temperature, nitrite, monosaccharides, ·total dissolved
organic carbon, particulate organic carbon and nitrogen, and
chlorophyll a represent "key" parameters. Further investigations
are needed to show the influence of the season on parameters
controlling bacterial biomass and activity.

     Biological and chemical exchange processes occur between
sediments and overlying waters, which may be of great importance for
both phytoplankton and benthos (Smetacek et al. 1976). The above
mentioned studies have shown that microbiological parameters in
sediments and overlying waters were closely related: saprophytic
bacteria revealed a positive correlation while total bacterial
biomass and activity were negatively correlated. High biomass and
activity in the sediment corresponded to low biomass and activity in
the water above. This seems to demonstrate that the microbiological
parameters in both systems are controlled by more than simply a
concentration gradient.

CARBON CYCLE AS INFLUENCED BY BACTERIA

     Based on the data derived from the above mentioned investiga-
tion during summer in the Kiel Fjord and Kiel Bight, bacterial
biomass production in sediments and overlying waters was estimated
(Fig. 2). In the sediment, the overwhelming portion of organic
carbon is present in the particulate form (POC). Dissolved organic
carbon (DOC) is two orders of magnitude lower. However, in the water
above the sediment, DOC represents 96% of the total organic carbon.
Bacterial carbon expressed as percentage of POC is roughly one order
of magnitude higher in the water than in the sediment (9.2 and 1.4%,
respectively). Natural free dissolved monosaccharides (FMS) account
for 0.8 and 0.2% of DOC (sediment and water, respectively). The
main components of FMS are glucose and fructose, which make up

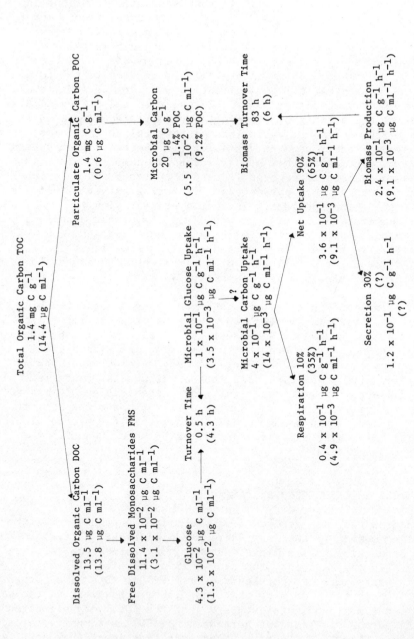

Figure 2. Diagram of carbon cycle as influenced by bacteria in sandy beach sediments and overlying waters (in parentheses) of the Kiel Fjord and the Kiel Bight (Baltic Sea; FRG).

80-90% of the total. Based on the concentration and actual uptake rates of glucose, a bacterial turnover time of 0.5 h (sediment) and 4.3 h (water) was calculated. If we assume that bacterial glucose uptake represents one quarter of the total carbon uptake, bacterial carbon uptake amounts to 4×10^{-1} µg C g^{-1} h^{-1} in the sediment and 14×10^{-3} µg C ml^{-1} h^{-1} in the water above. The assumption for this calculation is derived from the observation that monosaccharides and amino acids as labile, easily degradable components of DOC, are present in approximately equal concentrations, from which glucose roughly represents one quarter. On an average, 10% (sediment) and 35% (water) of the organic carbon taken up by bacteria (glucose uptake) is respired. If we assume that another 30% of the carbon uptake is secreted by bacteria in the sediment (slime production on particle surfaces), bacterial biomass production amounts to 2.4×10^{-1} µg C g^{-1} h^{-1} in the sediment compared to 9.1×10^{-3} µg C ml^{-1} h^{-1} in the water above. This represents a biomass turnover time of 83 and 6 hours (sediment and water, respectively).

The assumptions made in the calculations illustrate our limited knowledge of bacterial carbon conversion under natural conditions. There is still no direct approach for measuring bacterial biomass production. Glucose represents only one of the variety of substrates utilizable by bacteria. Again, one may argue that glucose is not the most suitable substrate to meet the specific carbon and energy requirements of the bacteria adapted to the type of sediment investigated. Other substrates may be respired and incorporated differently from glucose. This means that the conversion factor for the extrapolation from bacterial glucose uptake to bacterial carbon uptake may be wrong. We know that bacteria in the sediment exhibit a high extracellular production; however, secretion rates are unknown. Furthermore, the bacterial production derived from the decomposition of particulate organic carbon (POC) could not be included in the calculations, since up to now, no method is available to estimate bacterial carbon conversion based on the decomposition of POC. Because of this, the calculations are to be regarded as conservative. In summary, for the calculation of bacterial biomass production, we have to rely on a number of assumptions and the validity of conversion factors. However, the comparison with other data measured independently may add some confidence to the calculations presented above. The comparison between microphytobenthic primary production and bacterial production demonstrates that between 30 and 50% of the primary production is fixed by bacterial secondary production. Compared to meiofauna carbon, daily bacterial biomass production turned out to be sufficient to provide a daily turnover of meiofauna biomass. The comparison with microphytobenthic primary production and meiofauna biomass reveals that the bacterial biomass production calculated is at least in the order of magnitude one should expect. Bacterial biomass production calculated for the water overlying the sediment (approximately 10^{-2} µg C ml^{-1} h^{-1}) correspond well to the upper limits of the values reported for bacterial production in the water column (for literature cf van Es and Meyer-Reil 1982). The high bacterial biomass production is a reflection of the relatively high nutrient concentration and bacterial load in waters overlying shallow water sediments.

CONCLUSION

The limited information available on bacterial biomass in sediments indicates that bacteria contribute only insignificantly to the sediment standing stock carbon pool. However, based on their activity in mineralization and biomass production, the cycle of nutrients in sediments is dominated by the activity of bacteria. In this respect, the relationships between bacteria and benthic fauna are of specific interest (cf. Gerlach 1978).

The rapid uptake of dissolved organic substrates by bacteria has been well-documented, although the utilization of different substrates in different types of sediments and the carbon conversion efficiency need further investigation. Comparatively little is known about the decomposition of particulate organic matter, a process by which (high molecular) material of the primary production becomes available for higher trophic levels. We require information on the spectrum of extracellular enzymes acting in different types of sediments and the influence of environmental factors on decomposition rates. In calculations of bacterial biomass production in sediments, we still have to rely on a number of assumptions and the validity of conversion factors.

In future research, time and scale has to be reduced in microbiological studies. Short-term investigations and the exploration of the sediment as a variety of microenvironments seem to be most promising.

ACKNOWLEDGMENT

Publication No. 376 of the Joint Research Program at Kiel University (Sonderforschungsbereich 95 der Deutschen Forschungsgemeinschaft).

REFERENCES

Ansbaek, J., and T. H. Blackburn. 1980. A method for the analysis of acetate turnover in a coastal marine sediment. Microb. Ecol. 5: 253-264.
Ayyakkannu, K., and D. Chandramohan. 1971. Occurrence and distribution of phosphate solubilizing bacteria and phosphatase in marine sediments at Porto Novo. Mar. Biol. 11: 201-205.

Boeye, A., M. Wayenhergh, and M. Aerts. 1975. Density and composition of heterotrophic bacterial populations in North Sea sediments. Mar. Biol. 32: 263-270.

Burns, R. G. 1980. Microbial adhesion to soil surfaces: consequences for growth and enzyme activities, pp. 249-262. In: R. C. W. Berkeley, J. M. Lynch, J. Melling, P. R. Rutter, and B. Vincent [eds.], Microbial Adhesion to Surfaces. Society of Chemical Industry, Ellis Horwood Limited, London.

Christian, R. R., and W. J. Wiebe. 1978. Anaerobic microbial community metabolism in Spartina alterniflora soils. Limnol. Oceanogr. 23: 328-336.

Corpe, W. A. 1974. Periphytic marine bacteria and the formation of microbial films on solid surfaces, pp. 397-417. In: R. R. Colwell and R. Y. Morita [eds.], Effect of the Ocean Environment on Microbial Activities. University Park Press, Baltimore.

Corpe, W. A., and H. Winters. 1972. Hydrolytic enzymes of some periphytic bacteria. Can. J. Microbiol. 18: 1483-1490.

Daatselaar, M. C. C., and W. Harder. 1974. Some aspects of the regulation of the production of extracellular proteolytic enzymes of a marine bacterium. Arch. Microbiol. 101: 31-34.

Dale, N. G. 1974. Bacteria in intertidal sediments: Factors related to their distribution. Limnol. Oceanogr. 19: 509-518.

Faubel, A., and L.-A. Meyer-Reil. 1981. Enzymatic decomposition of particulate organic matter by meiofauna. Kieler Meeresforsch. Sonderh. 5: 429-430.

Geesey, G. G., and R. Y. Morita. 1979. Capture of arginine at low concentrations by a marine psychrophilic bacterium. Appl. Environ. Microbiol. 38: 1092-1097.

Gerlach, S. A. 1978. Food-chain relationships in subtidal silty sand marine sediments and the role of meiofauna in stimulating bacterial productivity. Oecologia 33: 55-69.

Gocke, K. 1976. Respiration von gelösten organischen Verbindungen durch natürliche Mikroorganismen populationen. Ein Vergleich zwischen verschiedenen Biotopen. Mar. Biol. 35: 375-383.

Gocke, K., R. Dawson, and G. Liebezeit. 1981. Availability of dissolved free glucose to heterotrophic microorganisms. Mar. Biol. 62: 209-216.

Griffiths, R. P., S. S. Hayasaka, T. M. McNamara, and R. Y. Morita. 1978. Relative microbial activity and bacterial concentrations in water and sediment samples taken in the Beaufort Sea. Can. J. Microbiol. 24: 1217-1226.

Gunkel, W. 1964. Die Verwendung des Ultra-Turrax zur Aufteilung von Bakterienaggregaten in marinen Proben. Helgol. Wiss. Meeresunters. 11: 287-295.

Hall, K. J., P. M. Kleiber, and I. Yesaki. 1972. Heterotrophic uptake of organic solutes by microorganisms in the sediment. Mem. Ist. Ital. Idrobiol. 29(Suppl.): 441-471.

Hanson, R. B., and W. S. Gardner. 1978. Uptake and metabolism of two amino acids by anaerobic microorganisms in four diverse salt-marsh soils. Mar. Biol. 46: 101-107.

Hargrave, B. T. 1972. Aerobic decomposition of sediment and detritus as a function of particle surface area and organic content. Limnol. Oceanogr. 17: 583-596.

Harrison, M. J., R. T. Wright, and R. Y. Morita. 1971. Method for measuring mineralization in lake sediments. Appl. Microbiol. 21: 698-702.

Jones, J. G. 1980. Some differences in the microbiology of profundal and littoral lake sediments. J. Gen. Microbiol. 117: 285-292.

Kepkay, P. E., R. C. Cooke, and J. A. Novitsky. 1979. Microbial autotrophy: A primary source of organic carbon in marine sediments. Science 204: 68-69.

Kim, J., and C. E. ZoBell. 1974. Occurrence and activities of cell-free enzymes in oceanic environments, pp. 368-385. In: R. R. Colwell and R. Y. Morita [eds.], Effect of the Ocean Environment on Microbial Activities. University Park Press, Baltimore.

King, G. M., and M. J. Klug. 1980. Sulfhydrolase activity in sediments of Wintergreen Lake, Kalamazoo County, Michigan. Appl. Environ. Microbiol. 39: 950-956.

King, G. M., and M. J. Klug. 1982. Glucose metabolism in sediments of a eutrophic lake: tracer analysis of uptake and product formation. Appl. Environ. Microbiol. 44: 1308-1317.

Krambeck, C. 1979. Applicability and limitations of the Michaelis-Menten equation in microbial ecology. Arch. Hydrobiol. Beih. Ergebn. Limnol. 12: 64-76.

Krambeck, C., H.-J. Krambeck, and J. Overbeck. 1981. Microcomputer-assisted biomass determination of plankton bacteria on scanning electron micrographs. Appl. Environ. Microbiol. 42: 142-149.

Litchfield, C. D., M. A. Devanas, J. Zindulis, C. E. Carty, J. P. Nakas, and E. L. Martin. 1979. Application of the ^{14}C organic mineralization technique to marine sediments, pp. 128-147. In: C. D. Litchfield and P. L. Seyfried [eds.], Methodology for Biomass Determinations and Microbial Activities in Sediments. American Society for Testing and Materials, Philadelphia.

Litchfield, C. D., J. P. Nakas, and R. H. Vreeland. 1976. Bacterial flux in some New Jersey estuarine sediments. Am. Soc. Limnol. Oceanogr. Spec. Symp. 2: 340-354.

Little, J. E., R. E. Sjogren, and G. R. Carson. 1979. Measurement of proteolysis in natural waters. Appl. Environ. Microbiol. 37: 900-908.

Maeda, M., and N. Taga. 1973. Deoxyribonuclease activity in sea-water and sediment. Mar. Biol. 20: 58-63.

Meadows, P. S., and J. G. Anderson. 1966. Micro-organisms attached to marine and freshwater sand grains. Nature, Lond. 212: 1059-1060.

Meyer-Reil, L.-A. 1977. Bacterial growth rates and biomass production, pp. 223-236. In: G. Rheinheimer [ed.], Microbial Ecology of a Brackish Water Environment. Springer-Verlag, Berlin.

Meyer-Reil, L.-A. 1978. Uptake of glucose by bacteria in the sediment. Mar. Biol. 44: 293-298.

Meyer-Reil, L.-A. 1981. Enzymatic decomposition of proteins and carbohydrates in marine sediments: Methodology and field observations during spring. Kiel. Meeresforsch. Sonderh. 5: 311-317.

Meyer-Reil, L.-A., M. Bölter, G. Liebezeit, and W. Schramm. 1979. Short-term variations in microbiological and chemical parameters. Mar. Ecol. Prog. Ser. 1: 1-6.

Meyer-Reil, L.-A., M. Bölter, R. Dawson, G. Liebezeit, H. Szwerinski, and K. Wolter. 1980. Interrelationships between microbiological and chemical parameters of sandy beach sediments, a summer aspect. Appl. Environ. Microbiol. 39: 797-802.

Meyer-Reil, L.-A., R. Dawson, G. Liebezeit, and H. Tiedge. 1978. Fluctuations and interactions of bacterial activity in sandy beach sediments and overlying waters. Mar. Biol. 48: 161-171.

Meyer-Reil, L.-A., and A. Faubel. 1980. Uptake of organic matter by meiofauna organisms and interrelationships with bacteria. Mar. Ecol. Prog. Ser. 3: 251-256.

Meyer-Reil, L.-A., W. Schramm, and G. Wefer. 1981. Microbiology of a tropical coral reef system (Mactan, Philippines). Kiel. Meeresforsch. Sonderh. 5: 431-432.

Moriarty, D. J. W. 1980. Measurement of bacterial biomass in sandy sediments, pp. 131-139. In: P. A. Trudinger, M. R. Walter, and B. J. Ralph [eds.], Biogeochemistry of Ancient and Modern Environments. Australian Academy of Science, Canberra, and Springer-Verlag, Berlin.

Novitsky, J. A., and P. E. Kepkay. 1981. Patterns of microbial heterotrophy through changing environments in a marine sediment. Mar. Ecol. Prog. Ser. 4: 1-7.

Oshrain, R. L., and W. J. Wiebe. 1979. Arylsulfatase activity in salt marsh soils. Appl. Environ. Microbiol. 38: 337-340.

Reichgott, M., and L. H. Stevenson. 1978. Microbiological and physical properties of salt marsh and microecosystem sediments. Appl. Environ. Microbiol. 36: 662-667.

Rheinheimer, G. 1977. Bakteriologisch-ökologische Untersuchungen in Sandstränden an Nord- und Ostsee. Bot. Mar. 20: 385-400.

Rheinheimer, G. 1980. Aquatic Microbiology. John Wiley and Sons, New York. 235 p.

Sayler, G. S., M. Puziss, and M. Silver. 1979. Alkaline phosphatase assay for freshwater sediments: application to perturbed sediment systems. Appl. Environ. Microbiol. 38: 922-927.

Smetacek, V., B. von Bodungen. K. von Bröckel, and B. Zeitzschel. 1976. The plankton tower. II. Release of nutrients from sediments due to changes in the density of bottom water. Mar. Biol. 34: 373-378.

Stevenson, L. H., C. E. Millwood, and B. H. Hebeler. 1974. Aerobic heterotrophic bacterial populations in estuarine water and sediments, pp. 268- 285. In: R. R. Colwell [ed.], Effect of the Ocean Environment on Microbial Activities. University Park Press, Baltimore.

van Es, F. B., and L. -A. Meyer-Reil. 1982. Biomass and metabolic activity of heterotrophic marine bacteria. Adv. Microb. Ecol. 6: 111-170.

Weise, W., and G. Rheinheimer. 1978. Scanning electron microscopy
and epifluorescence investigation of bacterial colonization of
marine sand sediments. Microb. Ecol. 4: 175–188.

Weise, W., and G. Rheinheimer. 1979. Fluroeszenzmikroskopische
Untersuchungen über die Bakterienbesiedlung mariner Sandsedi-
mente. Bot. Mar. 22: 99–106.

Westheide, W. 1968. Zur quantitativen Verteilung von Bakterien und
Hefen in einem Gezeitenstrand der Nordseeküste. Mar. Biol. 1:
336–347.

Wood, L. W. 1970. The role of estuarine sediment microorganisms in
the uptake of organic solutes under aerobic conditions. Ph.D.
thesis, North Carolina State University at Raleigh.

Wright, R. T. 1973. Some difficulties in using ^{14}C-organic solutes
to measure heterotrophic bacterial activity, pp. 199–217. In:
L. H. Stevenson and R. R. Colwell [eds.], Estuarine Microbial
Ecology. University of South Carolina Press, Columbia.

ZoBell, C. E. 1938. Studies on the bacterial flora of marine bottom
sediments. J. Sed.. Petrol. 8: 10–18.

SYNTHESIS OF CARBON STOCKS AND FLOWS IN THE OPEN OCEAN MIXED LAYER

Bruce J. Peterson

The Ecosystems Center
Marine Biological Laboratory
Woods Hole, Massachusetts   02543

"One aim of the physical sciences has been to give an exact picture of the material world. One achievement of physics in the twentieth century has been to prove that that aim is unattainable...There is no absolute knowledge. And those who claim it, whether they are scientists or dogmatists, open the door to tragedy. All information is imperfect. We have to treat it with humility."

J. Bronowski, "The Ascent of Man"

At the conclusion of the NATO ARI Conference on Heterotrophic Activity in the Sea a working group composed of Barry Hargrave, Peter Williams, Bo Riemann, Richard Newell, Angelo Carlucci, Alain Herbland and Bruce Peterson attempted to diagram the carbon stocks and flows in the mixed layer of the open ocean (Fig. 1). During the preceeding week we had heard a great deal on these topics and on many other aspects of heterotrophic activity in the sea as well. We chose to depict the open ocean mixed layer because that habitat allowed us to make maximum use of the information presented during the conference and because it is probably the best studied habitat in the sea.

Unfortunately there was not time during the conference to discuss this attempt at depicting the carbon cycle. The figure with associated notes and description was, however, circulated to the conference participants with a request for corrections and comments. Eighteen responses (over 50% of the participants) were obtained which is very likely more exchange than would have occurred in a discussion. The most relevant points are arranged below in alphabetical order by author's last name. One general point which

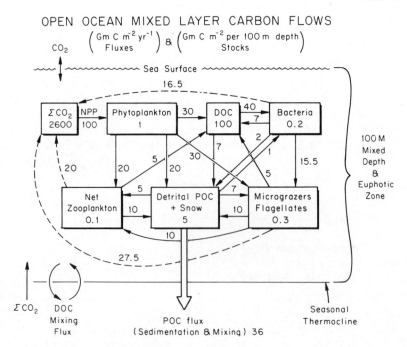

Figure 1.   Pools and flows of carbon in the open oean mixed layer.
Units for fluxes are g C m^{-2} yr^{-1} and for pools are g C m^{-2} in the
100 meter mixed depth or g C per 100 m^3.

many people wished to emphasize was that the diagram should not be
taken as truth.   The numbers for both stocks and fluxes can vary
widely and probably a better approach would be to give a range of
values which would be rather wide in many cases.   We chose to keep
the diagram as simple as possible.   A second important point is that
the net zooplankton biomass (and hence turnover) is probably too low
and many respondents supplied valuable information on that point.

A revision of the figure is included at the end of the
discussion section.   Once again it would be a mistake to conclude
that this is an accurate or sufficient description of any real
planktonic ecosystem.   Any of the fluxes could well be the topic
for future research.   The usefulness of such a figure is that it
provides a focus for discussion and debate.   It should <u>not</u> be viewed
as a state-of-the-art description or as a concensus achieved by the
participants at the conference.

ASSUMPTIONS AND DIMENSIONS

1.   The diagram presents estimates or best guesses of mean annual
values for stocks in grams carbon per square meter to a depth of 100
meters.   This 100 meters is presumed to be the euphotic zone depth

and also the depth to the seasonal thermocline. The fluxes are all
in grams carbon per square meter per year for the entire 100 meter
mixed layer.

2.   We assumed that 100 g C of <u>net</u> <u>primary</u> production was to be
distributed throughout the food web for both respiration and export.
Thus we avoided the problem of how to measure phytoplankton respiration and centered our attention on the subsequent heterotrophic
processes.

3.   Pool sizes show no annual net change. This does not deny an
annual cycle but it does mean that the standing stock on day 1 equals
the standing stock on day 366.

DESCRIPTION OF THE DIAGRAM

1.   Of the 100 g C net primary production 30 g goes to DOC via
excretion or leaching after death; 30 g goes to micrograzers, 20 g
goes to detrital POC and 20 g to net zooplankton grazers. The
standing stock of 1 g C m^{-2} turns over 100 times per year or once
every 3.6 days.

2.   The DOC pool receives inputs from phytoplankton excretion and
leaching (30 g), net zooplankton release and sloppy feeding (5 g),
microzooplankton excretion (5 g), and bacterial leakage and lysis
(7 g). Of this 47 g, losses are to bacterial uptake (40 g) and to
detrital POC by aggregation and adsorption or uptake by detrital
bound bacteria (7 g). There would be additional minor loss by
mixing downward. The DOC pool turns over in just over 2 years.

3.   Bacteria take up 40 g of DOC and 1 g of carbon derived from POC.
The bacteria lose 7 g of DOC via excretion or leaching losses after
death, respire 16.5 g, lose 15.5 g to microzooplankton predation and
2 g to the detrital POC pool through death. Bacterial carbon turns
over 205 times per year or once every 1.8 days.

4.   The net zooplankton consume 20 g of phytoplankton C, 10 g of
micrograzer C and 5 g of detrital C. Losses are to DOC via excretion and sloppy feeding (5 g), fecal pellets (10 g), and respiration
(20 g). Turnover of zooplankton carbon occurs 350 times per year or
about once per day which seems very high. [See individual responses
below.]

5.   The micrograzers receive inputs of carbon from phytoplankton
(30 g), bacteria (15.5 g), and detritus (7 g). They lose carbon to
respiration (27.5 g), predation (10 g), excretion of DOC (5 g), and
the detrital POC pool (10 g). Microzooplankton carbon turns over
175 times per year or once every 2 days.

6.   Detrital POC & snow receives inputs from phytoplankton death
(20 g), net zooplankton fecal pellets (10 g), micrograzers (10 g),
DOC aggregation (7 g), and bacterial death (2 g). Outputs are to
consumption by net zooplankton (5 g) and microzooplankton (7 g),
bacterial uptake (1 g) and sedimentation (36 g).

7.   A return upward mixing flux of total CO_2 (ΣCO_2) is included to
balance the losses of POC and DOC.

DISCUSSION

Responses to Letter

   Gilles Billen: I think that the figures in the diagram themselves are not the most important point, but rather the line of
argument and the feelings leading to them. I am myself not aware of
the basis for postulating such an important role of micrograzers and
flagellates.

   I wish to emphasize the different relative roles of zooplankton
versus bacteria in the recycling of primary production in coastal
and in oceanic areas (see Chapter 15). It seems clear from an
examination of the literature that net zooplankton grazing consumes
a much higher fraction of primary production in oceanic than in
coastal systems. Comparison of heterotrophic activity estimates
with the values of primary production suggests that bacteria play
only a minor role in primary production recycling in oligotrophic
oceanic waters, while their role is predominant in eutrophic coastal
waters. I suggest that this could be a general ecological feature:
microorganisms dominate in the material flows in "rich" environments,
while macroorganisms prevail in "poorer" ones. The same is observed
in sedimentary environments.

   John Davies: I think your 70/30 split of POC/DOC is about right.

   I can't see how zooplankton can consume their own weight of
carbon each day. Estimates of zooplankton feeding in a bloom are
usually about 50% of their body weight/day for fast growing animals
feeding maximally.

   The transfer efficiencies that you have used are high (e.g., net
zooplankton 15/35 = 43%, microzooplankton 25/52 = 48%). These seem
high for a "steady state" system where the biomass of "non-growing"
adults must be fairly high. Your zooplankton biomass does appear to
be small. If you quadruple zooplankton standing stock and reduce by
4 the "turnover" time you maintain the status quo.

   Ian Dundas: I must say that if somebody forced me to put down
numbers on such a diagram, I probably would be willing to accept

something like what you have put together. I do have some problems
with the turnover of net zooplankton being twice as fast as for
micrograzers.

Frank van Es: One point of particular interest to me is the
ecological conversion coefficient, or yield factor, that you used
for bacteria. This figure, a "common knowledge", is essentially
derived from laboratory culture work with rapidly growing bacteria
such as E. coli, B. subtilis and A. aerogenes, and from ^{14}C uptake
experiments using easy substrates, and may be in serious error.

Barry Hargrave: I have circulated the enclosed figure to
various people at the Bedford Insititute of Oceanography. No one
criticized the configuration or pathways of interconnection. Either
they hadn't thought about it, or had nothing better to offer!

A. Herbland: The diagram is very different from the classic
description in the textbooks. The role of the net zooplankton
becomes of minor importance, and inversely, the·role of the DOC and
POC is increased. I agree with the new concept for the open ocean
and especially for the tropical and subtropical open ocean, where a
permanent thermocline allows the development of a kind of "continuous culture" in the mixed layer, with very small organisms.

In the present configuration, the microzooplankton turns over
more quickly than the phytoplankton, and overall, the net zooplankton has the shortest turnover time. As you suggest, the biomass of
net zooplankton is probably higher.

I am not sure that this diagram is unconsciously not the result
of a workshop on microbial activity. What would be the configuration of same carbon flows elaborated after a workshop with zooplank-
tologists? I think it would be different.

My conclusion will be that of Yentsch (1980).

"...we naturally wish to exaggerate what we have created
and to oppose contradiction in a competitive manner. (...).
The problem in incorporating this in the development of any
theory of phytoplankton growth is to separate the truth from
exaggeration."

David Moriarty: When you say bacterial C turns over every 1.8
days, this should be distinguished from turnover of bacterial cells
or generation time. The generation time from the values in the
diagram (24.5/0.2) is 3 days. This is closer to the likely real
values, which I think will be around 5 days. If you knocked the
bacterial biomass up to 0.3, which is reasonable, the values would
look better to me.

With the net zooplankton, you have neglected losses to death
and predation. In the fluxes shown, there is no room for population
growth or turnover. Carbon into new animal tissue is additional to
losses in respiration, fecal pellets and sloppy feeding. The biomass
may be low - perhaps 0.2 is closer - so combining these additional
points should bring down the C turnover time (again different from
population turnover).

Richard Morita: None of the values indicated in the diagram
allows for the microbes in and on other organisms.

Richard Newell: I think the diagram could have some value in
the NATO communication as long as it did not then become entrenched
in the literature as "what the experts say must be going on!" I feel
a little sensitive about this because some of the data obtained with
chemostats on nitrogen-enriched media have given the firm impression
that energy flow through the bacteria under natural conditions (of
very low nitrogen) "should" approach a particular value. As Billen
pointed out, if the C/N ratio of a substrate is high, then microzoo-
plankton/bacteria will need to process more carbon to attain a given
nitrogen production than when the C/N ratio is low (as in nitrogen-
enriched brews).

Larry Pomeroy: I have gone through the box model of the carbon
flows in the mixed layer, and it seems perfectly reasonable and
internally consistent. For some, the model may be quite revolution-
ary. It actually gives short shrift to the net zooplankton.

Jonathan Sharp: I would alter the phytoplankton transfer to
detrital carbon. I really think 30% is too high for excretion to
DOC and I think, on an annual scale, much more goes to POC. I
prefer to assign 11 rather than 30 to the phytoplankton to DOC
transfer and 39 rather than 20 to the phytoplankton to detrital POC
plus snow transfer.

On bacterial supply, I would give less from DOC and more from
POC. I would alter the description of carbon flow through the net
zooplankton compartment so that the 5 g DOC loss is due only to
excretion; sloppy feeding is not consumed, it is part of the phytoplankton excretion. I would raise the zooplankton biomass to 0.5 g.
This is partially due to the jelly-like organisms usually missed by
net sampling due to disintegration. This then gives turnover 20
times a year or once every 5.2 days. (Sic.).

John McN. Sieburth: Due to the spatial and temporal changes
through the photic zone (and the fact that primary production just
does not occur on some sunny days (Burney et al. 1981c), I am very
leary about putting hard figures on any of this. I think that:
(1) we can't homogenize the photic zone, (2) we still don't know
the compartments, and (3) we don't even know the physical-chemical
nature of DOC!

Peter Wangersky: My big concern is that such estimates tend to become fixed in the literature as if they were handed down graven on stone tablets.

The one thought I would pass on is that the sediment trap work suggests that utilization in the upper layers must be less than we had thought, because of the rapid passage of large particles down the water column, and there may be a greater transfer from particulate to "dissolved" state further down in the water column, as the big particles are broken up. I suspect this takes place deeper than 100 m, and that the primary mechansim for getting the inorganic nutrients back into the mixed layer may well be the storm "event".

Bill Wiebe: I would refer to Larry Pomeroy's paper in R. J. Livingston's 1974 book as a starting point. Pomeroy presents a very similar model. I recognize that the flux measurements presented in your diagram differ some and perhaps the arguments about certain fluxes differ also but both are attempts to examine carbon flow in open ocean systems. I would bet that much of the data for the two modelling approaches is the same. Your model has one different flux, sedimentation and mixing through the seasonal thermocline.

The reason I said that such a model would have been helpful 10 years ago is that, as Larry has pointed out, there were problems with conventional models, i.e., the fishery models that demanded nearly 100% consumption of phytoplankton. I believe that if we are really going to understand open ocean food webs, we should focus on time and space scales that produce heterogeneity. There is more and more evidence that organisms congregate or accumulate in sharply defined layers of water at least periodically. Study of the dynamics of these phenomena should tell us a lot more about carbon transfer and organism strategies than spending more time on "big picture" models.

Peter J. leB. Williams: The zooplankton biomass is probably too low. I sense it should be greater than, or comparable to, the bacterial + microzooplankton biomass, say $0.2 - 0.5$ g C m^{-2}. That would increase the turnover time to five days which would seem better.

Dick Wright: I would question the figures for bacterial standing stock and flow to grazers. If the model stands as is, it implies a 1.8 day turnover time for all the open ocean bacteria. This, it seems to me, is too rapid. I think the standing stock for bacteria can be revised upwards.

The flux to micrograzers is sheer speculation as I'm sure you know. I suspect that more of the bacteria die of starvation and join the detrital POC or make new DOC, but who knows?

Revision of the Diagram

In response to the suggestions offered, I have revised the
carbon flow diagrams shown below. There was nearly unanimous agreement that the standing stock of net zooplankton should be revised
upwards. By changing the biomass from 0.1 to 0.5 g C, the rate of
carbon flow through the new zooplankton becomes more reasonable.

On a more difficult second point, agreement was not so good.
Jonathan Sharp felt that phytoplankton excretion of DOC should be
reduced by a factor of three with consequent increase in the flow
from phytoplankton to detrital POC. This detrital POC is then
available for bacterial utilization and this bacterial activity
remains similar but the carbon is supplied to bacteria via a
different pathway. These carbon alternative flows are shown in
parentheses in Figure 2 for comparison with the flows assigned
earlier.

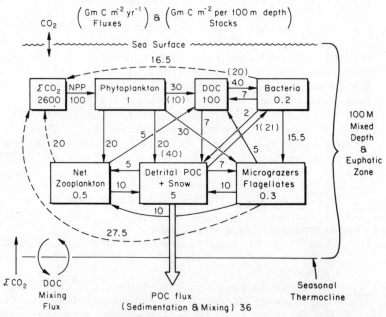

Figure 2. Revised diagram of pools and flows of carbon in the open
ocean mixed layer. Units for fluxes are g C m^{-2} yr^{-1} and for pools
are g C m^{-2} in the 100 meter mixed depth or g C per 100 m^3.

CONTRIBUTORS

Dr. Farooq Azam
Institute of Marine Resources
University of California, San Diego A-018
La Jolla, California 92093

Dr. Robert Bertoni
Istituto Italiano di Idrobiologia
28048 Pallanza, Italy

Dr. Gilles Billen
Laboratory of Oceanography
Universite Libre de Bruxelles
Avenue F. D. Roosevelt 50
1050 Brussels, Belgium

Dr. Angelo F. Carlucci
Scripps Institute of Oceanography A-018
La Jolla, California 92093

Dr. John Davies
Department of Agriculture and Fisheries Marine Laboratory
Aberdeen, Scotland, United Kingdom

Dr. I. Dundas
Department of General Microbiology
University Bergen
Allegt. 70, Bergen, Norway

Dr. Geoffrey Eglinton
Organic Geochemistry Unit
School of Chemistry
University of Bristol, Cantock's Close
Bristol BS8 1TS, United Kingdom

Dr. Gloria Cruz Ferreia
Departamento da Quimica
3800 ACEIRO, Portugal

Dr. Åke Hagström
Department of Microbiology
University of Umeå
S-901 87 Umeå, Sweden

Dr. Barry T. Hargrave
Marine Ecology Laboratory
Bedford Institute of Oceanography
Dartmouth, Nova Scotia
Canada B2Y 4A2

Dr. A. Herbland
Antenne O.R.S.T.O.M.
Center Oceanologica de Bretagne
B. P. 337, Cedex, France

Dr. John E. Hobbie
Marine Biological Laboratory
Woods Hole, Massachusetts 02543

Dr. Holger Jannasch
Woods Hole Oceanographic Institution
Woods Hole, Massachusetts 02543

Dr. William J. Jenkins
Woods Hole Oceanographic Institution
Woods Hole, Massachusetts 02543

Dr. Carl C. Barker Jørgensen
Zoophysiological Laboratory
University of Copenhagen
Universitetsparken 13
DK-21 Copenhagen Ø, Denmark

Dr. David M. Karl
Department of Oceanography
University of Hawaii
Honolulu, Hawaii 96822

Dr. Cindy Lee
Woods Hole Oceanographic Institution
Woods Hole, Massachusetts 02543

Dr. William Li
Marine Ecology Laboratory
Bedford Institute of Oceanography
Dartmouth, Nova Scotia
Canada B2Y 4A2

Dr. Lutz Meyer-Reil
Institut fur Meereskunde an der Universitat Kiel
Abteilung Marine Mikrobiologie
Kiel, Federal Republic of Germany

Dr. David J. W. Moriarty
Division of Fisheries and Oceanography
CSIRO, P. O. Box 120
Cleveland, QLD 4163, Australia

Dr. Richard Morita
Department of Microbiology
Oregon State University
Corvallis, Oregon   97331

Dr. R. C. Newell
Institute for Marine Environmental Research
Prospect Place, The Hoe
Plymouth PL1 3DH, United Kingdom

Dr. Bruce J. Peterson
Marine Biological Laboratory
Woods Hole, Massachusetts   02543

Dr. Lawrence R. Pomeroy
Institute of Ecology
University of Georgia
Athens, Georgia   30602

Dr. Francis A. Richards
Department of the Navy
Office of Naval Research
Branch Office, London, Box 39
FPO New York   09510

Dr. Bo Riemann
University of Copenhagen
Freshwater-Biological Laboratory
51 Helsingørsgade -DK 3400 Hillerød, Denmark

Dr. Jonathan Sharp
College of Marine Studies
University of Delaware
Lewes, Delaware   19958

Dr. John McN. Sieburth
Graduate School of Oceanography
University of Rhode Island, Bay Campus
Narragansett, Rhode Island   02882-1197

Dr. Frank B. van Es
Biol. Res. Eurs Estuary
Department of Microbiology
Groningen University
Kerklaan 30
9751 NN, Haren, The Netherlands

Dr. Hans van Gemerden
Department of Microbiology
Groningen University
Kerklaan 30
9751 NN, Haren, The Netherlands

Dr. Peter Wangersky
Department of Oceanography
Dalhousie University
Halifax, Nova Scotia
Canada B3H 4J1

Dr. William Wiebe
Department of Microbiology
University of Georgia
Athens, Georgia   30602

Dr. Peter J. leB. Williams
Department of Oceanography
The University
Southampton SO9 5NH, United Kingdom

Dr. Richard T. Wright
Department of Biology
Gordon College
Wenham, Massachusetts   01984